INSTITUTE FOR
Research On Teaching

Algorithmization
in Learning and Instruction

Algorithmization in Learning and Instruction

L. N. Landa
Institute of General and Educational Psychology
Academy of the Pedagogical Sciences
Moscow, USSR

Felix F. Kopstein
Scientific Editor

Virginia Bennett
Translator

Educational Technology Publications
Englewood Cliffs, New Jersey 07632

Library of Congress Cataloging in Publication Data

Landa, Lev Nakhmanovich.
　Algorithmization in learning and instruction.

　Translation of Algoritmizatsiia v obuchenii.
　1. Education—Mathematical models. I. Title.
LB1027.L2413　　371.1'02'0184　　73-11044
ISBN　0-87778-063-3

Official English language translation, published by agreement with "Mezhdunarodnaja Kniga," Moscow, USSR, and Professor L.N. Landa, Institute of General and Educational Psychology, Academy of the Pedagogical Sciences, Moscow. First published in Russian, Moscow, 1966.

Printed in the United States of America.

Library of Congress Catalog Card Number: 73-11044.

International Standard Book Number: 0-87778-063-3.

First Printing: January, 1974.

Preface to the English Edition

The joint preparation of a preface to the English edition of this book is most easily explained through a quotation from an informal account circulated to friends and colleagues by the editor upon his return from a meeting with the author in Moscow. It read, in part:

> It may be of general interest that an American and a Russian psychologist can come to a close, intimate, and complete understanding. . . . Of course, it helps if both sides arrive with the same fundamental cybernetic orientation. Nonetheless, it was remarkable how often in our discussions it turned out that (without prior contact) we had solved similar theoretical and practical problems in identical ways.

While this is a spontaneous and personal statement, it might be taken as a reflection of less personal and broader issues.

Science and similar intellectual endeavors have never been respecters of political and cultural boundaries. What is true for two individuals is, no doubt, to some degree true for English-speaking and Russian-speaking psychologists, cyberneticists, instructional technologists, etc., in general. After all, we are addressing the same intellectual, scientific problems, and, as we shall see in the first Part of this book, there is normally but a limited number of possible solutions for any class of problems. Hence, it is not particularly surprising that a "style of problem solving"—a way of thinking about and of approaching certain classes of problems—is of far greater consequence than the problem solver's national origins.

Readers will find a good deal of internal evidence (sometimes underscored by the editor) of "correlated thinking" in the U.S. and the USSR. An inspection of the bibliography will demonstrate a thorough awareness on the author's part of the relevant scientific literature in the English language. It is the editor's belief that possibly or even probably there is a lesser familiarity with the comparable scientific literature in the Russian language among English-speaking readers. We use the phrase "correlated thinking" here deliberately, because we wish to skirt the issue of primacy. Science is not a sporting contest, but a cooperative problem solving enterprise. Everyone benefits equally from new discoveries and new insights so that it matters very little who "thought of it first."

Of far greater importance may be the *kind* of approach to instructional science and technology represented by the present volume. It seems to us that in both of our countries psychologists and instructional technologists (pedagogical or educational methodologists) with few exceptions have been reluctant to adopt a cybernetic view of learning and instruction. Everywhere it has been difficult for those who espouse this view to make themselves heard. This book is one beginning at the task of exploring and explaining some ramifications of cybernetics for behavioral science and instructional technology. We hope it will suggest new insights and stimulate new research and development approaches to old problems.

As everyone probably knows by now, it was Norbert Wiener who coined the term "cybernetics" (see N. Wiener. *Cybernetics*, Cambridge, Mass.: The M.I.T. Press, 1948) and defined it as the science of "control and communication in the animal and the machine." Today we know that this definition is somewhat idiosyncratic and that other definitions of cybernetics are now given as the science of computation, of evolution or evolutionary processes, of problem solving, etc. Since psychologists have long maintained that problem solving is a facet of learning, and since most (possibly all) formal instruction has as its ultimate objective the endowment of a student with some sort of problem solving capability, it seems surprising that so few behavioral scientists have viewed learning and instruction as a "control and communication" process.

Since the original Russian publication of this book in 1966 the burgeoning of such technological developments as computer-administered instruction (CAI) have fostered a recognition of the interactive communication between a learner (student) and an instructional agent (teacher, tutor, device, book, etc.) as a key issue. It is fairly obvious that the instructional agent communicates subject-matter content (information) to the student. There is now a recognition that communication from the student to the instructional agent is also of importance. The question is: What should be the content and the structure (organization) of communication over the feedback channel? What should be the *structure* rather than the content of the entire interactive, instructional interchange? What are the implications of different message structures for the decision structures of the interacting communicators (teacher and student)? What are the implications for their respective internal decision—or regulation and control—structures? What types of control and what over-all structure of control over these processes and whose hands will steer them toward an optimal effectiveness and efficiency? Questions such as these demand thoroughgoing formal, logical (theoretical) analyses before hypothetical answers are formulated. One must *clearly* understand the question before proceeding to answer it. This denies in no way the absolute necessity for *ultimately* testing logical implications, i.e., hypothetical propositions, against experimental data. This book is a beginning at both tasks.

It should be noted that this English translation appears approximately seven years after its original publication in 1966. Thus it is no longer an up-to-date presentation of the author's perceptions of the state of these problems and issues. In the interim he has pursued them vigorously so that a great many have become more clearly and precisely defined. Also, since the author's first publications in this problem field (1959, 1961) and particularly subsequent to the original publication of this book (1966) a great deal of relevant work has been done by others. In the USSR as well as in other European countries a substantial number of research, development, and application studies have been published. They have extended the algorithmic approach to a variety of subject-matters, including mathematics, physics, chemistry, biology, foreign languages, and others. Literally hundreds of research studies on this topic are now available. Faced with the choice of preparing a second edition and the further delays in English publication that would be entailed thereby, it seemed preferable to reach a compromise with obsolescence. This decision was made easier by the fact that the basic approach, i.e., primarily the first part of this book, requires no significant revision to this day.

This book reflects only one aspect of the author's approach to a wider range of problems. Many of these problems are here touched upon only superficially. Thus only passing mention is made of the relationship between algorithmic and heuristic processes, the design of *adaptive* instructional algorithms with reference to diagnostic assessments of students' reasoning processes in terms of hypothetical systems of intellectual operations, etc. Some of these issues are addressed in a second book, whose English translation is now in preparation. Others have come to light and have been dealt with in more recent research which has been published so far only in Russian.

It is particularly important that readers should not arrive at an impression of the author as a singlemindedly rigid proponent of complete algorithmization for any and all instruction. Quite the contrary is true. The author's scientific concerns have been and continue to be the relationships between algorithmic and heuristic (independently creative) processes. In this book, however, the concern is with the role of algorithms (or rather algorithmic prescriptions) in the processes of learning and instruction.

Possibly, the manner in which this English translation came to be prepared will be of interest to readers. Some years ago the editor became aware of this book through references to it in European scientific literature. Eventually the German translation was obtained and served to convince him of its value and of the desirability of an English translation. Interest in such an enterprise proved to be difficult to arouse. Both of us wish to express our appreciation to the publisher for his willingness to take on the risks of this English translation, and for his steadfast support during some difficult periods.

The procedure in preparing the translation was as follows: The translator first prepared an essentially literal translation of each chapter. This was deemed necessary, because of the extreme precision of expression required by the content of this book, and especially its first part. For many words and phrases the most accurate English cognates had to be established first. The initial expectation had been that the editor would be able to recast this first and relatively *literal* translation into a more *literate* English form without distorting the originally intended meaning. It became clear very early, however, that serious distortions and misrepresentations were creeping in.

Thanks to a travel grant to the editor from the U.S. National Science Foundation plus additional support from his parent institution, the Human Resources Research Organization, he was able to make a trip to Moscow. There, sitting side by side, we reviewed and revised the initial versions of approximately the first third of this book. For the remainder of the book we arranged a procedure according to which subsequent chapters were mailed to the author, who then reviewed the initial version as prepared by the editor from the basic version of the translation. The author indicated all necessary changes and provided notes on all problematical renderings before returning the chapter to the editor. Thus every chapter in this translation has been reviewed and approved by the author. Even so the inescapable and fundamental differences in connotative meaning between Russian and English terms defeated us in some instances. Where this has been the case we have sought to elucidate the intended meaning through comments.

Differences of language also created another kind of problem. It happens that the content of the experimental instruction (described largely in the second part of the book) was a portion of the Russian grammar and syntax as they are taught to native Russian speakers. No amount of recasting could have preserved the critical features of the Russian examples, nor would such recasting have illustrated adequately the critical features of the instructional procedures. Thus the decision was made to retain the Russian examples to the extent to which the critical instructional features were clear and would illustrate the critical point. We hope that this decision will prove valid and that readers will be served well by these examples. Similarly, the editor has resisted strenuously the temptation to excise a great deal of elaborative discussion, because it tends to elucidate the author's fundamental meaning and intentions.

It remains for us to express our appreciation to all those people and institutions who helped us in some way during the unexpectedly long and difficult task of preparing this English translation. We have already mentioned the U.S. National Science Foundation, the Human Resources Research Organization (Hum-RRO), and the publisher. Next we want to express our appreciation to Dr. Virginia Bennett, the translator, for her stamina in achieving a precise first rendering of difficult material and under difficult circumstances. Dr. John Stelzer and particularly Mr. Edward H. Kingsley have helped immeasurably by clearing up problematical points of logic and mathematics, and by offering excellent advice in these areas. The editor's secretary, Mrs. Doris Stein, provided the same splendid clerical support during particularly difficult periods which the editor has come to expect from her routinely. Finally, we mutually tip our hats to each other and hope that the English translation of this book will help to foster the vital spirit of international scientific, intellectual interchange.

F.F.K.   L.N.L.
Alexandria, Virginia, U.S.A.   Moscow, U.S.S.R.

July, 1973

Introduction

Summed up in this book is research on the problem of teaching students some sufficiently general methods (ways, means, procedures) of thought (reasoning) which the author conducted from 1952 on. This research rested on the achievements of pedagogical psychology and represented, on the one hand, the application, and on the other, the future development of, a number of approaches and principles elaborated first of all in Soviet science. Originally, the research was conducted on material related to students' solution of geometrical problems by proof. This research showed that one of the important reasons for their inability to solve these problems was the fact that students often do not know those operations which must be carried out in order to find the solution, or they have not mastered these operations, since they are not formed in them (here, we mean operations which are known or can be found in science). If one were to designate the system of operations for the solution of a certain class of problems as a method of solution (the notion of method will be examined in more detail in the first chapter), then one may say that students are unable to solve problems well because they do not know the methods of solution. They do not know how they must think as they solve, and how, and in some cases, in what order they must act and operate with the conditions of the problem. This, in turn, is provoked by the fact that in school, as an analysis of the way mathematics was taught there showed, students are not taught these methods in many instances. Their attention is not directed at the operations which "make up" the solution of the problem, and these operations are not specially practiced. The teacher is often concerned only with offering the student knowledge about the *content* of what is studied, and is considerably less concerned with giving him the means of *operating* on or with this content and with how to *think, reason, reflect* as he assimilates the specific content and applies his knowledge about it. In this connection, the students themselves usually just think about the content of *what* they have to learn, and they do not think about *how* to learn. The problem of *what* to learn is much more dominant than the problem of *how*, and with the help of *which methods* one must learn and master what is being studied.

Of course, such a situation is not accidental, and it is not simply the fault of the teachers. The fundamental reason for the defect indicated consists of the fact that the problem of establishing procedures of thought and of reasoning (as well as the problem of ways to instruct the students in these procedures) is not sufficiently elaborated in the psychological and pedagogical sciences themselves. For the procedures of thought which are taught the students to be conscious, purposeful, and accessible to every teacher, one must first of all create the effective means to break down intellectual activity into the components which make it up—intellectual operations which are sufficiently elementary in a specific sense. One must create methods to specify the structure of operations which is best for different conditions, and one must also work out the methodology for their operation-by-operation formation.

The research which we mentioned at the outset of this introduction was directed at the solution of these problems as applied to the instruction of geometry.

During the research, an attempt was made to determine the operations which "make up" the process of thinking through a geometrical problem (by proof) and the process of searching for its solution. The system of these operations was formed as a specific system of rules and as a sufficiently general procedural prescription which shows what must be done with the conditions of the problem in the process of its solution, and how to think as one searches for the proof.[1] The reduction of the process of searching for proof into separate, sufficiently elementary operations and instructing students in them showed that, in a relatively short time, it was possible to raise sharply the efficiency of instruction and to teach students how to solve problems of a kind they had not been able to solve up until then.

The work described was completed at the end of 1954 and was defended in 1955 as a doctoral (*cand. phil.*) thesis [684].[2] At the same time, or somewhat later, efforts began in the United States on the analysis of the psychological and logical structure of intellectual activity during the solution of problems by proof (and certain other problems) and on the formulation of methods for the search for proof with the aim of simulating it on electronic computing machines. There, the corresponding methods were given the name "heuristic methods." Work to program the process of proof for computers based on the study of how this process takes place in substantive human thought was designated as "heuristic programming" (see, e.g., Newell, Shaw, and Simon [140], [141], [192], [193]). On the problem of heuristic programming and heuristic methods of thought in general, see also, [28], [130], [131], [134], [135], [147], [148], [150], [187], [188], [191], [724], [865]. Operations in the search for proof which were analogous, to some extent, to those which we taught students in our experiment were formalized, and the process of searching for proof was simulated on a computer. Programs were written and called "The Logic Theory Machine" and "The General Problem Solver," on the basis of which a general purpose computer could prove a series of logical and geometrical theorems. The first results of these studies, as far as we know, were published in American scientific journals in 1956-1957.

After the completion in 1955 of research to discover sufficiently general procedures (methods) of thought for the solution of geometrical problems, and experiments in teaching them to students, the question arose as to whether these methods were specific only to mathematics. If not, then would it not be possible to establish the common features of methods of thought for the solution of problems from different subject-matter fields

(e.g., mathematics, grammar, physics, chemistry problems, etc.)?
Would it not be possible to teach the students the solution of
mathematical problems, for example, in such a way that they
would simultaneously learn how to solve grammar problems, and
vice versa, to teach the solution of grammar problems in such a
way as to simultaneously cultivate mathematical thought?[3] The
establishment of such general procedures (methods) would be
extremely important, since it would make it possible to understand more profoundly and in more detail the nature of man's
general intellectual abilities and to find a way to form them
consciously and purposefully. There are grounds to assume that
general intellectual abilities are none other than the mastery of
general methods of thought and of reasoning, and the ability to
reflect on and solve problems which have a different subject-
matter content.

In order to answer the question of whether the methods were
specific only to mathematics, it was necessary to investigate
methods of solution for some non-mathematical problems. Grammar problems were chosen as the object of investigation. The
process of solving grammar problems was analyzed in exactly the
same way as the process of solving geometrical problems. It turned
out that the procedures for solving both kinds of problems had
much in common and that one may also formulate sufficiently
general procedures for the solution of grammar problems quite as
for the solution of geometry problems. But an essential difference
between these procedures was also brought to light. The procedures in the search for proof in the solution of geometry
problems were precisely methods of *search*, i.e., they had a
heuristic character; and, therefore, the way to accomplish them
depended essentially on the type of problem, its complexity, etc.
These methods do not completely and univocally determine all of
the students' actions as they search for the solution, and therefore,
they do not guarantee—as applied to each specific problem—its
inevitable solution. For grammar problems, we succeeded in
designing procedures which completely and univocally determine
the students' actions, and when they are applied correctly,
guarantee the inevitable solution of real problems encountered in

practical work in school. In other words, these methods have an essentially algorithmic character.

It is known that the study of methods of this kind occupies an important niche in the study of mathematics. Analysis has shown that such methods are also taught during grammar instruction (although, until recently, they were not called algorithms). At the same time, one type of algorithmic method was established, namely *procedures for the identification* of grammatical phenomena which in many cases are not taught or are incorrectly taught as grammar is studied. Often neither the teacher nor the students know about these algorithmic procedures; most of them are not in the methodologies.

It turned out that the students' lack of knowledge (and lack of mastery) of this type of algorithm is one of the main reasons which causes difficulties in the students' mastery of grammar. It is one of the basic reasons for their mistakes. A good knowledge of the rules of grammar does not change things under these conditions. Just as a good knowledge of geometry theorems still does not ensure the solution of mathematical problems—for in order to solve a problem one must be able to specify which of the theorems is applicable in the given situation—neither does a good knowledge of grammar rules itself ensure correct grammar usage. The whole difficulty consists in the determination (identification) of which of the known rules must (may) be used in the given instance. But this is precisely what many students are unable to do. When they encounter a word where they must specify the correct spelling, or a sentence which must be punctuated, they often do not know which operations must be performed with this word or sentence in order to choose the necessary rule and solve the problem of spelling. In other words, as with the study of mathematics, they have a knowledge of the content of the academic material—they know the theorems and rules—but they do not know general methods for the solution of problems. They have not mastered the uniform reasoning procedures. And, just as in mathematics, the teacher often does not teach these methods to the students. This is not surprising, since many of these methods have not been established and formulated in science itself.[4]

In order to verify the hypothesis that the main reason for difficulties in the mastery of grammar and for mistakes in usage stems from the students' lack of knowledge of uniform methods for the solution of grammar problems, we developed and formulated such methods (applicable to a number of topics) and began the experiment of teaching them to students. The results of the experiment turned out to be similar to those which were obtained when students were taught general methods for the solution of geometry problems. Students began to master grammar better and with considerably more ease, and the number of grammar mistakes decreased by five to seven times over a relatively short period of instruction.

By 1959, experimental work on teaching students methods of grammatical reasoning and algorithmic procedures for the solution of grammar problems was ended and some articles on the algorithmic approach to educational processes [271], [272], [275], [688] and a book were written about the results of the research. When the book was completed, however, it became clear to the author that the manuscript had to be revised considerably.

The reason for this was that during the second half of the 1950s, there was a rapid growth of work in the field of cybernetics in our country. The ideas of cybernetics became very widespread and also attracted many representatives from the humanities. In the process of experimental work, it became obvious that many of the assumptions from which we proceeded as we developed and formulated uniform methods of thought and taught them to the students could obtain good interpretations within the concepts of cybernetics. Moreover, the concepts and methods of cybernetics and also mathematical (symbolic) logic made it possible to interpret pedagogical and psychological phenomena with greater depth by considering them as a particular instance from a wider circle of phenomena and processes studied in cybernetics. They also made it possible to use certain concepts, ideas, and methods of cybernetics and mathematical logic for a deeper and more precise statement of psychological-pedagogical problems themselves. At the end of the 1950s, however, the cybernetic approach to psychological-pedagogical phenomena was so unusual and strange for many people who worked in the field of pedagogy that any attempt to apply the ideas and methods of cybernetics to pedagogy encountered sharp protests. At the beginning of the 1960s, when the above mentioned book was finished, the situation had begun to change somewhat. It became clear that the moment had arrived when it was possible to set forth the basic propositions and ideas of the book in such a way as to present them from a broader, cybernetic point of view. But in order to do this, it was necessary to rewrite at least the theoretical part of the book.

The work turned out to be time-consuming and difficult, since, at that time, a great deal of data had accumulated in the most diverse fields of science which were directly related to the topic of our research. An understanding of these data made it possible to define the original theoretical propositions much more accurately and to state some problems which we did not envision when we began our experimental research. In particular, these were the problem of the specification of the conditions of expediency for the teaching of one or another algorithm to the students, the problem of methods for the selection and evaluation of an optimal strategy for search and identification, and several others. Undoubtedly, in an approach to the solution of these problems, various defects may come to light later on, and these approaches will have to be improved and supplemented. But it is known that, in science, the statement of problems and the formulation of methods are often no less important than the solution of these problems and the demonstration of how the corresponding methods must be applied.

How are the problems touched upon in this book connected with programmed instruction? The elaboration of the ideas underlying the research described in this book was begun before programmed instruction made its appearance. The research itself was conducted without any contacts with programmed instruction. This, however, does not disaffirm the internal connection of the material of this book with the principles of programmed instruction. Proposed in the research are means to formulate effectively methods for the students' intellectual activity which could be used successfully in programmed instruction. Moreover, a number of the propositions from which we proceeded basically coincide with many of the approaches of the theory of programmed instruction.

However, this is not a book on programmed instruction. The basic problem of programmed instruction is to construct algorithms of instruction. The basic problem of our research was to teach algorithms. If algorithms of instruction are programs by which the teacher must be guided (in them, he is shown which actions he must perform in the teaching process and, in particular, how he must control the students' actions), then the instruction in algorithms is the instruction in programs by which the students' themselves must be guided (in them, they are shown which actions they must perform with the object of the actions, depending upon the purpose, conditions, and results of the actions). Algorithms of instruction are possible in which the teaching of some kind of algorithm to the students will not be presupposed. It is also possible to teach algorithms without acting in accordance with some algorithm and without using methods of programmed instruction. More will be said about this in Chapter Three. Here we note only that an algorithm of instruction can be devised which will direct the teacher's activity well, but this activity will control (direct) the students' activities poorly. In other words, a poor, but sufficiently detailed methodological elaboration can direct the teacher's activity well, but when the teacher uses this elaboration, he directs the students' activities poorly.

Programmed instruction, in the form in which it exists now, does not include the direct control of the students' intellectual activity as a necessary indicative feature.[5] This is confirmed by the fact that programmed textbooks may be designed in such a way that they will not include instruction in algorithms and general methods of thought. Actually, many of the programmed textbooks and programs for instruction with the aid of machines control only the presentation of the academic material (subject-matter to be learned, *Ed.*) and check the results of his intellectual operations (whether the student answered the questions correctly or not, and whether he did or did not solve the problem). They do not directly control (steer, regulate) the internal processes of intellectual activity nor the formation of methods of intellectual activity in students. Of course, control of internal, covert processes can be accomplished only through the control of external behavior. But the character of the control actions intended only to control external behavior differs from those aimed at directing the internal, psychological mechanisms (processes) which determine external behavior. The significance of research on the problem of teaching students algorithms for the theory and practice of programmed instruction consists of the fact that this research points out one of the possible ways to improve programmed textbooks and teaching machines. It indicates what must be included in programmed textbooks and programs for teaching machines, and how they must be changed in order that, along with indirect control of the students' intellectual activity, a direct, purposeful control may be realized. At present our collaborators in the Laboratory for Programmed Instruction at the Institute of General and Educational Psychology of the Academy of Pedagogical Sciences of the USSR are attempting to use the means of programmed instruction to form in the students methods of thought which have a sufficiently general character. They are also attempting to teach them algorithms and some heuristic methods for the solution of problems and for a more effective, more direct control of their intellectual activity than is usual now.

Related to the fact that our book is not directly about programmed instruction (this is evident from the correlation of the problems of our research with the basic problem of programmed instruction), we hardly touch upon the problems of programmed instruction as such. We limit ourselves—where it is expedient—to just a few considerations and remarks. The field of programmed instruction is an independent field of research, and it requires its own theory and its own methods for the solution of problems which arise within it.

We also shall not state the basic notions and ideas of cybernetics in our book, since there is an abundant literature on this topic.[6] We shall note only that the possibility of applying the ideas of cybernetics to the field of psychological-pedagogical phenomena is based on the fact that instruction (and education) may, in a specific sense, be considered as a form of control. In a word, it is precisely the control of the formation and development of the intellectual processes and characteristics of an individual. The intellectual processes, in principle, are just as controllable as physical, chemical, biological, and other processes, although they are also considerably more complex and therefore more difficult to study. The problem consists of studying the laws of control of these processes and, on the basis of their cognition, mastering them more fully. The elaboration of the theories and methods for the control of intellectual processes is one of the most important tasks of pedagogy. The circumstance that instruction is a form of control provides a basis for examining the laws of control in the field of instruction as a particular instance from among more general laws of control.

This, in turn, makes it possible to hope for the possibility of the use of the ideas and methods of cybernetics both to raise the effectiveness of the pedagogical process and to perfect pedagogical theory.[7]

Of course, the theory of instruction does not and could not be considered to be the same thing as cybernetics. The sphere of application of cybernetics to the field of pedagogy (as to other sciences) is limited by the abstract character (abstraction) which is the core of cybernetics. The answer to many questions of pedagogy (e.g., ones like the question about the aims of education, about the content of instruction, and many others) are specified not by cybernetics, but by the needs of society and by its interests and politics. Education has its specific object, its specific traits, and laws, and to study them only from a cybernetic viewpoint would be insufficient and incorrect. But the analysis of the process of instruction from the viewpoint of the principles of control and of the demands made on a "good control system" makes it possible to establish certain essential defects in the theory and practice of instruction and to point out the way to eliminate them.[8]

A characteristic of cybernetics is that it studies mainly the control of complex systems, and devises methods to optimize this control. If one examines educational control from this viewpoint, then it is perfectly clear that the student (and, in general, any person) is a complex system. Moreover, he is, evidently, one of the most complex systems which one can have to control. It is for this reason that problems of optimizing control which must be solved during instruction (not to mention education) are very difficult.

Among the concepts used by cybernetics, the concept of an algorithm is of the greatest interest to us in this book. This concept arose in mathematics and is now applied widely where there is a question of creating a program for the control of different objects and processes. Algorithms have significance for instructional theory because, when we teach a student an algorithm for the solution of problems, we not only provide him with the means to control those objects which he will transform with the help of that algorithm, but also with the means to control himself, his own intellectual operations, and his own practical actions. An algorithm, as a means of control after it is mastered, becomes for the person a means of *self-control*, and the means for him to regulate independently his practical and intellectual activities.

Since the basic subject-matter of this book is the teaching to students of some uniform reasoning procedures and sufficiently general methods of thought which have an algorithmic character, the concept of an algorithm will be examined in it. This is all the more important, because the concept is not always used correctly in pedagogical and psychological literature. The concept of an algorithm is rather complex and has a long history. In order to reveal its content and show to what degree and in what sense it may be rightfully applied in pedagogy and psychology, in the first section of the book, we had to go beyond the bounds of psychology and pedagogy and examine the concept of an algorithm on a broader plane and touch upon its use in other sciences. Especially instructive for psychology and pedagogy are the application and development of this concept in the theory of automatic control. Here, as is known, the problem of instruction and self-instruction has become at the present time quite important (in connection with the problem of simulating the process of instruction of a person and of building self-instructing cybernetic systems).

Considerable difficulties arose for the author as he was writing this book. These were in connection with his aim to make the material set forth accessible to different categories of readers. The problems of instruction interest not only psychologists and pedagogues, but also engineers, mathematicians, specialists in the various fields of cybernetics, and representatives of many of the fields in the humanities. And that which is known and accessible to an engineer or to a mathematician is not always known and accessible to a psychologist and a pedagogue or vice versa. It is impossible to write a book which would take into account, in equal measure, the level of knowledge and the character of inquiries of the representatives of all the indicated specialties and professions. We chose pedagogues and psychologists as the basic addressees of this book, and we tried to set forth a number of problems related to cybernetics, mathematics, and mathematical logic in a form which would be sufficiently accessible to an unprepared reader. Nonetheless, it is not impossible that some places in the book may be difficult for individual readers. To elucidate potential difficulties we have relied heavily on explanations, notes, and commentaries which can be skipped over during reading without serious detriment to immediate understanding.

A great many problems from various fields of science are touched upon in the book. In order not to hamper the comprehension of the material and so as not to disperse the readers' attention, we tried, whenever possible, to avoid historical digressions and not to set forth the history of the problems examined. Since we tried to avoid an excessive enlargement of the book's scope, we do not provide a detailed analysis of the points of view of different authors on questions examined. We limit ourselves to an indication of literature on these questions. In the bibliography of the book, it is chiefly the works of contemporary authors which are indicated.[9]

As we end the introduction, we would like to note that everything in this book is by no means indisputable. Certain problems are just stated, where the solution of part of them is not conclusive, even for the author himself. And if a discussion arises on the questions touched upon in this book, the author will do nothing but welcome it. It is clear already that additional and specialized research is needed for some problems. Other approaches are undoubtedly possible for certain problems. But if the author succeeded in directing the reader in the search for these approaches, he will consider his task completed.

In conclusion, I would like to express my sincere gratitude to B.V. Biryukov and B.V. Gnedenko, who helped in the editing of this book, as well as to the Academician A.I. Berg, psychologists L.M. Wekker, N.I. Zhinkin, L.B. Itelson, P.A. Shevaryov, mathematicians V.G. Ashkinuze, A.I. Markushevich, N.M. Nagorny, V.B. Orlov, Yu. A. Shikhanovich, engineer A.V. Shileiko, linguists and methodologists L. Yu. Maksimov, V. Yu. Rozentsveig, and A.V. Tekuchev, with whom the author often discussed the problems touched upon in this book and whose views were of considerable help in the work on the manuscript. Also of essential help to the author was his collaboration with A.R. Belopolskaya, who successfully developed some algorithms for translation from German to Russian.

I would also like to express my gratitude to my colleagues in the didactic sector of the Institute of the Theory and History of Pedagogy of the Academy of Pedagogical Sciences of the USSR, and heads of the sector, B.P. Yesipov and M.A. Danilov, as well as my collaborators in the Laboratory for Programmed Instruction of the Institute of General and Educational Psychology of the APS USSR who took part in evaluating the manuscript of the book and who made a number of valuable comments. My cordial thanks to the teachers, Ye. V. Akimova, N.A. Arutyunyan, N.S. Beryozovsky, L.A. Bobrovskaya, Ye. V. Vishnevetskaya, M.I. Gorelovaya, N.P. Gorovaya, G.A. Domogatskaya, I.S. Zabuga, S.F. Ivanova, S.I. Nikitina, I.N. Okoyemovaya, I.I. Orlova, G.L. Ryabininskaya, A.F. Solovyova, and V.A. Edelstein. These teachers took part in the experimental work and in the discussion of separate problems stated in this book, and they gave the author considerable help in the practical verification and realization of the ideas developed in the book.

L.N.L.
Moscow, U.S.S.R.
January, 1966

Notes

1. The necessity for developing sufficiently general methods in the search for proof was realized long ago. Some rules for such a search were formulated, in particular, by the mathematicians Hadamard [38] and Polya [750].

2. The basic content of the dissertation was published in addition to [684] in two articles, [686], [687], appearing in 1959 and 1961. (See page 16 for note on [ ], *Ed.*)

3. The problem of general traits in formal-logical reasoning in different subject-matter was stated a long time ago (i.e., all logical laws have to do with the form of reasoning irrespective of its content); on the psychological plane, its significance has been particularly emphasized recently by Kabanova-Meller [648], [649] and Shenshev [845].

4. Some teachers discover these methods independently, teach them to their students, and attain great success in their work; but we are speaking here not of individual teachers, but about the general state of affairs with respect to the level at which this given problem has been elaborated in science.

5. By direct control, we mean influence on the students' intellectual operations by way of special directions, rules, and any other prescriptions which appeal directly to these operations and which have a direct influence on their flow. In addition to direct control of intellectual activity, it is possible to influence the flow of intellectual operations indirectly through a particular selection and organization of the content of instruction, of the educational material, of the type of exercises, etc. With instruction which is correctly presented, both these means of control of intellectual activity must be combined.

6. See: [40], [42], [46-51], [53-55], [57], [60], [61], [69], [74-76], [78], [86], [87], [91], [94], [96], [98], [100], [108], [109], [112-116], [120], [121], [132], [137], [143-145], [151], [153], [155-158], [162], [164], [171-173], [177-180], [183], [196-198], [201], and others. For the application of cybernetics to pedagogy and

psychology, see, for example: [208], [231], [256-259], [262], [272], [274], [309], [314], [324], [355], [375-377], [381-383], [389], [390], [397], [406], [408-410], [422], [426], [429], [437], [439], [440], [446], [447], [450], [454], [470-472], [479], [480], [494], [508], [511], [516].

7.   The fact that a controlled system in instruction (the student) is active and that it is not only controlled but a *self-con-trolled* system does not change the essence of the matter although it introduces a definite specific into instructional control in comparison with other types of control. In instruction it is especially important that the program of control ensure the quickest elaboration of programs for *self-control* in the students in order that the control be directed toward freeing the student more quickly from the need for this control. But the formation of abilities for self-control must also be controlled, and in this sense, one may speak of the student as a controlled system.

8.   The cybernetic analysis of the instructional process in no way signifies a "destruction" of everything of value which has been amassed by the theory and practice of teaching through many centuries. Quite the contrary, when the ideas of cybernetics are applied in pedagogy, it is very important to use as much as possible of all that is meaningful and progressive in the pedagogical heritage.

9.   In examining this bibliography, the date of this book's original (Russian) publication must be kept in mind. The author has pointed out that since that time the relevant literature has grown considerably—and not only in the USSR. (*Editor.*)

Table of Contents

Preface to the English Edition   vii

Introduction   xiii

PART ONE: Theoretical Foundations

Section One: Algorithms and the Control Process

Chapter One. Instructing Students in Reasoning Procedures: Algorithmic and Non-Algorithmic Methods   7
    1.   Training Students in Reasoning Procedures   —
    2.   Characteristic Traits of Algorithmic Methods   10
    3.   The Features of Non-Algorithmic Methods and
       Their Role in Problem Solving   21
Chapter Two. Quasi-Algorithmic Prescriptions   33
    1.   The Concept of Quasi-Algorithmic Prescriptions   —
    2.   Relativity of Elementary Intellectual Operation:
       Concept and Criteria   37
    3.   Formal and Content Aspects in Quasi-Algo-
       rithmic Prescriptions   44
Chapter Three. Algorithmic Prescriptions, Algorithmic
Descriptions, and Algorithmic Processes   50
    1.   Basic Concepts   —
    2.   Methods for the Description of Algorithmic
       Processes   54

xxix

3. Algorithmic Description of Teaching Activity   60
4. The Algorithmic Description of the Students' and Teacher's Activities in the Process of Instruction: Its Role in Programmed Instruction   64

Chapter Four. Algorithms as Determined by the Characteristics of the Controlled Object and the Information Available About It   73

1. Classification of Algorithms According to the Conditions and Types of Control   —
2. Control with Complete Information Available   76
3. Control with Incomplete Information Available   86
4. Peculiarities of Learning and Self-Instruction in Man and Machine   90

Chapter Five. Algorithms of Transformation and Algorithms of Identification   105

Section Two: Some Theoretical Problems in Teaching Algorithms

Chapter Six. Expediency of Instructing Students in Algorithms   113

1. Classification of Thought Problems   —
2. Conditions Under Which the Teaching of Algorithms Is Expedient   132

Chapter Seven. Teaching Algorithms in School   148

1. Implications of Teaching Algorithms   —
2. Some Ways to Teach Algorithms and Their Influence on the Development of Students' Intellectual Abilities   159
3. Problems in Whose Solutions the Order of Operations Can Vary   166
4. Individualization in the Teaching of Algorithms   169
5. Interrelation Between the Concept of "Algorithm" and the Concepts of "Ability" and "Skill"   170
6. On Teaching Algorithms of Identification   174

Section Three: Logical and Psychological Problems in Constructing Algorithms of Identification

Chapter Eight. Indicative Features of Phenomena and Their Logical Structures   191
    1.   The Concept of Indicative Features   —
    2.   Selection of Indicative Features   195
    3.   Logical Structures of Indicative Features: Concept and Classification   203
Chapter Nine. Deficiencies in Presenting Logical Structures of Indicative Features in Textbooks   214
    1.   Definitions Where No Logical Operator Is Mentioned   215
    2.   Definitions in Which Logical Connectives *"And"* and *"Or"* Are Expressed by Different Grammatical Conjunctions   216
    3.   Definitions with Incorrectly Described Structures of Indicative Features   220
Chapter Ten. Methods of Identification as a Function of the Logical Structure of Indicative Features   224

Section Four: Mathematical Methods in the Design and Evaluation of Algorithms of Identification

Chapter Eleven. The Importance of Quantitative Methods of Analysis and of the Evaluation of the Efficiency of Algorithms   243
Chapter Twelve. Optimization of the Process of Search and Identification   257
    1.   Identification Under Alternative Choice   259
    2.   Search in Multiple Choice   277
Chapter Thirteen. The Stochastic Mechanisms of Intellectual Activity in the Process of Perception and Identification of Phenomena   318

PART TWO: Research Findings

**Section One: Organization of Experimental Instruction
and Research**

Chapter Fourteen. Objectives and Methods of Experimental Instruction   341
   1.   Objectives of Experiment   —
   2.   Materials and Methods of Experiment   344
Chapter Fifteen. Forming in Students the Concepts of
Indicative Features, Their Logical Structure, and Corresponding Reasoning Methods   348
   1.   Students' Mistakes Due to the Lack of Definite
       Logical Knowledge and of Corresponding
       Abilities   —
   2.   Design and Implementation of a "Logic Lesson"   352
Chapter Sixteen. Studying the Topic "Types of Simple
Sentences"   387
   1.   Substantiation of Instructional Methods   —
   2.   Description of the Procedures and Lessons
       (Basic Fragments)   397
Chapter Seventeen. Studying the Topic "Compound-
Coordinate Sentences"   445
   1.   Substantiation of the Methods of Instruction   —
   2.   Description of the Course of Lessons (Basic
       Fragments)   472
Chapter Eighteen. Some Aids and Technical Means for
Teaching Algorithms   513
   1.   "Exercise-books for Independent Study" as a
       Means of Forming Algorithmic Processes Operation-by-Operation   —
   2.   The Teaching Machine "Tutor I"   526
Chapter Nineteen. Breaking Down Intellectual Activity
into Elementary Intellectual Operations in the Course of
Experimental Instruction   537

Section Two: Results of Experimental Instruction and Research

Chapter Twenty. Comparative Analysis of the Results of Instruction in Experimental and Control Classes 549
   1. Quantitative Data —
   2. A Qualitative Analysis 555
Chapter Twenty-One. Characteristics of the Way Students in Experimental Classes Assimilated Algorithms 585
   1. The Causes of Student Errors in Experimental Classes —
   2. The Students' Attitudes Toward the Methods of Experimental Instruction 592

Conclusion 600

Bibliography 616
   I. Problems of Logic and the Theory of Knowledge —
   II. Mathematics, Mathematical Logic, and the Theory of Algorithms. Cybernetics and Certain of Its Applications 619
   III. The Application of Cybernetics and Logic in Pedagogy 634
   IV. The Physiology of Higher Nervous Activity. General and Pedagogical Psychology. Didactics and the Methodology of Instruction. Linguistic Foundations for Teaching a Native Language 664

Author Index 705

Subject Index 709

A Glossary of Logical Symbols Used in This Book 715

About the Author, Editor, and Translator 717

Algorithmization
in Learning and Instruction

PART ONE

Theoretical Foundations

Part One: Section One

Algorithms and the Control Process

Chapter One

Instructing Students in Reasoning Procedures: Algorithmic and Non-Algorithmic Methods

1. Training Students in Reasoning Procedures

The guidance of students' reasoning processes has always been one of the most important problems in pedagogy and psychology. The question of how to teach people so that they not only acquire knowledge, but also learn to think has long held the interest of scholars from different disciplines.

A person needs knowledge not so much for its intrinsic value, but mostly in order to solve problems arising in practical and theoretical activities. In order to solve problems through the application of acquired knowledge, the problem solver must have mastered appropriate methods of reasoning.

Problem solving in the broadest sense of the word comprises a large part of any activity—practical and theoretical. In order to learn to do something well (make things, construct, invent, write grammatically, prove one's statements, find the causes of things, etc.), one must learn how to solve problems: logical, physical, chemical, design, grammatical, etc. But the solving of problems requires reasoning ability. This is why teaching students the ability to think in the process of teaching them to solve problems is the most important aspect of preparing them for practical and theoretical activities.

Problems which a person must be able to solve in the course of his activity are extremely diverse. It is impossible to teach in school the solution to all problems which one might encounter in life; they are practically boundless. Nonetheless, school must

7

prepare students for a future in which they must be able to solve the most varied problems. There is only one way to do this: while teaching students the specific solutions to concrete problems, there must be developed in them sufficiently general reasoning procedures, i.e., general approaches to the solution of any problem that will enable them to search for a solution in any novel situation. The formation of such sufficiently general methods of thinking is, therefore, one of the most important tasks of a school.[1] The development of effective reasoning capabilities is an important part of and a means to the general psycho-educational development of students, especially with reference to the growth of their intellectual capabilities.

Methods of reasoning formed in the context of one subject by appropriate instruction can be transferred to another subject. Thus, the more that general reasoning approaches are shaped in students in the course of their instruction in each subject, the greater will be the rate of growth of their problem-solving skills; the higher will be the level of their general intellectual development; and the better will be their preparation for the diverse aspects of practical and theoretical activities they will encounter in life.

The task of teaching students sufficiently general reasoning procedures is not only a methodological problem but also a didactic and psychological one of basic importance. Its solution is related to the formation of those basic cognitive, or intellectual, functions which are important for the mastery of any subject and provides preparation for any kind of activity. It is a basic didactic problem also because the teaching of general ways (methods) of reasoning presupposes the development of general ways (methods) of instruction. The latter can be applied to the teaching of different topics within one subject as well as to the teaching of different subject-matters.

Awareness of the necessity for teaching students general reasoning procedures in science and in practical affairs has existed for a long time, but its solution has progressed slowly. In order to create a constructive theory of instruction in general reasoning processes, one must understand the nature of these processes. This requires the explanation of what general methods of reasoning are, what their psychological nature and internal mechanisms are, and what takes place in man's consciousness when he reasons and solves problems.

The elucidation of the nature of general reasoning procedures is a most important and at the same time difficult task. So long as the nature of reasoning has not been established, one cannot place the teaching of these methods on a scientific foundation, and purposively direct the process of their formation.

It is well known that the word "method" is used extensively in science. Thus, we speak of the method of dialectics (in philosophy), of parallel translation (in geometry), and the method of unique distinction (in inductive logic), of situation analysis (in psychology), and of the method of treatment (in medicine), and so on. Although these methods differ from one another in the degree of their generality, and in their character (in some instances these are general principles of the approach to reality or to its particular domains; in others, they are the specific instructions for actions during the solving of problems belonging to a rather narrow category), they are not by accident called by one and the same word, "method." When one speaks of methods, one usually has in mind certain prescriptions or instructions on the modes in which a person is to act in order to attain defined goals (particularly instructions as to which operations to carry out in order to solve particular problems). Sometimes these modes of action are themselves called methods.

For example, when it is said that a person *knows* a certain method, it is understood that he acquired or discovered a certain prescription, remembers it and can, when necessary, be guided by it and even explain it to other people. When it is said that a person has *mastered* a certain method, it means that he is able to carry out a certain system of operations; but, in this instance, knowledge of a prescription is not necessarily included in the notion of "mastering a method." A person may be able to carry out defined operations and solve defined problems without knowing their corresponding prescription and without awareness of his operations. In this book, the word "method" (and "procedure," *Ed.*) will be used in both senses of the word. But the first section will, for the most part, be concerned with methods defined as prescriptions.

We will attempt to arrive at a more substantial exposition of the notion of method. Our starting point in so doing will be the concept of the algorithm. This is especially advisable, since the basic subject-matter of this book is devoted to the application of methods of an algorithmic character to pedagogy and psychology.

2. Characteristic Traits of Algorithmic Methods

Algorithms are one of the kinds of methods underlying much ordinary activity, and not just intellectual activities. In other words, the notion of "algorithm" can be applied not only to an activity which is carried out through intellectual operations, but also to activity which is carried out through physical operations. Moreover, algorithms can direct operations for a machine. In this book, however, we shall for the most part examine general methods underlying the intellectual activity (or the general reasoning procedures—which is the same thing) of people.

The distinction between mental and physical actions (operations) lies in the fact that physical actions transform real physical objects, i.e., change "things," while mental or intellectual operations change the percepts and concepts of these things. For example, if a real rotation of some three-dimensional figure (made, for example, from wood) is a physical operation, then the internal rotation of the image of that figure is an intellectual operation. It is easy to see that intellectual operations represent a transformation not of a material thing, but of its image. At first there was one image of the figure in the observer's consciousness (with one side turned toward the observer), then there was a second image of the figure (with another side turned toward the observer). These are different images. Let us note that, although the intellectual operations are transformations not of material objects, but of their images (particularly, of concepts), these internal transformational mechanisms are purely material processes taking place in the brain. The intellectual operation by which such transformations are completed are actions bringing about real changes of those images and ideas toward which they are directed. A person's mental and physical actions do not exist in isolation from each other. Many physical actions are generated by mental actions and are regulated by them. On the other hand, the mental actions themselves are formed on a basis of physical actions, and in the course of activity interact with them and after are supported by them. The immense significance of intellectual operations lies specifically in the fact that, because of them, man can carry out an intellectual experiment with objects from the exterior world that permits him to obtain knowledge of the results of his future influence on these objects prior to the point of a real execution of the corresponding physical activities.

A detailed analysis of the notion "algorithm" is necessary for us, not only in order to understand the nature of reasoning methods, but also in order to teach algorithms and to make use of algorithms for greater control of the reasoning processes of students.

The notion of algorithm arose in mathematics. By algorithm is usually meant *a precise, generally comprehensible prescription for carrying out a defined* (in each particular case) *sequence of elementary operations* (from some system of such operations) *in order to solve any problem belonging to a certain class* (or type).

These characteristics of algorithms do not constitute a precise, mathematical definition, but nonetheless quite clearly reveal the essence of the notion. An example of an algorithm can be given in terms of the well known procedure of dividing natural numbers or the prescription for finding the greatest common denominator of two natural numbers.

That algorithm can be formulated as follows:[2]

In order to find the greatest common denominator of two natural numbers x and y one must:

1. *Write* both numbers x and y as products of prime factors ($x = \prod x_i$; $y = \prod y_j$).

where:

Π signifies the product of a finite series of numbers (analogous to Σ for the sum),

x_i and y_i are the prime factors of the numbers x, y and the indices
i, j   take on as many values as there are factors in x, y respectively.

2.   *Find* the smallest factor of the first number (min x_i).

3.   *Ascertain* whether at least one factor in the expansion of the second number equals the smallest factor of the first number (does y_j = min x_i exist?).
If not, then go on to instruction 4.
If yes, then go on to instruction 5.

4.   *Cross out* the smallest factors from the expansion of the first number and go on to number 7.

5.   *Eliminate* the factor from the expansion of the second number which is equal to the factor of the first number (eliminate y_i = min x_i).

6.   *Cross out* that factor from both expansions.

7.   *Ascertain* whether at least one uneliminated factor remains in the expansion of the first number.
If no, then go to instruction 8.
If yes, then go back to instruction 2.

8.   *Multiply* all of the eliminated factors and finish according to the algorithmic prescription.

The product of the eliminated factors will thus be the greatest common denominator.

This algorithm can be expressed a great deal more clearly in the form of the block diagram shown in Figure 1.1.[3]
One can solve the problem of finding the greatest common denominator by another method, by means of another set of

Figure 1.1

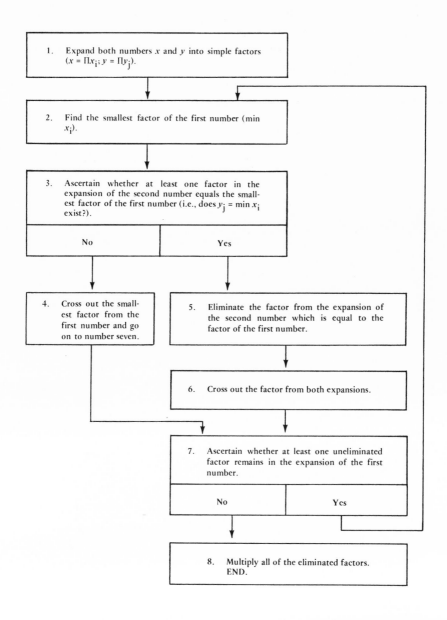

1. Expand both numbers x and y into simple factors ($x = \Pi x_i$; $y = \Pi y_j$).

2. Find the smallest factor of the first number (min x_i).

3. Ascertain whether at least one factor in the expansion of the second number equals the smallest factor of the first number (i.e., does y_j = min x_i exist?).

   | No | Yes |

4. Cross out the smallest factor from the first number and go on to number seven.

5. Eliminate the factor from the expansion of the second number which is equal to the factor of the first number.

6. Cross out the factor from both expansions.

7. Ascertain whether at least one uneliminated factor remains in the expansion of the first number.

   | No | Yes |

8. Multiply all of the eliminated factors. END.

operations whose order can also be described in the form of a particular system of instruction.*

In order to find the greatest common denominator of two natural numbers, one must:

1. *Ascertain* whether the first number is less than the second.
    If so, go to instruction 2.
    If not, go to instruction 3.

2. *Subtract* the first number from the second and consider the result as the second number. Go to instruction 1.

3. *Ascertain* whether both numbers are equal.
    If so, go to instruction 5.
    If not, go to instruction 4.

4. *Subtract* the second number from the first and consider the result as the first number. Go back to instruction 1.

5. *Conclude*: The first of the numbers is the sought for greatest common denominator of the two initial numbers—and thus—the problem is finished as prescribed by the algorithm.

The block diagram of this algorithm is shown in Figure 1.2.

The operations which lead to the solution of the problem are, as we see, of a rather general character. They may be applied to any number whose greatest common denominator has to be found. The special character of these operations lies also in the fact that they are organized within a particular system and are carried out in strict sequence (although these systems, even in resolving one and the same class of problems, can be, as we have seen, different). But within each system the sequence is strictly defined.

The discovery and formulation of algorithms was one of the most important tasks of mathematics from the moment it became a science. In this connection, the academician A.N. Kolmogorov

---

*The diagram shown is a somewhat modified version of what is known as Euclid's algorithm.

Figure 1.2

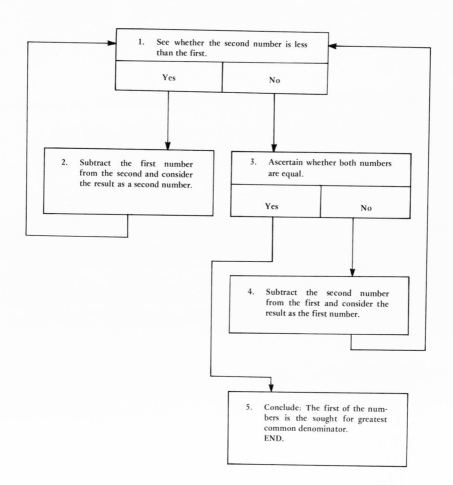

[110] * writes: "In all cases where possible, the discovery of algorithms is the natural goal of mathematics." In the process of its development, mathematics has tried to find all the most general, effective algorithms for the solution to problems which would permit the uniform solution of broader classes of problems by one means, i.e., by the use of one and the same system of operations.

The successes in the elaboration of general ways of solving problems even led, in due course, to the idea that it might be possible to completely formalize mathematics and create a universal algorithm having such a high degree of generality that it would be possible to solve all mathematical problems by one method.[4] Leibnitz, in particular, tried to create such an algorithm. As is known, this attempt did not achieve success. Moreover, it was later proved that it is impossible to completely formalize mathematics (or, for that matter, the other sciences).[5] The existence of classes of problems for which it is impossible to construct any algorithm (what is known as an algorithmically unsolvable problem)** has been proved.[6] Later, algorithmically unsolvable problems were discovered by Markov [126], Post [195], Novikov [138], and other mathematicians.

The notion of algorithmically unsolvable problems should not be confused with the notion of a non-algorithmized (at a given level of knowledge and practice) problem. If, in the first case, there is the impossibility of devising any algorithm to solve a certain broad class of problems (actually, it is sometimes possible

---

*All references referred to with bracketed numbers are listed at the end of this volume.

**The Russian term "mass problem" or "mass quantity problem" amounts to a mathematical idiom whose accurate rendering into English presents some difficulty. The translators of Markov [127] have suggested that "general problem" may be more meaningful to English readers. The intent is to convey the notion that it is impossible to construct an algorithm that deduces the solution for every class of problem. Thus, the literal term "mass quantity problem" might be rendered as "universal deducibility problem." It is herein called simply "algorithmically unsolvable problem." (*Editor.*)

to devise algorithms for some subclasses of this class of problems), in the second case it is a matter of problems that have not as yet been properly specified and formalized. This has made it impossible to formulate univocal prescriptions for a uniform process for solving them. Examples of such problems will be given later in this book.

Despite the existence in mathematics of algorithmically unsolvable problems, the search for algorithms to solve problems (in those cases where it has not been proved that algorithms are impossible) plays a very important role in the development of mathematics. Defining the types of problems for which algorithms are possible and those for which they are not has a great gnoseological significance. It clarifies the degree of generality among different types of problems and methods of reasoning. This also has a great practical significance, since it indicates the pointlessness of searching for algorithms in those instances where it has been proven that a certain problem is algorithmically unsolvable.

Algorithms are usually characterized by the following basic properties: specificity, generality and resultivity.

SPECIFICITY. This property resides in the requirement that the prescriptive directions in algorithms must be strictly defined. Directive instructions must indicate precisely the nature and conditions of each action, exclude chance components in the choice of actions, be uniformly interpretable, and be unambiguous. Thus, they must refer to sufficiently elementary operations for an addressed system—person or a machine—to carry them out unequivocally.

The specificity of an algorithm is expressed in the fact that problem solving by algorithm is a strictly directed process, completely guided and not admitting of any arbitrariness. This is a process which can be repeated by any person (or machine, if the algorithm is programmed into it) and will lead to identical results, if the two data sets are identical.

GENERALITY. This property is reflected in the fact that *any* member belonging to the defined (problem, *Ed.*) class may take the part of any other member as the initial datum of a problem which is solved by an algorithm. Thus, the algorithm of the division of numbers is applicable not only to the numbers 243 and 3, or 150 and 5, but to any two natural numbers. Therefore, algorithms can be considered as general solution methods, because they make possible the solving of not just one particular, given problem with one particular set of initial data, but of the most varied problems drawn from some class. This class can contain an indefinitely large—and in deductive sciences usually infinite—number of specific problems distinguishable by their data sets.

RESULTIVITY. This property is reflected in the fact that an algorithm always converges on a specific sought-for result, which is always obtained in the presence of the appropriate data set. This property of an algorithm, however, does not assume that algorithms result in the obtaining of the desired result with all data sets belonging to the defined class. It is possible that the algorithm will be inapplicable to certain sets of data; and, in that case, the process of carrying out the algorithm will either halt suddenly, or it will never end.[7]

It was said earlier that an algorithm is a prescription for carrying out, in a defined sequence, a certain system of elementary operations for the solving of all problems of a given class. It is clear from the characteristics of the basic features of an algorithm that, although the algorithm is a prescription for the execution of a certain system of operation, by no means is every directive for the carrying out of operations an algorithm.

In order to further clarify the basic properties of algorithms, let us take note of the following:

In the process of solving a problem according to a certain prescription, one always has: (1) the prescription itself, consisting of specific instructions (directions, rules, commands) for carrying out specific actions (operations) on specific objects* (it can perhaps, for example, be written down on paper); (2) a certain system-operator (for example a person or a machine) to whom these instructions are addressed and who performs the specified operations on these objects; (3) the objects toward which the

---

*Entities, phenomena. (*Editor.*)

operations are directed and which are transformed by these operations. The solution of any problem lies ultimately in the transformation of a certain object or objects from one state to another (final) state.*

In the process of solving a problem by any prescription (not necessarily having an algorithmic character in the above described sense), there are two sorts of relations: (1) that of the prescription to the executing system engaged in performing the corresponding operations, and (2) the relation of a system and of its operations to the real object at which these operations are directed and which is transformed under their influence. In this connection, one may speak about the verb "determine" in two senses. In one sense the prescription determines (evokes, actualizes) the actions of the problem solver. In the other sense his operations determine (evoke, are conditional for) the transformation of the object.

A prescription univocally (entirely, strictly) determines actions directed to the solving of problems, if each instruction of this prescription always elicits from the solver identical operations in identical situations.[8] Conversely, a certain prescription ambiguously (non-strictly, incompletely) determines the actions aimed at the solving of problems if even a single one of the instructions can elicit from identical or different solvers different operations in identical situations.

Each operation directed at an object engenders the transition of the object from one state to another.[9] The transformation can consist of a single transition, or it can consist of chains of transitions (a one-step or many-step transformation). The transition of an object from state to state will be univocal (also called strictly determined), if a given operation applied to the object, which is found in a given state, always transforms it to the very same other state. Conversely, the transformation of an object from state to state will not be univocal, i.e., will not be strictly determined, if the exact same operation applied to the object which is in the very same state, in some instances provokes its

*If a reasoning problem is being solved, then the objects of transformation are the images, concepts of material objects, their properties, and relations.

transformation to some one particular state, and in other instances to a different one.[10]

Now, let us agree on indicating the state of objects by lower case letters and the operations applied to the objects (found in given states) by capital letters. The passage from state "a" to "b," we will indicate by $a \Rightarrow b$; if, in so doing we wish to say that the transformation takes place under the influence of operator A, then we will write $a \overset{A}{\Rightarrow} b$ or else $A(a) \Rightarrow b$. We will also make use of the expressions of the form $(a \overset{A}{\Rightarrow} b) \dot{V} (a \overset{A}{\Rightarrow} c)$ or, in a more shortened version, $a \overset{A}{\Rightarrow} b \dot{V} c$, understanding them, correspondingly, as indications that one and only one of the transformations $a \Rightarrow c$ is taking place.

In general, the expression $q_0 \overset{A}{\Rightarrow} q_1 \dot{V} q_2 \dot{V} \ldots \dot{V} q_n$ is read: the application of operator A to state q_0 (of the object under consideration) transforms it in one and the same situation to one and only one state from a finite number of n states $q_1, \ldots, q_n$.*

At each step of the algorithmic process, the object liable to change can be found in one of several states $q_1, q_2, \ldots, q_n$. The character of the transformation of the object at that step is determined by the character of the operator being applied and by the state of the object. An algorithm is considered to have been formulated where a set of instructions precisely specifies all of the actions which must be taken with respect to the object in all of its possible states to transform it to the required final state. Here, all of the possible states of the object are considered known, and the exact reactions required of the solver are specified for each one of them.[11] Algorithms require identical actions in response to identical situations (states) independently of what kind of real reasons resulted in the object's transforming into one or another of the multiple possible (specified) states, and whether or not the

---

*The symbol $\dot{V}$ as it is used here must not be confused with the strict (partitive) disjunction in mathematical logic. On the difference of meanings of the partitive "or" in natural language and the strict disjunction in logic, see below, in Chapter Eight.

transformation into that state was strictly or not strictly determined.

What has been said above permits us to contrast more precisely algorithmic methods with non-algorithmic ones and to explain the properties of the latter.

3. The Features of Non-Algorithmic Methods and Their Role in Problem Solving

Non-algorithmic methods in contrast to algorithmic methods do not completely nor unambiguously determine the actions of the problem solver. The instructions have some degree of uncertainty, and evoke different actions from different problem solvers (or from one and the same problem solver) in one and the same situation. Thereby, such occurrences as the following can take place:

(1) The instructions entering into the prescription are such that they do not evoke (do not actuate) in the problem solver any kind of action whatever, since they are not comprehensible to him;

(2) Instructions entering into the prescription are such that each of them (or at least one) can be carried out by means of different actions, which lead to different transformations of the initial objects. Thereby, in turn, the following may happen:

(a) Specific actions by means of which the given instruction can be carried out are not indicated.

(b) The specific actions by means of which the given instruction can be carried out are indicated, and the problem solver can choose one of them, and so on.

Let us give some examples of prescriptions of a non-algorithmic type.

EXAMPLE 1. Let us suppose that the student who could not solve a certain mathematical problem were given the following prescription about what to do in order to solve it.

1.   Read the text attentively.
2.   Think about how the given data are interrelated.
3.   Draw all of the conclusions arising from the given data.

This prescription will not be an algorithm, because the operations indicated here are not elementary and are not comprehensible to all. It is precisely because the student may have to read the problem statement carefully and not know how to think and draw conclusions that he is unable to solve the problem. Yet instead of being shown "how," he is told: be more attentive, think, draw conclusions, etc. In order to turn this prescription into an algorithm, it is necessary to break down each of the instructions into such simple, elementary instructions that the student will comprehend them and will be able to execute them. Only then can the prescription evoke (actualize) the necessary operations and through them engender the kind of reasoning process which will undoubtedly assure the solving of the problem.[12]

EXAMPLE 2. In a certain prescription indicating the actions which the problem solver ought to carry out in the course of looking for proofs of geometry theorems as part of his homework, there is the following instruction: "juxtapose the isolated element of the figure with other elements and draw conclusions about its properties." An instruction of this type is non-algorithmic because different students can carry it out in different ways. It is possible to juxtapose the isolated element of the figure with different elements; and, as a result, different conclusions may be drawn.

EXAMPLE 3. In a prescription which specifies the course of a game, there are such instructions:

1.   Go to the book shelf which has three books on it.
2.   Take the book in the middle.
3.   Open it to a page whose number ends in 5.
4.   Find the first word on the page.
5.   Note its first letter.
6.   If the letter belongs to the first half of the alphabet, then carry out action A with the book and end your actions at that point.

7. If the letter belongs to the second half of the alphabet, then carry out action B with the book and end your actions at that point.

If one were to assume that all of the operations indicated in that prescription were sufficiently elementary, and the persons to whom it is addressed were able to carry out those operations, then that prescription still would not be algorithmic, because it has one ambiguous instruction—"open the book to a page whose number ends with 5."

Different individuals carrying out that instruction could open the book to different pages, and having performed different actions with the book, arrive at different results. The total process of activity, thus, is also not completely determined: the third instruction involves uncertainty, since it can be carried out in different ways.

In some instances, when a precise number of different actions is known, by means of which one or another instruction can be carried out, the degree of uncertainty of that instruction can be measured. Thus, for example, if the book has four pages whose number ends in 5, and the probabilities of opening to each of the pages are identical, the degree of uncertainty of the instruction, "open the book to a page whose number ends in 5" is equal to the $\log_2 4 = 2$ bits. If, for example, there are eight such pages in the book, then the degree of uncertainty of the same instruction in that case will be equal to the $\log_2 8 = 3$ bits.

It is clear from what has been said that non-algorithmic methods, which do not completely determine the action of the problem solver in the process of solving problems, require an independent search for appropriate actions from him and an independent making of decisions. Incomplete specification of activity by the corresponding prescription, lack of knowledge (or imprecise, incomplete knowledge) of which operation should be carried out in one or another case in order to solve the problem is evidently that essential peculiarity which makes it possible to call such an activity independent.

The notion of independent activity is used here not in the

sense that a person *performs* certain actions himself (in that sense, algorithmically prescribed actions are also independent), but in the sense that he himself *chooses* actions, *finds* them, and independently *decides* on the character and course of action without the availability of some kind of ready-made prescription on how to act. The highest degree of independent activity is creative activity, which demands not only independent decisions about how to act, but also the discovery of new and original ways of acting.

Using the notion of algorithm, one can characterize independent activity (including reasoning activity) as activity which is not determined by corresponding prescriptions or is not completely determined.

From the examination of different non-algorithmic methods, one may conclude that the degree of uncertainty as to actions inherent in a prescription can be different in different cases. Sometimes in an ambiguous prescription there is only one instruction not completely specified, and therefore only a small "part" (i.e., one step) of the activity is independent or creative. It is another matter when several or even a majority of the steps are independent or creative. Also character and degree of specificity of each step may be dissimilar. This means that the degree of independence or creativity which is necessary for carrying out the corresponding step is dissimilar as well.

If, for instance, the set of operations is indicated by means of which an instruction can be executed, then it is easier to make a choice of one operation from among the number indicated, than in a case where this number is not indicated, and the choice must be made from memory (here we do not have in mind a case where an algorithm is "imprinted" in the memory). It is more difficult to find the necessary operation when there is a higher degree of non-specificity of an instruction than when there is a lower degree of non-specificity (in the first case, the area of choice is greater). The degree of non-specificity of an instruction can thus be one of the indices of "the degree of independence" which is necessary for the implementation of the given instruction. It is more difficult to find and employ an unlikely operation (with the given degree of non-specificity of the instruction) than a more probable one, and so on.

Some difficulties are indicated above. They are, as we said, difficulties in independent activity or, in defined situations, difficulties in creative work, and in some cases they are measurable. There is nothing impossible in setting up criteria which will allow a judgment to be made (at least by certain parameters) to what extent one or another problem requires independent thinking, and to what extent it is creative.

Wherein lies the necessity for elaborating and using methods of non-algorithmic character along with algorithms in the process of solving problems?

It lies in the fact that for many problems it is impossible to know or foresee all situations and operations that will be necessary for the problem solution. (The actual design of an algorithm presupposes the isolation of a complete system of such situations and operations.) It is also impossible to know the required sequence of operations, the connection between specified operations and specified situations, etc. For the solving of many problems the issue is not simply the application of some knowledge and means of action to a concrete situation. It is the learning of what is not yet learned and the discovery of the unknown. It is precisely because it is impossible to take everything into account beforehand and to foresee everything ahead of time in solving many problems, that it also is impossible to devise the algorithms for their solution. One can only indicate certain modes of arriving at the solution which partially direct the actions of the problem solver, but do not completely determine them. The problem solver who uses such methods must, during the course of solving the problem, find and carry out those actions which the situation requires, the properties of which will become revealed or be discovered in the actual process of solving.[13]

As an example of non-algorithmic methods which do not completely determine the actions of the problem solver, one may point to the methods of criminological situation-analysis. Thus, the author of one of the books on the logic of legal investigation, Starchenko [30] writes: "In order to carry out an investigation of a case quickly while 'hot on the trail,' it is necessary to have at least hypothetical answers to such questions as *where* to look for evidence, *how* to look for it, and most important, *what* to look for, i.e., what facts in the given instance can fulfill the role of evidence in the case." But it is precisely for such questions that the investigator has no answers at first. It is only in the process of the investigation itself that he can obtain the necessary facts. It is impossible to construct a general, universal algorithm for criminological situation-analyses, since it is impossible to foresee all of the objects, their properties, and connections which could turn out to be essential for the exposure of the crime. Some new circumstance, which was not known earlier and never previously encountered, could turn out to be essential and, therefore, could not be taken into account in advance and included in the algorithm.

However, as the essential conditions for the solving of specific problems progressively are precisely revealed, an exhaustive list of them is compiled (where this is possible), and the set of conditions and their interrelations become defined. Thus, it becomes possible on the basis of methods of non-algorithmic character to construct algorithmic methods and to transform prescriptions containing indefinite instructions into prescriptions which completely specify the process of solution.[14]

It is easy to show, from the example involving the books on the bookshelf introduced above (Example 3), how a method of non-algorithmic character can be changed into a method of an algorithmic character. To do this, one must substitute for instructions characterized by non-specificity those which have specificity. Thus, the indefinite instruction, "open the book to a page whose number ends in the number 5," can be replaced by the definite instruction, "open the book to page 15." Now the instruction will be completely specific (algorithmic).

The development of science, especially in recent years, has given many examples of the transforming of methods of a non-algorithmic character into algorithms. That which not long ago was considered as not susceptible to algorithmization, it turns out, can be algorithmized. Many kinds of activity for which only methods of non-algorithmic nature were once formulated (and which therefore required intuition, creativity, etc.) can now be described algorithmically. Such precise and unambiguous prescriptions have been created for solving what in the recent past were particularly creative problems that it is now possible to transfer their solutions to electronic computers. Among such examples are diagnostic problems.

In his day, S.P. Botkin* outlined the basic characteristics of a "method of diagnosis," which as Osipov and Kopnin [18] write, remain correct to this very day, without any essential changes. Here is the way these authors describe "the method of diagnosis" in today's language: "The first and very important part of the method is a detailed study of the objective symptoms observed in the patient. The second part is a detailed study of the case history and of the subjective symptoms. The third part is 'an analysis of the facts revealed by the examination of the facts' (Botkin) which includes anatomical and functional diagnostics as well as a diagnosis of the illness. The fourth part consists of 'a diagnosis of the patient' which should be strictly distinguished from 'a diagnosis of the illness.' The fifth part is the prognosis of the disease."

The description of the stages given is not difficult to formulate into a prescription as to what the physician must do to supply a diagnosis:

(1) study the objective symptoms observed in the patient;
(2) study the case history and the subjective symptoms;
(3) analyze the facts brought out by the examination and diagnose the illness;
(4) give "a diagnosis of the patient";
(5) carry out a prognosis of the disease.

Here the "method of diagnosis" is formulated as a prescription

---

*Sergei Petrovich Botkin (1832-1889). Noted Russian diagnostician, professor of the Imperial Academy of Medicine and Surgery, Medical Consultant to the Medical Board of Russian Internal Affairs, and President of the National Health Commission of Russia. (*Translator.*)

(system of instructions) about the activities of a physician. This prescription, however, is not an algorithm, since each instruction appeals not to elementary but to rather complex and compound operations, and the degree of ambiguity for each instruction is quite high. At this time algorithms that completely specify the solution of diagnostic problems for the diagnosis of a series of diseases have been developed. In a series of studies (see [41], [44], [63-66], [77], [118], [194], and others) the logical and mathematical principles of the approach to the construction of such diagnostic algorithms have been described.[15] The significance of the transition from non-algorithmic methods of diagnosis to algorithmic ones consists not only in the fact that a machine can now solve a diagnostic problem. Knowing the algorithms for solving diagnostic problems is also of the utmost importance for physicians, since, acting according to an algorithm, they are able to carry out an examination of the patient more systematically, take into consideration those symptoms they might have "let slip by" or "not noticed," and remember such illnesses which might otherwise not have "come to mind." The strict specification of a diagnostic process considerably lessens the possibility of diagnostic mistakes as a consequence of various subjective reasons (the peculiarities of a physician's memory, the character of his knowledge, his type of reasoning, etc.), thus greatly raising the "reliability" of the diagnosis.

We have introduced only one example of the transition from non-algorithmic methods to algorithmic ones. It would have been possible to cite many more examples (for example, the design of algorithms for optimal planning in economics, algorithms of machine translations in mathematical linguistics, and several more). The algorithmization of the processes of reasoning activity in many different fields is a progressive tendency in modern science. The role of algorithmization is extremely great now. Algorithmization has enormous importance for theory as well as practice, opening up to science great new possibilities.

Notes

1. "General methods of reasoning" does not mean universal methods. The general character of methods is always relative, since it is connected with those classes of problems which are solved by means of these methods.

2. A prescription (algorithm) is made up of instructions which are sometimes called rules.

3. The instruction presented here was devised by us as a result of the expansion of separate operations of the rule, given in an arithmetic textbook of the secondary schools. "In order to find the greatest common denominator of several numbers, it suffices, after having expanded them into simple factors, to multiply among themselves those factors which are common to all of the given numbers" [660]. Let us note in this connection, that the word "rule" in science has many meanings. Those instructions are called rules which make up the prescriptions-algorithms, and the prescriptions similar to what was just shown plus several other instructions and statements. More will be said in detail about several varieties of rules in one of the following chapters.

4. The notions "means" and "method" are used here in one and the same sense. We use two different words solely out of stylistic considerations.

5. In 1931 K. Goedel proved that even a theory of the elementary arithmetic of natural numbers cannot be completely formalized, i.e., that it is impossible to construct such a finite and uncontradictory system of axioms and rules from which one could infer all propositions which are true in content for elementary arithmetic.

6. On the notion of deducibility problems, see, for example, Gastev [79], Glushkov [84], Uspensky [165], [166].

7. However, as Markov [127] points out, "the possibility of a resultless breaking off can be excluded, without limiting essentially, the general sense of 'algorithm.' "

8. Here we are not going to define more precisely the meaning of "identical situations" and "identical operations," assuming

that we can discern (identify and differentiate) operations as well as situations. We can, for example, distinguish "open" from "close" and are always able to say about certain actions with a door whether it is being opened or closed. It is necessary, however, to note that the task of identifying and differentiating operations is not always as simple as in this instance, and often requires precise defining of which operations to consider "identical," and which "different." But inasmuch as a more detailed examination of this question does not enter into our task, we will consider that we are able to identify and differentiate operations and situations in the instances in which we are interested.

9. Here, by state, we mean a finite number of distinguishable and statable characteristics of an object (in a particular case, this may be one unique characteristic), which are important for its examination from a defined viewpoint, and by means of which the object is described in terms of specified goals. In addition, an object usually has other characteristics which are not included in the examination and are not stated and, therefore, do not enter into the notion of state in the sense just indicated. Real objects also have properties which can be generally unknown at a given level of knowledge.

10. The question of types of transformation from state to state in general has been examined specifically in Ashby's book [179]. Here we will note only that the univocal transformations from state to state are explained by the fact that certain of the system's properties, or processes taking place in it, had not been taken into consideration earlier and included in the initial state. Also, all kinds of outside accidental factors which cannot be taken into account can influence the transformation from state to state.

11. Inasmuch as it is not always possible to allow for all of the possible states in the transforming of actually existing objects, in which an object can pass to one or another step of transformation, the number of possible states in some cases turns out to be uncertain, and it is impossible to set up a precise algorithm in such situations. More will be said on the

causes making the algorithmization of the solution to all problems impossible.

12. It must be pointed out that "algorithms" similar to the one given here are often addressed not only to students (by teachers) but also to teachers (by methodologists). When, for example, in instruction on how to give a lesson, the teacher is told that he must "arouse the students' interest" or "provoke an active attention," but the kinds of actions which constitute "arousing interest" and "provoking attention" are not made clear, then such instructions are of little use. After all, the teacher to whom such instructions are addressed often really does not know how this should be done. Instructions of that type are characteristic of all abstract, non-specific guidance. Yet, complete control demands concretely detailed guidance expressed in unambiguous instructions.

13. What was said about the fact that the solution process in many cases cannot be completely determined by algorithm, but is directed only by the solver's knowledge and mastery of methods of an algorithmic character, does not mean that it is impossible to teach such methods deliberately. They were taught and are being taught (an experiment in such teaching is described by Landa [686]). Without a doubt, however, the teaching of non-algorithmic methods is more complicated than the teaching of algorithmic methods. It is also less reliable, since it is never possible to guarantee that, as a result of teaching, the student will successfully solve all problems requiring the application of corresponding non-algorithmic methods.

14. These algorithmic methods (algorithms) will still not be all-embracing, since it is impossible to exhaust the process of cognition. They can be suitable for solving those problems whose conditions and methods of solving are known at the present moment. These algorithms are a result of the development to a particular level of the knowledge of a certain domain of phenomena.

15. A peculiarity of many diagnostic algorithms lies in the fact

that these algorithms process the initial data of diagnostic problems (symptoms of the illness) into syndromes of not one illness, but several possible illnesses, with an indication of the probability of each one of them. In other words, as the outcome of the algorithm, there is a table with indications of the probability of several diseases that are possible with the given symptoms. But, with identical initial data, there will always be identical tables as the outcome of the algorithm.

Chapter Two

Quasi-Algorithmic Prescriptions

1. The Concept of Quasi-Algorithmic Prescriptions

In this book, we have attempted to apply the concept of algorithm to a number of pedagogical and psychological phenomena. As we have already said, the process of instruction is one kind of control. This suggests why the concept of algorithm is extremely important for pedagogy and psychology. Algorithms can serve as one means of effectively controlling the processes of learning and teaching. The concept of a (mathematical) algorithm implies the process of transforming the objects of certain standard types (let us say symbols of a certain alphabet) and the complete formalization of the transformational process. However, in pedagogy and psychology, the formalization[1] of the process of transformation can rarely be made complete. Psychology and education (pedagogy) frequently deal with objects of other-than constructive type, and their transformation rarely can be completely formalized. That is why the concept of a (mathematical) algorithm seems to be not directly applicable to the fields of pedagogy and psychology. In many instances, we do not know how to construct (i.e., devise, design, *Ed.*) algorithms in the precise mathematical sense of that word[2] when applying them to problems being examined in these fields of knowledge. Nonetheless, we are able to construct prescriptions having a number of the essential traits of algorithms (namely, specificity of actions, generality, and resultivity), but not possessing certain of their other features. In order to distinguish a prescription of the latter type from prescriptions which are algorithms in the precise mathemati-

33

cal sense of the word, we shall call such prescriptions *quasi-algorithmic prescriptions,* or, for brevity, simply *algorithmic prescriptions.*[3]

The decision to introduce the concept of quasi-algorithmic prescriptions is determined by the following considerations. The notion of a (mathematical) algorithm is linked to two requirements. In pedagogy and psychology (and also several other sciences) these requirements are fulfilled in a special way and often with some degree of approximation. The first of these is the requirement for the univocal recognizability (discriminability and identifiability) of those objects on which the operations prescribed by the instructions of the algorithm are to be carried out. The second requirement is for *the isolation and clear indication of the final set of the operations* used in the given algorithmic system.

These requirements are fulfilled in the following way—with various specifications of the notion of algorithm—within the theory of algorithms.

First, the sphere of objects of some standard type (these are usually letters or symbols of some alphabet) is fixed ahead of time. With reference to these objects, it is stipulated that there must be no difficulties in distinguishing and identifying them, since they will be perceived clearly and unambiguously and grasped intuitively in the same way by all. The unambiguity of these objects is considered to be absolute and unconditional *a priori.*

Second, a finite set of operations on these objects is also fixed ahead of time. Further, the elementarity and realizability of these operations are also considered to be absolute and unconditional and are assumed to have been given beforehand.

The distinction between the concept of quasi-algorithmic prescriptions and a (mathematical) algorithm is the following. In accordance with the character of problems arising in pedagogy and psychology, the sphere of unambiguously recognizable objects with which one has to deal cannot be completely fixed in advance, or standardized in some way. The lack of ambiguity (specificity, distinguishability, and identifiability) of objects in this case is particularly relative. It depends on the intellectual level of the students, on the skills and artistry of the experimenter, teacher, or methodologist, on the level of knowledge attained by the person, and so on. In the practical situation of teaching people, the ambiguity of an object, or its absence, is for the greater part—and this will be shown later—something which can be established only through observation and experiment.

Analogous with this are operations taking place in the process of teaching people. They are too diversified to be presented in some kind of complete, definitive list. Their elementarity or simplicity also has a relative character. (We will discuss this later in more detail.) Instructions to be included in algorithmic prescriptions often presuppose an appeal to the understanding of what a person is doing, to a clear evaluation of the meaning, the content of language expressions, and so on. One can say—what this signifies will be explained below—that in contrast to algorithms in the strict mathematical sense, quasi-algorithmic prescriptions permit instructions to be directed not only to operations on formal symbols, but also to operations on material content (i.e., operations in the empirical world, including its semantic manifestations, *Ed.*).

This should be understood as follows. An algorithm (in the mathematical sense) presupposes operations on unambiguously recognizable objects of some standard type (usually some sorts of symbols) according to completely precise, univocally understood rules. However, these operations are not necessary per se, but only to attain certain practical or cognitive goals. Beyond the symbols with which one operates is a defined content extraneous to the symbols, a content which is represented with the help of the symbols. In the case of algorithm in the mathematical sense, we abstract from that content. Such an abstraction is not arbitrary on our part. It facilitates our acting according to an algorithm, since it does not divert the attention onto the meaning of the operations and the basic sense of the content beyond the symbols which are transformed. With such an approach, applications of algorithms are included within the general "fabric" of cognition by means of the fact that usually material interpretation of symbols is made only for the initial data of an algorithmically solved problem and for the obtained results.

Such an approach is more easily realized in deductive fields of knowledge, i.e., in mathematics, logic, certain areas of theoretical physics, and the like. This is so, because in order to apply algorithms in the mathematical sense it is necessary to establish a sufficiently defined correspondence between the symbols on which the algorithm operates and the content represented by them.

It is considerably easier to establish such a correspondence in the deductive sciences, especially in mathematics, which deal with idealized abstract objects (see, for example, [3], [7]) than in the experimental-descriptive sciences. The latter deal not with abstract things such as numbers, points, or functions, but with processes and patterns having direct expression in concrete, sensory realities.[4] People have long since overcome this difficulty by addressing themselves to their meaning while solving problems in these fields. Insofar as such a solution always essentially presupposes the use of symbols (even though they are symbols of a natural language such as words, grammatical sentences, and so on), they are considered as having a definite meaning. The application of definite rules of transformation to them presupposes the taking of this meaning into account. Quasi-algorithmic prescriptions, while largely retaining the properties of specificity, generality, and resultivity, at the same time permit operating with objects not only of a symbolic nature but as well with whatever lies *beyond* those objects, i.e., with their content and meaning.

It is evident from what has been said that the notion of a quasi-algorithmic prescription is less precise (in the mathematical sense) than the notion of an algorithm. Appealing to the meaning content of the object of an operation brings with it a rejection of what is characteristic of modern mathematical logic. That is the distinction among levels of consideration, one of which refers to symbols as such (syntactical level) and the other takes into account the meaning (significance, materiality) of the symbols (the semantic level). The above-mentioned rejection, by the way, is closely linked with the rejection of considering only the unambiguously recognizable objects of some standard type which is characteristic, as we noted above, of the mathematical theory of algorithms. Further, this leads to the view that the objects of transformations by quasi-algorithmic prescriptions can be any things, phenomena, or processes, so long as they possess, within the necessary limits, the properties of constancy and identifiability. Such an approach leads to the replacement (in our judgment) of the concept of algorithm by the concept of quasi-algorithmic prescription. The character of such prescriptions specifies or stipulates the approximation (by comparison with mathematical algorithms) of their properties of specificity, generality, and resultivity.

What is essential to note is not alone the fact that these properties are realized only approximately, but also that a quasi-algorithmic prescription may contain a reference to the semantics of the operations and to the meaning of the objects on which the actions are carried out. At the same time, quasi-algorithmic prescriptions—just like algorithms are applicable to the solving of various problems—do permit the varying of their initial data, do specify or stipulate the orientation of the prescription toward obtaining the desired result, and so on.

Since an algorithmic prescription permits instructions appealing not only to the formal but also to the semantic operations having to do with meaning content, this concept is pertinent above all to the widening of the sphere of what may be considered an elementary operation. It is largely in this connection that we speak not of algorithms, but of quasi-algorithmic prescriptions (in the sense outlined above) in the fields of education and psychology.

2. Relativity of Elementary Intellectual Operation: Concept and Criteria

If one may consider the most varied operations as elementary, then what are the criteria for elementarity of operations? How can one determine whether an operation is elementary? In mathematics that question is always answered by concurrence. Specific complexes of operations are introduced (which are different in different algorithmic systems) and these operations are declared elementary.[5] In other fields of knowledge, particularly in the cybernetics of non-living systems, such a means of solving the problem of elementarity or non-elementarity of an operation may have only a very limited application, inasmuch as cybernetics has to do with real and diverse controlling systems.

In these fields, the question of whether an operation is elementary must be solved in each separate instance on the basis of a detailed analysis of the inherent capabilities of the system which produces these operations and for which an algorithm is established. Therefore, acknowledging specific operations as being elementary must in that case depend on the detailed study of the properties of real controlling systems and on drawing conclusions from these properties. Thus, the *notion of elementary (simple) operations is relative.* One and the same operation may be elementary for one system and non-elementary for another. It all depends on the complexity and the structure of the system. One may characterize an operation as elementary only after having correlated it with the peculiarities of the system which produces those operations. The elementarity of operations, therefore, is always correlative with the system which carries out these operations.

Thus, certain operations which are elementary for a man, for example, operations to distinguish simple images, are not elementary for a machine. In order to instruct a machine how to distinguish objects, it is necessary to break down the act of distinguishing, which is so elementary for a man, into even more minutely elementary acts (operations).

What has been said about the relativity of the concept of an elementary operation has significance not only for the comparison of such systems as man and machine. It is quite clear that many operations which are elementary for a grown man are non-elementary for a child. Insofar as the process of teaching a man is characterized by the fact that as he learns he *develops* his capabilities, then the operation which was formerly non-elementary on one level of development becomes elementary on another. This is why the question of whether an operation for a given system (or for a given level of development of a system) is elementary must be resolved in the final analysis by experimentation.

What is the experimental way of solving the question of whether one or another operation is elementary or not? What is the criterion of the elementarity of an operation?

That criterion is the capability of a controlled system to carry out uniformly and faultlessly a given operation in response to a corresponding instruction within a quasi-algorithmic prescription.[6] The task arising for the compiler of a quasi-algorithmic prescription consists, in that case, in the following.

Let us suppose that we must construct an algorithm for controlling some system. In order to ascertain the degree of fragmentation to which the instructions given in the prescription should be broken down, one must know which operations are elementary for the given system. If we know this (for example, we know which operations the given machine or person can produce), the instructions should be adapted to the possibilities of that machine or person. An algorithm in that instance should cover instructions appealing only to those operations which the machine or person is capable of carrying out. But there could be instances in which we would not know which operations are elementary for the given system. Not knowing that, we would not know to what degree of fragmentation the instructions must be broken down in the algorithmic prescription. *The only way of solving the problem in this case lies in experimentation.*[7]

An experiment could be constructed in the following manner. At first an hypothesis is formed about which operations are elementary for the given system. Then, proceeding from the hypothesis, a prescription is formulated which is tried out afterward, one part at a time. If, in response to the instructions contained in the prescriptions, the system (for example, a person or a group of people) uniformly and correctly performs the corresponding operations, then the hypothesis was correct. The given operations are indeed elementary for the given system, and the prescription is algorithmic. If the system is not able to carry out the instructions, "doesn't know how" to do the necessary operations or does them differently in different instances (or incorrectly altogether), then there are the following implications. The given operations are not elementary for the system. The hypothesis was incorrect. More fragmentary instructions must be put into practice, and the operations must be broken down into more elementary ones.

Thus, a prescription is quasi-algorithmic (that is, appeals to the elementary operations of a given system) if it results, according to the accepted criterion, in a uniform and correct execution of the instructions which make up the prescription.

In finding out which operations are elementary for a certain system, it is important, however, to determine in each instance not only the upper limits of the elementarity of an operation, but also the lower levels. That is, one must determine whether or not the operations toward which the algorithmic prescription is directed are too elementary. This, too, can be discovered only through experimentation. Suppose a system is given certain instructions, and it carries them out uniformly and without mistakes. In order to determine whether these instructions are excessively detailed or directed at too elementary operations, one must gradually render these instructions more complicated. Rendering the instructions more complex must be continued until such a time as the system will no longer be able to carry out the instructions uniformly and without mistakes. The instructions of the final level which the system executed correctly will determine the lower limit of fragmentation of the instructions and the elementarity of the operations.

What was said about determining the degree of fragmentation of the instructions and of the elementarity of the operations has tremendous significance in particular for programmed instruction. In order for programmed instruction to be correct and effective, one must know precisely the level of development in knowledge and in operations of the students for whom the educational material and the corresponding activities are programmed. It is necessary to discover and analyze what they have mastered and what they have not mastered, and which operations at a given level of their development are elementary for them. This analysis, including the formulation of an hypothesis and the conduct of a delicate psychological experiment also determines "the level of fragmentation" that must be attained in breaking down complex processes into elementary operations.

Let us illustrate what has been said by two examples. Suppose that for a certain group of people it is necessary to set up a prescription for carrying out a certain activity. This activity includes the following operations: pushing a button, turning a handle to the right, and assessing the situation according to some criterion. Will the prescription which includes instructions for carrying out the specific sequence of these operations be algorithmic? It is impossible to give a simple, univocal answer to that question so long as it has not been discovered whether the operations are elementary for the given group of people. In regard to the first two operations, they are undoubtedly elementary for normal adult people who understand the English language. In response to the instructions: "push the button," "turn the handle to the right"—they all univocally carry out the necessary actions.* The third operation is more complex. If the criterion by which one must assess the situation is indicated precisely enough, and if the people know this and are able to apply it identically (correctly), then one may consider the given operation as elementary and the instruction for carrying it out as "algorithmic." If these conditions do not exist, i.e., if certain people from the given group cannot make an evaluation in response to the instruction so as to assess the situation according to the given criterion, or if they do it incorrectly, or assess the situation differently (i.e., the evaluation will be equivocal), then one may not consider the given operation elementary for everyone, and the corresponding instruction will not be "algorithmic."

An analogous situation exists when setting up other algorithmic prescriptions, including those that have to do with the realization of intellectual activities. Suppose, for example, that it were necessary to resolve the question of whether or not one could include, when formulating an algorithm for solving a certain grammatical problem, the instruction, "choose the sentence's subject." If it is known that the students for whom the

---

*However, the second operation could be non-elementary for a small child. He might not know or might not be sure of what "to the right" means.

quasi-algorithmic prescription is constructed can uniformly and faultlessly choose the subject in the sentence, then one may consider that operation elementary and not break down the corresponding instruction for that operation into more fragmentary ones. If it is known, however, that the students cannot discriminate the subject uniformly and faultlessly, then the given instruction must be broken down into a number of still more fragmentary ones appealing to yet more elementary operations. In the case where it is not known whether the students can or cannot pick out the subject, an experiment must be set up. If, in response to the instruction, "pick out the subject," the subject will be correctly discriminated in the corresponding sentences, then one may consider that operation elementary; if not, then it should not be considered elementary.

It is clear from what has been said that the notion of a quasi-algorithmic prescription is not one that can be defined at the same level of strictness as the notion of an algorithm. At the same time, when applying it to the experimental-descriptive fields of knowledge, one may consider it sufficiently precise to the degree that one may for each practical instance evaluate against experimental criteria the problem of whether or not the given prescription is algorithmic. It is important to note that, although it is rather precise, this notion is nonetheless relative. This means the following. The same prescription may be either algorithmic or non-algorithmic, depending on the system to which it is directed, the level of development of this system (for example, the level of a person's development), and the control over whose actions must be achieved with the help of the prescription.

With respect to a single person, the question of whether a certain prescription is algorithmic can be answered rather simply. However, determining the algorithmic character of some prescription intended for a group of people is a problem which cannot be solved simply. In classroom teaching, it usually happens that certain operations are (or have become) elementary for some students even of the same class group, while for others they are not. Therefore, the prescriptions directed at those operations are algorithmic for the former and non-algorithmic for the latter. This leads to the recognition that the prescription which is directed at a group of students cannot determine the process of activities for all of the students of a given group univocally and on an identical level.

What is the criterion for the algorithmic character of prescriptions under conditions of large-scale, class-group teaching? This criterion can only be statistical. One may, for example, admit that a certain prescription is algorithmic if it determines the process of activities of 85 percent of the students in a given group. One may admit, for example, that it is algorithmic if it determines the process of activity of 95 percent of the students in a given group, and so on. Precisely which criterion should be accepted in a situation depends on a number of conditions which require additional analysis and study.

One of the most important means of raising the "determining power" (power to govern the actions of a student, *Ed.*) of a quasi-algorithmic prescription in large-scale teaching consists of the following. Before calling certain operations elementary, it must be determined whether they actually are elementary for everyone. If not, then these operations must be specially formed. A highly effective method of forming single intellectual operations (reasoning operations) is the method of forming them by stages, which was developed by Galperin and applied in a series of investigations (see, for example, [311], [312], [519], [572], [573], [578], [581], [599], [612], [656], [723], [730], [731], [767], [769], [811-814], [834], [835], [854]).

Successful instruction for each student is possible only when, as it is conducted, the level of development of his knowledge and of the formation of his operations is taken into account. Obviously, in large-scale, class-group teaching, this is impossible from a practical viewpoint.[8] During a lesson, the teacher in this situation cannot treat every student individually, taking into account the level of his development, his individual peculiarities, and whatever method is optimal *for him* (the student). The teacher ordinarily has to orient himself toward the "average" student. In this respect programmed instruction opens up completely new perspectives. It makes possible the combining of class-group teaching with individualization and the adjustment of the level and character of teaching to the level of development of knowledge and the formation of operations in each individual student. Programmed instruction also opens up perspectives for adapting instruction to the individual characteristics of the student (at least to some of them), permitting in a number of cases the implementing of a flexible transition from one level of teaching to another, according to the degree to which a student has mastered the knowledge and the operations. Thus, programmed instruction first creates the possibility in mass instruction of using the prescriptions which will take into consideration the level of development of knowledge and operations of *each student*, and which—if needed—are directed toward operations which are elementary *for him.*

3. Formal and Content Aspects in
Quasi-Algorithmic Prescriptions

In speaking of operations to which algorithms must be directed, we emphasized that these operations may deal not only with form, but also with content. The important thing is that they be elementary and that a person be able to carry them out correctly and uniformly.

One example of an elementary semantic operation dealing with content could be a person's evaluation of the truth-value of a certain type of phrase. For example, if one were to ask whether the phrase, "He got off the plane at Tashkent and immediately went home to his Moscow apartment," is truthful, then everyone would say no, since a person cannot be both in Tashkent and Moscow at the same time. The impossibility of finding an object in two different places simultaneously is a distinctive semantic axiom which people use in their reasoning and in their actions. The application of that axiom can undoubtedly be considered as some elementary semantic operation. Different people carry out this operation identically and, therefore, the instructions appealing to that operation may be considered algorithmic (in the sense of "algorithmic prescription"). This instruction univocally and completely specifies the corresponding semantic action of a person and leads to identical results on different occasions.

It is easy to be convinced of this, if one sets up such a prescription:

1. Read the sentence.
2. Ascertain whether it is true.

> If yes, then cross out the first city in it.

> If no, then cross out the second city in it.

It is unnecessary to conduct special experiments in order to be convinced that all normal, literate adults knowing the English language and basic geography will see the senselessness of that sentence and perform exactly the same action (cross out "Moscow"). Consequently, the given prescription univocally determines not only the actions of the problem solvers, but also the change which they bring about in that sentence. Since people normally execute such univocal, specific transformations not only with a specific given sentence, but with any sentence of this kind, a prescription similar to the one cited may be considered algorithmic. Ways of determining the elementarity of some semantic operation would be the same as that described above for any other operations; i.e., a person must carry out the operation identically and correctly in response to the algorithmic instruction.

It follows from what has been said above about algorithms that algorithmization in the mathematical sense presupposes the formalization of problem solving and permits an automatic execution of actions not only in the sense that these actions may be handed over to a machine and carried out by it,* but also that when a person carries out these actions, they *may* be accomplished "purely mechanically," i.e., without penetrating into the meaning content of the operations and the significance of the symbols with whose assistance the transforming operations are executed. What has been said, however, does not mean that the solving of problems by algorithm *demands* without fail an automatic execution of operations and does not permit awareness of their meaning (in the instance where a person is solving a

---

*Any problem for which an algorithm may be constructed can be solved, in principle, with the help of a machine. See, e.g., Glushkov [85].

problem by algorithm). B.A. Trakhtenbrot [163] says: "The formal character of elementary operations [he is speaking of a mathematical, computing algorithm] lies in the fact that they *can be* carried out automatically" (the italics are mine—L. Landa). But neither Trakhtenbrot nor other mathematicians say that it is *obligatory* for operations to be carried out automatically and only automatically with no allowance for any awareness of what kind of operation it is, or its meaning-content, and why it is necessary to execute them. In other words, algorithms guarantee the *possibility* of an automatic execution of the operation, but they do not *require* it. They do not exclude being aware of and comprehending operations and their *conscious* execution.[9]

If mathematical algorithms which are already computerized may be consciously applied and carried out, then what was said applies to an even greater extent to algorithmic prescriptions which in many cases are directed specifically at the semantic operations having to do with content. Moreover, the execution of algorithmic instructions applied to semantic operations without a definite level of awareness about the object of the activities and the operations themselves, i.e., purely automatically (mechanically) is impossible. The reason why semantic operations are semantic is that they cannot be fulfilled formally, i.e., without attributing symbols to their object-meaning and without taking into account their meaning.

What has been said does not signify that machines cannot solve semantic problems. However they can solve them (at least today) only after these problems have been formalized beforehand. At the present time, studies on the formalization of semantics have assumed greater and greater scope (see, for example, [10], [22], [35-37]).[10] Studies which describe understanding via "semantic factors" [101] and tests on the discernment of the meaning of a phrase by a machine (see, e.g., [88]) are of great interest. It has already been shown in practice today that it is possible to solve semantic problems (which have been formalized in advance) with the help of algorithms. It has also been shown that many semantic phenomena may be described strictly and analyzed precisely.

The possibility of formalizing and consequently—in the mathematical sense of the word—of algorithmizing semantic processes has a great gnoseological as well as practical significance for both pedagogy and psychology. One may assume that a deeper penetration into the essence of semantic processes and a precise description of them will permit the development of ways for more effectively forming these processes in students.

Quasi-algorithmic prescriptions appealing not only to the formal but also to the semantic operations having to do with content require in the latter instance the implementation of specifically human modes of action. They include the cognition of objects, of actions, and of the operations to be carried out on these objects. It is precisely for this reason that teaching students to act on the basis of quasi-algorithmic prescriptions can serve as a means of forming conscious activities and mastering specifically human modes of reasoning. Therein lies the great significance for pedagogy and psychology of quasi-algorithmic prescriptions.

Notes

1. For a concept of formalization, which has been defined more precisely in modern mathematical logic, see, e.g., Klini [107] and Kaluzhnin [103]; the methodological aspects of formalization are more thoroughly examined in the article by Biryukov and Konoplyankin [56].

2. For the mathematical concept of an algorithm see, e.g., the article of Uspensky [166], which has an extensive bibliography.

3. We note that often the term "quasi-algorithmic prescription" (or "algorithmic prescription") leads to stylistic difficulties. In those instances where such difficulties will arise, we will replace that term by the word "algorithm." We emphasize, however, that the word "algorithm" will be used here, not in the strict mathematical sense, but as a synonym for the term "quasi-algorithmic prescription." For example, instead of the expression, "an algorithmic prescription for a mode of

teaching," we will simply say "a teaching algorithm" or "an instructional algorithm." Instead of the expression "an algorithmic prescription for an identification procedure," or, "identification of an algorithm," we will simply use "an algorithm of identification," and so on.

4. This, however, does not mean that in deductive sciences the establishing of such a correspondence—and, therefore, the formalizing of their contents—does not meet with difficulties. There are difficulties, and they have a fundamental character as the famous results of Goedel attest, which prove the incompleteness of formal arithmetic. (See Note No. 5, Chapter One.)

5. Of course, in mathematics, declaring that defined operations are elementary is not a purely arbitrary matter. It is always based on defined theoretical considerations and on relevant experience.

6. The highly automatic execution of this operation is also psychologically important in the sense that it is carried out (if it consists of a series of even more simple operations) as *a single* operation and as a *unique act.*

7. It is assumed that an experiment on ascertaining whether a prescription has an algorithmic nature is carried out under conditions normal for the functioning of a system (for example, the person with whom the experiment is conducted is not sick or tired, etc.). Inasmuch as the very notion of "normal" conditions is not sufficiently specific for these purposes, what is chosen as a criterion of the algorithmic nature of an instruction does not have to be a completely faultless execution of that instruction. It can be a correct execution in only a certain percentage of cases in repeated trials (for example, 95 percent, 90 percent, etc.). The choice of a criterion can be determined by the most diverse reasons, but in each case it must be indicated.

8. In the experience of progressive teachers, there are many examples of combining group instruction work with individual (drill and practice, *Ed.*) work, particularly examples of conducting classes where different groups of students are given

assignments of different degrees of difficulty (see, for example: [590], [598], [659], [714], [761], [808]). But, here, adjusting the character of teaching to the level of knowledge and operation of *each* student has not succeeded.

9. The concept of the conscious execution of an operation is not very well defined. In speaking of the consciousness of executing the operations of an algorithm, we have in mind the person's comprehension of the objects at which operations are directed, i.e., a comprehension of the meaning-content of the operations, his understanding of the sense of the problem being solved by means of that algorithm, his understanding of the significance of the algorithm and of its structure, the comprehension of the completed activity, though such a type of understanding and comprehension is not foreseen by the algorithm, etc. We also have in mind different sorts of associations that may arise during the course of carrying out the operations. To solve a problem by an algorithm (in the mathematical sense) a similar comprehension and understanding are not necessary. But a person as a reasoning being tries to comprehend any activity carried out by him, even that which can be done mechanically.

10. A good bibliography on semantics can be found, e.g., in Schaff [33].

Chapter Three

Algorithmic Prescriptions, Algorithmic Descriptions, and Algorithmic Processes

1. Basic Concepts

Up to this point we have treated algorithms as prescriptions covering a certain system of rather elementary operations, which when carried out permit the solving of specific classes of problems.

It will be completely understandable that algorithms in themselves solve no problems whatever. The problems are solved in the process of executing the operations prescribed by the algorithm or corresponding to some algorithm. It is perfectly natural to distinguish precisely a *prescription* for carrying out a specified system of operations from the *system of operations* itself. If a specific type of prescription for carrying out a system of operations is called an *algorithm* (or an *algorithmic prescription*), then it is expedient to call the very execution of the system of operations and the process of solving the problem by an algorithm—or in correspondence with an algorithm—an *algorithmic process.*

Uspensky [165] characterizes the algorithmic process in the following way: "An algorithmic process is the process of applying an algorithm to some object." He adds: "When using that notion in the fields of psychology and pedagogy, it requires further differentiation."

The necessity for such a differentiation arises from the fact that one must distinguish between execution with and without awareness, of operations (including those prescribed by an algorithm). As we know, a person often solves specified problems without knowing the algorithm which is the basis of the solution (this is even more so in the case of animals). Even a person is often unaware of the operations carried out by himself. This is particularly evident from the fact that he cannot name them and also cannot describe the way the problem is solved. In other words, a person often solves a problem in a way which is not an *application* of some algorithm, which he may or may not know. However, there may be a strict order in his actions, a strict pattern which can be determined and precisely described algorithmically.

The situation here is analogous to one which takes place, for example, in the processes of verbal communication. Grammatically, people speak correctly often without knowing or remembering the rules of grammar (and children always speak correctly without knowing grammar). Their speech is not an *application* of the rules of grammar, but it is *in correspondence* with grammar rules and can be grammatically described. Moreover, a prescription on how to speak can be created on this basis.

Shevaryov has done special research on actions whose characteristic trait it is that the people who carry them out are unaware of the corresponding rules, but conform to them (see, for example, [604], [616], [623], [646], [810], [840], [841]). Shevaryov singled out a class of associations which are distinguished by the fact that each one of them corresponds to a particular rule, but the "explicit recall of the rule (as a general statement) does not enter into the functioning system of associations" [840]. Shevaryov called such associations generalized, or "conforming to rules."

It is natural to designate as algorithmic not only a process of applying an algorithm to the solution of a defined class of problems where the algorithm is familiar to a person (or a machine), but also to the process which displays such a high regularity that an exact mapping into an algorithm is possible. In other words, we are referring to a process to which an algorithm can be put in exact correspondence.

Therefore, here and further on, when we say an algorithmic process, we will mean a process which is either the application of some algorithm, or a process which is not an application of an algorithm, but which may be described algorithmically. In the latter case, it is a process which could *correspond* to some specified algorithm or algorithmic prescription (one that proceeds in correspondence with that algorithm).

The proposed interpretation of the notion of an algorithmic process corresponds to the problems and methods of cybernetics. "The basic method of cybernetics," writes Lyapunov, "is the method of algorithmically describing the functioning of controlling systems" [120].

Algorithmic descriptions of the most diverse systems and processes have been accomplished. Thus, Lyapunov and Shestopal [122] algorithmically described the work of Watt's regulator, the process of forming conditioned reflexes in animals, and the activities of a controller for a railroad. They considered as elementary such operations or acts as the changing of gears in a steam engine—which correspond to the weight distribution of the mechanism (in terms of a parallelogram), changing the weight distribution of the parallelogram into the distribution of the joints, altering the amount of steam going from the boiler to the cylinder, the jerking back of the hand as a result of the influence of an irritating stimulus, the chemical action of an organism (e.g., the excretion of saliva) in response to an action of the irritant, the command of the controller for sending out a train or for transferring a train to a siding, and so on. Braynes, Napalkov, and Svechinsky [61] algorithmically described the process of establishing specific forms of behavior in animals. Pushkin [146] algorithmically described the activity of an assistant stationmaster. Zhinkin [99] described the vocal communication system of monkeys. In Landa [271], [272] the process of solving grammatical problems is described algorithmically. Also, an algorithmic description of a series of industrial and other processes has been made.

Obviously, one must distinguish an algorithm that is a *description* of an algorithmic process from an algorithm that is a *prescription* for its execution. If an algorithmic prescription directly *controls* an algorithmic process and "evokes" it, then an algorithmic description merely *states* how the algorithmic process comes about and how it proceeds.

Wherein lies the significance of an algorithmic description of processes? It lies in the fact that such a description is an essential condition of cognizing the algorithmic process toward the purpose of a subsequent control over them or in order to model them. One could even say that some process (if it is algorithmic) is actually known only when its algorithmic description is given. Conversely, an algorithmic description of a process may be given only when it is known in sufficient depth.

The task of providing algorithmic descriptions of processes arises when it is necessary to determine the structure of these processes (i.e., to establish the algorithm of the systems' workings). When it is necessary to give a description of an algorithmic process, the question always arises as to the degree of fragmentation into which one must break down that process and which acts (operators) and logical conditions should be considered elementary. It is impossible to answer that question in a general form, since everything depends on the level of description. Thus, for example, in the macro-analysis of a process, one "level of fragmentation" of a description is necessary, and in micro-analysis, a different one.[1] Within the framework of macro- and micro-analysis, there may also be different levels of fragmentation of descriptions.[2] However, in those instances where the direct purpose of a description is the development of a prescription for controlling that process, the necessary level of fragmentation of descriptions can usually be determined rather precisely. Here, the criterion of whether the level of fragmentation of the description has been determined correctly is the capability of the controlled system uniformly to produce those processes prescribed by the algorithm.

Cognizing and describing algorithmic processes is necessary for the active intervention of a person in these processes (often proceeding naturally), in order to control them directly or indirectly. Often, one may construct an algorithmic prescription only after the algorithmic process has been discovered and described. Therefore, bringing to light the structure of algorithmic processes is one of the more important tasks of science.

The interrelation between an algorithmic prescription and an algorithmic description implies that each prescription for the execution of an algorithmic process may at the same time also be considered as its description. But not every description of an algorithmic process performs the role of the prescription by which it is fulfilled. One may give, for example, an algorithmic description of certain processes taking place in the brain (Braynes, Napalkov, and Svechinsky [61] have made such an attempt), but the resulting description does not constitute a prescription which "governs" the brain in its work.

Therefore, algorithmic prescriptions and algorithmic descriptions are not one and the same. They have different roles.

2. Methods for the Description of Algorithmic Processes

Let us turn to the methods of describing algorithmic processes, and exhibit some examples of such descriptions. We begin by taking certain aspects of human activity as the object of the description.

If we first observe the work of an operator at a control panel which has many lights, levers, knobs, etc., then at the beginning it is very difficult to see some sort of pattern in the operator's flow of activities. It is difficult to answer why at a particular moment the operator pushed a certain knob and not another, or turned a lever to the right instead of to the left, and so on.

In order to gain an appreciation of this flow of actions and describe precisely the regularity of the activity (without such a description it is impossible to set up a program for studying and teaching this activity), it is necessary to dismember or break down the complete activity into the sum total of its component actions. Then it must be brought to light how these elements are linked, and the whole as a specific composition must be recreated. Insofar as actions are always evoked by specific conditions and are directed at specific objects, one must also include these objects and conditions in the description as some of the elements. However, in the description, objects may often be united with actions as, for example, in the phrase "push the knob."

Suppose there is an operator who, before beginning his work, must verify whether the apparatus with which he is working is in order. The problem is solved in the following manner. First of all, he must verify whether the apparatus is plugged in. If not, then he must do this. If it is plugged in, then he must flip the switch and see whether the red light goes on. If it does, then the apparatus is in order, and he may begin working with it. If it does not light up, then the apparatus is not working, and a technician must be summoned to repair it. Here, the link between conditions and actions is "rigid," and when a specific condition appears (for example, the red light goes on or does not) a specific action must be executed (either work with the apparatus is begun, or a technician is summoned). The activity would have a non-algorithmic character if the person, when responding to the lighting up of the red light, sometimes performed one action and sometimes performed another.

The first and most widespread method of algorithmically describing processes (including different aspects of activity) is the verbal description. In essence we have already given an algorithmic description of the activities reviewed by us. We showed how the operator must act under specific conditions. If one wishes, one may make these instructions more imperative by writing each of them on a separate line and numbering each of them.

For example:

1. Verify whether the apparatus is plugged in.
   If yes, then proceed to instruction 3.
   If not, then
2. Plug it in.
3. Flip the switch.
4. See whether the red light has come on.
   If yes, then proceed to instruction 5.
   If not, then proceed to instruction 6.
5. Begin work.
6. Call the technician.

Operations 5 and 6 are final. The activity prescribed by the algorithm ends with them, depending on the presence or absence of the preceding conditions.

Lyapunov and Shestopal [122] proposed a method of symbolically describing control processes which have an algorithmic character. To make such a description possible, it is necessary, as we already said, to break down the process of activity (particularly its control) into a series of elementary acts and to establish the conditions which provoke these acts. These acts (operations) are called *elementary operators* and are designated by the capital letters A, B, C. The conditions which must be taken into account are called *logical conditions,* and are designated by small letters, a, b, c. When necessary to indicate a permanently logically false condition, ω is also used.* The sequence of checking the conditions and the carrying out of the actions is written down in the form of a *logic diagram* which represents the expression made up of operators, logical conditions, and numbered arrows arranged in a specific manner. The operators as well as the logical conditions are called *members* of the logic diagram.

Let us set up a logic diagram of the above described activity, and for this purpose, let us introduce the following designations:

Logical conditions:
a—The apparatus is plugged in;
b—The red light is on.

Operators:
A—Plugging in the apparatus;
B—Flipping the switch;
C—Beginning work;
D—Calling the technician.

A period next to a capital letter will signify a cessation of activity. It signifies the end of the algorithm's work.

*The ω is a notational convenience. It signifies a permanently negative logical condition. It functions as a concise indicator of a logically required loop. Put another way it produces a double negation—a negation of a negative logical condition (e.g., it will be *not* true that the apparatus is *not* plugged in). Thus it serves to change the material (not the formal) truth value of some statement from F to T (or - to +). If ω were not used, loops in the formulas would require infinite sequences of segments of the algorithm (i.e., the requisite logical conditions for completing the algorithm would never be met). (*Editor.*)

Then the process of activity is algorithmically described thus:

$$a \overset{1}{\uparrow} \overset{3}{\downarrow} Bb \overset{2}{\uparrow} C. \overset{1}{\downarrow} A \omega \overset{3}{\uparrow} \overset{2}{\downarrow} D.$$

Rules for reading logical diagrams are the following:

First of all, one must look at the letter at the left. In the given instance, this letter designates a logical condition. Consequently, according to the algorithm, one must ascertain whether that condition has been fulfilled, i.e., whether the apparatus is plugged in.

If it is (if the logical condition is fulfilled), then disregard the arrow and go on to the next letter (fulfill operator B—flip on the switch). If not (if the logical condition is not fulfilled), one must pay attention to the arrow and see where it points.* It points to letter A (plug in the apparatus). After that, one must go on to the following letter (in this instance, inasmuch as this is always a false—that is, an unfulfilled—logical condition, one must transfer to the member of the logic diagram indicated by the arrow which directly follows the Greek letter ω). That arrow indicates that one must carry out operator B—flip on the switch, and so on, until there is a member of the diagram which is not followed by any other members which could be acted on. (There is no such member in our example.)

Let us summarize what was said. The action according to an algorithm begins at the far left of the diagram.

The principle of action is the following:

First instance: the letter considered is a logical condition.

---

*It should be noted that arrows occur in pairs of which one member (e.g., $\overset{1}{\uparrow}$) points "from" (e.g., a) and the other (e.g., $\overset{1}{\downarrow}$) points "to" (e.g., A) so that the numerical superscripts serve as pair indices and arrow direction expresses "go to" (up) and "here" (down). (*Editor.*)

Here there may be two variants:

(a) If the logical condition is fulfilled, then, disregarding the arrow, one must move on to the next letter on the right.

(b) If the logical condition is *not* fulfilled, then one must pay attention to the arrow and go on to the letter which it indicates.

Second instance: The letter considered is an operator.

In that instance, the operator must be fulfilled, and then one must go on to the next letter on the right.

Action is carried on until it meets an operator with a period (sign of cessation) or an operator which is not followed by a member which must be acted on. After fulfilling one such operator, the action ceases. Let us note that algorithms may exist in which the action they direct continues infinitely.

The third way of describing an activity which has an algorithmic character is its graph.

Here is a graph of the activity under examination:[3]

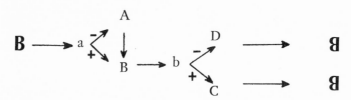

The arrows on the diagram indicate the sequence for verifying the conditions and executing the actions (operators). The plus and minus signs indicate the presence or absence of a condition. The letters with periods represent the concluding actions.

The conditions as well as the actions may be designated in a graph, not only symbolically, that is, by letters, but also by ordinary words.

Along with the means already indicated for algorithmically describing processes, still others have been developed (e.g., Markov's description of algorithmic processes [127] in the form of "normal algorithms" and description in the form of a diagram of "Turing machines" [200]).* However, it is inconvenient to apply them in education and psychology. For psycho-educational purposes, it may be more convenient to describe algorithmic processes in the forms mentioned in Chapter One, particularly block diagrams, than in the above notation.

As we already noted above, in order to describe algorithmically a concrete activity, one must first break it down into elements (logical conditions and operations) and then synthesize this activity as a combination of these elements.** It is important to underline the general methodological significance of this scientific method.

The method guiding research toward the discovery of elements "making up" a whole, their functions, and forms of connections permits a structural analysis and structural descriptions of phenomena and processes. Once the basic elements of a whole are found, and their functions and the ways they are connected are determined, the description of these phenomena and processes in the precise language of logic becomes possible at once. Let us note that describing the whole and the concrete through the combining of elements does not mean that the whole is reduced to the sum of its parts, but that it becomes comprehensible only through an analysis of its component elements, their functions, and the ways they are connected.

Not only can one describe algorithmically a controlling activity expressed in external physical actions, but one can do this also for psychological and particularly cognitive activity as expressed in the internal intellectual operations. Examples of an algorithmic description of intellectual activity while solving problems (based on material from students' solving of grammatical problems) will be given in the second part of this book.

---

*Popular exposition of these algorithmic systems is given, for example, in Trakhtenbrot's brochure [163].

**If one considers the symbols as representing logical conditions and operations, and the ways they are linked, then any concrete algorithm can be presented as a word in that alphabet.

3. Algorithmic Description of
Teaching Activity

In the preceding segment of this chapter, we introduced an example of an algorithmic description of a controller's overt activity, and we indicated the possibility of algorithmically describing people's intellectual activity while solving problems.

May one algorithmically describe the activity of the teacher (teaching activity) and create on the basis of that description an algorithmic prescription for an instructional procedure (a teaching algorithm)?

Obviously, it is possible if that activity in some instances has an algorithmic character. But, in order to do that, one must break it down into elements and present it as a specific combination of those elements. Teaching activity (like many others) cannot be completely algorithmized, insofar as it is impossible to foresee all the logical conditions and their combinations which might confront the teacher in the course of his or her work and which might require that he or she introduce new methods for influencing students which had not previously existed. The instructional process is an evolving one; as the years go by, new conditions and new kinds of instructional actions appear. But the activity of the teacher when solving often-encountered, typical pedagogical problems can be algorithmized, and therefore, can be described algorithmically.

Teaching, like any other activity, is made up of specific teaching activities (operations) which are applied in a sequence depending on the logical conditions and the chosen plan of instruction (throughout, we will consider the objectives of teaching as given). From the foregoing it is clear what must be done to carry out an algorithmic description as the basis for an algorithmic prescription for an instructional procedure (a teaching algorithm). First of all one must determine conditions which are important for choosing among instructional actions (as an algorithmic description, these will be logical conditions). Second, one must determine the instructional actions themselves (in an algorithmic description, these will be the operators). Third, one must determine the way in which they are linked, i.e., the plan of instruction (in an algorithmic description, this will be the logic diagram of the algorithm).

As an example, let us develop one teaching algorithm, having constructed a program of activity for the teacher in the process of teaching.

Let us suppose that we have to acquaint the students with a concept that is new to them (for example, with the concept of a "circle"). This can be done in various ways. Let us choose one of the possibilities. It can be expressed in the form of the following algorithmic prescription:

In order to give the students a notion of a circle one must:

1. Give a definition and a sketch. ("A circle is a closed curve lying in a plane, all points of which are equidistant from one point. This point is called the center of the circle."*)

2. Suggest that the students identify the characteristics indicated in the definition and name them.

If a given student identifies all of the distinguishing characteristics, go to instruction 6.

If he has not identified all the characteristics, go to instruction 3.

3. Sketch a figure having all of the characteristics named by the student, but not corresponding to the actual definition of a circle and its geometrical form.**

*Here we proceed from the assumption that the teacher gives the definition introduced in the secondary school geometry textbook by Nikitin [732]. In this definition, four characteristics of a circle are given: a circle is the kind of line which (1) is curved, and (2) closed, and (3) on a flat surface, and (4) all of its points are at an equal distance from one point which is called the center.

**Vorobyov called such a device the counter-image [563]. The essence of this device can be clarified by such an example as this: Let us suppose that, in defining a circle, the student indicates the following characteristic: "A curved closed line on a flat surface" and does not indicate the characteristic "all points are equidistant from one point called the center." Instead of simply pointing out his mistake to the student, the teacher draws a figure (in this case, an ellipse), having all of the characteristics indicated by the student, but not corresponding to the true definition of a circle and its geometrical figure. Inasmuch as a student distinguishes intuitively between a circle and that which is not a circle, he sees immediately that he did not indicate all of the necessary characteristics, and that his definition is incorrect.

If a student did not correct his mistake, go to instruction 4.

4. Call on another student.

5. Suggest that this student correct the mistake.[4]

6. Suggest the naming of logical connectives which link the characteristics.

7. Write down all of the characteristics together with the connectives. This concludes the steps prescribed by this algorithm.

Therefore, the teacher goes on to exercises in which the students apply the concept of a circle and verifies their mastery of it. Space does not permit the algorithmic description of this stage of work here.

It is easy to see that the above algorithmic prescription is nothing but a specific program of actions to be carried out by the teacher,* which, in turn, is aimed at actualizing (evoking) specific actions from the students and their regulation. In order to describe this program of actions in the form of a graph, where the conditions and operators will be designated by symbols and/or in the form of a logic diagram of an algorithm, let us introduce the following notation:

Logical conditions:

a—The student identified all of the characteristics indicated in the definition;

b—The student corrected his mistake.

Operators:

A—Formulating a definition of the concept and representing a circle by a sketch;

---

*We do not say that this is the best program. It is one of the possible programs close to the ones which are applied practically in a number of cases during the course of teaching. The aim of the present example is not that of recommending certain teaching programs, but of showing how they may be described. An understanding of methods of description can be derived from good programs as well as from poor ones.

B—Suggesting that the student identify the characteristics indicated in the definition and that he name them;

C—Suggesting that the pupil name the logical connectives linking the characteristics;

D—Applying the device of a counter-image;

E—Calling on another student;

F—Suggesting that he correct the mistake;

G—Writing down the characteristics along with the connectives.

The graph of this algorithm will have the following appearance:

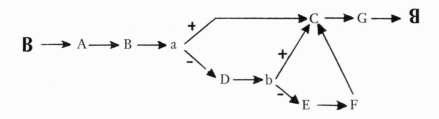

The logic diagram of this algorithm will be as follows:

$$\begin{array}{ccccccc} & 1 & 3{,}4 & & 1 & & 2 & 3 & 2 & & 4 \\ A & B & a\uparrow & \downarrow & C & G. & \downarrow & D & b\uparrow & \omega\uparrow & \downarrow & E & F & \omega\uparrow \end{array}$$

It is easy to see that, depending on which logical conditions will be encountered in a situation (the student identified all of the characteristics or not all of them, corrected his mistake or did not), the teacher's sequence of actions will be different. There is a "rigid" link in the algorithm between conditions and actions (each condition demands a completely specified action), but there is no rigid link between the actions themselves. Since combinations of conditions may be different in different instances, the combinations of actions also will be different. But the very same

combination of activities always corresponds to the same combination of conditions.[5]

Let us consider what combinations of conditions and
corresponding actions there could be.

Significance of logical conditions		Sequence of actions						
a	b							
0	0	A	B	D	E	F	C	G
0	1	A	B	D	C	G		
1	0	A	B	C	G			
1	1	A	B	C	G			

1. Conditions a and b are both violated (i.e., the student did
not identify all of the characteristics, and the student did not
correct his mistake);

2. condition a is violated, but condition b is fulfilled;

3. condition a is fulfilled, but condition b is violated;

4. both conditions a and b are fulfilled.

The dependence of combinations of actions on combinations
of conditions may be indicated in the table (1 signifies the
fulfilling of the corresponding condition; 0, that it is not fulfilled).

The possibility of algorithmically describing the activities of
students and teachers has tremendous significance for programmed
instruction. Let us stop at this point.[6]

4. The Algorithmic Description of the Students' and Teacher's Activities in the Process of Instruction: Its Role in Programmed Instruction

The most important principles of programmed instruction as
stressed by almost all the authors writing on this problem are the
significant augmentation of the proportion of independent effort

and the individualization of instruction. But in order that students be enabled to assimilate knowledge effectively in the process of individual work, it is necessary that it be directed in a special way. Specific conditions and a special organization of instruction are needed.

A teacher, within his human limitations, working with a class cannot possibly guide, and guide sensitively, the process of individual study by *each* pupil. He cannot keep track of and take into account the character and degree of each student's assimilation of knowledge, his skills and habits, and his individual peculiarities, and test him appropriately. Under conditions of mass instruction the teacher cannot simultaneously receive and process information from all students concerning the state of their knowledge as well as skills and habits and at the same time analyze the information. He cannot simultaneously make decisions on optimal procedures for teaching different students. He cannot simultaneously impart or communicate different material to different students (one student may have one type of deficiency and another others). He cannot simultaneously direct study activity at several different tempos. He cannot simultaneously apply different procedures corresponding to individual characteristics of different students, and so on.

Providing conditions for an effective assimilation of knowledge by means of independent work, broadening the sphere of its application in the learning process, and individualizing instruction are possible only in those instances where special devices and equipment are designed and used. These devices permit a flexible adaptation of the course of teaching to the dynamics of assimilation of knowledge, skills, and habits by each student. They control and regulate automatically or semi-automatically the process of learning and teaching.[7] Such devices and equipment are programmed textbooks and teaching machines. Their significance lies in the fact that they—in contrast to ordinary textbooks from which people study independently—permit not only the imparting of knowledge and the controlling of its assimilation, but also changes in the course of instruction in accordance with the course of assimilation. Thus, they adapt the instruction to the dynamics of assimilation and at the same time simulate the process of tutoring the individual student. It is important to note here that, with programmed textbooks and teaching machines, the changes in the course of instruction and its adaptation to the dynamics of knowledge assimilation take place automatically or semi-automatically without the direct participation of a person-teacher.*

The flexible and automatic (semi-automatic) adaptation of the course of instruction to the dynamics of knowledge assimilation is impossible without breaking down the educational material into small parts (steps) and without constant, uninterrupted control over the character of the student's assimilation of that material. It could be pointed out that the principles of programmed teaching as usually indicated (assimilation of knowledge through independent effort, active participation in instruction, breaking down of instructional material into small steps, continuous control over assimilation, direct reinforcement, and individualization) are not independent. They are closely connected with the requirement for assimilating knowledge in the course of independent work and for flexibly adapting the course of instruction to the dynamics of assimilation, which itself necessarily presupposes individualization. The automation of the instructional process (although partial) is not an end in itself in programmed instruction, but it is the most important means for solving the tasks before us. Under conditions of mass instruction, i.e., of class-groups, there is no means of ensuring an effective control over the students' independent work other than to hand over certain functions of the teacher to special teaching machines, i.e., by means of at least partial automation of the instructional process.

In speaking of partially automating the teaching process, it should be emphasized that complete (or sufficiently complete) automation, though possible in principle, in many instances is unnecessary, unsuitable, and may even bring harm. The teacher has been and remains the principal factor in the educational process. He continues to be responsible for the major functions of

*Only the possibility for adaptation exists; it may not be realized. (*Editor.*)

organizing the group and educating the students. The influence of his personality cannot be fully replaced by any textbook or machine. Moreover, programmed instruction as only one of many methods of instruction, can never replace the entire series of forms for collective educational work which always did and always will play an important part in the educational process. Programmed textbooks and teaching machines automate only parts of the teaching process and at some stages. However, they do ease the teacher's work significantly and make him more productive and creative. Freed from many aspects of mechanical, uncreative work, the teacher is able to give considerably more time and attention to the general administration of all educational work, including those complex and difficult problems to which, up until this time, he has been unable to "devote any time."

Developing methods of programmed instruction and introducing partial automation into the pedagogical process produce new and considerably greater requirements in didactics, psychology, and the methods of individual disciplines. When methods were written for the teacher, the methodologists did not have to break down the activities of teachers and students into elementary operations and set up algorithmic prescriptions. There was no need to foresee all of the possible details and variations of the pedagogical process and to specify in sufficient detail the activities of the teacher and of the students. The teacher himself, as the most complete and flexible controlling system, as an intelligent and thinking being—often an experienced specialist—filled the gaps found in methodological instructions himself, found ways of acting in those situations which the methodologist did not foresee, dealt with those mistakes which were not mentioned in the methodological texts, made concrete those recommendations which were not sufficiently concrete, and so on. In brief, the teacher could in many cases adjust quickly (though not always successfully) to the real course of the students' assimilation of knowledge, skills, and habits.

The situation changes in a fundamental way, however, when instruction must be carried out (at least partly) not by the teacher, but with the help of programmed textbooks and teaching machines. The programmed textbook will not "know" how to react to a mistake which has not been taken into consideration beforehand. The teaching machine itself will not find the means of acting in situations which were not foreseen by the programmer.[8] Thus, breaking down the educational process (activity of the students and the teacher, and the educational material itself) into elements and setting up prescriptions that will determine the course of this process are necessary conditions for programming the teaching process. It is a condition without which programmed instruction is impossible. But devising such prescriptions for the course of instruction consists of its algorithmization.[9] It is clear from what has been said that programmed instruction is inseparable from algorithmization. Moreover, any teaching program is none other than that teaching algorithm which, by breaking down the teaching process into sufficiently elementary components (steps, operations), ensures the automatic (or semi-automatic) control over the assimilation of knowledge, skills, and habits by students. A major component of the process of automatic (or semi-automatic) control over teaching and studying is the availability in programmed instruction of feedback (i.e., student to teacher, *Ed.*) that makes possible a continuous exchange of information between the student and the teaching machine. This makes possible a sufficiently flexible and continuous adaptation of the instruction to the process of learning.

Two meanings should be differentiated in which one may use the expression "algorithmization of instruction." On the one hand, algorithmization of instruction may be interpreted as the algorithmization of the student's activities; and, on the other, the algorithmization of the teacher's activities. Algorithmization in the first sense is that of teaching algorithms to the students. In the second, it is the construction and use of teaching algorithms. It is clear that to teach algorithms and to use teaching algorithms are not the same things. It is possible to teach algorithms without following a teaching algorithm, and it is possible to proceed from some teaching algorithm and not to teach algorithms. Clearly, not every teaching algorithm presupposes the teaching of algorithms. A teaching algorithm can be so constructed that it does not foresee the specific teaching of algorithms to students and the forming of algorithmic processes in them. Programmed instruction directly represents the algorithmization of teaching in the second sense, i.e., the algorithmization of the instructional activity itself, although it does not exclude the teaching of algorithms to students. Algorithmic processes, too, may be formed in pupils through methods of programmed instruction. Moreover, "good" programmed instruction must necessarily form algorithmic processes where this is possible and beneficial. In other words, algorithms of teaching must include, in necessary cases, the teaching of algorithms.

Can it be said that programmed instruction is always better than unprogrammed instruction? The question cannot be answered unequivocally. Everything depends on the kind of instruction with which programmed instruction is being compared. To what teaching problems is it applied? On what material is it based? By what algorithm is it carried out? One and the same subject can be taught many different ways. The choice of material and the method of teaching (on which basis the instructional algorithm is subsequently constructed) depend on the theoretical basis of programming, how the creator of the program envisions the mechanism of the students' processes of assimilating knowledges, skills, and habits, and on the conception of ways of raising the effectiveness of this assimilation.

In itself, the algorithmization of instruction and good automatic or semi-automatic control of instruction still do not solve the problems of quality and effectiveness in such instruction. Much depends on what the goal and the material are, as well as what means of control are being used in the course of instruction, and what processes are being formed in the students as this takes place. In this connection it must be said emphatically that unprogrammed instruction with a proper choice of material and methods can turn out to be more effective than programmed instruction based on an unsatisfactory algorithm. From all this, however, it does not follow necessarily that the role of automatic control in the teaching and learning processes is insignificant. Existing data indicate that increasing the proportion of independ-

ent study and improving channels of student-teacher communication can improve instructional effectiveness to some degree even
though instructional content and method (in the broad sense of
the word) are not changed significantly.[10] It is important,
however, to keep in mind (and this follows directly from all that
has been said above) that such a way of perfecting instruction does
not in itself solve all of its problems. The greatest progress in
raising the effectiveness of teaching is possible only when
algorithmization and automation of teaching are based on an
in-depth psychological analysis of the patterns in which students
assimilate knowledge, skills, and habits, and on a fundamental
improvement of content and teaching methods. Improving the
feedback in the teaching process and controlling with greater
frequency the way the student learns and remembers material or
solves a problem do not suffice. Nor does simply forcing him to
work more intensively. He must be taught how to study correctly.
He must be provided with general and more efficient methods of
thinking. Such skills as, for example, independently finding the
solution to a problem, carrying out the search process, transferring
solution methods from one context to another, and so on, must be
formed in the pupil. Attaining such goals solely by means of
algorithmizing the teaching process and improving the feedback is
impossible. The satisfactory solution of teaching problems must
proceed from correct psychological and didactic positions (and
take into account all aspects of learning and teaching).

Notes

1. On macro- and micro-methods of approaching phenomena in
   cybernetics, see Lyapunov and Yablonsky [123].
2. Newell, Shaw, and Simon attempted to describe a person's
   thinking while solving problems at the level of "information
   processes" [193]. "Our theory," write these authors, "is a
   theory of the information processes involved in problem
   solving and not a theory of neural or electronic mechanisms
   for information processing" (p. 163).

3. When graphs are applied to a visual portrayal of algorithms, one often speaks of graphs of algorithms. On describing algorithms with the aid of graphs see, for example, Kaluzhnin [103]. In the diagrams, we will make **B** signify objects which are given to the input of the algorithm (in the present case, this object is the apparatus for verifying the specified features and with which the specified activities are produced); the symbol **ᙠ** represents those output objects which are produced at the conclusion.

4. In order not to create a cumbersome prescription and so as not to complicate the records (which would hamper making clear the essence of the matter), we constructed an algorithm for a case where the second student called upon always rectifies the mistake of his predecessor. In reality, he might not correct it, and then another student would have to be called upon. This is not difficult to provide for in the prescription. In such a case, it must be indicated in the algorithm to what point the operation of calling on another student must be repeated, and what operation must take place if none of the students called upon can correct the error of his predecessor and give the correct answer (this could be, for example, telling the students the correct answer, asking a leading question, indicating some plan of action, and so on).

5. One may therefore say of the algorithm that it is simultaneously *rigid* and *flexible.* It is rigid because under the same conditions, it presupposes the execution of the same actions. It is flexible because in different situations, it presupposes executing different actions (and in this sense, it is adjusted to the conditions).

6. For more about this in detail, see Landa [278].

7. The notion of automatic control is used here in the sense in which it has been used in the theory of automatic control and also in cybernetics. By automatic control is meant controlling certain processes with the help of special auto-

matic equipment without the direct participation of a person. "Automatic" in that sense has nothing in common with the notion of automatic used in psychology and pedagogy (automatic as opposed to conscious). With the help of automatic teaching equipment, one may completely embody the principle of consciousness in teaching.

8. When the development of self-adjusting teaching machines is achieved, they will be able to solve (at least in part) even these problems.

9. For more detail on the correlation of the notions of "programmed" and "algorithmized" teaching, see Landa [278]. Here, we will note only that in a case of teaching with the help of self-adjusting teaching machines, there will be no need for installing complete teaching algorithms. Efficient teaching algorithms may be found by the machines independently in the process of teaching, and the program (algorithm) installed in such a machine will, to a considerable degree, be the program (algorithm) of discovering the teaching algorithm, i.e., the program (algorithm) of teaching oneself how to teach.

10. This fact demonstrates the relative mutual independence of such factors as the organization of instruction and the form of control over learning (especially the character of the feedback). It permits, if well arranged, a significant amelioration of control over the work of each student and assures him quick knowledge of the results of his actions.

Chapter Four

Algorithms as Determined by the Characteristics of the Controlled Object and the Information Available About It

1. Classification of Algorithms According to the Conditions and Types of Control

The use of algorithms to solve various control problems has shown that the character of algorithms depends a great deal on the features of the system to be controlled with its help. It depends also on the character of information about this system and the control process which is found in the instructions of the controlling system. In this connection, the question arose in science as to what should be the kind of classification of algorithms which reflects these dependencies. Particular attention is given to this problem in the theory of automatic control, where the problem of synthesizing the automata (computers) with specific properties is a practical problem.

The data of the theory of automatic control are of exceptional interest for psychology and pedagogy, since one of the important goals confronting this theory is the creation of automatic equipment to simulate man's activity, and especially his intellectual activity. The creation of control equipment in many cases is one of the important indices of how well and profoundly the character and organization of human activities are known. A precise knowledge of them is of paramount importance for the instructional process. Without such knowledge it is impossible to have fully controlled, conscious, and highly effective instruction.

An important thesis in the theory of automatic control is that of classifying the types of controlling systems according to the algorithmic types which underlie their functioning (Artobolevsky, Bernstein, Bulgakov, Gavrilov, Lerner, Meyerov, Sukhov, Feldbaum, Filippovich, Khramoi, and Shorygin [43]). The algorithms themselves may be classified into different types according to different characteristics.

One of these is the sufficiency of the algorithm for the functioning of the system. Thus, for example, if a system operates according to a certain algorithm* and "copes" with its functions in such a way that no additional control of the system by other systems is necessary, then, in the theory of automatic control, such an algorithm is called a *functional algorithm* (Artobolevsky, Bernstein *et al.* [43]). A functional algorithm is an algorithm by which the operation of a system is carried out. If, as a result of disturbing influences from outside the system, this system cannot execute the functioning algorithm which has been prescribed for it, then one must exert particular influence on the system for the purpose of insuring the execution of the functioning algorithm. This can be done only in that instance where some other kind of system (of a higher level) will be established over the first system and will control it. The algorithm determining the controlling influences of the second system over the first is called *a controlling algorithm* [43].

Let us note that the concepts "a functional algorithm" and a "controlling algorithm" are, generally speaking, relative. If the operation of the basic system is examined in relation to those systems which it transforms (controls), then the algorithm by which the system operates (functional algorithm) is at the same time a controlling algorithm. On the other hand, the system which supervises this system and exerts controlling influences in relation to it operates itself according to a specific algorithm. This algorithm as controlling algorithm in relation to the first system is a functional algorithm in relation to the second system. Therefore, one and the same algorithm viewed in one perspective is a functional algorithm, while viewed in another perspective is a controlling algorithm.

*We are speaking here of algorithms, not of algorithmic prescriptions, since it is precisely that term which is used in the theory of automatic control.

The necessity of classifying algorithms into two types by the characteristics outlined here derives from the fact that the control of phenomena and processes is often carried out by means of not just one controlling system, but by a whole aggregate of systems which have a hierarchical structure.

If a system of a lower rank cannot cope with its functions, then a system at a higher level is engaged, which begins to control the first system. It, one might say, begins to control the controlling. If, then, at some given moment, the second system also ceases to cope with its functions, a third system is engaged at a yet higher level, which begins to control the controlling of the second system, and so on.

It must be said that we also have to deal with such conditions in instruction. For example, if a student knows the method for solving a certain problem and solves it independently, then he acts in compliance with a certain functional algorithm, and no interference on the part of the teacher in the process of his activity is necessary. If some difficulties arise for him, and he cannot succeed in solving the problem independently, he turns to the teacher for help. The teacher is, in this sense, a controlling system of a higher level, which joins in the process under specific conditions. Guided by certain rules, the teacher begins to influence the student, to control his thinking and actions in order to lead him to the solution of the problem. In many cases, the teacher acts in accordance with a specific controlling algorithm of which he may or may not be consciously aware. It can happen, however, that the teacher does not know how to help the student in the case of some difficulties. He turns to a methodologist who is (or at any rate, should be) a controlling system at a still higher level. Then, if the methodologist cannot answer the question, the teacher (or the teacher and the methodologist) have recourse to scientific literature and to scientists, and try to find the solution through them. If science cannot supply the answer to the given question, then it begins to be considered as a problem requiring further research.

Algorithms can be classified by another characteristic. The basis may be a goal that is set for the controlling system operating on the principle of a particular algorithm. Such a goal could be, for instance, the stabilization of the controlling parameter value, maintaining it at a certain level or within specified limits (for example, in controlling a production process, maintaining some exact temperature in the furnaces; and, in teaching, maintaining a specified activity level or the attention of the students in the class, etc.). A goal could be the changing of the controlling parameter value in response to a previously given function of time (for instance, increasing the pressure in the controlled aggregate during a definite time period to a specified magnitude, changing the dosage of medicines for a patient as he takes them, increasing the difficulty of problems presented to students according to their mastery of specified skills and habits, and so on).

Of greatest interest to us is the possible classification of algorithms according to the features of the controlled object (more precisely, the laws determining its functioning), and also according to the character of information acquired by the controlling system which, on the basis of this information, executes the control process.

Here, two basic possibilities exist. First, when the controlling system possesses all the information necessary for control; let us call such control: *control with complete information.* Second, when the controlling system does not have all the information necessary for control at its disposal; let us call such control: *control with incomplete information.* Let us examine the specifics of these types of control.

2. Control with Complete Information Available

With such control, the controlling system, for example, a person controlling some other system (henceforth, we will say controlling a certain controllable object):

(1) Knows all of the existing features of the controlled object and the laws of its functioning;

(2) Knows all of the influences under which the object might fall, and also how they change its state;

(3) Knows the existing ways of obtaining information on the state of the controlled object; and

(4) Knows the methods of successfully controlling it (namely, knows in which instances what operators must be applied, and what will be the results of applying each operator.[1]

If all of these data are available, an algorithm may be devised—the exact prescription for when and under what conditions; precisely what must be done to attain the goal which has been set for successfully solving the problem.

Here, several varieties of this algorithm may occur, depending on whether the transitions—strictly or non-strictly determined—are realized in the process of transforming objects while applying the algorithm. The fact that the character of the transitions of an object from state to state is not strictly determined does not contradict the fact that the transformation as a whole may have a specific character. But, for this, the reaction must be specifically anticipated for every possible transformation of the system, and the sum totals of these reactions must ensure the "arrival" of the object with identical initial states at the exact same final state. After such a state is identified as final, the transformation of the object according to the algorithm is concluded.

We indicate the dependence of the structure of the algorithm by what type of transition takes place in the controlled object in the process of its transformation.

Algorithms of Controlling Objects with
Strictly Determined Transitions

If control is accomplished on a basis of complete information and the controlled object in the process of transformation uniformly changes from state to state, then under these conditions, it is always known:

(1) Precisely which operator must be applied to what state in the controlled object, and

(2) to what state the controlled object passes as a result of applying the given operator to the given state.

If, for example, it is known that the controlled object is in state a, then operator A must be applied to it, but if it is in state b, then operator B, and so on. It is also known that if operator A is applied to state a, then the object passes on to state b and only state b; if operator B is then applied to state b, the object passes on to state c and only state c, and so on. As we see, each subsequent state is the single valued function of the preceding state and of the operator applied to that state. Thus, b is the single valued function of a and A, c is the single valued function of b and B, and so on.[2]

We will designate by $a \rightarrow T (A)$ the instructions of the type: "If the controlled object is in state a, then one should apply operator A to it."* We will indicate by $a \overset{A}{\Rightarrow} b$, the fact that if operator A is applied to state a, then the controlled object passes from state a to state b.

Let us assume now that we must transform a certain object from state a to the given final state d. The algorithm $\mathfrak{B}$ for the solving of this problem may be described thus:

$$\mathfrak{B}: \quad \left\{ \begin{array}{l} a \rightarrow T (A) \\[1em] b \rightarrow T (B) \\[1em] c \rightarrow T (C) \end{array} \right.$$

If it is necessary to represent the state into which the object passes when specific operators are applied to it, then it can be done in the following manner:

$$a \overset{A}{\Rightarrow} b, \quad b \overset{B}{\Rightarrow} c, \quad c \overset{C}{\Rightarrow} d.$$

This notation signifies that if A acts as operator on state a of the object, then it passes to state b; if, subsequently, B acts as an operator on state b, then the object transforms to state c, and so on. In this diagram, a is the initial state and d the final. If we are interested in the sequence of states through which the controlled object passes at the application of the given algorithm, then it may be described thus:

---

*T (A) signifies: "Operator A should be applied."

$$a \Rightarrow b, \quad b \Rightarrow c, \quad c \Rightarrow d.$$

or	a	b	c
	$\Downarrow$	$\Downarrow$	$\Downarrow$
	b	c	$d.$

Many algorithms for switching on instruments, an algorithm for starting up a motor vehicle (on the condition that it is in good repair), and certain others could be examples of algorithms accomplishing a similar type of transformation.

We examined the process of control whose characteristic feature is the fact that the application of a certain operator to a certain state of the controlled object transforms that object to one sole state which is known in advance (for example, $a \overset{A}{\Rightarrow} b$). Let us go on now to examining algorithms for controlling objects which do not have strictly determined transitions from state to state.

Algorithms for Controlling Objects Which
Do Not Have Strictly Determined Transitions

If, in controlling the objects just described, a certain operator A is applied to a certain state a of the controlled object and always transforms that object to one specific state (and that state is known to the controlling system), then in controlling objects which do not have strictly determined transitions, the application of a certain operator A to a certain state a of the object can transform it to different states from among a certain finite number of known states $b, c, d, e, \ldots, \dot{n}$, where it is not known in advance into which from among the set of possible states the controlled object will pass. However, the probability of such a transition is known (or can be determined). We will also consider such control as control with full information. Here that set of states into which the object may pass is known, given that a specific operator acts on a specific state, and the transition probabilities to each state also are known, i.e., the distribution of the probabilities of transition into the different states.

Above we represented strictly determined transitions from state to state by formulae like $a \overset{A}{\Rightarrow} b$. Transitions which are not strictly determined are to be represented by formulae like $a \overset{A}{\Rightarrow} b \lor c$, where $\dot{\lor}$ stands for the strict "or." The probabilities of transitions into each of the states could be shown in parentheses next to the letters designating corresponding states. Thus, when applying operator A to state a of a certain object causes it to pass to state b in 0.8 cases, but in 0.2 cases to state c, then this could be symbolically represented in the following manner:

$$a \overset{A}{\Rightarrow} b \ (0.8) \ \dot{\lor} \ c \ (0.2)$$

The transition probabilities may also be represented in the form of a table.

	b	c
A(a)	0.8	0.2

where A(a) signifies the application of operator A to state a.

Transitions which are not strictly determined, but whose probabilities are known, are called stochastic. In life, we are constantly dealing with instances of stochastic transitions from state to state. The simplest example would be a situation which we encounter in telephoning. Having dialed the number, i.e., having acted upon the initial state a of the telephone by operator A, we may, with a definite probability hear either a ringing sound (the party's line is free) or short buzzes (the line is busy). If the state of the telephone when emitting ringing sounds is designated by letter b, and the state of the telephone when emitting short buzzes is designated by the letter c, then we get $a \overset{A}{\Rightarrow} b \ \dot{\lor} \ c$.*

Thus, here, the transition from one state to another is multifarious. On different occasions, the controlled object with the action of one and the same operator, in the same initial state passes from this state into different states. Having carried out an extended observation of this process, one may calculate what the

---

*At present, it is not important for us that these states b and c themselves are conditional on the "states" of the subscriber.

probability will be that when operator A acts on state *a*, the controlled object will pass into state *b*, and what the probability will be that it will pass into state *c*.

A large number of examples of transitions which are not strictly determined (stochastic) could be given from the field of learning and instruction. For example, in the process of solving geometric problems, it is often necessary to carry out supplementary constructions. In drawing some sort of supplementary construction or in order to gain additional cues, the student, as a rule, expects that he will be able to obtain them with only a certain degree of subjective probability. The supplementary construction (i.e., the transformation of the original geometric object) may engender the most varied results, some among them being unsuitable. An analogous situation also takes place in many other instances. Let us say that the teacher applies some measure A to a student who is in a state *a*, with the goal of transforming him from state *a* to state *b*.* As a rule, the result of exerting a certain influence on some state of the student is rarely univocal. With certain probabilities, the very same influence, in some instances leads to one result (state)—the desired one—and in other instances to undesirable ones.[3]

The skill of the teacher consists, in particular, in the ability to increase the probability of obtaining the desired states in the student and in lessening the probability of obtaining undesirable states. We note that, at the present time, the probabilities of different results when applying specific influences on students who are in one fixed state are not being studied, or are being studied very little. Nevertheless, such studies would have great significance for developing pedagogical theory and practice. The availability of quantitative data on the probability distribution of various results in applying specific influences to specific initial states of students is an essential condition for the formulation of precise pedagogical laws and developing efficient strategies of instruction.

*In speaking of the states of the student, we have in mind the meaning for state of which we spoke earlier; see Chapter One.

Comparing the features of objects with strictly and non-strictly determined transitions leads to the conclusion that different types of controlled objects require different types of controlling algorithms. Somewhat further on, we will show that they also require different designs of controlling systems.

Let us turn to algorithm ℬ, designed for controlling an object with strictly determined transitions from state to state (cited earlier in this chapter). The conditions for applying each operator were indicated in this algorithm: $a \to T(A); b \to T(B); c \to T(C)$. However, there is no need whatever to give an algorithm in such a form for the type of control being examined. If the influence of operator A on initial state a leads to state b, and if the influence of operator B on state b always leads to state c, and so on, then, in the algorithm, it suffices to show the initial state and the sequence of operations A, B . . ., without showing the conditions of their application; for it is known that the necessary conditions appear as a result of applying the preceding operator.

If it is necessary to construct an algorithm providing for the object's transitions from state to state which are not strictly determined, then it is impossible to show an algorithm indicating only the initial state and the sequence of operations. It is known that the result of a preceding operation might be multiform. The preceding operator when applied to the corresponding state might transform the controlled object into various states (for example, into b or c). The character of the next operator to be applied, then, depends on exactly in which state the object arrived at as a result of applying the preceding operator. Subsequently, if, in the first instance, one may speak of a strictly determined sequence of operations themselves (operation B always followed operation A, and operation C always followed operation B), then, in the second instance, there is no such rigid sequence. The only thing rigid here is the link between conditions and operations [i.e., a link like $a \to T(A)$]. The sequence of the object's states that results from applying specific operations is not rigid, since situations like $q_0 \Rightarrow q_1 \dot{V} \ldots \dot{V} q_n$ can take place. It is clear that these two types of algorithm are not identical.

How the Design of Controlling Systems
Depends on Types of Algorithms

The fact that realizing algorithms of different types requires different designs of controlling systems is very important. Thus, for example, to realize an algorithm providing for the control of objects with strictly determined transformations requires neither the recalling of the conditions for each operation, nor the analysis of its results (for it is known in advance that operator A transforms the controlled object from state *a* to state *b*). Here one must remember only the sequence of operations and know the initial state of the object, as well as what operation must conclude the algorithmic process. In realizing an algorithm of the second type, which provides for control of objects without strictly determined transformations, not only is it necessary to remember the conditions for each operation, but also to have information about what the results of each operation have been. Without receiving such information, and analyzing it, the controlling mechanism (for example, a man's brain or the controlling block of an automatic system) will not be able to decide which operation must be executed during each successive moment. In other words, the controlling system for realizing such an algorithm must, without fail, be a system with feedback.

It is not only through theoretical examination of the problem that the dependence of the type of algorithm and the design of a controlling system based on the type of process which must be controlled and on the character of information about this process come to light. Today, different types of controlling systems are being designed which put into practice the different types of algorithm and execute control at different levels.

Let us introduce a diagram for control without feedback (Figure 4.1) and one with feedback (Figure 4.2) as they are described in the theory of automatic control (cf. Feldbaum [169]).

It is evident from Figure 4.1 that the controlling mechanism A on the basis of the externally received program (w) sends controlling signals u to the controlled object B; and the controlled object, in receiving these signals (influences), changes in a manner

Figure 4.1

Control Without Feedback

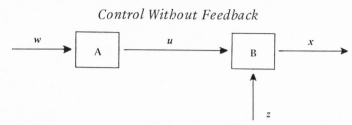

corresponding to them, and at the end yields the value x. Not only does controlling mechanism A, which is sending signals u, exert influence on controlled object B, but so does the "exterior world" from which come the perturbating influences z.

The system described is not a system with feedback, since the controlling influences u are not determined by states x of the controlled object (these states are not analyzed, and information about these states does not influence the character of the controlling influences of u). Control without feedback is possible in the case when the value x is determined rather completely by the controlling influences u and by the states of the controlled object B, and the perturbating influences z are insignificant and may be disregarded.

When these conditions are not observed, then, for control, the controlling system must know at every moment what happened at the output of the controlled object as a function of the preceding influence. For this, one must continually analyze x, and the information from x must continually participate in the elaboration of the controlling signal u. Let us show the diagram of the simplest form of system with feedback.[4]

Figure 4.2

Control with Feedback

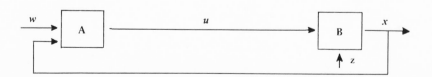

As we see, the controlling mechanism A "proceeds" in its operation, not only from information included in the influence w, but also from the output value x of object B. It is not difficult, having made the system somewhat more complex, to include block C in it, which will measure the disturbing influences and send information about them into controlling system A (Figure 4.3). In that case, when the controlling signals are synthesized, u will not only be influenced by value s at the beginning of the controlling system and by the data on the significance of quantity x, but also by data on the disturbing influences z. Such a system, which has been made more complex, is depicted in Figure 4.3.

Figure 4.3

Control with Complex Feedback

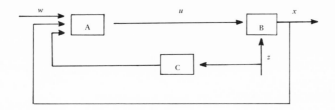

The diagrams show clearly that the conditions in which the process of control takes place and the character of information received by the controlling mechanism determine the design of a control system.

All of the control processes thus far examined were constructed on the basis of complete information, both on the type of processes occurring in the object under control, and the factors which influence it, as well as on the methods of controlling that object. Owing to this, it is possible in all cases to devise algorithms which give a system accurate prescriptions for carrying out the control process; that is, for the way of acting under specific conditions.

However, quite often the control process must be executed under conditions when the person setting up the control program for himself, for other people, or for an automatic control system, does not have some of the information needed for control. In these cases, it is not known in advance how to act under one or another condition. Here it is impossible to construct an algorithm of the system's functioning or control in advance.

This algorithm must be *found, discovered* in the very process of control. But discovering the algorithm is itself a process which the system discovering the algorithm (a person or automaton such as a computer) must be able to execute. Thus, the problem arises of creating algorithms through which other algorithms can be discovered. Such algorithms of discovering algorithms are essentially *search algorithms.*

Let us note that the term "search algorithm" can have a double sense. It is known that there exist problems (e.g., fine-tuning a radio receiver) which can be solved only by a search process. In that case, the algorithm for solving the problem is the search algorithm. But the words "search algorithm" could have the meaning: "an algorithm for finding another algorithm" (and namely an algorithm for solving a given class of problems). A search algorithm in this sense is distinct from a search algorithm for finding a particular solution value (e.g., from among a range of values).[5]

Insofar as the necessity for search arises in conditions where the control process is carried out with incomplete information, let us examine in more detail the situation of control with incomplete information.

3. Control with Incomplete Information Available

Control with incomplete information is very interesting, since many controlling systems (including man) must often work (and do work) precisely under these conditions. In mathematics there has been for a long time a predominance of problems in which people (or computers) had all the necessary information on the problem and on the process of solving it (for example, information on all of the conditions of the problem, on intermediate data, and so on). With the development of cybernetics and present-day computational methods, the situation has begun to change. Lately, more and more attention is being given to problems with incomplete information in the new fields of mathematics.

The reasons why controlling systems are often forced to carry out the control process without having all necessary information are extremely diverse. Among them, for example, one may mention the following:

(a) The structure of a controlled object is unknown as well as the function describing the relations of its initial and terminal states;[6]

(b) The perturbating factors influencing the controlled object or the character of their influence is unknown;

(c) The conditions under which one must apply one or another operation (to use one or another operator for processing information, to exert one or another influence on the object of control, and so on) are not known;

(d) The operations leading to the required transformation of the state of the object are unknown, and so on.

We must keep in mind here the fact that the absence of necessary information can be linked with lack of cognizance of certain processes at a particular moment, as well as with the impossibility of foreseeing all those factors which could exert all kinds of random influences on the object under control.

Let us describe some of the situations arising in control processes with incomplete information, making use of the designations which we introduced earlier in this chapter.

In discussing control processes with complete information, we were dealing with instances where for each step of control:

(a) It was known precisely which operator must be applied to what state of the controlled object in order to attain the goal (i.e., the rules like $a \rightarrow T(A)$ were known).

(b) It was known precisely into what state the controlled object would transform (or the set of possible final states was known from among which one and only one would realize that transformation as well as the probability of that transformation), if on a certain precisely known state, e.g., a, some precisely known operator will act, e.g., A (i.e., dependencies like $a \overset{A}{\Rightarrow} b$ or $a \overset{A}{\Rightarrow} b_{0.8} \lor c_{0.2}$ were known).

In control with incomplete information, we are dealing, accordingly, with instances when:

(a) Perhaps the state into which the controlled object would transform is unknown, if some known operator, e.g., A were to act on some known state, e.g., a. In other words in this instance the following situation takes place:

$$a \overset{A}{\Rightarrow} ?^7$$

(b) Perhaps it is not known which operator must act upon some known state, e.g., a in order to transform the object into another state, e.g., b. This situation may be represented thus:

$$a \overset{?}{\Rightarrow} b.$$

(c) Perhaps it is not known upon which state of the controlled object the given known operator, e.g., A must act in order to transform it to another known state, e.g., b:

$$? \overset{A}{\Rightarrow} b.$$

(d) Perhaps it is not known what chain of transformations must be executed in order that some final state n, may be reached, starting from an initial state a. Here we confront the situation:

$$a \Rightarrow ?, \ldots, ? \Rightarrow n.$$

(e) Combinations of all these different cases can also take place.

As we said earlier, *a priori* it is impossible with incomplete information to devise an algorithm for solving problems. The necessary dependencies for constructing an algorithm must be found beforehand, or perhaps discovered during the actual processes of control. To do that, a system effecting control must possess the ability on the basis of the algorithm put in it (program), to seek out independently that regime of control which will lead to the required solution—and, in so doing, find frequently the best (in a defined sense) solution. Such systems are called *purposive*.

Man could serve as one of the examples of a perfected purposive system. Very complex problems are also presented to non-living (e.g., electronic) automatic purposive systems.

In speaking of the necessity for creating equipment which would control systems with a wide range of changing conditions of control, Li Yao-tsu and Vandervelde [119] write: "For such types of systems [they are discussing complex systems which must be automatically controlled] it is advisable to have the kind of purposive system where there would be no need to know in advance, either the surrounding circumstances or the characteristics of the object, and which would nevertheless ensure the attainment of the required dynamic properties . . . ."

Purposive systems are systems which do not change the control algorithm installed in them, but, during the process of searching, perfect the regime of the controlled object's operation in conformity with previously given criteria.

An example of such a system is the "optimizer" devised at the Institute for Automation and Telemechanics in Moscow under the direction of Professor A.A. Feldbaum (see, e.g., [168]).

The essential activity of an optimizer consists of the following: Let us suppose that, in the process of controlling a complex system, one must obtain certain products as the output, which have precisely specified properties. Certain factors controlled by the controlling equipment—a, b, c, . . .,n (for example, pressure in the system, temperature, the presence of prescribed properties of certain substances, etc.)—influence the development of these properties. But precisely how these factors influence the final product is not known. It is also not known what value must be attributed to each of the factors in order to obtain the necessary properties. If the properties of a final product are designated by r, then $r = f(a, b, c, . . ., n)$ where function f is unknown. The only way of effecting control under such conditions is to test different values of factors, creating various combinations from them and evaluating the outcomes of testing the different values. Some algorithm can be constructed which will indicate how the tests must be carried out and how to carry out search process for the necessary values of the factors. Similar search algorithms underlie the performance of several types of optimizers. These algorithms, like all other algorithms, are "rigid" programs of actions, but they are programs for search actions. The solution of control problems through the help of algorithms of this type is carried out in the process of continual search.[8]

4. Peculiarities of Learning and Self-Instruction in Man and Machine

We examined purposive systems which continually adjust themselves to changing conditions in the process of control, but which in the process of such adjustment (tuning) do not change the algorithm installed in them.

There are, however, systems—and among them man must again be noted[9]—which during the control process *discover, change, or perfect the very algorithm of control.* They are realized through an original search algorithm or on the basis of methods using random processes. These are known as search-improving (or learning or self-instructing) systems. The controlling mechanism in these systems consists of at least two parts, and one of them (the controlling mechanism of a higher rank) in the process of searching, discovers (or changes or perfects) the algorithm of the other part which directly carries out the control.[10]

One may indicate the following general features of the process of self-instruction in a man or a machine as they occur at the present time (see, e.g., Artobolevsky, Bernstein *et al.* [43]):

(a) The process of self-instruction takes place in a closed loop.[11]

(b) During self-instruction, the characteristics of the learning system, its behavior, and its algorithm change.

(c) As a result of changing the algorithm, the system begins to do what it had earlier not been able to do.

(d) The capability for specific modes of behavior attained as a result of self-instruction is acquired. It was not placed into the system in advance by its designer (with respect to man—by nature).

Various ways of learning can exist, depending on the interrelations of learning systems with other systems.

One of the essential differences in the method of learning consists in the fact that by looking for efficient—or approximately efficient—modes of behavior the learning system can obtain an evaluation (reinforcement) of its actions from other systems, or it can appraise its own actions on the basis of criteria installed in it (or developed in the course of previous behavior).

We noted earlier the processes of self-instruction having a number of common traits in man and machine (we will discuss essential differences later).

In the theory of automatic control, when speaking of learning or self-instructing machines, up until recently, the technical community for the most part had in mind the process of self-instruction, i.e., a process of quite independently devising (without a person-teacher) new programs (algorithms) of behavior on the basis of initial programs installed in the machine by the constructor or the programmer.[12]

The "installation" of the initial programs *per se* was not usually considered part of the process of instructing a machine, and it was not considered instruction. A completely different situation exists with respect to instructing people. The process of transmitting knowledge, skills, and habits from some persons to others (this is connected with the social nature of man and of his learning), i.e., the process of "installing" initial programs in the consciousness of those learning, must be included necessarily in the notion of instructing a person.

In programming machines (including their preparation for self-instruction), the initial information ("knowledge") and the program of action ("skills and habits") in the precise sense of the word are *installed* (or inserted) into the machine at the beginning; in instructing people, these things are *transmitted* to the learners during the course of the instruction itself.[13] But if in the human case the knowledge, skills, and habits are transmitted to the learners in the course of instruction rather than just simply being "inserted" into their heads, a special problem arises with respect to the instruction of people. This is the question of how the learners *assimilate* it. (Assimilation represents a special kind of activity through which exterior information is processed into percepts and concepts and consolidated in memory. The percepts and concepts, in turn, serve as a basis for man's reasoning and acting, and his direction and regulation of his behaviors.) The problem of assimilation (or a process analogous to assimilation) with reference to machines arose comparatively recently. Hitherto in cybernetics, learning (or self-instruction) in machines meant essentially the process of independently developing new programs on the basis of initially installed information and programs. Therefore, the very process of program "installation" and "assimilation" virtually did not enter into the notions of learning in and/or instruction of a machine.*

The comparison of the notions "learning in a machine" and "learning in a human being" shows how these notions do not correspond. A more precise definition of how one must understand the expression "learning in a machine" avoids that confusion which sometimes arises when the concepts of cybernetics are used to analyze man's activities, in particular his learning.

The fact that learning in man not only includes instructing oneself within the closed loop of man and his environment but most of all the assimilation of experience amassed by humanity raises a very acute question. What is the interrelation of learning as an assimilation of "ready-made" (i.e., previously digested and formulated information) knowledge and programs of activity, and of learning as instructing oneself? Obviously, when teaching students ready-made knowledge and ready-made programs of action (including algorithms), the strategy of teaching must be to carry out that teaching so as to form in them the ability of instructing themselves. They must be given the ability independently to develop new knowledge and new modes of activity. In

---

*Of course, the process of man's installing programs in a machine can also be called the process of instructing a machine ("man instructed a machine how to act"). But obviously the word "instruct" will in this case be used in a figurative sense.

order to do this, the students must not only be taught algorithms for solving a specified class of problems, but also search algorithms; and, in general, methods of a higher order, which permit the independent discovery of other algorithms and methods.[14]

We spoke earlier about the kinds of conditions in which the necessity for search arises. These conditions are the absence of necessary information either about laws of the controlled object and processes taking place in it, or about the very means of control. This is natural, because if the controlling system lacks, for example, information as to which operator should be applied to a certain state a of the controlled object in order to transform it to state b (i.e., the situation $a \overset{?}{\Rightarrow} b$ takes place), then this question can be answered only by means of an experiment and by *attempting* to apply different operators to a. Thus, the search process, the discovery of new algorithms on the basis of search, obligatorily includes an experiment and trials as a means of carrying out this experiment. What has been said applies to people and animals as well as to electronic computers.

Thus, Artobolevsky, Bernstein *et al.* [43], in discussing the fact that a specific mode of behavior attained as a result of learning must not be installed ahead of time in a computer (automaton) by the constructor, state: "In a contradictory situation [i.e., if the mode of behavior were installed in the computer by the constructor] a simple reproduction of the mode installed in the computer occurs. Because the discovery of new modes of behavior that were not installed ahead of time in the automaton is possible only as a result of learning by the automaton from its own *trial* and *errors* [the italics are mine— L.L.], the method of automatic search is the only appropriate one for this purpose.[15]

It is absolutely clear that, depending on precisely which information is lacking for the controlling system, trials seek different goals and have a different character. Thus, for example, if situation $a \overset{A}{\Rightarrow} ?$ takes place, i.e., it is unknown to which result the applying of operator A to state a will lead, then the trial consisting of applying the operator will have the goal of studying the character of the transformation of the object under the influence of that operator. If the situation $a \overset{?}{\Rightarrow} b$ takes place, i.e., it is not known what operator should be applied to initial state a in order to transform it into another state b, then the trials will be directed at finding the necessary operators and studying their influence on the specified state.

Search trials with the subsequent analysis of their results on the basis of information appearing through the feedback channels are the most important means for evolving new forms of behavior in animals and man. This applies also to the solving of creative problems by man. Pavlov repeatedly pointed out the role of search trials in the formation of behavior (see, e.g., [745]). This question when applied to man's reasoning has been much studied by psychologists (see, e.g., [633], [634], [685], [775], [822], and others).

Owing to the presence in man of a second signal system, his trials can be covert, i.e., they can be realized on an interior plan and on an interior model of the exterior world which is constructed on a basis of percepts and concepts. The real, overt experiment in human activity is usually preceded by a covert experiment directed by the personal experience of the person and his knowledge. Before beginning to act overtly in one or another manner with objects of the exterior world, man usually "plays through" solving the problem on an interior model.[16] All that has been said, of course, does not mean that the process described does not lend itself to being modeled cybernetically. There already exist some analogs of a mental experiment in cybernetic machines today. But for a really thorough modeling of these processes in machines, it is necessary, evidently, to learn how to simulate specific features of the subjective reflection, i.e., percepts, and concepts, of the objective world.

Strategies of search by means of testing trials may be very different, depending on the type of problem and the character of information the controlling system has. Both psychological data and certain results of theoretical and technical cybernetics testify to this (see, e.g., [80], [81], [167], and others). Thus, for example, a search can be carried out by scanning, when the whole field is searched step by step. If, before beginning the search, the controlling system has some information about a narrower field in which the object may be found, then the original field is considerably narrowed and the trajectory of the search which differs essentially from the consecutive testing of the whole field "step-by-step" can be observed. In this connection the methods of search which Gelfand and Tsetlin called methods of local and non-local investigation are of great interest [80].

Looking for new search strategies (or algorithms of search) has as the goal the improvement of the search process, i.e., making it optimal. Increasing the efficiency of search, in particular diminishing the elapsing time, has enormous importance for control processes. A more efficient search algorithm solves problems arising in the course of control (including the discovery of new algorithms) much more quickly, while wasting less time, means, and energy.

Returning to the question of the role of search trials in the process of solving mental problems, two types of trials must be pointed out:

(a) Trials which are realized according to a program known ahead of time (according to a known test algorithm), and

(b) Trials whose algorithm has not been given to a person or is not known to him.[17]

Although both kinds of trials lead to the discovery of new regularities, or the obtaining of new results, or the constructing of new algorithms, only the second kind of trials, by their psychological nature (excluding, of course, instances where they take place at random) can be called genuinely creative trials (one may also say that these are trials which are realized in the process of creative search).

We must obviously distinguish two notions: "Discovering something new" and "the creative process." One can arrive at the discovery of something new, previously unknown (not only to a given person, but also to mankind on the whole) not only by way of creative trials or creative search, but also by way of investigative trials executed according to an algorithm.[18] On the other hand, in the process of creative search, something new can be discovered which is new only for the given person.[19] Thus, the discovery of something new *in itself* is not an indication of creativity, as people often think (the discovery of something new is often set forth as an indication of the creative process). The discovery of something new and the creative process are different, though they are usually closely linked.

The majority of contemporary self-adjusting and self-instructing machines that engage in search and discover during the process of search for new regularities (including new algorithms) work according to specific search algorithms which represent a "rigid" prescription on how to conduct the search process, what kinds of trials are to be used, and in what sequence they must be carried out. However, at the present time, attempts are already being made to simulate, in particular, the non-algorithmic search process. Simulation is carried out by introducing a random generator of trials and a choice (usually by providing the machine with the probabilities associated with different types of trials) into the search process. It is precisely this principle which is used, for example, in machines which create music.

The presence of random trials (trials which are not carried out according to an algorithm) is the common factor linking both man's creative search and the "creation" of a machine.[20] But a person's random trials are based upon specific knowledge and intuition, representing the probability estimate associated with objects, situations, and modes of action. These are almost always trials *directed* in a specific manner. The trials of the majority of machines existing up until recently were evoked by the generator of random processes. Therefore, not only were these trials random, but also "blind." It is only recently that machines pursuing a goal oriented search, operating on the basis of accumulating experience and probability estimates have begun to be developed (see, e.g., Ivakhnenko [102]). But in a machine without goal orientation, there are only two extremes: It acts either according to a rigid algorithm, or on the basis of "blind" and purely random trials. Rarely does a person—unlike such a machine—resort to purely random trials. In many cases, the presence of prior knowledge and intuition excludes the necessity for a blind search.[21]

All that has been said does not signify that, in principle, it is impossible to simulate any creative mechanisms, or to make any scientific and artistic discoveries with the help of machines. At present, however, we are not able to simulate even the basic mechanisms of human creativity. For the time being, it is difficult to say to what degree this will succeed in the near future and on what progress in this domain is contingent. It is quite possible, however, that attempts at simulation of specific and basic peculiarities of human creativity will not have succeeded by the time when we shall be able to simulate biological processes, and—what is especially important—the subjective reflection of the exterior world, brought about by the specific organization of informational (psychological) processes in the brain. Evidently, the difficulties in simulating the mechanism of man's creativity arise from the fact that creative processes are linked with the activity of highly organized material in the brain which possesses a number of properties that other types of material do not possess, or not to the same degree. These properties are (a) great plasticity of the processes taking place in the brain, (b) the ability to synthesize signals (which is a cardinal trait), (c) the ability to form different types of associations, (d) the ability to dynamically transform percepts and concepts by means of psychic actions, and (e) a number of others.[22] At present, these particularities do not submit easily to simulations (so long as only the separate parameters of psychic activity are modeled). Meanwhile, these qualities of the nervous and psychic process play an enormous role in the mechanism of non-algorithmic search (creative trials), and have an important influence on its results.

Let us note that the indicated qualities of the nervous and psychic processes determine the specific character not only of man's creative acts, but also the specific character of his assimilation and application algorithms. This is reflected first and foremost in processes of generalization which arise during the course of mastering the algorithms. The generalization of the operations of which algorithmic processes are composed is not given *a priori* to a person. It evolves during the process of mastering the algorithm. Once formed, however, it ensures the transfer of the algorithm to new circumstances and its application to new specific content. As we will see later (Part Two, Section Two), many features of attempts to apply algorithms (including mistakes in their application) are linked with the character of the generalizations formed during the course of mastering the algorithms.

From everything that has been stated, it is clear that, despite their "rigidity," algorithms[2 3] make it possible to solve problems of different classes, including problems of discovering new, previously unknown, patterns and methods of behavior.[2 4] On the basis of algorithms, one may carry out a study of phenomena (by means of investigative operations), their transformation, and also the design of other algorithms. Algorithms may serve as a program of learning and self-instruction. The perfection of controlling systems and their acquisition of new experience may also take place on the basis of algorithms. Algorithms, despite their rigidity,[2 5] may ensure a flexible adaptation of control systems to changing processes by changing programs. Owing to the fact that algorithms can form hierarchical systems entering into interaction with one another, the rigidity of algorithms is not an obstacle to changing them during the control system's process of functioning.

The important gnoseological significance of research in the field of algorithmizing a broad sphere of problems of control lies in the fact the research makes possible the profound study of the structure of many processes which until recently were considered resistant to precise scientific analysis, description, and modeling. The thought that something besides man could solve intellectual problems leading to the discovery of something new seemed, even recently, to be blasphemous. Creativity was considered man's absolute prerogative. Any attempt to break down the creative process into elements and to represent it as some combination or a certain aggregate of those elements was considered to be mechanistic.

The development of cybernetics has shown that many processes (by no means all) which were considered (and until now actually were) creative, have an algorithmic nature, and that we simply did not know the underlying algorithms. This is what Trapeznikov said on this subject in his introductory lecture at the

International Conference on Automatic Control [161]: "Computers aiding the researcher can be self-instructing. They can acquire habits of a simple type to carry out narrow tasks. However, in principle, it is also not impossible that they devise to some degree 'creative' skills, i.e., automatic improvement of search in complex fields. At present, the limits for developing such systems are not evident. The creation of a computer, which on the basis of the aggregate of empirical data and of testing different hypotheses could create complete theories that would explain the data of an experiment in some field must not be considered impossible. Meanwhile, such activity is being called genuinely creative." Further on, Trapeznikov notes that, even for specialists, it was a revelation "how complex and profound those processes are, which earlier could only be carried out by the human brain, and which now lend themselves to automation, thanks to the application of computing machines."

The precise description and determination of the structure of creative processes—many of which have in reality an algorithmic character—have tremendous significance for pedagogy. Their simulation shows that creative processes may not only be "cultivated." Creative activity can be *taught*. There is no doubt that further research in analyzing the structure of intellectual activity, its precise description, and its modeling will have great importance not only for the theory, but also for the practice, of instruction.

Notes

1. Feldbaum [169], in particular, points out the significance of such factors for the design of control systems.
2. A detailed function could be specified in the form of a table such as this:

	a	b
A	b	p
B	q	c

The table signifies that the application of operator A to state *a* transforms the object to state *b*; the application of operator A to state *b* transforms the object to state *p*, and so on.

3. Here it is important to note that for model learning situations there are a number of states into which the student may transform as a result of applying specific instructional influences to a momentary state. The transition probabilities, in principle, need to be studied and even defined. However, for non-model situations, it is very difficult to do this, inasmuch as the states into which the student may transform cannot be foreseen in practice.

4. Control systems without feedback are called systems of automatic rigid control (SARC); systems with feedback are called systems of automatic regulation (SAR).

5. Of course, the search of an algorithm may also be considered as the solution of a certain problem (namely the problem of finding the solution algorithm). However, the terminology which has been proposed is convenient and will not lead to confusion.

6. In other words, the regularity with which the controlled system transforms under specific influences into one or another state is not known. Often even those states into which the controlled object could change under the influence of different exterior influences are also not known.

7. Here we are not considering a case where the states (from a set of possible states) into which the object may transform are known, while the probabilities of the corresponding transformations are not known, since of greatest interest for us now are precisely those situations in which the factors about which we are talking are not known. Let us also note the following: In speaking about the fact that the state into which the controlled object passes as a result of the influences of a specified operator upon it is unknown, we certainly do not mean that the controlling system knows absolutely nothing about these states. As a rule, the character of the states and the type of states are known in their general form. For example, perhaps it is not known exactly how a

person will react to a rebuke or a scolding, and into what particular state he will transform (the reaction can be most unexpected). It is known, however, that he will not evaporate, dissolve, or fly off into thin air, etc. His particular reaction is unknown, but the approximate character of the possible reactions and the set of phenomena with which it could be connected are known. What has been said, however, does not exclude the fact that, in the development of science, there are instances where even the character of the state into which an object could transform as a result of specific influences upon it is unknown.

8. At the present time, systems with stochastic search processes are being developed, which carry out trials on the basis of a random process (see, e.g., [102], [176]).

9. One of man's characteristic traits is the fact that he, as it were, contains different systems within himself, and thus can work, depending on conditions, either as an "automatic system of rigid control," or as a "system of automatic regulation," or as a "self-instructing system." The immense resources of man, the flexibility of his physiological organization, and primarily his nervous system give him the possibility for carrying out control at the most diverse levels, switching, when necessary, from one level to another.

10. Self-instructing systems may have controlling mechanisms of different ranks. The controlling mechanism of a higher rank can change the operation of the algorithm of the control mechanism in a lower rank.

11. Concerning the learning in computers, Artobolevsky, Bernstein *et al.* [43] write: "The process of learning in an automatic machine is the model of man's becoming acquainted with the world around him, which also takes place within a closed loop: man—environment." Let us note, however, that the latter statement is true only under those conditions when by "learning" is meant only "self-instruction" and that included in the notion "environment" are the materialized products of human cognition (e.g., books). For man learns (instructs himself) not only in the process of

direct interaction with his environment or the aggregate of material objects, but also by way of reading books. If the materialized products of human cognition are not included in the notion "environment," then learning in the closed loop "man-environment" must be considered as only one of the special situations in man's self-instruction.

12. Experiments on man's instructing a machine have been developed only comparatively recently.

13. Let us note that the initial forms of man's behavior (unconditioned reflexes) are also installed (by nature) before the beginning of his learning process. Man inherits them ready-made. But these programs of behavior which are installed by heredity, despite their immense significance for the life of the organism, occupy a very small place in the common fund of man's ways of behaving. Man acquires the basic fund of ways of behaving in the process of learning with the help of other people and in the process of self-instruction. The majority of contemporary machines acquire a basic fund of ways for behaving through the ready-made programs installed in them.

14. Let us note that a search algorithm by which a machine discovers new algorithms is also the result of experience amassed by man, which is placed in the machine in the form of an initial program. The machine then carries out the process of self-instruction on the basis of that program.

15. Let us note that trial-and-error as a method of discovering new laws and as a fact of behavior (there is no difference— whether machine, animal, or man) and behaviorism as a theory of behavior interpreting this fact in a specific manner are different things. Obviously, one and the same fact can be interpreted differently.

16. Attention is called to the parallel of these propositions to those of G.A. Miller, E. Galanter, and K.H. Pribram (*Plans and the Structure of Behavior*, N.Y.: Holt, 1960), whose TOTE (*T*est *O*perate, *T*est, *E*xit) process appears to be, at least, very similar. (*Editor.*)

17. An example of trials which are carried out according to an

algorithm given previously, or an algorithm made up independently, could be tests of applying a theorem in the process of investigating some sort of geometric proposition. In order to prove the equality of two segments, for example, one can draw up an index of all known theorems on the equality of segments and a special plan for testing (checking) their suitability for the given conditions. The search of "appropriate" theorems carried out according to this plan in a specific sequence will also be a search according to an algorithm. The experiment of introducing elements of algorithmization into the process of search for theorems while teaching geometric proofs has been described, in particular, in Landa [686]. The algorithmization of search, as we already said, underlies programs recently developed to prove mathematical and logical theorems with the help of general purpose digital computers (see, e.g., [140], [141], and others).

18. The discovery of something new by means of algorithmic search (we will thus designate an investigation carried out by an algorithm) is already a fact today. Machines into which the algorithm is placed, containing a large number and variety of trials, are capable, in principle, of discovering things which man has not yet discovered, or which are difficult for him to discover. Thus, machines find the optimal solution to many problems, bring to light complex regularities of processes, and so on. It is interesting that a machine, working according to a program designed by Bongard [59], discovers laws by which series of numbers are organized and also solves a number of other problems connected with the discovery of new patterns which were not known earlier.

19. The distinction of something new in the social sense (new for science, with respect to knowledge amassed by mankind) and something new in the psychological sense (new for the individual) is natural.

20. The significance of random trials and the role, in general, of chance in creativity, especially in scientific discoveries, was noted long ago (see, e.g., [8], [14], [23], [24], and others),

although the corresponding facts could be interpreted differently.

21. The most important feature of man's creative processes consists also of the fact that owing to the extraordinary plasticity of his anatomical and neuro-physiological organization, man is capable of producing such new actions (physical and psychic) as have never taken place before and, in principle, could not have been programmed in advance. This is ensured, in particular, by the structural features of the human organs of movement. Vecker [556] writes on this subject: "The basic feature of the human hand as a performing organ in comparison to a performing organ of any automatic mechanism (including a mechanical hand) is its high functional plasticity, permitting the solution of an infinite class of motor problems on one and the same morphologically unchanging design. Discovering the freedom of executing an infinite diversity of motor solutions, the design of the hand as a performing organ does not fix in itself any special program of movements. Neither are the program of separate performing operations nor their algorithmic sequences incorporated in the structure of the hand."

22. The concrete psychological analysis of the specifics of intellectual reflection and, in particular, of means of organizing informational processes in man's brain is contained in publications [4], [31], [68], [694], [695], [699], and others.

23. What has been said refers also to the search algorithms insofar as they are also "rigid," in the sense that they completely determine the search process.

24. Behavior is understood here in the broad sense which cybernetics gave this term (see, e.g., Klaus [106], and the afterword of Bazhenov, Biryukov, and Spirkin to Klaus' book).

25. The fact that algorithms are not only "rigid," but also flexible (and in precisely which sense), was discussed previously, in Chapter Three.

Chapter Five

Algorithms of Transformation and Algorithms of Identification

We have examined the problem of how types of algorithms depend on the types of problems which must be solved with their help, and how they depend on the character of information about the controlled object and about the actual process of control which the controlling system has at its disposal. There is, however, yet another indicative feature by which one may classify algorithms into types. This is the purpose for applying the algorithm, i.e., the transformation of an object, or its identification. Let us examine this indicative feature more thoroughly.

Kaluzhnin [103] says that in algorithms which are defined by graphs (and graphs are one of the general means for representing algorithms)[1] there are objects with a double nature: operators and identifiers. The operators (or elementary operators) transform the information entering into the algorithm. The identifiers, as Glushkov [84] characterizes them, "identify—perform characteristics of the information processed by the algorithm, and depending on the results of the identification, change the sequence in which the elementary operators follow one another." If, in other words, operators are mathematical objects which correspond to certain operations *for transforming* information, then identifiers are mathematical objects *for identifying* information by revealing its indicative features. Depending on which indicative feature the information processed by the algorithm has, different operations are applied to it for transformation.

Similar to the way in which certain elementary acts of

105

transforming information are considered in the theory of algorithms to be operators, so certain elementary acts of identifying information are considered identifiers. Identification achieved through identifiers is, in the theory of algorithms, considered as a *single-action* process which cannot be broken down into any other more elementary acts ("steps," operators). If we turn to the operation of real control systems, we see that the identification of objects, phenomena, or their features is rarely a single-action process. Most often, it is realized not by means of one operation, but with the help of a *whole system* of identificational acts (operations). Analysis shows that this system of identificational operations may have an algorithmic character and may be described algorithmically. Hence, it follows that not only can there be algorithms for transforming information (and, in general, material objects), but also algorithms for identifying it. If each algorithm of transformation contains the identificational process as its component, then the identificational process may not include any transformation of the initial objects.[2] As we shall see later, there are, however, cases where the process of identification includes the transformation of an initial object.* These are special cases, though rather frequent. Hence, transformation appears here as a means of identification, as its condition, and is a prerequisite for it. The solution of a problem in these cases is not the transformation of the initial object, but attributing it to a specific class.

The task of identifying an object is always particular, in relation to the task of transforming it. However, when the process of identification is complex, it becomes a separate and often very difficult task.

For example, the identification of an illness (diagnosis) is necessary in order to cure it (treatment is the transformation of the state of the patient). However, the identification of an illness, i.e., its diagnosis, is usually a separate problem. A diagnostic algorithm is none other than an identificational algorithm. Very

---

*By "initial object," the author means any entity *to be* processed according to an algorithm; for example, a sentence. (*Editor.*)

often, they include operations of transformation (for example, the patient is given specific medicinal substances which transform his state in such a way as to define his illness by the character of the patient's reaction to them). But these transformations have as their goal diagnosis, i.e., they serve as a means of identifying the state of the patient and not as treatment. Transformation for the purpose of treatment is a completely different transformation from that of identification.

Therefore, algorithms can be divided into two types from the standpoint of the goal attained with their help: algorithms of transformation and algorithms of identification. However, algorithms of transformation include operations (or even algorithms) of identification, and the latter may include operations (and even algorithms) of transformation.

How can one distinguish such algorithms from one another? Seemingly this can only be done by examining the nature of the intended goal and the final result of the application of the algorithm. If such a result is a judgment as to the initial object's belonging to a certain class (of the type $x \in A$, where x is the initial object, A a certain class, and $\in$ the sign that the object belongs to a certain class), then the given algorithm is an algorithm of identification. In the opposite case, the algorithm is a transformational algorithm.

Let us dwell at somewhat greater length on the problems of identification which are solved by means of transformation. We introduced examples of this kind of problem (from the field of medical diagnosis). A large number of problems of identifying substances in chemistry (including those being solved in school) are also problems of identification by means of transformation. Yet another example could be the transformations to which mechanical systems are subjected in order to discover why they are out of order or out of tolerance. Other examples could also be given.

Identification through transformation is often used in instructional practice. The simplest example from the field of instruction in grammar is the device, frequently used in school, of identifying the case of a noun (nominative or accusative*) by substituting a pronoun (e.g., "John had George for a friend" and "He had him for a friend"). In any process of identification which is carried out by transformation, i.e., by some constructive (though perhaps only purely covert intellectual) activity, the most important operation is that of comparing the transformed object with certain indicative features given by a definition or any other kind of theoretical statement.

The type of these theoretical statements is the following: If some object possesses some given indicative feature or features, then it belongs to a certain class. For example, if the patient has symptoms $a, b, c, \ldots, k$, then he is suffering from illness A, and so on.[3] It is easy to see that these theoretical statements point out the features indicating that phenomena belong to specified classes (e.g., to a class of specified illnesses).

Many examples of identification via transformation deriving from other subject-matters could be introduced. For example, in arithmetic, in order to determine (identify) whether some number can be divided by 9, it is transformed (the sum of its digits is sought); in physics, to determine the boiling point of a substance, it is heated to that point, and so on.

That which in an algorithm of identification through transformation is one of the last operations (i.e., matching obtained and specified characteristics of objects) is the essence of the algorithmic process in pure identification algorithms (i.e., in algorithms that identify objects without transformation). Any process of identification amounts to the progressive establishment of indicative features in objects which then are compared to features characteristic of objects belonging to certain classes. If the indicative features of the object under examination correspond to the distinctive features of objects of a certain class, then the given object belongs to that class. If they do not correspond, then it does not belong to that class.

Since the process of identification may be broken down into elementary operations which can be applied to varying initial data, and since the process can be strictly determined by a definite

*In English, objective case. (*Editor.*)

prescription, it can be algorithmized just like processes of transformation.

We shall ask: In general, wherein lies the significance of a problem of identification? Why does the process of identification require special consideration?

In a general form, the significance of this problem consists in the fact that the overwhelming majority of human or animal actions are applied not simply to particular real objects, but to objects as members of certain classes of objects. It is biologically and socially inefficient to devise a special form of behavior for each separate object, process, phenomenon, influence, etc. It is much more expedient to devise generalized forms of behavior applicable to objects as representatives of whole classes of objects. Only in this way is there a possibility of transferring learned behavior from one object to another without passing through a stage of learning and without devising a special form of behavior with respect to each novel object. Only after the process of identification, i.e., the process of including an object in some class of objects, has been successfully accomplished can a pattern of behavior be transferred to a previously unencountered stimulus object belonging to that certain class.

Let us suppose, for a moment, that animals and humans could not identify objects, i.e., attribute them to certain classes. Life would be impossible. An animal which eats grass, in connection with each new blade, would be confronted with the decision of whether or not "this" was edible. It is only owing to the fact that each new blade is identified as an object belonging to the class "grass" that the animal transfers actions onto it which are applicable to objects of the given class. Animals presumably recognize objects only on the level of perception (i.e., on the basis of their external resemblance). Man is capable of recognizing objects on the level of reasoning as well. Here, the external resemblance of objects is not necessary. Possessing the ability of bringing to light the internal indicative features of objects which are hidden from direct observation, and of operating on them, man is capable of attributing to one and the same class even those objects which do not externally resemble each other but which do resemble each other in terms of internal hidden features.

From all that has been said, the significance of identificational processes for instruction follows. Any transformations
which students must carry out in the process of assimilating
knowledge, skills, and habits include as components—and often as
a special task—the identification of their membership in a specified
class. When the process of identification is not carried out, or is
carried out incorrectly, it is often impossible to accomplish its
transformation (or that transformation will be incorrect). The
special teaching of identificational processes together with the
determination of the possibility of their algorithmization is
therefore an important problem, the solving of which has a
considerable significance for the theory and practice of instruction. The greater part of the remaining sections of this book will
be devoted to examining approaches and methods of solving this
problem. However, before examining the problem of teaching
students identificational algorithms in particular, it is necessary to
examine in more detail the more general problem of how
algorithms (algorithmic prescriptions) may be applied in instruction. We shall now proceed to this question.

Notes

1. On this question, see Freidman [342].
2. Glushkov [84] notes: "No changes of a word take place at
    nodes representing identifications" (i.e., in a graph).
3. The expression "if a certain object x has the indicative
    feature a, then it belongs to class A" can be written in the
    form of symbolic logic thus:

$$a(x) \Rightarrow \epsilon \ A.$$

Here $a(x)$ signifies that object x possesses the feature a.
The arrow corresponds to the implication "if . . ., then"; and
the symbol ϵ signifies the membership of x in the class (set)
A.

Part One: Section Two

Some Theoretical Problems in Teaching Algorithms

Chapter Six

Expediency of Instructing Students in Algorithms

1. Classification of Thought Problems

As stated previously, in the course of his daily activity, man must solve a great many of the most varied practical and theoretical problems. In solving these problems, he acts as a controlling system, transforming specified objects or information about them. In order that the solving of problems be successful, man must assimilate specific methods for problem solving. But to do this he must be taught them, and he must master them. In those cases where man is not given methods in prefabricated forms, or where these methods are altogether unknown, he must be able to discover them independently. This, in turn, requires instruction in ways ("methods") of independently discovering methods.[1]

In "entrusting" a machine with the solution of specific problems, we either equip it with a specific solution algorithm (particularly if a search for a specified goal is involved), or with an algorithm for discovering another algorithm.[2] The latter will permit it to discover the problem-solving algorithm for the purpose of later applying it. The question arises: What should people be equipped with and what must be "installed" in them when they are confronted with specific problems? In particular, should they be equipped with algorithms for solving problems and search algorithms through special instruction in them? After all, people are the most completely self-adjusting and self-instructing systems. They are capable of working out algorithms independently or of finding non-algorithmic ways of solving problems, even in those cases where algorithms do not exist or are unknown.

In order to answer that question, it is necessary to examine the kind of problems people have to deal with in the course of their practical and theoretical activity and the mastery of which kind of methods is required to solve these problems.

Here, the following questions arise:

1. Can all problems be solved by means of algorithms?[3]

2. If not, precisely which problems cannot be solved by algorithms?

3. Conversely, do there exist problems which cannot be solved without carrying out some algorithmic process?

4. If there exist problems which can be solved by algorithms, is it always expedient to solve them in this way? When is it expedient, and when not?

5. If it is expedient to solve certain problems by algorithm, is it expedient to teach this algorithm? When is it expedient and when not?

Let us pause first of all at the question of which problems it is expedient or necessary to solve by algorithm and which it is not. The general classification of basic types of problems from the standpoint which interests us can be presented in the following manner.

Let us examine more closely the most basic among these types of problems.

Problems for Which Algorithms Cannot Be
Or Have Not Yet Been Constructed

We have already discussed these problems in Chapter One. Therefore, we will not pause here. It is completely obvious that problems for which algorithms cannot be or have not been constructed cannot be solved by algorithm, and that it is impossible to teach such algorithms.

Problems Which Must Be Solved by Algorithms
(or Algorithmic Procedures)

There are a great many such problems. These are problems which require the execution of a specific system of operations to

Figure 6.1

Schematic Classification of Problems

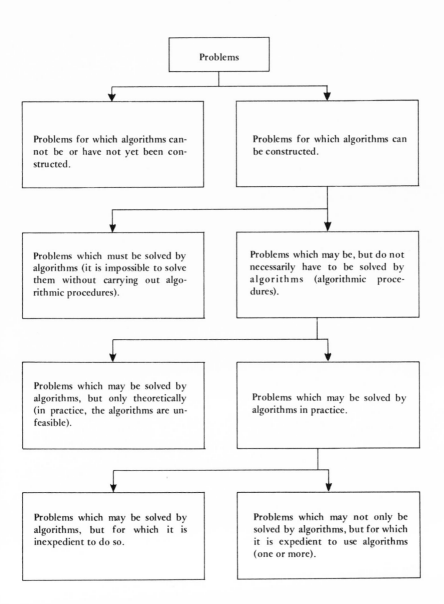

Problems

Problems for which algorithms cannot be or have not yet been constructed.

Problems for which algorithms can be constructed.

Problems which must be solved by algorithms (it is impossible to solve them without carrying out algorithmic procedures).

Problems which may be, but do not necessarily have to be solved by algorithms (algorithmic procedures).

Problems which may be solved by algorithms, but only theoretically (in practice, the algorithms are unfeasible).

Problems which may be solved by algorithms in practice.

Problems which may be solved by algorithms, but for which it is inexpedient to do so.

Problems which may not only be solved by algorithms, but for which it is expedient to use algorithms (one or more).

solve them, and this must be in strict sequence. A problem is impossible to solve when the necessary system is not carried out, or its necessary sequence is destroyed. Industrial and practical activity provide many examples of such problems (turning on instruments, controlling machines, regulating industrial processes, and so on).[4] A great number of such problems can be found in study activity as well (for example, when studying mathematics, chemistry, and other subjects).

A number of examples of grammatical problems, the correct solving of which can only be brought about by carrying out operations in specific sequence were cited, in particular, in the works of Granik [240], [592], Polyakova [752], [753], and Vlasenkov [223].

The necessity for strict sequence of operations when solving these types of problems is connected with a specific stage-by-stage transformation of those objects, which are carried out by means of those operations. As soon as it is necessary to solve a problem where each subsequent act of transforming the initial object is directed at the result of the preceding act, the sequence of these acts becomes critical. It is impossible to solve these problems without knowing the algorithms for solving them and/or without carrying out a specific algorithmic procedure—hence the significance of such algorithms for instruction. Some of these problems may be solved by one single algorithm and one possible system of operations. Others may have different solution algorithms, i.e., may be solved by different systems of operations. But inside each system, the sequence of operations must be strictly specified. An important question when solving problems having different solution algorithms is: Which algorithm is the most efficient under precisely which condition? But this is a special question and it will be discussed in more detail later.

We examined problems which cannot be solved by algorithms and problems which cannot be solved without algorithms and without carrying out an algorithmic process. Let us go on to examine problems which may be, but do not necessarily have to be, solved by algorithms; and, first of all, problems for which algorithms may be constructed, but where the algorithms are so complex as to be unfeasible in practice.

Problems Solvable by Algorithms Beyond Practical
Realization (the Algorithm Is Unfeasible in Practice)

The most famous example of this type of problem is the problem of finding the best move in a chess game. Since there are a finite number of positions in chess, and in each position a finite number of possible acts (moves), it is possible, in principle, to construct an algorithm for finding the best move which, in part, represents an exhaustive search algorithm that considers all possible moves in each position on the board and, hence, all possible games (see, e.g., Orlov [142]). However, the number of possible positions and moves in playing chess is so great that it is impossible to carry out this algorithm in practice. It has been calculated that if this algorithm were to be implemented in a computer carrying out millions of operations per second, then finding a best move in the mid-game would require approximately 10^{69} seconds (the age of our planetary system is estimated to be 10^{18} seconds, *Ed.*).

However, not only problems of finding the best move in chess games belong to this class of problems. Geometric problems of proof are examples from scholastic practice. One can, in principle, devise algorithms similar to those for sorting out moves during chess games for solving school geometry problems. However, it would be impossible to look for a solution on the basis of such an algorithm, since this would also require reviewing the immense number of variants. It is clear that it would also be impossible to teach such an algorithm. The solution is to devise methods which would narrow considerably the scope of searching for a solution and would permit solving problems without going through exhaustive testing of *all* possible variants of action. These methods are the so-called heuristic methods. Heuristic methods presume a support of accumulated experience in problem solving and the use of this experience for specifying the direction of the search toward the greatest probability of solution. Heuristic methods may differ from one another by the degree of specificity of the instructions entering into the corresponding prescriptions. However, the degree of specificity of the instructions could be so great that heuristic methods could be formalized and presented in the form of algorithms which even a computer could follow.

We have pointed out examples of certain problems for which one may, in principle, construct algorithms, but these algorithms are found to be unfeasible in practice. Let us now examine problems for which one may construct algorithms which are feasible in practice. However, it should be clear from Figure 6.1 on page 115 that it is not expedient to solve every problem which can be solved by algorithm in this way. Since this class of problems is very important, let us examine it in more detail.

Problems Solvable by Algorithms
But Not Expedient for Algorithmic Procedures

The first example concerns problems which are difficult to formalize, i.e., problems for which it is difficult to create univocal prescriptions which completely specify the solution process, such as, for example, certain grammatical problems. Let us examine one of them, namely, the problem of specifying the usage of the prefixes *pre-* and *pri-*.*

In the standard textbook on the Russian language for high school, Barkhudarov and Kryuchkov [532] set forth the following rules of usage for these prefixes:

"In order not to make mistakes in correct usage of the prefixes *pre-* and *pri-*, one must understand their meanings.

"The prefix *pri-* has two basic meanings. First, it signifies a rapprochement or a union with something, also a proximity to something, for example, *pri*yekhat (v.), to arrive; *pri*bezhat (v.), to come running [up to] ; *pri*kleit (v.), glue onto; *pri*kupit (v.), to buy some more [in addition to what has already been bought] ; *pri*brezhny (adj.), coastal [by, at the shore] ; *pri*morsky (adj.), seaside, maritime [by, at the sea] ; *pri*gorod (n.), suburb [near the

---

*While the example here is drawn from Russian grammar and aimed at readers whose primary language is Russian, the example holds, as well, for instruction in Russian as a secondary (foreign) language. A parallel in English, e.g., in terms of the prefixes *ante-* and *anti-*, can be readily imagined. Here and subsequently the Russian grammatical examples have been retained unless it was thought that they would tend to confuse rather than illuminate the point under discussion. (*Editor.*)

city]. Second, it signifies incompleteness of actions, for example: *pri*lech (v.), to take a nap; *pri*sest (v.), to take a seat; *pri*tvorit (v.), to shut.

"The prefix *pre-* also has two basic meanings. First, it signifies a high level of quality or actions. For example, *pre*dobry (adj.), exceptionally kind, good; *pre*mily (adj.), exceptionally nice, kind; *pre*uvelichat (v.), exaggerate; *pre*vysit (v.), to exceed, to overstep. Secondly, it has the meaning similar to that of the prefix *pere-* [trans-, transition, passage], for example, *pre*rvat (v.), interrupt; *pre*lomleniye (n.), refraction.

"In some words the meaning of the prefixes *pre-* and *pri-* is difficult to explain, for example, *pre*sledovat (v.), pursue; *pre*nebregat (v.), neglect; etc. The correct usage of these prefixes must be memorized."

In these rules, the conditions for using each prefix are shown, but what must be done, and how to judge in order to specify the use of one or another prefix in a word is not directly indicated. Polyakova [752] formulated two possible ways for solving the problem of which prefix must be used for the word, based on the prefixes indicated in the textbook. We cite them in Polyakova's formulation:

"1. *Method of Assumption.* If it is *pri-*, then one of the meanings of that prefix should be present: either union, rapprochement, proximity to something, or incompleteness of action—and the meaning of *pre-* could not be present: either a high level of quality or action, or the prefix *pre-*, which resembles *pere-* in meaning. Let us see which meaning fits. This is the one that fits. Here, with this meaning, the prefix *pri-* (*pre-*) is used.

"2. *Method of Defining the Lexical Meaning of the Word.* What idea is expressed in the given word? The given word signifies. . . . Here, the prefix *pri-* (*pre-*) is used.

"This second method leads to a direct selection of the meaning of the word.

"It must be said that very often it is difficult to establish the meaning of a word directly. Therefore, it is necessarily recommended that the various meanings be compared in advance, and that the most appropriate be chosen on the basis of that comparison."

Although Polyakova did not formulate methods of solving the problem in the form of a prescription exactly directed at the actions of the students, it is easy to formulate such prescriptions from what she has said. For example, for the first method, the following could be done.

In order to establish which prefix (*pre-* or *pri-*) should be used in the word, one must:

1.   Assume that it has the prefix *pri-*.
2.   Ascertain whether the word signifies a union.
        If yes, then write the prefix *pri-*.
        If no, then,
3.   Ascertain whether the word signifies rapprochement.
        If yes, then write the prefix *pri-*.
        If no, then,
4.   Ascertain whether the word signifies proximity to something, and so on.

An analogous prescription may be set up for the second method of specifying the correct usage of the prefix.

If the question arises—is such a prescription algorithmic?—it must be answered negatively. This prescription is not algorithmic, since it is applied to the kind of semantic operations which are not sufficiently elementary, and therefore cannot, in all cases, be carried out univocally even by the very same student. One or another result of carrying out the operations depends entirely on the particularities of the student's individual experience, even on random associations which he might have while perceiving the word he must use. In this connection, having correctly defined the meaning of one and, in that case, correctly attributed the prefix, the student could, following the same prescription, incorrectly define the meaning of another word and make a mistake. Examples of such mistakes can be found in any methodological work devoted to the topic.

Therefore, the given prescription does not fully specify the process of the student's solving the problem, and the actions which it prescribes do not ensure the accurate use of the prefix in all cases. However, as Polyakova indicates, teaching this prescription *lessens* the number of mistakes in comparison with the method of instruction where students are *not* shown how to judge (reason) when solving problems of a given type.

If the prescription cited above is not algorithmic, then is it possible to construct algorithmic prescriptions for this and similar problems which would completely and univocally specify the actions of students and ensure the correct use of prefixes in all cases? Since the number of words for which prefixes must be specified is finite, a corresponding algorithmic prescription *could* be constructed. However, it would include the operation of referring to a dictionary. An algorithm which includes the operation of referring to a dictionary ensures the correct use of prefixes in all cases. But it is inexpedient to act according to such an algorithm. It is impossible to have a dictionary on hand at all times, and it is inconvenient to look in the dictionary for every word. Therefore, students are taught grammatical rules and prescriptions similar to the ones cited, so that they will not have to turn to a dictionary every time.

Let us note that a greater expediency in solving problems other than by algorithms (algorithmic prescriptions) and based on non-algorithmic types of prescriptions can be found not only in the case where it is necessary to carry out non-elementary semantic operations, but also in some cases where it is necessary to carry out operations of a non-semantic character. An example would be connected with the isolation and appraisal of character-istics of objects which are sensorily perceived. Thus, in order to teach students how to correctly use roots with alternate vowels, Polyakova [753] set up such a prescription for one of these types of roots (*gor-gar, zor-zar*): "First, one must establish where the accent stress falls; then one must single out the root; then one must find the doubtful vowel. The conclusion is that if the stress falls on the vowel of the root, then in the roots of *gor-gar- a* is always written under stress. In the roots of *zor-zar-* the vowel of the root is always *o* under stress." This prescription is also not algorithmic. The third instruction (the first two do not interest us) does not univocally specify the process of problem solving. The vowel which seems doubtful to one student does not seem doubtful to another. Moreover, one and the same vowel seems doubtful in one word to the same student, while in another word he has no doubts about it whatsoever. As in the preceding case, an algorithm for solving this problem could be devised which would propose consulting a dictionary, but proceeding by such an algorithm would also not be expedient.

We examined examples in which it is possible to construct algorithmic prescriptions which could be carried out in practice. To be guided by them, however, is inexpedient, since they engender a number of difficulties or inconveniences and demand a considerable amount of time, and so on.

Let us examine an example of a different sort, where following an algorithm may lead to an inefficient solution of problems.[5]

When teaching identical transformations* in algebra, a tendency to algorithmize strictly all of the students' activities becomes distinctly apparent. This is done exemplarily in the following procedure. For algebraic expressions examined at each stage of teaching identical transformations, a special canonical form is chosen and stated. Teaching the technique of identical transformations amounts to the fact that students get accustomed to reducing each algebraic expression to that canonical form, and to carrying out operations on the algebraic expressions used in this form. If, for rational algebraic expressions, this approach is no longer expedient in all cases (there are problems which can be solved more economically without reducing the expression to a canonical form), then in passing to identical transformations of irrationalities, the notion of canonical form of an algebraic expression loses its precise mathematical sense. At the same time, a strict algorithmization of the transformation process also loses its meaning. But the procedure indicated above continues to act by inertia in the practice of teaching. Here, in practice, one uses, in the role of a "canonical form" of fractional-irrational algebraic

---

*The phrase "identical transformations" refers to sundry operations on both sides of the equality sign within an algebraic expression, which do not disturb this equality. (*Editor.*)

expressions, the presentation of these expressions in the form of fractions with rational denominators. As a result, a steady habit of considering one of the transformations—freeing the fractions from irrationality in the denominator—as something universally necessary and of carrying it out every time before undertaking any operation on irrational algebraic fractions is produced in the students. They are often totally unable to carry out the operations on irrational fractions in attempting to bypass this transformation. They have not been taught how to do this. With the ordinary system of instruction, the possibility of solving all problems on identical transformations of irrationalities is attained by some uniform method indicated in advance (i.e., by means of a certain uniform algorithm). However, this method turns out, as a rule, to be much longer and cumbersome than applying, in each case, individual methods corresponding to that case. Such individual methods do not permit complete algorithmization (because expanding into multipliers in the field of irrational expressions turns out to be equivocal), and, evidently, it is for this reason that it is often not taught to students.

In order to carry out identical transformations of even uncomplicated irrational expressions, the student must not only freely master the technique of carrying out elementary operations, but he must also be able to evaluate in each particular instance the possibility and the expediency of applying some other transformations. All this requires the establishment of special skills for orientation.[6] Of course, since the identical transformation of irrationalities carried out in secondary school embraces only a finite number of different cases, in principle, it is possible to enumerate all such cases with an indication of the best means of solving each one of them. However, such an algorithm itself would by its formulation be infinitely cumbersome, and teaching it would obviously be a pedagogical absurdity. On the other hand, it is evident from what has been said, that algorithms which are simple in their formulation (and the formulation of the traditional algorithm for transforming fractional-irrational expressions is very simple) could be too complex to be realized and could lead to inefficient solutions to the problem.

Let us examine yet another group of problems for the solving of which algorithms may be devised, but which it is inexpedient to solve algorithmically. These are problems which are easier to solve by means of trials, where the number of possible trials is relatively small. Such problems are often encountered in ordinary life and in industrial as well as in scholastic activities. The simplest example of it is an algebra problem: the problem which arises when solving a simple equation with one unknown.

As every schoolboy knows, in order to solve an equation, it is necessary to shift those expressions containing the unknown to one side and to shift the free expressions to the other. In principle, it does not matter at all which expressions are transferred to which side. But let us suppose that when transferring expressions with unknowns to one side and free expressions to the other, and in reducing like expressions, we want the coefficient to be positive in the unknown. It is not difficult to establish an algorithm which, when followed, could specify in different cases to which side it would be preferable to transfer the free expressions and to which side the expressions having unknowns. However, it is inexpedient to set up such an algorithm and use it. It is much more economical to mentally try out bringing the unknown over to one side and see with which sign the unknown turns up. If it is positive, then the trial was correct. If it is negative, then the trial was incorrect, and the unknown must be transferred to the other side.[7]

Thus, trials with the subsequent evaluation of their results by some criterion are, in many cases, more economical than solving problems by algorithm. It is precisely for this reason that man often uses trials as being the more efficient method. It must be said that, in many cases, man solves problems partly by trials and partly (in specific stages) by algorithm. A special feature of man, as we already said, is his ability to switch quickly and flexibly from one program of solution to another. He is a "multi-programmed" system.

It is not just in the solving of mathematical problems that we encounter examples where trials are more efficient and where people act according to a trial-and-error procedure rather than by algorithms. Thus, for example, people studying a foreign language for the purpose of understanding and translating a foreign text often carry out an analysis of the sentences, not on a basis of algorithms for grammatical analysis (even in cases where they know them), but by means of trials in which the success of each trial is evaluated on the basis of semantic criteria.[8] Such a method is certainly not always successful (often it does not yield the correct comprehension and translation). Sometimes, however, it does permit attaining one's goal in a more economical way.

An objective prerequisite for the possibility of solving problems by trials is the problem solver's knowledge of the final outcome (state of the object) which must be obtained at the end. From this viewpoint, problems may be divided into two types which may, for convenience, be called *problems with previously known final outcomes* (specific expectations, *Ed.*) and *problems with unknown final outcomes*.

To the first type of problems, for example, belong the problem of unlocking a lock if the key is lost (the final state of the lock which must be obtained as a result of the solution is precisely known here), the problem of maintaining the temperature of an oven within the limits of *a* and *b* degrees (the temperature which must be obtained at the end of the action is known in advance here), the problem of synthesizing a certain chemical substance with previously given properties, the problem of proving a certain mathematical proposition, and many others.

To the second type belong such problems as adding one number to another, for example, 18191 and 39944 (the number representing the sum of the addition is not known in advance), or the problem of how many parts an object that is being disassembled will yield, and so forth.

It is possible and in many cases expedient to solve problems with previously known final results by means of trials (it is possible because in knowing the final result to be obtained, one may see precisely whether the trials led to this result or not). But it is often not expedient, or it is even impossible to solve by trials problems with previously unknown final results. Indeed, if it is not known in advance what number would result from the addition of 18191 to 39944, then attempting to solve the problem by means of trials is senseless, since the criteria for the correctness or incorrectness of the results are lacking. An analogous situation occurs when solving many problems of pronunciation in English. It is impossible to specify by trials whether the "gh" in such words as bough, cough, draught, drought, taught is silent or pronounced "f," since we have no direct criteria as to whether the trial is correct or incorrect. It is precisely that which should serve as a criterion for the correctness of the trial which is unknown.

However, in similar situations (i.e., in solving problems where the final outcome is unknown) it is usually possible to find (discover) independently an algorithm of solution if one exists. But a knowledge and use of either oblique criteria for evaluations, or other persons' evaluations of the correctness of the trials is indispensable. In a number of cases, models of the final product which man encounters in the process of living and acting may serve as criteria (for example, a person encounters the correct spelling of words in a book). In such cases, the process of independently discovering an algorithm takes place in the following way. A person constructs some kind of hypothesis concerning a course of action, and attempts to apply these actions (this may be one action or a whole chain of actions). If the trial leads to a positive outcome (evaluated on the basis of criteria at his disposition or by other people), then these actions are kept in mind and are accepted as a possible algorithm. This algorithm is then verified by solving problems with different conditions. If the experiment did not lead to a positive outcome, then another hypothesis is constructed and another trial takes place which is also evaluated on the basis of criteria at hand or by other people, and so on, until the person finds the actions which lead to the correct outcome.[9] It is by exactly such means that some students, for example, independently discover grammatical algorithms if they are not taught them.

As we said earlier, many grammatical problems are problems with unknown final outcomes. Therefore, finding algorithms by means of trials with an evaluation of the results by a *direct* criterion, i.e., using the required final outcome as criterion, is often impossible, since the final outcome is unknown. What guides the student when searching for an algorithm as to the correctness or incorrectness of his hypothesis and his trials is, in these cases, the evaluation of the teacher. More rarely it is self-evaluation deriving from a comparison of the results of his trials with some oblique criteria. If the student, for example, wrote a word correctly, then this fact itself corroborates the hypothesis from which he started, or indicates success of the trial. If he made a mistake, then the hypothesis is discarded. The teacher's indication of a mistake or its independent detection through comparison of obtained outcomes with some kind of model is a stimulus for searching for another hypothesis and carrying out other trials. Thus, as a result of testing various hypotheses and trials against teachers' evaluations or self-evaluation, some students (usually the more capable ones with flexible reasoning) succeed in independently finding (discovering) an algorithm for solving problems of one or another type.

Here is one more example of a problem for which it could turn out to be inexpedient to use algorithms. These are problems where the probabilistic solution is more advantageous than completely carrying out an algorithmic process. One model of such problems is the following. Let us suppose that action A must be carried out if, and only if, a certain object x has characteristics a, b, and c. That is, the situation a (x) & b (x) & c (x) ⟷A exists.* In order to specify whether or not action A should be applied to the object, the presence in it of all three characteristics must be verified. The verification procedure itself could be prescribed by some algorithm. However, it can be observed generally that, while knowing an algorithm for solving a problem, people often do not verify all characteristics, but only some of them. The greater the probability that with a and b present c is also present, the smaller the risk of carrying out action A on only two characteristics instead of three. But the expediency of a probabilistic solution depends not only on the probability of characteristic c given

---

*The symbol & signifies the logical conjunction, and the double arrows ⟷ signify "if and only if." The symbols of Symbolic Logic will be introduced more systematically and explained below, in Chapter Seven and Chapter Eight.

characteristics a and b, but also on the value of the mistake if characteristic c is not present, and action A should not have been carried out. If the negative value of the mistake is not great (for example, if the value of the mistake is less than the value of the time spent carrying out operations to verify characteristic c), then it is more advantageous to accept a probabilistic solution based on a verification of only some of the characteristics. But, if the value of the mistake is great, then it is more advantageous to verify all characteristics, i.e., to carry out all the operations prescribed by the algorithm. Although the decision to carry out all the operations of the algorithm or to accept a probabilistic solution usually takes place intuitively, in principle, it is entirely possible to construct mathematical criteria for decision making. Because these criteria take into account the probability of a mistake within a probabilistic solution, the value of a mistake, and the economy of not fully carrying out the algorithm, they show in which cases it is more expedient to accept a probabilistic solution.

The following relations could be the basis for a criterion. Let us designate probability (frequency) of mistakes in a probabilistic solution by the symbol p (mis); the value of a mistake in a probabilistic solution (in fixed arbitrary conditional units) by the symbol v (mis); the economic value (time and means) of not completely carrying out the algorithm by the symbol v $(econ)$. It will be obvious that a probabilistic solution is more advantageous under the condition p $(mis) \cdot v$ $(mis) \leqslant v$ $(econ)$ and unadvantageous if the contrary. If the expression "the probabilistic solution is advantageous" is designated by the letter B, then what has been said may be written in the following formulas:

$$(p\ (mis) \cdot v\ (mis) \leqslant v\ (econ)) \Rightarrow \mathrm{B} \qquad (1)$$

$$(p\ (mis) \cdot v\ (mis) > v\ (econ)) \Rightarrow \overline{\mathrm{B}} \quad [10] \qquad (2)$$

The meaning of the formulas is easy to understand from the following example. Let us take the very same situation where in order to carry out action A, it is necessary to verify whether the three characteristics a, b, and c are present. Action A is carried out if, and only if, the object has all three characteristics (*a* (*x*) & *b* (*x*) & *c* (*x*) ⇔ A). Correspondingly, verifying the three characteristics requires three operations. In carrying out all three operations, the problem is correctly solved, and the problem solver has no losses whatever. Let us suppose, however, that verifying characteristic *c* is a costly or complex procedure (we often meet such cases when solving identification problems, e.g., when it is necessary to establish a medical diagnosis or identify an equipment malfunction). Instead of carrying out the entire algorithm, i.e., verifying all three characteristics in our example, might it not be more expedient to act in a probabilistic way on the basis of only two characteristics, e.g., *a* and *b*? The answer to that question depends on the probability of a mistake, the value of the mistake and the advantage which will be derived from the person's carrying out one operation less in the process of solving the problem. Let us suppose that the probability of a mistake *p* (*mis*) equals 0.3 (a person is likely to be mistaken three times out of ten); the value of the mistake *v* (*mis*) is equal to two conditional units; the advantage *v* (*econ*) obtained from the person's verifying in one less operation within the process of solving the problem is equal to one conditional unit. Now it is easy to compare the losses which a person has with a probabilistic solution and the advantage he obtains with it. If the loss is less than the advantage, then it is expedient to follow a probabilistic solution. If it is more, then it is inexpedient. The losses in a probabilistic solution are equal to *p* (*mis*) · *v* (*mis*) = 0.3 · 2 = 0.6 conditional units on the average. The advantage which is obtained in a probabilistic solution is equal to one conditional unit. Since 0.6 is less than 1, it is evident from formula (1) that following a probabilistic solution is more advantageous under the given condition than to act according to an algorithm. Under other conditions (for example, if *v* (*econ*) were equal to 0.5), it would have been more advantageous to solve the problem by algorithm.

There can be varying degrees in the probabilistic character of a solution, where the degree of probabilistic character is not directly related to the number of verified characteristics (more precisely, of their proportionate representation among all of the characteristics being examined). Thus, in some cases, the verification of one characteristic out of three could ensure a greater probability of a correct solution than in other cases where two out of three characteristics are verified. The probability of a correct solution depends not so much on the number of verified characteristics. Rather it depends on the character of their relations and hence the probability, in the presence of some specific characteristics in the object, of there being other characteristics as well. Ultimately, it depends on what contribution the verification of one or another characteristic makes in adopting the correct solution.

Adopting the correct solution on the basis of probabilistic criteria and some subsequent evaluation takes place not only when a person knows the algorithm for solving the problem, but also when it is inexpedient for him to act according to an algorithm for specific reasons. Probabilistic approaches also underlie the solution of problems where an algorithm does not exist or is unknown. Under those conditions, the choice of actions (A, B, C, . . ., N) may be specified only by the probabilistic evaluation of their success, and by the probabilistic prognostication of their outcomes. It is obvious that, from among all probable actions, a person will give first choice to those which will lead to the problem solution with the greatest probability.[11] In ordinary language, it is often said that man *intuitively* senses how it is best for him to act and which action it is most expedient to perform.[12]

Probabilistic evaluations may, therefore, apply to evaluations of situations and their separate characteristics, as well as to evaluations of particular actions from the standpoint of success in attaining the goal through these actions. One would expect that probabilistic bases of reasoning are more universal than approaches based on strictly deterministic processes. In this sense, strictly deterministic processes may be considered a special case of probabilistic processes, i.e., a case where the probabilities of success of various actions are either equal to one or to zero.

As certain data show (see, e.g., Dodonov [610]), a probabilistic evaluation of a situation[13] often precedes its detailed analysis on the basis of theoretical propositions known to a person.

Moreover, a probabilistic evaluation guides that analysis through a specified course.

There are facts in psychology and pedagogy which show that probabilistic techniques of reasoning may be guided in a specified manner, forming those probabilistic evaluations and actions in a person which are necessary for solving problems. The work of Vishnepolskaya is of great interest in this connection [559], [560]. The work shows how, in changing the frequency of orthograms which are given to students to write down, one may change at will the tendency toward one or another type of mistake.

According to Vishnepolskaya's data, unaccented *o* occurs in the Russian language more than twice as often as unaccented *a*. Unaccented *e* and unaccented *i* have approximately the same relationship. The number of words with the prefix *pri-* encountered by children in texts of literary works exceeds by almost four times the number of words with the prefix *pre-*. There appears a tendency in Russian children, in all doubtful cases, when it is not possible to depend on a rule to write the orthogram which predominates in the Russian vocabulary. Where students hear an unaccented *a-o*, they write *o*, where *e-i*, they write *e*, where *pre-/pri-*, they write *pri-*. Often, children make this rule absolute. However, it is possible to change the character of the "tendency" in dictating texts where the relative frequencies of occurrence of the orthograms mentioned are reversed. Correspondingly, the types of mistakes also change.

We examined cases in which it is possible to construct algorithms (algorithmic prescriptions) for solving one or another class of problem, but where solving these problems by means of an algorithm is inexpedient. Before going on to a detailed examination of problems for which it not only may be, but it actually is expedient and in many cases necessary to solve by means of algorithms, we pose the following questions. If it is expedient to solve some problems by an algorithm, does it follow that it is always expedient to teach this algorithm? If not, then what are the conditions of expediency for teaching algorithms? Let us dwell on these questions in more detail.

2. Conditions Under Which the Teaching of
Algorithms Is Expedient

In order to answer the question raised, let us examine the ways one may teach a person how to solve problems for which there exists an algorithm and which are impossible to solve without executing the algorithmic procedure.

There can be at least five ways of teaching a person how to solve such problems:

(1) teach algorithms of solution;
(2) teach search algorithms for discovering other algorithms;[14]
(3) teach general methods of searching which are of non-algorithmic character;
(4) teach separate rules of action, showing the students which operations *may be* applied in the solution process;
(5) teach neither algorithms nor methods of non-algorithmic character, but place a person in a problem situation and confront him with problems, counting thereafter on his ability to discover independently an algorithmic process (algorithmic procedure) in the course of self-instruction.[15]

Let us clarify the features of each of the approaches in the example. For simplicity, let us take the case where a student must be taught how to turn on some equipment, and for this purpose it is necessary to execute a specific algorithmic procedure. In order not to introduce new situations into the discussion, let us modify a little the example which we examined in Chapter Three.

Suppose that in front of the student there is some panel which has buttons a, b, c, switch e, knob f, and lights m (the apparatus is defective), and n (the apparatus is operative). Suppose that in order to turn on the apparatus and verify that it is in working order, it is necessary to execute the following algorithmic procedure:

(1) Place switch e in an up position;
(2) Turn knob f to the right;
(3) Press button a;

(4)   Press button b;

(5)   See which of the lights turns on.

If light m goes on (the apparatus is defective), then turn knob f to the left (turn off the apparatus) and summon the technician.

If light n goes on (the apparatus is in working order and turned on), then press button c and begin use of the apparatus.

The *first* method of instruction consists of directly forming the necessary algorithmic process within the student by explaining the rules for using the apparatus to him or showing him how to use it. As a result of training in applying rules or models of action, the student masters the algorithm which is the algorithm for solving the problem of turning on the apparatus and verifying that it is in working order.[16]

A *second* method of instruction is to supply the student with some search algorithm without explaining the algorithm of solution to him, pointing out, for example, what sequence of actions he must try to perform with the apparatus in order to find the unknown rules for turning it on and for verifying that it is in working order (i.e., discover the algorithm of solution). In that algorithm, for example, there could be indications of the type: At first try pressing button a, then b; if nothing happens, then try placing the switch in the up position, then push button a, and so on. In an algorithm to search for another algorithm, all possible operations and their sequences must be foreseen. In carrying out these operations, the student necessarily discovers which sequence of operations leads to the goal, i.e., uncovers the algorithm of solution which he can then apply to any equipment of the same design.

The *third* possible method consists of neither explaining the algorithm of solution to the student nor the search algorithm, but of supplying him with some methods of non-algorithmic character. The student is given instructions containing one or another uncertainty, on the basis of which, however, it is possible to discover the algorithm of solution. These instructions could be: "Specify the purpose of the elements of the apparatus which are on the exterior of the panel"; "try out different sequences of actions with them"; and so on. Obviously indications of this type are non-algorithmic.

If, in using the search algorithm, the student will necessarily discover the algorithm of solution sooner or later, this will not necessarily be so in using a method similar to the one above. Instructions analogous to the ones cited above do not univocally specify the character of the operations, and being to some degree uncertain, do not completely determine the student's actions. In carrying out these instructions, he must show independence and perhaps even creativity in trying to find the operations which he must execute in order to realize the directions and solve the problem. It is, of course, quite possible that, in carrying out the directions, the student will not "light upon" certain necessary operations, will not notice some important element on the panel, will not try out a certain sequence of operations (precisely the one leading to the goal), and so on. This is the reason why he may not discover the algorithm and may not solve the problem.

The *fourth* possible method of instruction consists of explaining to the student the rules of action which *may be* carried out with the object of finding the solution, without telling him which sequence of actions unfailingly leads to the solution. These are rules of the following type: One may push button *a,* one may push button *b,* one may turn the knob *f* to the right, and so on.

A clear example of the type are the rules for playing chess.[17] The rules of chess show how one *may* move in order to win the game (where each rule is precise: the knight moves in such a way, the bishop in such a way, and so on), but they do not show how it is *necessary* to move in order to win the game.[18] Many other rules have an analogous character (for example, rules referring to actions which may be applied to proving a theorem can have such a character). All these rules are not given in the form of an algorithm, but in the form of some calculation.[19]

The difference shown between an algorithm and a calculation consists of the fact that if an algorithm shows what *must* be done to solve a problem, then a calculation shows what *may* be done to solve it. An algorithm is a set of *prescriptive* rules, whereas a calculation is a set of *permissive* rules.[20] Applying the rules of some calculation to the initial object of that calculation, we engender new objects. For example, in a chess game (which may be considered as a calculation) in applying specific rules (authorized moves) to a specific situation, we produce another kind of situation. When solving geometry problems, from one kind of geometric proposition and construction, we produce other propositions and constructions, and so on. Since calculations contain permissive and not prescriptive rules, different people may apply different rules to one and the same situation, and thereby produce different successor situations from one and the same situation. This is precisely what happens in various chess-game strategies and in the different ways of proving theorems as well as in other instances. It is evident that a calculation may be turned into an algorithm, if a strict sequence for applying the rules is given. However, in many cases, doing this would be inexpedient.

The *fifth* possible way of instruction consists of not explaining to the student either an algorithm or a non-algorithmic method, nor furnishing any rules for possible actions. Instead, the student is placed in a problem situation. He is confronted with problems, and it is expected that he will discover the algorithmic procedure quite independently while instructing himself. If a person knows neither the algorithm of solution, nor the algorithm of search for another algorithm, nor a method of non-algorithmic character for problem solving, nor rules telling of possible permissible actions, then only one route is open to him. That route is to discover independently the algorithmic procedure (algorithm) via trials and analysis of obtained results.[21] If the trials are successful, the student will discover the algorithmic procedure or even the algorithmic prescription. If they are unsuccessful, he will not discover it. As in the two preceding cases, there is no guarantee of success here.

A large number of problems which cannot be solved without executing a specific algorithmic procedure confront man. What must he be taught in order to be able to solve these problems?

First of all, he must be taught the ability to carry out search trials, so as to arrive independently at a solution of the problem and, moreover, when possible, to discover the corresponding algorithms. Teaching trials assumes a working out of a large arsenal of actions, the forming of diverse links (associations) among them, drawing conclusions which ensure the transfer of knowledge and actions from one field to another, being able to pass from action to action if the preceding trial was unsuccessful, and several other operations and abilities. The ability to carry out test trials lies at the basis of any scientific or practical experiment, any scientific or artistic discoveries, and underlies every creative activity.

It is very important, also, to teach general methods of non-algorithmic character. These methods may have a different degree of generality, and on the basis of each it is possible to solve—if the method is sufficiently general—diverse problems from different domains. A person must be able to use different methods—methods of different levels of generality; and when it is necessary to solve a problem, he must be able to specify which of these to apply. Here it is essential to have the ability to choose a method which is more or less general, and which, therefore, maximally narrows down the field in searching for a solution or shrinks the trial space.

The ability to find a solution to a problem by way of trials, and the ability to choose the correct methods of solution (non-algorithmic methods are meant) permit the solving of problems for which the algorithm of solution is unknown. These abilities are also necessary for independently discovering algorithms. Teaching them is an important task of the school.

Whether or not students should be taught specific algorithms, whether specific algorithmic processes for solving one or another type of problems should be formed in them, depends on a number of factors. In each case, they must be specially considered and evaluated.

Stemming from the fact that any algorithm is always a method for solving a particular class of problems and is inapplicable outside that class, one must first of all judge the *significance* of those problems which will be solved by the algorithm. If these problems have no great scientific, practical, or general educational significance, then there is no point in spending time teaching students the algorithms of their solution.

If, however, these problems are significant in one or another context, then answering the question of whether or not it is expedient to teach the algorithm is influenced by the *difficulty (or ease) of the algorithm.* If the algorithm is very complicated, and it is easier to solve the problem on the basis of some non-algorithmic method or search trials, then obviously there is no point in teaching the algorithms. The time and energy spent on teaching the algorithms are not justified. In some cases the algorithms can be so complex that, as discussed above, it is impossible to apply them in practice. Therefore, one of the conditions for expediency in teaching algorithms is their low level of complexity.

However, even this condition is insufficient. In order that it be expedient to teach an algorithm, it is necessary that the problems to be solved by means of this algorithm *be encountered often enough.* Thus, if the time spent in forming an algorithmic process is great and the problems to be solved by it will be encountered rarely (and the total time spent on solving the problem by search trials is less than the time spent teaching the algorithm and solving the problem by algorithm), it is obviously inexpedient to teach this algorithm.

What has been said, however, is correct only on condition that no solution, an incorrect solution, or time spent on independent search for a solution do not bring a great deal of harm.[22] This is evident from the following examples. Let us take two problems, the first of which is a grammatical problem in using the endings of nouns and the second an industrial problem for controlling some equipment configuration. If a person who is uncertain about or does not know some grammatical algorithm ponders for a long time over some grammatical form (e.g., *who* or *whom*) and finally decides on the incorrect form, this, as a rule, hardly has any serious consequences. If, by contrast, he is going to try lengthily different variants of a control process and thereby exceed, for example, the permissible pressure in a tank, the harm could be very considerable. The relative values of not solving the problem versus a long search for the solution or even an incorrect solution can be different, therefore, in different cases. This is why in deciding whether it is expedient to teach a person one or another algorithm, it is important to take into consideration not only the number of problems which have to be solved by the algorithm, but also the value of no solution or an incorrect solution which is always possible if the problem is solved without knowing the algorithm.

In discussing conditions that influence the expediency or inexpediency of teaching a person one or another algorithm, we have indicated such factors as the complexity of the algorithm, the number of problems which must be solved by this algorithm, and the degree of harm caused by not solving a problem or incorrectly solving it. But, in order that considering these factors may serve as a basis for deciding whether it is expedient to teach one or another specific algorithm, these factors must be somehow evaluated numerically and correlated with each other. This is an extremely difficult problem, and at present, one can consider only approaches to solving it. However, sometimes approximate evaluations of these factors may be obtained. Having such evaluations at one's disposal, it is possible to propose criteria (also, of course, sufficiently approximate) for evaluating the expediency of teaching algorithms under one or another condition. Setting up such a criterion might be done in the following manner.

Let us suppose that we are dealing with problems where losses occurring from an inability to solve a problem or solving it incorrectly are insignificant and may be disregarded (this happens, for example, when solving grammatical and several other problems in school). Since, in that case, it is possible to discount the losses occurring from the inability to solve the problem or from an incorrect solution of it, a determination of expediency or inexpediency may be carried out, for example, on the basis of a comparison between the average time spent in solving the problems when an algorithm has been taught and when it is not taught and must be independently discovered. The quality of the problem solutions in both cases as well as all other things are assumed to be equal.

Let us clarify this in the following way. Suppose that students must be taught to solve a grammatical problem of a certain type. This can be done either by teaching the students an algorithm of solution of these problems, or by some other means which do not propose a special teaching of the algorithm. Experience shows that some of the students actually do discover the necessary algorithmic procedure, but the process of this discovery is often very long and painful, leading to the goal via a large number of mistakes. In so doing, the students arrive at insufficiently general procedures which lead to mistakes in writing, against which the teacher has to struggle for a long time (and often unsuccessfully). Let us evaluate approximately the expediency of teaching algorithms to the students under the indicated conditions. For that, we shall designate:

$n-$ the mean number of problems of a given type which a student solves during the time of instruction in school;*

$a-$ the mean number of problems of a given type which a student solves up until the moment of independently discovering the algorithmic procedure;

t_1- the mean time spent by the student who has not been taught the algorithm on studying the material;**

t_2- the mean time which the student who has not been taught the algorithm spends on correcting mistakes and in supplementary study of the subject-matter.

$s-$ the mean time which the student spends solving one problem without knowing the algorithm;

g_1- the mean time which the student spends studying the material when the algorithm is taught;[2][3]

g_2- the mean time a student spends in correcting mistakes and on supplementary study of the subject when an algorithm is taught;

$p-$ the mean time the student spends in solving one problem with the help of an algorithmic procedure he discovered himself.

*All of the discussions here and later on are based on the average student.

**The special training exercises in solving problems are not included here in the meaning of studying the material. The time spent in solving these problems is taken into account.

Let us examine the case where the student is not taught the algorithm. The mean time spent on assimilating the subject-matter and solving the problem will equal:

$$t_1 + t_2 + a \cdot s + (n - a) \cdot \text{p}^{\,24} \tag{3}$$

The meaning of this formula may be verbalized in the following manner: t_1 is time spent by the student in studying the material corresponding to the program. Since, after studying the material (and solving a specific number of problems), the student often still makes mistakes, time t_2 is that time spent in correcting mistakes, re-examining the subject-matter and on supplementary study of it, and so on. Total time spent on basic and supplemental study of the material is thus equal to $t_1 + t_2$. Since the student in this case solves problems without having been taught the algorithm, he spends $a \cdot s$ time on them. If the student in the process of solving them discovers the algorithmic procedure independently, he presumably solves the remaining $n - a$ problems by means of this procedure, spending $(n - a) \cdot \text{p}$ time on it. Thus, the total expenditure of time on assimilating the material and solving the problems in the case where the student is not taught the algorithm is equal to:

$$t_1 + t_2 + a \cdot s + (n - a) \cdot \text{p.}^{25}$$

Let us examine the case where the student is taught the algorithm. The mean time he will spend on mastering the material and solving the problem is equal to:

$$g_1 + g_2 + n \cdot \text{p.} \tag{4}$$

If the statement "it is expedient to teach the algorithm" is designated by A, then the basic strategies of teaching the methods which interest us may be written thus:

$$(t_1 + t_2 + a \cdot s + (n - a) \cdot p \geqslant g_1 + g_2 + n \cdot p) \Rightarrow \text{A.} \tag{5}$$

$$(t_1 + t_2 + a \cdot s + (n - a) \cdot p < g_1 + g_2 + n \cdot p) \Rightarrow \overline{A}. [26] \qquad (6)$$

The implication of these formulae is easy to understand: If the total expenditure of time on teaching the algorithms plus the subsequent time for solutions of problems by algorithm are less than the total expenditure of time on "non-algorithmic" teaching plus the subsequent time for solving of problems without knowing the algorithm (assuming identical outcomes), then it is expedient to teach the algorithms. If it is more, then it is inexpedient.

Here we have uniformly started from the optimistic hypothesis that students learn to solve problems correctly, even in the case where they are not taught the algorithm. However, such a supposition is far from being justified in all cases. Very often, students do not succeed in discovering a good and sufficiently general algorithm. Often, for example, they do not succeed in learning how to write grammatically correct sentences. In the latter case, when specifying the teaching strategy, it is necessary to take into consideration not only the mean time spent in teaching according to one or the other strategy, but also the quality of assimilation which is attained in teaching according to each of the strategies. Correspondingly, one must change the formulae cited above, taking into consideration the value of losses in them due to the lower quality of assimilation. One must take into consideration the negative value of mistakes which students will make *after* finishing the course of instruction. (The values of losses due to mistakes cannot be expressed as an expenditure of time, but it is possible to express the expenditure of time in terms of a loss value. Thus the amounts of time characteristically lost as a consequence of mistakes can be converted to a specific loss value.)

Does it not follow from the fact that people who are able to use an algorithm of solution for a specific class of problems often solve them not on the basis of the algorithmic process, but on the basis of intuition by way of accepting probabilistic solutions and so on, so that it is generally inexpedient to teach algorithms? These questions must be answered negatively, not only because there exist a great many problems which it is expedient to solve by algorithmic procedures, but also because solving problems by such a method is one of the conditions for forming intuition itself.

After all, in order to quickly form a good, correct intuition, it is usually necessary at first to solve problems on the basis of good methods, in particular, by means of algorithms if they exist. The better a person masters an algorithm for solving problems, the quicker its successive execution becomes overlearned and automatized. In other words, mastery of algorithms through overlearning is one of the conditions for emerging intuition. A person's intuition will be better developed if the underlying procedures are more strict and precise and the amassed *experience of correct solutions* is greater. But this experience is formed most successfully in the course of solving problems on the basis of precise methods.

The important conclusion deriving from the problems examined in the first part of this chapter consists in the fact that the problem of the expediency or inexpediency of teaching one or another algorithm cannot be solved *a priori*. This is always a *problem* that must be solved on the basis of *estimation* which presupposes the taking into account of the totality of factors and some evaluation of them. Methods of estimation are applied at present in all fields of human activity where there is a problem of choosing one best means of acting relative to specified criteria from among a certain number of possible means. The problem of determining the expediency of teaching algorithms must be examined as a problem of choosing the optimal solution. Therefore, it is related to that type of problem which is studied by "Operations Research."[2 7] One of the goals of further work in this field should, therefore, focus on the expediency of teaching algorithms as a problem of Operations Research. For these purposes it is necessary to bring to light all (or at least the basic) factors influencing the expediency or inexpediency of teaching algorithms and to work out methods for their precise quantitative evaluation.

Notes

1. In other words, in any case a person must learn some initial methods. He has to be taught them.

2. On the differentiation between search as a method of solution and search as a means of discovering an algorithm, see Chapter Five.
3. It is not important for us now whether they are solved by the conscious application of some algorithmic prescription or by carrying out some algorithmic procedure without realizing total composition and structure of the operations which make up the process of solution. It is a question of the principal possibility of constructing an algorithmic prescription for solving some class of problems and carrying out some algorithmic process (algorithmic procedure).
4. Thus, for example, without fulfilling the specific sequence of operations, it is impossible to start a vehicle; it is impossible to carry out certain important functions for controlling an airplane, train, etc. Destroying the necessary sequence of operations (such a sequence may not be the only one) could lead to a wreck or a catastrophe.
5. V.G. Ashkenuze called our attention to this example and permitted its use in this book. We cited the example as Ashkenuze interprets it.
6. The problem of patterns and the special features of orienting activity in the process of solving problems and forming habits has been studied by Galperin, Zaporozhets, and their colleagues (see, e.g., [574-576], [632], [739], [762], [769]). They developed special devices for forming orienting activities. Let us note that teaching of search and identification algorithms which we discussed in Section One (Chapters One through Five) is nothing other than the forming of the ability to orient oneself in specific situations, i.e., to carry out orienting activities.
7. If a person has a strict system of trials (i.e., he always carries them out in a specific order), then one may say that he acts according to a specific algorithm of trials.
8. For example, if the trials are directed at establishing the links between words in a sentence.
9. Often a person does not construct a hypothesis, but simply produces different trials, comparing their results with some

kinds of criteria, or with the evaluations of other people.

10. The line over the letter is a sign negating the corresponding expression ($\overline{B}$ signifies "not B," i.e., "It is not true that it is B.").

11. The actions which are based on the probabilistic evaluation of their outcome may be called "probabilistic actions," or "probabilistic moves." In referring to probabilistic evaluations of possible (expected) outcomes, we do not mean to suggest the evaluation of such outcomes by a specific number (e.g., action A in a certain situation leads to success with the probability of 0.6, action B with the probability of 0.3, and so on). A probabilistic evaluation appears most often in the form of an expectation which can have a different power, in the form of a feeling of certitude which can have different levels, and of some other "feelings," the power and level of which serve as a distinctive reflection of the degree of probability of the outcomes of specific actions.

12. The notion of intuition is very equivocal. In science, as well as in ordinary speech, this word connotes the most varied phenomena. "Acting by intuition" can mean, both, actions proceeding from unrealizable characteristics and actions based on a probabilistic evaluation of expected outcomes and some other processes. In examining the solutions accepted on the basis of intuition, it is usually necessary to take into account the nature of that process which underlies the intuition.

13. Dodonov calls it an evaluation on the basis of direct and practical generalizations. But, as we see it, its meaning consists of having a probabilistic character.

14. Teaching algorithms of solution, and search algorithms, and forming algorithmic processes may be carried out in two forms: In the proper form of an explanation of algorithmic prescriptions to the students, and in the form of a demonstration on how to act in specific situations. These forms are often combined in a specific way.

15. If this algorithmic process is realized and formulated by the student (even though merely for himself) in the form of some

system of rules or instructions on how to act in specific circumstances, then one may speak of his finding not only the required algorithmic process, but also the corresponding algorithmic prescription.

16. In speaking of mastering an algorithm, we mean throughout not the learning of the algorithmic prescription as such by heart, but the mastery of the corresponding algorithmic operations, i.e., forming the ability to execute the algorithmic process. In this sense, mastering (and being able to use) an algorithm should be distinguished from knowing an algorithm, i.e., knowing an algorithmic prescription. This does not necessarily include the ability to use the corresponding algorithmic operations. However, more will be said about this in detail later.

17. If it is unreasonable to teach the ability to turn on some equipment by informing a person of such rules, then in many other cases, it is quite reasonable.

18. Shevaryov [841], in particular, indicates the difference between these rules in the psychological and logical scheme.

19. Without going into detail on the meaning of calculation, we refer the reader to Yanovskaya's article [184].

20. Compare with Uspensky [166].

21. Trials were also necessary in the preceding two instances, but there, they were directed in a specific manner (and they were determined), by rules concerning possible actions (these rules are contained in corresponding methods of non-algorithmic character which are explained to the students). In the latter case, the trials are not specified externally by anything; they can be directed only by the knowledge and experience of the person. Let us note that if trials in the preceding two cases were not determined completely by external directives, then in the case where a person is guided by search algorithms in his actions (the second method of teaching), his actions are determined completely by algorithmic prescriptions.

22. We said above that not knowing the algorithm for solving the problem and the necessity, consequently, of solving the problem by means of search trials always leaves open the

possibility of no solution or an incorrect solution of the problem. It is also clear that solving a problem by algorithm (if it is not very complex) requires less time than solving it without the algorithm. The more difficult the problem, the greater is the probability of not solving the problem, or of a long search, or of an incorrect solution.

23. In speaking of teaching algorithms, we mean not simply conveying algorithms to the students in a ready-made form, but also teaching their independent discovery. Under this circumstance, the influence of such instruction on the development of independent creative reasoning of the students and on their general mental abilities will be no less than the influence of instruction where the students are not taught algorithms, and they have to discover them quite independently. Moreover, in many cases, the influence of such instruction on the development of independent creative reasoning will be greater, since any specially organized instruction is, in general, more effective than spontaneous instruction.

24. It is assumed that instruction is conducted until such time as the student learns how to solve problems of a corresponding type without mistakes or attains a given level of solution, i.e., until various students arrive at approximately the same level. However, an estimate can be made, which deals not with the difference in the time needed to attain identical results from instruction, but with the difference in the number of mistakes which the students make with different methods of teaching, holding time spent on instruction constant. The results of teaching can also be judged by other parameters.

25. If the student in the process of solving problems did not discover the algorithm, then $n - a = 0$, and formula (3) takes the form of:

$$t_1 + t_2 + a \cdot s, \text{ where } a = n.$$

26. These formulae indicate that, in pedagogy as well as in other sciences, it is possible to engage in a quantitative evaluation

(even though very approximate) of the optimality of one or another teaching strategy and of the choice of strategy based on quantitative estimates. It is unlikely that in the near future methods of estimating strategies of teaching will become as precise as those of several other fields of human activity. However, specific quantitative substantiation of teaching strategies undoubtedly is possible in pedagogy even now.

27. On the subject and methods of analyzing operations, see, e.g., Ventsel [71], and Morse and Kimball [133].

Chapter Seven

Teaching Algorithms in School

1. Implications of Teaching Algorithms

Earlier we examined the different types of problems, including ones which it is impossible or inexpedient to solve via algorithms. We also saw that it is not always possible or expedient to teach algorithms and that there can be problems and conditions under which it is more expedient to teach non-algorithmic methods.

Let us pass on now to examining problems where it makes sense to solve via algorithms and where it is also expedient to teach these algorithms with premeditation. These are problems (often with an unknown final outcome) which students must solve in large numbers. They are the kinds of problems for which it is difficult to discover independently a good solution algorithm and where not knowing the algorithm leads to considerable difficulties. There are many such problems in practice exercises in school.

If one were to analyze, for example, the mathematical problems which students have to solve in school, one would readily note that many of them are of the kind which are impossible or difficult to solve without knowing the appropriate algorithms. The process of solving them consists entirely of applying a specific algorithm to the data of the problem. Multiplication, division, raising numbers to powers, finding the root, many transformations in algebra, fulfilling basic geometrical constructions, finding the values of trigonometrical functions by tables, etc., are of this type. All are problems which are solved by means of specific algorithms, and it is essentially these algorithms that the students are taught (although all instruction is not and should not be restricted to teaching algorithms). However, algorithmically solvable problems exist not only in mathematics, but also, in great numbers in other subject matters. The algorithms for some of them have been found, but for many others they are not known and remain to be discovered.

The circumstance that algorithmically solvable problems exist not only in mathematics has a great significance for the practice of teaching. Usually it is much easier and simpler to solve a problem when the person knows the algorithm than to solve a problem when the algorithm is unknown. Moreover, solving via algorithm is incomparably quicker and more reliable. Think of how much time and energy it would require for someone to divide one large number by another if he did not know the algorithm of division (as in ancient times) and how easy and simple it is to solve this problem now, even for a sixth grader, when the algorithm is known.

The discovery of algorithms of solution for mathematical problems produced a radical change, one could say a revolution, in the practice of teaching mathematics. It led to the practice of teaching algorithms and thus facilitated and accelerated to a great extent the assimilation of this subject and particularly the material in those of its fields where teaching of algorithms plays a major role.

There are ample reasons to believe that the discovery and formulation of algorithms (algorithmic prescriptions) for solving problems in non-mathematical subjects will facilitate and accelerate to the same degree the solution of these problems and teaching the processes of solution, as the discovery, in its time, of the algorithm of division facilitated and accelerated the process of teaching the ability to divide.

The discovery and formulation of algorithms in some instances is a very simple task. In others, it may take decades or even centuries. In our daily life we deal ordinarily with simple, uncomplicated algorithmic prescriptions and algorithmic processes (procedures). The algorithm for using a vending machine could serve as an example of one of these. Formulating such an algorithm does not provoke difficulties. It emerges from the design of the machine and is a description of the overt and obvious operations performed by its customers. Although such an algorithm is extremely simple, without knowing it, or without carrying out even one of the operations (for example, not putting in the money, or not pushing the knob, etc.[1]), the desired result will not be attained. We also will not get the result in the case where we destroy the sequence of operations, for example, if at first we push the knob, and then put in the money. If we assume that someone does not know the rules for using the machine and did not see how others were using it, we would still expect that after a small number of trials he would discover those rules independently.

The situation, however, changes radically when we have to deal with a much more difficult problem, for example, with a problem of learning how to control some kind of complicated machine or apparatus. Here, also, trials can lead to an independent discovery of the algorithm, but this process will be long, costly, and in some cases even dangerous.

Practical problems exist for which efficient algorithms have not yet been found or have been found only very recently. For example, there are the problems of troubleshooting malfunctions in mechanical and electronic equipment.[2] In order to repair the machine which is out of order, the mechanic must first of all find the reason for the fault. Sometimes there can be dozens of reasons for each type of machine. In what order must the mechanic examine the functional units of the machine so as to discover the reason for its malfunctioning in the shortest time? Until recently there were no efficient algorithms for solving such problems, and troubleshooters proceeded solely from intuition and experience. However, solutions based simply on experience and intuition are often far from optimal. They engender much wasted time and lead to considerable economic losses. Only recently have efficient algorithms for solving these problems been found through mathematical analysis. The use of these algorithms has produced great economic savings.

In the process of instruction, we uncover the same picture. In comparatively simple cases the operations for solving one or another problem are known, and are usually taught without calling the corresponding prescriptions algorithms. If, for some reason or other, the teacher does not provide the students with the necessary algorithm, they usually discover it themselves.

The case is somewhat different when it is necessary to teach the solutions for more difficult problems. In these cases, only the most capable students and the most experienced and talented teachers succeed in independently discovering the appropriate algorithms. These algorithms are often not formulated by the teacher as general prescriptions for carrying out the sequence of actions for solving all problems of a given class. We are not even referring to the fact that algorithms discovered during empirical trials are not always adequately efficient.

The difficulty of discovering algorithms of intellectual activity as compared with algorithms of overt activity consists, first of all, in the fact that the mechanisms of specific kinds of intellectual activity are unknown in many cases. Secondly, intellectual processes are covert and thus hidden from observation. Therefore they are difficult to uncover, to isolate, and to describe. In consequence, solution algorithms for even not very complicated intellectual problems often turn out to be unknown both to the problem solvers and to the teacher.

Discovering the solution algorithms for different types of problems in widely different subject-matters provides the possibility not only of more quickly and successfully teaching students to solve these problems, but it also facilitates the mastery of a subject-matter as a whole. This is clearly important at a time, such as the present, when students tend to be overworked in school. Therefore, teaching algorithms (in those cases when it is expedient) should have a major role in the teaching of any subject-matter.*

However, even in teaching mathematics, where algorithms play a major role, the instructional process should not and cannot

---

*It goes without saying—in the presence of those conditions discussed in the preceding chapter.

under any circumstances be reduced to teaching algorithms. Instruction in algorithms is only one of the aspects of the teaching process, though a very important one.

In teaching algorithms to students and in forming algorithmic processes[3] in them, one may follow different approaches. One of these is to provide the students with the algorithm in a prepared form, i.e., to begin by teaching algorithmic prescriptions. This way, however, is often not the best, although in a number of cases and under special teaching conditions, for the purpose of economizing time, it is expedient to provide ready-made algorithmic prescriptions. But even under these conditions, for intelligent assimilation and application of the algorithm, the students must understand well the content of the material on which they must operate according to the algorithm. They must know the patterns of the phenomena, their essential characteristics, etc. In general, it is much more worth while, from a pedagogical viewpoint, when the student discovers the appropriate algorithms himself (if, of course, this task is not beyond his powers), or with the help of the teacher, than when he receives them ready-made.[4] It is expedient to follow this approach not only in teaching mathematics, but also in teaching other subject-matter, particularly grammar. In studying grammatical phenomena, one should start, not by telling students the algorithm for recognizing these phenomena and for operating on them, but by acquainting them with the content of these phenomena, their peculiarities, and relations to other phenomena, in brief, by forming the concepts. These concepts, of course, are formed in the process of activity, but the operations which the students perform at this time do not necessarily have an algorithmic character. The task of devising an algorithm for identifying and operating upon a phenomenon arises only during the process of studying the phenomenon and forming a conception of it. Often, however, it appears as a special problem at the stage of instruction which follows, when the students have already become familiar with the content of the phenomenon and have gained an appreciation of its special features, and when it is necessary to learn how to operate freely with this phenomenon. It is here that the necessity for an effective and efficient algorithm arises. The more perfect, easy, and simple the algorithm, the more easily and quickly one can form the necessary abilities and skills in the students, and the greater will be the possibilities of freeing their consciousness from the need to think of *how* to write (in the sense of grammatical correctness), leaving them to concentrate their attention on *what* to write.

The operations carried out in the process of *forming* a concept of a certain phenomenon are not necessarily the same as the operations which are carried out in the process of *applying* that concept. This is explained particularly by the fact that the number of characteristics which are established in the process of cognizing a phenomenon (i.e., in forming a concept of it) are usually greater than that number of characteristics on the basis of which the identification of the phenomenon takes place. Also the characteristics themselves differ among themselves in some cases. The concepts which are formed in the process of studying phenomena and the concepts on the basis of which their subsequent identification takes place are generally different, although closely connected. The distinction between these concepts was correctly noted in the textbook on logic edited by Gorsky and Tavanets [15]. In contemporary logical reasoning, the concept, as the authors point out, fulfills a double function. The first function of the concept is to be the condition for understanding judgments. It fulfills this role only when it presents a precise thought about the characteristics of an object which distinguish the given object from other objects. For these purposes the concept must incorporate within itself a small number of those of the object's characteristics that distinguish it from other objects. However, distinguishing one object from another object is only one of the functions of a concept. The other, no less important, function consists of the concept's ability to reflect, more or less fully, the result of what has been cognized. The concept as an outcome of the object's cognition is not a mere image of its distinctive features. It is a complex representation summing up a long series of preceding judgments and conclusions which characterize the essential aspects of the object. A concept as a result of cognition is a cluster of many previously obtained items of knowledge about the object which are compressed into one single concept.

In the methodology of teaching different subject-matters, including Russian, the question about how to acquaint students with different phenomena, how to establish their essential features, show their meaning, make them interesting, etc., has been well elaborated. In other words, the problem of how to provide students with a specific sum of knowledge about a subject-matter so as to form in them a concept of its essential characteristics, i.e., as a totality of cognition, has been much discussed. In methodology, the question of how to teach different, including grammatical, rules—rules which are none other than specific algorithms of transformation—has also been much discussed.

Much less explored is the question of how to form concepts that could serve as the basis for identifying objects and solving problems. Also unexplored is the question of how, on the basis of these concepts, to design more economical, efficient, and effective algorithms for their application (these are algorithms to identify phenomena), and how to teach these algorithms. Although many progressive teachers often do teach students algorithms of identification (practice itself generates the necessity for discovering and teaching such algorithms), a scientifically elaborated theory of devising and teaching algorithms of identification does not exist in pedagogy. It is for this reason that instruction in algorithms of identification, when it takes place in practice, often proceeds intuitively and unconsciously, executed spontaneously, and not always adequately. It is worth mentioning also that it is often only after many years of work, a large number of mistakes and failures, and many trials that a teacher arrives at a recognition of the necessity for breaking down intellectual activity into operations when identifying phenomena, and for teaching these operations with premeditation. These failures and mistakes could have been prevented and the path to pedagogical expertise made easier, if teachers had had at their disposal a specific theory for designing algorithms and teaching algorithms of identification.

In this book, we will not deal specially with the first stage of studying the phenomena, namely establishing their peculiarities (characteristics), establishing relations between them, and so on. We only underline the great importance of that stage and refer the reader to literature on that question. *The chief subject of our investigation is the elaboration of a theory for devising identification algorithms for the purpose of instruction as well as the means for teaching them.*[5]

Does not the teaching of algorithms lead to rigidity and stereotypy in the reasoning of students? Does not the danger arise of repressing their creative forces with that kind of instruction (one may say: "creativity should be cultivated, and we teach algorithms")?

First, it is not only creativity which must be cultivated. Forming different habits which must each proceed in the most automatized way plays a very big role in teaching. These skills are important not only in themselves, for without them many kinds of activity cannot exist. They are a necessary component of any creative process. For example, there can be no profound understanding and creative comprehension of a literary work, if a person reads poorly and if all his efforts and attention go into "penetrating the words" and into the technique of reading itself. No creative process is possible if these separate components of the process are not highly automatized.

Second, learning of algorithms by students should not be reduced to the acquisition of prepared algorithms or learning them by heart. An effective instructional presentation of algorithms necessarily presupposes independent discovery, design, and formulation of algorithms, and these are, as a rule, psychologically creative in character. Teaching algorithms can be an effective way of cultivating qualities of creative thinking.

Third, what was said above about teaching of algorithms does not mean that it must replace the cultivation of quickness of wit, insight, and generally fostering in students the ability to find a solution in those instances where there is no algorithm, or it is unknown. We stated only that if (when) it is possible to construct algorithms for some problems, and if (when) solving these problems by an algorithmic procedure is more efficient than by some other means, then it is inexpedient not to try and find, as well as teach, the appropriate algorithms.

Today, many algorithmically solvable problems (e.g., in teaching grammatical skills) are taught "non-algorithmically." This takes up so much extra time and effort of students and teachers that little time remains in school for solving problems of a creative character, and for developing the higher intellectual capabilities. The cultivation of creative thinking is threatened not by the fact that the teaching of algorithms will constitute a larger proportion of the educational process, but by the fact that, at present, in the period of rapid development in science and technology, an insufficient proportion of the instruction (i.e., in non-mathematical subjects) is devoted to it.

The teaching of algorithms is necessary for yet another reason. If, for the solving of some problem, it is required that there be some number of sequential operations (i.e., a specific algorithmic procedure be carried out), then not knowing and not carrying out, or carrying out one of these operations incorrectly leads, as we already said, to a mistake. If the student is taught these operations in the framework of an algorithm, then he masters the correct method of solution and the correct way of reasoning and acting relatively easily and quickly. If, however, he is not especially taught these operations, he will have to discover them for himself, embarking on the path of "trial and error." Since finding the correct, complete, and economical systems of operations for solving different classes of problems is often a matter of difficulty, it is natural that many students cannot discover them independently and that there are many defects in the operations they carry out. All this engenders difficulties in mastering the knowledge in question and leads to mistakes in solving problems (ungrammatical usage, incorrect understanding of specific questions, the inability to act efficiently in executing practical tasks, and so on).

Research conducted by us over a number of years has demonstrated[6] that a weak mastery of material, difficulties in learning, and mistakes are provoked in many cases by the fact that students do not know or have not mastered a number of important algorithms whose application is necessary for solving problems of a specific class. This, in turn, is related to the fact that in corresponding divisions of pedagogical science, not enough attention is paid to study of these algorithms. They are not formulated for all classes of problems (e.g., in native and foreign language instruction). In many cases the teacher does not know the set and structure of operations, which must be taught to ensure the mastery of specific material and the solving of problems, and does not intentionally teach these operations (algorithms).

Lately, the problem of teaching students intellectual operations has attracted the attention of psychologists, pedagogues, and methodologists. In many works of Soviet scholars,[7] the significance of developing capabilities for rigorous reasoning in students is shown. The efforts of many teachers are also devoted to this subject.[8]

Wherein, however, lies the drawback of elaborating this aspect of instructional methodology in a number of methodological and didactic investigations? The process of reasoning often has not been broken down into a specific number of sufficiently simple, elementary operations. (This is related, in turn, to the fact that, for the moment, in science, the methods of analyzing the fine mechanisms of reasoning hidden from external observation are in general poorly developed.) Further, in research, the most efficient sequence of operations for different conditions is often not indicated, which is the result of the fact that general principles of determining, devising, and describing algorithms for the purposes of instruction, as well as the criteria of the efficiency and economy, have not been established. Finally, the problem of designing a *system* of algorithms, determining the interrelations among them, comparing them according to degrees of generality and some other criteria has not been solved, although recently some efforts devoted to this question have begun to appear.[9]

In general, different devices (techniques) of thinking being pointed out in methodological writings often have a particular character and do not equip students with sufficiently general methods of reasoning which could make possible for students an independent attainment of knowledge and a successful application of it under the *most varied* circumstances. Bringing to light general systems of the intellectual operations that underlie the assimilation of specific knowledge and the solving of specific problems has not yet become a guiding principle in the design of instructional methodology. That is why algorithms of intellectual activity are formulated at present only for a very limited range of problems in comparison with those for which they could be discovered and then taught more efficiently with the help of these algorithms. This slows down the process of instruction as a whole and engenders many difficulties and mistakes for the students.

Hence, the elaboration of algorithms of mental activity for the purpose of teaching them to students is an important task of psychology and didactics. This task is general-didactic as well as general-psychological, since it concerns the most general principles of studying the algorithms, designing them, and teaching them.

Algorithmic prescriptions which may be used for teaching students differ essentially from algorithms designed for computers. This difference is contingent, first of all, on the concept of an elementary operation. Any algorithm presupposes a breaking down of the process of solving problems into sufficiently elementary operations. However, as said previously, the concept of an elementary operation itself is relative.

Operations which a machine can carry out are specified by its design; they depend, also, on the physical properties of the material of which it is made. Operations which a person may produce depend on the physiological properties of the highly organized brain and on its "design." Because of this, an operation which is not elementary from a "machine" standpoint may be considered quite elementary from a psychological and pedagogical viewpoint. It is also important to take into account the fact that the brain is the kind of self-perfecting system that develops its properties in the process of activity. That is why an operation which may be non-elementary at one level of a person's development and instruction can and must become elementary for another, higher level.

An algorithm that is placed into a machine in the form of a

program relates to its capabilities and presupposes the breaking down of operations to such a degree that the machine can execute them. An algorithmic prescription which is formulated for a person must be correspondingly related to his particular characteristics (physical and intellectual). It must take into consideration the capabilities which have been formed in him at a specific moment in time. The degree of fragmentation of operations prescribed by the algorithm designed for a person must be specified by the level of development of his reasoning, in particular by the character of the intellectual operations which have taken shape in him. Devising an algorithm for a problem-solving person, therefore, must be guided by a psychological evaluation of that person and must take into consideration the level of development of his intellectual processes.

In further discussions of algorithms, we will have in mind only algorithms (algorithmic prescriptions) established for people and, more specifically, for students. It is natural, therefore, that all these algorithms will represent the kinds of prescription or rules of action that indicate the sequence of operations which are elementary for a *given level* of the students' development. Here it is presupposed that each student *is able to perform* the operations which are indicated in the rules for action, that he knows how to execute them, and that they are formed in him beforehand, or are in the course of being formed.

2. Some Ways to Teach Algorithms and Their Influence on the Development of Students' Intellectual Abilities

Earlier, in discussing the role and place of teaching algorithms and of forming algorithmic processes, we also touched on the question of the means by which these algorithmic processes can be formed. We noted in particular that algorithmic processes may be formed by informing the students about algorithmic prescriptions, by demonstrating the operations which make up the algorithmic process, and also in the course of independent search trials. We add that they may also be formed by organizing the instructional material and designing exercises and problems so that the necessary systems of operations will be formed in the students in the course of solving these problems.[10]

These ways of forming algorithmic processes do not exclude each other. Moreover, the forming of algorithmic processes proceeds more successfully when these different ways are combined with each other. In this case, there can exist many variations in designing the educational process. For example, one may begin by telling the students about an algorithmic prescription, who then, while training to act according to the prescription, develop the corresponding algorithmic operations. One can deal with this also in the reverse order, i.e., begin by solving the problems and developing the corresponding operations in the actual process of solving the problems, so as to make the students aware of them in the very process of solution and, if necessary, to bring them to an independent formulation of the algorithmic prescription.[11] One may tell the students about the whole algorithmic prescription, or one may do it piecemeal. There can also be other variations for designing the educational process. Which variant to choose depends on the specific tasks and the circumstances of instruction. Let us examine several of these variants.

Suppose, for example, we are confronted with the problem of teaching a person in a very short time to carry out some sort of activity in which the speed with which it is carried out is of no importance. In this case, it may be expedient to begin the instruction by telling about the whole algorithmic prescription or about a whole system of algorithms. We have such a situation when, for example, a person has to be taught quickly to understand and translate a foreign text with the help of a dictionary. Knowing several algorithms for analyzing and translating a foreign text,[12] a person will be able to correctly understand and translate the text with their help, but he will do this slowly, carrying out a great number of logical and grammatical operations. He will approach understanding a foreign text and translating it into his native language by the method of "reconstruction," by way of many (often very complex) logical discourses about the grammatical form, i.e., discourses about the language. Grammatical operations formed by acting on an algorithmic prescription can, by subsequent training, be automatized, but it is quite obvious that attaining an understanding of a foreign text by means of complex logical discourses is undesirable in principle and in a specific sense unnatural.[13] When teaching a language in normal circumstances, unnecessary logical discourse about a language should be excluded when possible. One should try to have the students think *in the language,* and not *about the language.* To be sure, in order to teach how to think in a language, it is necessary to some extent to teach the ability to think about that language, but *only* to some extent. In order to attain the goal of "learning how to think in a (foreign) language," one must follow in teaching, the route which is impossible in the above-mentioned "emergency" (translating with dictionary) situations. This is the route of forming and establishing the algorithmic procedures step by step, and bringing each of the operations entering into these procedures to automatization before a transition to other operations included into the algorithm. It is precisely this route that it is expedient to follow in the ordinary circumstances of teaching a foreign language. We followed this route in those experiments on teaching a native language which will be described in the second part of the book.

We have repeatedly noted above different possible ways of forming algorithmic processes. But any methodology (technique) of instruction requiring students to remember long prescriptions is always undesirable. Wherever possible, one must try to eliminate the need for having to remember prescriptions at all. The forming of the necessary algorithmic procedures should take place in the process of independently solving problems. As to the prescriptions, they should appear mainly as a means toward awareness of the algorithmic process and a way of controlling it consciously and at will. What has just been said is related to the following problem.

As mentioned earlier, one must distinguish between *knowing* an algorithm and *mastering* it. Knowing an algorithm is knowing the operations which have to be executed in order to solve problems of a specific type and knowing the conditions of their application, i.e., knowing the prescription. Mastering the algorithm is the ability to perform the operations quickly and easily, i.e., the ability to carry out the algorithmic process. One may be able to carry out an algorithmic process, i.e., master the algorithm without knowing it, and conversely, one may know an algorithm and have a poor mastery over it. The situation here is completely analogous to the one which generally takes place in the process of reasoning and speech. One may be able to carry out complex logical discourses, for example, and prove theorems, and not know the rules for constituting the discourse. Conversely, one may know the rules for constructing such discourses and not be able to prove comparatively simple mathematical propositions. Analogously, one may be able to speak and write grammatically without knowing the rules of grammar, or know the rules of grammar well, yet speak and write with mistakes.

There are many examples of how people carry out algorithmic processes without knowing the algorithms, and moreover, without being aware of all the operations which "make up" the algorithmic process.[14] Such a situation was encountered, for example, when a task to replace human operators controlling certain industrial processes with automatic systems arose. For an automatic system to replace a human operator, it is necessary, first of all, to discover the algorithm by which that operator is guided and to transfer this algorithm to the automatic device. But discovering appropriate algorithms turned out in many cases to be a difficult affair. Human operators who execute control well in practice often cannot describe exactly how they do it. They do not realize the algorithmic process by which control is executed. Phenomena of this type are encountered even more frequently in the sphere of intellectual activity. People very often are able to solve specific problems without being able to explain how they think in the process of solving, and what operations they perform. In carrying out algorithmic processes which could, in principle, be described algorithmically, they do not know and cannot express the algorithms in accordance with which they act.

The fundamental task of teaching algorithms is to convey the *mastery* of them, i.e., the forming of algorithmic processes. Knowing an algorithm is but a means of attaining that end. But this means is extremely important. Psychological research has shown (see, e.g., [549], [577], [593], [596], [597], [625], [626], [632], [640], [643], [649], [650], [686], [711], [716], [720], [753], [768], [776], [790], [792], [799], [809], [817], [855], [862]) that forming knowledge (i.e., concepts, *Ed.*) in students not only of objects and phenomena of the external world, but also about how to act with them is of great importance. Forming actions (including intellectual ones) on the basis of *knowledge about the actions* considerably facilitates and accelerates the development of habits and skills, and at the same time improves their quality.

It must be emphasized that knowing about actions permits one to *control* these actions consciously and at will. If one does not know how to think in order to solve some problem or other, the requisite intellectual operations are actualized (i.e., elicited or cued, *Ed.*), for the most part, involuntarily and only as a response to the spontaneous influence of the objects of reasoning. In the presence of such knowledge, on the other hand, one can actualize and control the intellectual processes *at will.* Knowledge of action appears as a controlling system of a higher level, regulating the course of intellectual operations and directing the flow of intellectual processes. It is natural that, in becoming more controlled, the intellectual processes are formed more quickly, more easily, and more unerringly. There also emerges the ability to apply operations at will to new conditions, i.e., the ability to make a broader *transfer* onto these circumstances emerges. Applying the operations to new objects or conditions now depends less on how much these conditions externally resemble those under which the operations were formed; while, with involuntary actualization, it is precisely this which is the decisive factor in many cases. Owing to the awareness of the operations, they can be more easily applied to conditions which are not externally similar to those in which they were formed. Rather, they may be similar internally and essentially, but not in terms of their phenotypic character. Such a transfer, in turn, facilitates a broader and quicker *generalization* of operations which is extremely important for the intellectual development of students. The wider the range of transfer, the greater the possibility for new "moves" of thought, for new conclusions and discoveries. It is known that underlying scientific discoveries or inventions there is often a transfer of known methods to new conditions in which they were not applied earlier.

We discussed the importance of teaching algorithmic prescriptions for becoming aware of physical and intellectual operations by means of which human activity is carried out, and for the undeviating and voluntary control of those operations. The teaching of algorithms also has importance for the *systematization* of operations. For example, certain individual rules of grammar prescribe the ways of acting under some definite, limited conditions. However, grammatical algorithms may unite in one method a few or even many rules. They systematize operations related to different divisions of the subject-matter. Thus, the rule for when to use a comma and when not to use a comma before "and" is presented in different divisions of instruction on grammar. This has led to the fact that many students, when encountering this word, do not know which of the rules to apply in the given situation. As a result, mistakes appear.[15] However, it is possible to devise a single algorithm which synthesizes all the rules related to the conjunction "and" and permits univocal specification of whether a comma should or should not be placed before the conjunction. Many other similar examples could also be cited.

Teaching of algorithms systematizes operations in yet another respect. For example, it is possible to construct an algorithm for identifying the subject of a sentence and to teach it to students. After the students master the system of operations underlying this algorithm, and these operations become automatic, identifying the subject may be considered as one single, elementary operation. Further, it is possible to construct an algorithm for identifying the types of simple sentences. In order to identify a simple sentence, it must be specified whether it has a subject. Thus, the algorithmic procedure for identifying a subject enters as an element into another algorithmic procedure of a higher order. In its turn, the algorithmic procedure for identifying types of simple sentences can enter (in a specified method of teaching) as an element in an algorithmic procedure for identifying complex sentences (as distinct from a simple sentence with homogeneous members) i.e., in an algorithmic process of yet a higher order, and so on.[16] Thus, instruction in algorithms which are correctly constructed can serve as a perfect means for systematizing the knowledge underlying the algorithms as well as the operations.

The great *general educational importance* of teaching students algorithms and methods of devising them independently must be especially underlined. Although each particular algorithm serves only to solve problems of a specific class, in devising the algorithms, the students penetrate into the structure of objects and phenomena of the external world and of the intellectual processes themselves. They get to know the significance of general methods of reasoning; they learn how to discover, analyze, synthesize, and apply the methods. All this undoubtedly plays a large role in the students' development.

Finally, this must also be noted: Insofar as the correct teaching of algorithms includes comparing them by a level of efficiency or optimality, teaching algorithms may serve as a good means of cultivating in students *habits of pondering the efficiency* of different modes of action and of choosing the most efficient from among them.

As we know, any algorithm presupposes a univocal link between conditions and actions. It is often thought that this is an absolute deficiency of algorithms (see, e.g., Alekseyev [206]). Such a statement of the question is, however, untrue.[17] It was shown earlier that there is a large number of problems which cannot be solved once the specific sequence of operations is destroyed.[18] However, there are also problems where the sequence of operations has no significance and where the necessary result can be achieved through operations which are executed in different sequences. If, in connection with problems of the second type, the rigidity of the algorithm is often its deficiency, then, in connection with the first type, rigidity is its merit. The rigidity of an algorithm must not be considered as its deficiency or its merit irrespective of the type of problem which must be solved by means of that algorithm. Any attempt to render absolute some of the qualities of algorithms can lead only to confusion in evaluating their pedagogical significance.

3. Problems in Whose Solutions the
Order of Operations Can Vary

Above, we examined for the most part problems which require the carrying out of a certain strict sequence of operations. Let us look in more detail into the problems of the second type, which can be solved by means of various sequences of the very same operations. Many mathematical problems belong to this type. For example, the area of a triangle is equal to half of the product of the bases multiplied by the height. But there are many ways of calculating the area of a triangle. One can multiply the height by the base and divide the product obtained in half. One can multiply the base by the height and divide the product in half. One can divide the base in half and multiply the quotient by the height. One can divide the height by half and multiply the quotient by the bases; and so on.[19] Obviously, in forming a procedure for solving this problem in mathematics, there is no point in presenting it in the form of an algorithm indicating one particular sequence of operations. It is much more expedient to show all of the possible sequences of operations. The means through which this can be done is the *formula*. This formula for calculating the area of a triangle is, of course: $A = ½ bh$. Using this formula one may calculate the area of a triangle by different means, i.e., through different sequences of operations (algorithmic processes and their corresponding algorithms).

The formula above as well as all other analogous formulae are not algorithms, since they do not prescribe one specific sequence of actions.[20] This is a theoretical proposition reflecting specific relations among mathematical entities. But some rule* corresponds to this formula, and to the rule corresponds some *set of algorithms* (algorithmic procedures) representing all the possible sequences of actions for solving the given problem. One may say

---

*The inexpediency of prescribing a strict sequence for operations in this kind of rule is linked with the commutativity and associativity of the operations of adding and multiplying.

that this set of algorithms is contained within the formula in a hidden form. If some kind of problem may be solved not just in one way but in several different ways, it is much more expedient to offer them not in the form of algorithms, but as a formula to which all possible solution algorithms will directly correspond. Shevaryov [841] particularly emphasizes the correspondence between theoretical propositions and rules for action as applied in teaching mathematics. He cites the following rule: When it is necessary to multiply two polynomials, each member of one polynomial has to be multiplied by each member of the other polynomial and the products obtained must be added together. Then he cites the theoretical statement "the product of two polynomials is equal to the algebraic sum of the products of each member of one polynomial by each member of the other polynomial." In reference to the latter, he says: "This is a *theoretical* proposition, theoretical in the sense that a certain fact is noted in it. But there are no indications as to which operations *must* be carried out, or it is *advisable* to carry out, and so on. At the same time, it is evident that this theoretical proposition and the rule cited above are equivalent. A rule issues from a theoretical proposition, and a theoretical proposition can be deduced from that rule."

The rule for multiplying polynomials cited by Shevaryov is equivalent to the corresponding theoretical statement but is not an algorithm. It indicates what to do in multiplying two polynomials, but it does not indicate the sequence of actions in multiplying, nor does it fully determine these actions (each member of one polynomial may be multiplied by each member of the other polynomial in a different order). This shows, by the way, that the notions of "rule" and "algorithm" are not entirely congruent. "Rule" is a broader notion (and in certain cases less specific). Every algorithm can be considered as a rule, but not every rule is an algorithm. From this, however, it does not follow that rules must always be given in the form of algorithms. It is expedient to state many rules in the form in which the rule cited by Shevaryov was formulated.[21] It is important only to be aware always of what kind of rule is being formulated. For, in instruction, it is not at all the same thing whether a rule will be formulated as an algorithm or as a rule which is not an algorithm.

Formulating mathematical statements as formulae in no way lessens the significance of teaching algorithms, since the different algorithms for solving one and the same problem are not always differentiated solely by the sequence of operations and, therefore, they cannot always be placed in relation with some one single formula.[22] In those cases where different algorithms are distinguished only by the sequence of operations, and all of them may be placed in a relation with some formula, teaching algorithms is often really inexpedient. In those cases, however, it is important to teach the students to see the different algorithms hidden in the formula and to progress from the formula to a more efficient algorithm under some conditions, and conversely to progress from possible algorithms to the formula. For it is possible to solve a problem on the basis of a formula only by progressing from the formula to an algorithm. Not the least among the reasons for students' being unable to solve problems efficiently is the inability to progress from the formula to an efficient algorithm and to specify independently an efficient means for acting on the basis of the theoretical statement. Irrespective of whether algorithms for solving some general problem are distinguished only by the sequence of operations or whether by the operations themselves, students must be taught to evaluate which algorithm from among a number of possible algorithms is the most efficient in the given case in order to choose precisely the most efficient one for the problem. One may show the importance of choosing an efficient algorithm from the following simple example. There are many ways of adding 18 to 29: One may add 20 to 29 and subtract 2; one can add 18 to 30 and subtract 1, and so on. Facts show that the students do not always act in a sufficiently efficient manner. In turn, the process of solving the problem becomes more complicated, and this leads to excessive time spent on the solution. In order to solve such problems and others like them, one must know not only the corresponding solution algorithms, but one must also be able to choose the most efficient one from among them for the given case.[23]

4. Individualization in the Teaching of Algorithms

We must tell about still one more obvious requirement for teaching algorithms: It consists of the necessity for carrying out, whenever possible, the individualization of instruction in algorithms. Inasmuch as different algorithms for solving one and the same problem may require accounting for different indicative features and carrying out different operations, the different algorithms may seem more convenient or easier (i.e., in a specific sense, optimal) for different people, depending on their individual peculiarities. This can be illustrated with a rather simple example.

In loading a film projector, it is very important to thread the film into the machine correctly, since otherwise the image may appear on the screen "upside down." There are various indicative features which permit specifying the correct position of the film before loading, and in this connection, there are several possible algorithms for recognizing the position of the film. Let us cite a group of potential indications from which one may specify the proper position of the film when loading a projector for 16mm films: 1. (a) relative to the operator the film must be turned to its dull (emulsion) side for loading, and (b) images must be right side up and not "upside down"; or 2. (a) the perforations must be on the left, and (b) the film must have the dull side turned toward the person loading the projector; or 3. (a) the perforation must be on the left, and (b) the image must be right side up. How should one act efficiently in order to specify the correct position of the film? By which indication is it best to orient oneself? This depends on the individual peculiarities of a person. It is easier for different people to orient themselves by some indications than by others.[24]

However, the individualization of teaching algorithms consists not only in choosing from all possible indications and actions those which best "suit" a given person and which correspond to his individual peculiarities, but also to what degree of fragmentation specifications of the operations entering into the algorithm should be broken down. Since operations are developed differently in different people, the level of detail in the algorithm's instructions must be different for different people.[25]

We are no longer speaking of the importance of individualizing the *means of teaching* algorithms as such. One and the same thing can be taught by different methods. The individualization of teaching must consist also of teaching each person by a procedure corresponding to his individual peculiarities.

Thus, for example, a person with good phonetic discrimination and a good ability for mastering foreign languages aurally can be taught successfully certain aspects of a foreign language mainly through oral-aural methods. A person who has trouble distinguishing by ear, and who needs to depend on visual perception in order to master the spoken word, cannot be taught successfully by an oral-aural method alone. This does not mean that the first person is more capable than the second. They simply have different abilities. It must be noted that many people who are considered dull, are, as a matter of fact, dull only with respect to a specific method of teaching. Teaching them by another method may suffice to have their "dullness" disappear.

Thus, the method of teaching must be adapted to the individual peculiarities of a person and be adequate for him. Under present-day conditions of mass teaching, it is very difficult to realize such individualization, if possible at all. Programmed instruction, however, offers this possibility. In elaborating the procedures for teaching one or another algorithm, it is very important to take into consideration those changes which the algorithmic process undergoes when being formulated.[26] These changes could consist of combining several operations into one operation, in automatizing operations, and in several other changes which will be discussed later. At present, let us only note that, in teaching an algorithm, one must keep in mind the structure of that final product (process) which must be obtained as a result of the instruction. Knowing the structure of the final process is the necessary condition for its purposeful and effective formation.

5. Interrelation Between the Concept of "Algorithm" and the Concepts of "Ability" and "Skill"

It may be asked: Why use the concepts of "algorithm" (or "algorithmic prescription"), when there exist the concepts of "ability" and "skill" and when one may speak not of teaching algorithms but of teaching abilities and skills? In answer, the following must be said.

The concepts "ability" and "skill" are not sufficiently specific in pedagogy and psychology. The popular notion of ability subsumes properties and processes of very different character. For example, one speaks of the ability to guess, the ability to add 3 and 5, the ability to carry on scientific research, and the ability to sew on a button, even though all of these processes are completely different. To the present time, there are heated arguments in psychology and pedagogy about where the difference lies between abilities and skills, and about what meaning must be attributed to each of these words. This question has not yet been solved, and even today different authors use these words in different and often very vague senses.

By contrast with the concepts of "ability" and "skill," the concepts "algorithm," "algorithmic prescription," and "algorithmic process" are considerably more precise. These concepts indicate prescriptions and processes of a highly specific type for univocally discriminating their characteristic features. An algorithmic process may be distinguished from a non-algorithmic process by these features. This ensures an accurate differentiation of the processes investigated and a precise specification of the subject of investigation.

But it is not alone the fact that the concept "algorithm," "algorithmic prescription," and "algorithmic process" are more precise than the notions "ability" and "skill." These concepts are of a different character.

The concept "algorithm" (or "algorithmic prescription") refers to *prescriptions* for carrying out specific operations. The concepts "ability" and "skill" refer to the method of *mastering* operations. The concepts of "ability" and "skill" are closer to the concepts of "algorithmic process," though they do not stop there. There are processes called abilities (such as abilities to guess, or the ability to carry on scientific research) which, by their mechanisms, are not algorithmic. On the other hand, there are algorithmic processes which are not abilities or skills. Thus, someone who does something for the first time according to a complex algorithmic prescription, undoubtedly executes an algorithmic process, although one can scarcely say that he *is able* to do this, and even more so, that he mastered the corresponding skill.

A person, as we have said, may only poorly master the operations which are indicated in the algorithm. He can only know them. When we speak of abilities and skills, we mean that someone must master specific actions (operations). The essence of the matter does not change if one does not completely perform the operations, if one does not know of what elements the ability or skill that are being applied in practice are made up.

The notion "algorithmic process" and the notion "ability" are neither identical nor opposite. Some (but not all) algorithmic processes engage certain abilities. Certain abilities do become apparent in algorithmic processes. As to skills, one may obviously insist that all skills are realized within algorithmic processes, although the opposite is not true. Not every algorithmic process is an indicator (and therefore a manifestation) of a skill. A skill is a higher stage of development of an algorithmic process when an algorithmic process reaches a high degree of automatization. However, establishment of a more precise relation between the concepts "algorithm" ("algorithmic prescription") and "algorithmic process" on the one hand, and the concepts "ability" and "skill" on the other will only be possible when the latter will have been made sufficiently precise.

Since it is important that students not only know certain algorithms (algorithmic prescriptions) but master them in practice and be able to apply them, an important task arises in the process of teaching algorithms. This is the conversion of the knowledge of an algorithm into the skill of using it, i.e., developing the ability in the students for applying algorithms. However, it should be clear from what has been said above that the problem of teaching algorithms does not amount to the same thing as the problem of teaching students abilities and skills. The tasks confronting the teaching of algorithms are considerably broader. Teaching algorithms assumes the development of a general logical culture in which are cultivated a number of qualities of creative thinking. It assumes their formation in connection with becoming proficient in the methods of independently devising algorithms. On the other hand, the problem of developing abilities and skills in students is not the same as the problem of teaching algorithms and algorithmic prescriptions. When developing abilities and skills, a number of specific problems arise including such a very important problem as that of automatizing the actions which make up the algorithmic process.

From what has been said it is clear that the application of the concept "algorithm" ("algorithmic prescription") to the field of pedagogical phenomena is neither accidental nor arbitrary. The need for its application derives from the necessity for describing a number of pedagogical processes more precisely, with more differentiation, and from the need to control them more fully.

We constantly encounter the situation in which concepts that have been formed in one scientific discipline are dispersed into a much wider circle of phenomena. Here, the process of a more general and a deeper knowledge of reality is expressed. Without this, it is impossible to create general theories. Transferring a concept from one field of science to another is an important consequence and at the same time a condition for uncovering general patterns for these different fields of science.

What has been said refers both to the concept of algorithm and to algorithmic prescription. If these concepts characterize methods of solution not only for mathematics, but also for other problems (including grammatical ones), then this means that mathematical reasoning is not something absolutely specific and sharply separated from the reasoning in other fields of pure and applied science. The models of logical reasoning are general, and it is very important to consider this when examining the pedagogical significance of teaching algorithms. The concept "algorithm" may be applied to various aspects of reasoning for precisely this reason—that it has common features with mathematical reasoning which are discovered and realized at specific stages in scientific cognition.

There is no need to discuss how important it is to reveal such commonalities, not only for creating general psychological and pedagogical theories of reasoning, but for the very practice itself of teaching. For, if mathematical and, for example, grammatical reasoning involve common intellectual processes, it follows that one may develop general methods for teaching different subjects. It follows that methods of reasoning may be transferred from one subject-matter to another, that one may generalize these methods and purposefully shape general mental abilities. In uncovering what is general, for example, in mathematical and grammatical reasoning, one may begin to teach mathematics in such a way as to facilitate and accelerate the mastery of grammar, and conversely, teach grammar in such a way as to facilitate and accelerate the mastery of mathematics.*

It follows from what has been said here that applying the concept "algorithm" ("algorithmic prescription") to different kinds of reasoning is a necessary prerequisite for investigating more general patterns of reasoning and for revealing the deeper connection between their various kinds.

6. On Teaching Algorithms of Identification

It remains to examine in more detail the problem of teaching algorithms of identification, whose significance we discussed in a very general way earlier, in connection with the characteristics of two types of algorithms.

Let us begin by examining some (Russian, *Ed.*) examples of the identification of grammatical phenomena. Let us take the following grammatical rule: "If the given sentence is of complex construction, consisting of two simple sentences (indicative feature *a*) and these simple sentences have no common part to

---

*Clearly, this does not amount to a resurrection of the doctrine of formal discipline, but asserts only that with appropriate instruction the existence of common components in the algorithmic processes will lead to a savings in effort and/or time to achieve criterion levels of mastery. To put it simply:

$$M \cap G \neq \emptyset$$

where M is the set of algorithmic processes converging on a mastery of mathematics, and G is the set of algorithmic processes converging on a mastery of grammar. (*Editor.*)

which both sentences refer (feature b), then these simple sentences are separated by a comma (action A)"[27] In symbolic language, this rule may be written thus:

$$a \ \& \ b \Rightarrow A. \tag{1}$$

Like all other grammatical rules, the given rule may be considered an algorithm of transformation which includes only one operation of transformation. In applying this rule to some sentence, we transform it in a specific way: At first it had no comma; then, as a result of our action, one appeared in it.*

Take another example: "If the given sentence is simple with homogeneous parts (feature c), and if the homogeneous parts are joined by co-ordinating conjunctions (feature d), and if these conjunctions are not repeated (feature g), then the homogeneous parts are not separated by a comma (action $\overline{A}$)." In symbolic language, this rule may be written thus:

$$c \ \& \ d \ \& \ g \Rightarrow \overline{A}. \tag{2}$$

This rule may also be considered as an algorithm of transformation, but transformation of a special kind. It is an identical transformation, for $\overline{A}$ signifies: Action A is not carried out. In identical transformations, the initial object does not change, rather, it is as though it is transformed into itself.[28]

Let us suppose now that in the following sentence, *"I'm not feeling well, and I'm getting ready to go to the doctor,"* it must be specified whether a comma should be placed before the conjunction *and* (i.e., which action, A or $\overline{A}$, must be carried out).

In order to solve this problem, one must first of all determine whether the given sentence is complex (does it have feature a) or simple with homogeneous parts (does it have feature c), for the choice of a rule depends first of all on this.[29] Without identifying the type of sentence in advance and without establish-

---

*It is implied that at first the sentence be given without the punctuation mark about which the question is asked.

ing to what class it refers (to do this, one must know, in turn, the features of objects belonging to those classes which are not indicated in the rules), it is impossible to specify which rule must be applied to it, i.e., how to transform it. The actions indicated in the grammatical rules do not relate to the separate grammatical objects (sentences, *Ed.*) but to whole classes (or types) of objects which are characterized by specific features.

It is evident from these examples that the process of identification is a necessary condition and prerequisite for the process of transformation. When identification is not carried out, the subsequent transformation cannot be carried out. The examples considered also show that the transformation process can be in multiple stages. In order to solve the basic problem of identification—specifying whether the object belongs to a class which has a rule of transformation—it is often necessary to use additional definitions which reveal the aspects of the features interesting us (as indicated in the basic rule). The necessity may arise within the given examples, in particular, of using the definition of a complex sentence and a simple sentence with homogeneous parts (and along with this, the definition of homogeneous parts). These auxiliary definitions can serve as a basis for auxiliary algorithms of identification.

What was said above can be illustrated somewhat differently. In the rules cited, it was asserted that *if* the sentence is complex (i.e., has the feature a) and also has certain other features, *then* the simple sentences entering into its composition are separated by commas. *If* the sentence is simple with homogeneous parts (i.e., has feature c) and also has certain other features, *then* its homogeneous parts are not separated by commas. But, in encountering some sentence x, we do not know what kind it is—complex or simple with homogeneous parts—and it is impossible to detect features a or c directly. In order to specify the type of sentence, i.e., the presence in the sentence of features a or c, one must establish the presence in the sentence of other features: α, β, γ, . . ., which are not indicated in the rule. Precisely which features must be verified in the sentence in order to specify whether it has features a or c (i.e., whether it belongs to the class of complex or simple sentences with homogeneous parts) is given by special theoretical statements (definitions) on the basis of which the identification is made. These theoretical statements have the form:

$$\alpha\,(x)\,\&\,\beta\,(x)\,\&\,\gamma\,(x)\,\&\,\ldots \Leftrightarrow a\,(x),\!^{30}$$

where the double-headed arrow signifies the bi-conditional "if and only if . . . then," i.e., that the sum total of the features α, β, γ, . . . in the object x is necessary and sufficient for a conclusion as to the presence in that object of feature a. The statement that object x has feature a in the part to the right $a\,(x)$ is equivalent to the statement that this sentence belongs to the class of objects having the given feature (for example, if the sentence has the feature "complexity," then it belongs to the class of complex sentences). This means that instead of $x \in A$ (x belongs to class A) we can always write $a\,(x)$, (x has the feature a).*

An important conclusion follows from what has been said: Features indicated in the formulation of definitions and rules can be of two types. Features of one type are such that their presence or absence in an object can be determined directly (e.g., the feature "a conjunction uniting homogeneous parts is repeated"). Other features often cannot be directly detected. In order to determine their presence or absence, it is necessary to carry out a special process of identification involving the aspects of those features which are not indicated in the rule (e.g., features of types of sentences).

It was noticed long ago that many students, though they know rules, laws, and theorems well, are often unable to solve problems. This phenomenon is usually explained in psychology and pedagogy by the fact that, although students have the knowledge, they are not able to apply it. Much research has been devoted to the problem of applying knowledge, and it has produced significant contributions to understanding its nature (see, e.g., [543], [547], [586], [587], [624], [627], [642],

---

*Henceforth, we will be using sometimes one, sometimes the other, form.

[647], [651], [652], [676-678], [680], [717-722], [737], [764], [777], [778], [782], [783], [807], [832], [859]).

Viewed in terms of the interaction of the processes of transformation and the processes of identification, apparently one of the reasons for the students' inability to apply knowledge is that, in mastering algorithms of transformation, they do not master algorithms of identification. They do not know what must be done with the situation in order to specify which rule must be applied to it. This inability may be due to two reasons:

(1) Students know rules of type a & b $\Rightarrow$ A; c & d & g $\Rightarrow$ $\overline{A}$, but when encountering some situation to which one of the rules must be applied, they do not know that in order to choose a rule it is necessary to establish whether this situation possesses this or that aggregate of features (from those pointed out in the rules). For example, in encountering some sentence which contains the conjunction *and* they do not know that in order to specify the punctuation before this conjunction they must verify which from among the above specified conditions: a & b or c & d & g must be fulfilled. Without this determination a deliberate choice between A or $\overline{A}$ is, of course, impossible.

(2) The students know the appropriate rules and they also know that to choose a rule one must establish which of the aggregates of features indicated in the rules are found in the given situation, but they do not know how to determine (discriminate) the presence of these same features. For example, when encountering some sentence which contains the single conjunction *and*, they know that in order to specify the punctuation before *and*, they have to verify which features it has: a & b or c & d & g, but they do not know how to identify the presence of these features or of one of their parts.[31]

The difficulty of finding solutions to problems is aggravated by the fact that often, many different rules can be applied to one and the same situation. In order to define which of them must be applied in one or another instance, a complex process of verification must be carried out for the purpose of determining whether the situation belongs to one or another class.[32]

The expression "the ability to apply knowledge" (specifically the ability to apply the knowledge of rules or simply rules) is equivocal.[33] One may speak of the ability to apply rules in at least two different meanings. One may mean the ability to progress within the given rule from conditions (antecedents) to actions (consequents) and to carry out the transformation indicated in the rule.[34] One may also mean the ability to choose the necessary rule from among a number of rules in order later to carry out that activity (transformation) indicated in the rule. These two meanings of the expression "the ability to apply rules" must be strictly separated. They are two different abilities requiring entirely different operations. The inability of students to apply rules is most often expressed in their inability to choose the necessary rule from among a number of possible ones. In turn, the ability to choose a rule assumes the ability to carry out a process of identification.

There are two basic types of processes leading to a choice of rules. In order to present them more clearly, let us examine a situation where different rules may refer to one and the same situation or object x (it might be a word, a sentence or even a physical body, a chemical substance, and so on), where the choice of a rule is specified by the presence in the object of definite features. Let the connection between features and actions be given by the following three rules:

$$a\ (x)\ \&\ b\ (x)\ \&\ c\ (x) \Rightarrow A,$$
$$d\ (x)\ \&\ e\ (x)\ \&\ f\ (x) \Rightarrow B,$$
$$g\ (x)\ \&\ p\ (x)\ \&\ q\ (x) \Rightarrow C.$$

The person solving the problem must determine which action of the three possible ones must be carried out with a given object. In order to do so, it is necessary to establish which, from among the aggregate of features $a, b, c; d, e, f; g, p, q$, the object possesses. The presence of one of the combinations specifies the choice of a rule. If we suppose that each of the features can be directly detected (i.e., no other special identification procedure is necessary for identifying the object[35]), the following two types of procedures may be followed:

Type 1

(1) Various features are distinguished in the object in the process of analysis.

(2) The features are fixed.

(3) These features are compared with features indicated on the left side (antecedents) of the rules.

(4) The choice is made of that rule whose features indicated on the left side correspond with the features established in the situation.

Type 2

(1) Direct perception of the situation leads to the assumption of the applicability of a specific rule to it.

(2) This rule is fixed.

(3) The features indicated on the left side of the rule are separated from the rule.

(4) The presence of these features is verified in the object. If the object has the given features, then this rule is chosen for subsequent application. If they are not present, then another assumption is formulated, and the applicability of another rule is tested.

In the process of the first type, a person proceeds from a detailed, though non-directed analysis of the object. He distinguishes its various features without a preliminary set (*Einstellung*) to apply a contemplated rule and only later sees which of the rules could be applied. When setting out to analyze the object, he does not know *a priori* which features to look for. In processes of the second type, one carries out an analysis of the rule chosen for examination (fixed) on the basis of a previously global (undetailed) perception of a situation and on the ensuing hypothesis about the applicability of a specific rule to it. Then one carries out a detailed systematic analysis of the situation starting from the features specified within the rule. In so doing, one knows in advance which features to look for in the situation, and to what to pay attention. With the second type of process, the person, therefore, does not wait for what the object will yield him. He carries out a purposeful search for the features which interest him.[36]

Usually, a person with well developed reasoning combines both types of processes. In analyzing objects, he proceeds as though from two approaches: From what the object "gives" and from what must be found in it. It is not unreasonable to think, however, that the first type of process is developed on the basis of the second. The greater the experience one has in a purposeful, systematic identification, the easier it is to carry out a search which is not externally determined.

If one observes the practice in schools, one notes that fundamental attention in instruction is given to having students thoroughly master theoretical propositions (definitions, laws, and theorems) and the rules (algorithms of transformation) which correspond to them. It is different with algorithms of identification. The essential deficiency in the organization of instruction, as noted above, is the fact that students are often not taught algorithms of identification at all, or they are not taught them explicitly, or they are taught insufficiently, unsystematically, or incorrectly.[37] These algorithms for many classes of identification problems have simply not been discovered or formulated. As a result, students who find themselves confronted with the necessity for identifying one or another phenomenon often do not know what to do, and how, and in what order to carry out which operations.[38] It is possible to lessen substantially the difficulties of study, to lower the number of errors in solving problems, and generally to raise the effectiveness of instruction. It requires that students be taught not only algorithms of transformation, but also algorithms of identification. Thus, forming the processes of identification in the students must be the special object of the teacher's job.

The great importance of teaching algorithms of identification to students demands research on this form of algorithm and into the methods for formulating and teaching them. Section Three of this book will be devoted to these questions. Its structure is defined by the following.

Since the process of identification is linked with that of determining and analyzing the indicative features of phenomena, it is necessary to examine the logical and psycho-pedagogical aspects of the problems connected with the concept of an indicative feature. In particular, the problem of the logical structure of indicative features must be examined, since the structure of the identifying process is specified to a considerable degree precisely by the logical structure of the indicative features. Further, since the verification of indicative features during identification may be carried out in different sequences (i.e., different algorithms of identification may be used), the question arises of how to specify the most efficient sequence for verifying indicative features and the most efficient algorithm for their identification. The latter problem cannot be solved on the basis of logical and psychological considerations alone, and by applying logical and psychological methods of analysis alone. Here, the use of quantitative methods is also required. Section Four of this book is devoted to examining the application of quantitative methods to the estimation of optimal algorithms of identification.

Let us note that the sequence of questions mentioned here reflects the sequence in formulating algorithms of identification. It proceeds along the following stages:

(1) Starting from specific logical and psychological consider-ations, indicative features are determined, on the basis of which the identification of various phenomena can be made;

(2) the logical structure of indicative features is specified and corresponding theoretical statements and rules are constructed;

(3) with the help of specific quantitative methods, the most efficient sequence for verification of indicative features in a phenomenon is defined for the purpose of identifying the phenomenon. The most efficient algorithm of identification is being sought.

In the first two stages problems of a quantitative character are solved. At the third stage, in solving the problem of designing an algorithm, specific quantitative estimates and evaluations are also applied.

Notes

1. One type of automat requires pushing the button after the money has been inserted.

2. These problems are a particular case of problems of establishing a technical diagnosis.
3. We are talking about forming algorithmic processes, since a person must also be taught the ability to apply algorithmic prescriptions and carry out algorithmic processes. Algorithmic operations which constitute the execution of the algorithmic process must, in the precise sense of the word, be *formed or developed*, since these operations are not innate and are also not installed in him in prepared form as opposed to machines which have the ability to carry out specific operations installed in them when they are built.
4. The question of the pedagogical value of each of the methods for teaching algorithms was experimentally explored in the dissertation of our associate Granik [241].
5. This is why this aspect of instruction is described in the detailed outlines of lessons which will be cited later. This, however, does not mean that teaching of algorithms to students should substitute for a profound acquaintance by them with the substantive aspects of the phenomena being studied, with their peculiarities, their cognitive and practical significance, and so on. Such a substitution is utterly impermissible. Of course, the acquisition of knowledge about the substantive aspects of the phenomena studied and the teaching operations for applying this knowledge do not necessarily have to be separated in time, i.e., appear as stages differentiated according to time for mastering the corresponding phenomena. The teaching can be so structured as to have the algorithms progressively being formed in keeping with the assimilation of the indicative features of the phenomenon, and no special stage of forming algorithms is necessary. It is precisely this route which was the predominant one in the experimental instruction, the course of which will be described later. However, in whatever way the process of teaching is designed, it is important that, when expedient, the teaching of efficient systems of intellectual operations and the teaching of algorithms be carried out.
6. Part of the research concerning the mechanics of the

intellectual activity of students when analyzing a foreign text and translating from a foreign language to their native one were carried out jointly with Belopolskaya [216], [688].

7. See, e.g., [520], [542-544], [547], [561], [582], [584-587], [591], [601-603], [605], [606], [615], [617- 619], [630], [636], [657], [667-669], [679], [705], [706], [727], [733], [756], [780], [785-787], [816], [826], [827], [839], [857], [858].

8. See, e.g., [529], [555], [589], [607], [620], [655], [683], [704], [714], [726], [751], [755], [806], [838], [852].

9. See, e.g., Granik [241], and Vlasenkov [223].

10. In doing this, one may design the instruction in such a way that, in the course of solving problems, not only are separate operations formed within the students, but also the full systems. If no special effort is made to form algorithmic processes, the operations sometimes are formed in the students who are trying to solve problems by the trial-and-error procedure in the process of "adjusting" to the other object or situation (Leontyev pointed out, in his time, certain peculiarities of forming operations in the process of "adjusting)."

11. In the latter instance, the algorithmic prescription assumes a special function—as means of realizing an already formed algorithmic process. The significance of such a realization will be shown later.

12. Belopolskaya [214], [215] is conducting some interesting work in that direction.

13. In the circumstances described, that route, though forced, is efficient, since it leads to rapid and successful attainment of the indicated goal.

14. *Knowing* an algorithm and *being aware of it* must be differentiated. Knowing an algorithm means knowing the prescription that can be acquired in finished form from others, or through an awareness of one's own algorithmic operations. Being aware of an algorithmic process is merely the awareness of one's own algorithmic operations. One of the factors that may foster awareness of these operations is a

familiarization with the algorithmic prescription.

15. For example, in order to write grammatically, it is not sufficient to know grammatical rules. One must also know which rules to choose and apply in one or another situation, i.e., know the rules for choosing rules and the rules for applying them. Many algorithms are such rules.

16. We consider that identifying a complex sentence does not have to depend necessarily on specifying the kinds of simple sentences entering into the composition of the complex sentence. But that will be another algorithm, and it will be discussed later.

17. A critical analysis of similar viewpoints is given in our article [276].

18. We have noted already that the necessity for a strict sequence of operations is linked with a specific "stage-by-stage quality" of those processes whose control is carried out by means of these operations. Besides that, the overwhelming majority of algorithms presuppose different sequences of operations depending on different conditions in which the problem solution takes place, i.e., they permit a flexible adaptation to these conditions.

19. In all there are 12 different sequences of operations possible for solving this problem.

20. Formulae of this type differ from those with brackets which specify an exact sequence of operations and correspond to algorithms.

21. The inexpediency of prescribing a strict sequence for operations in this kind of rule is linked with the commutativity and associativity of the operations of adding and multiplying.

22. In discussing different solution algorithms for one and the same problem, two possibilities must be distinguished. Algorithms may differ only in terms of the order (sequence) of operations (as when calculating the area of a triangle). They may differ also in terms of the character of the operations themselves. In the latter instance, students have to be taught (at least sometimes) not just one single algorithm

for solving a problem, but many.

23. When taking into consideration such natural criteria as speed and reliability of getting a result (reflective of the degree of difficulty), it is not difficult to set up an algorithm specifying which from among the number of possible algorithms for adding numbers is the more efficient in different instances. Many people who are good at "mental arithmetic" use this algorithm consciously or unconsciously. Some teachers also use it.

24. There can also be different methods of identifying one and the same indicative feature. Thus, dullness (i.e., of the emulsion side) can be recognized by the sense of touch as well as by sight. Though the discrimination ability of sense organs can be improved and must be trained, for some people one method is easier and more convenient and for others the other one. The two methods may be of equal worth only for some people.

25. This corresponds entirely to what was said above about the relativity of the notion of an elementary operation. The same operations may or may not be elementary for different people.

26. This circumstance is completely missing when installing algorithms into machines (under the condition, of course, that these are not self-instructing machines).

27. Sometimes they are also separated by other punctuation marks, but in order not to complicate the analysis, we are not going to examine the additional conditions calling for still other punctuation marks.

28. Algorithms which transform an object into itself (i.e., leave it unchanged) are called identical algorithms (by analogy with identical functions in mathematics). The absence of transformation is considered here as a special case of transformation.

29. It is evident from the examples cited that among the indicative features indicated "on the left" there are those whose verification is necessary, first of all, for solving the question of the applicability of one or another rule. Verifying the features in the object for choosing appropriate rules may

be called the process of identifying the applicability of rules. The class in which the object is included in the given instance is the class of objects to which one or another rule is applicable.

30. The conjunction of the indicative features in the left side of this expression represents the complex feature of the situation (sentence). In formulating rules and definitions, complex features of other kinds also are possible (α, β, γ, ... may be complex features in the expression cited). Specific theoretical statements containing complex features of other kinds will be examined below.

31. Students know, for example, that it is necessary to verify whether the sentence is complex or simple (i.e., whether or not it has indicative features a or c), but they do not know how to detect this feature itself.

32. Many of the students' mistakes are evoked by the fact that in knowing the rules but not knowing the algorithms of identification, they incorrectly relate situations to specific classes and therefore do not apply to them those rules which are appropriate. In psychological and methodological litera- ture, facts bearing witness to students' inability to correctly relate objects to one another class have been shown many times.

33. The ability to apply knowledge of theoretical statements also includes progressing from the theoretical statement to the rule, i.e., to the prescription about corresponding practical actions.

34. These transformations are carried out easily in grammar, e.g., placing punctuation marks. In other cases, carrying out the actions shown in the rule are not at all easy, so that they must be studied sometimes at length.

35. Analogous to the procedure required to identify types of sentences (complex or simple with homogeneous parts) on the basis of indicative features not shown in the rule.

36. The difference between the two types of processes appears clearly when solving geometrical problems of proof. With the so-called "synthetic" (or progressive) method of solution, the

problem solver starts out from what is given. He analyzes the concepts indicated in what is given and the corresponding geometrical figures, trying to uncover their properties and deduce the different consequences from what is given. He tries to "squeeze" everything possible out of the data. With the so-called "analytical" (or regressive) method, the person starts out with what has to be proved and verifies (looks for) only those features which will help to carry out the proof. With the first method, one starts out, to a considerable extent, from what the geometrical objects "give" one. With the second, one starts from what has to be obtained for the proof.

37. Teachers often do not even have a clear idea of the existence and significance of algorithms of identification. For example, when students make grammatical mistakes, they are often referred to the textbook and are required to learn and repeat the corresponding rules. Meanwhile, students make the large majority of their mistakes not because they do not know the rules, but because they do not know how to identify phenomena and because they do not know the algorithms of identification (and choice) for the rules. It is not surprising that frequent repetition and study of rules usually yield no result.

38. Not one of the students to whom the question was posed of which action must be carried out in order to distinguish a complex sentence from a simple one with homogeneous members could cite the complete set of operations. Also, a number of teachers and methodologists who were presented with this question were unable to do so. It is not surprising, therefore, that instruction in syntax is often so imperfect, and the mastery of punctuation in grammar entails many difficulties.

Part One: Section Three

Logical and Psychological Problems in Constructing Algorithms of Identification

Chapter Eight

Indicative Features of Phenomena and Their Logical Structures

1. The Concept of Indicative Features

As we have seen, the process of identification consists of detecting specific indicative features in an object, comparing them with the indicative features given in definitions and other theoretical statements, and attributing the object to a specific class on this basis. However, this means of identification is not unique. A person identifies many things without using definitions and other theoretical statements and without knowing them (for example, he distinguishes a chair from a table without knowing the definitions of these objects). However, examining such processes of identification does not fall within the scope of this book.[1] Here, and later, we will deal only with the processes of identification via the matching of indicative features given in specific theoretical statements against observed features. Teaching these theoretical statements and developing the ability to recognize specific objects, phenomena, and processes in this way occupies a major niche in practical school activity as well as generally in life, and it is a very important task of teaching.

We spoke of the fact that it is possible and in many cases expedient to algorithmize the process of identification.[2] But, in order to construct good, efficient algorithms for identifying specific objects, one must first construct a good concept-definition for these objects, i.e., find those indicative features that will permit quick and easy identification.

In philosophy, logic, and in ordinary speech, the word "indicative feature" is used in different senses. Indicative feature often means some properties, qualities, and states of an object, as well as relations among objects such that from their presence or absence one may draw a conclusion about the presence or absence of other properties, qualities, and relations, and about the existence of some other objects, and so on. For example, it is said that the presence of smoke coming out of a chimney indicates (is an indicative feature) that there is a fire on the hearth in that house; that tracks in the snow are an indicative feature that a certain species of animal has passed by; that a lowering of the atmospheric pressure is an indicative feature of an approaching storm; that the equality of angles at the base of a triangle is an indicative feature of an isosceles triangle; that the reddening of metal is an indicative feature of its being hot; and so on.

In another, broader sense, any properties, qualities, and relations of an object to other objects, i.e., everything in which some objects resemble others or are distinct from them, are called indicative features. This includes everything which could be characteristic of an object and by which, consequently, we can distinguish objects from one another.

In the first sense, "indicative features" refers only to those features which serve as a basis for corresponding conclusions. In order to distinguish these features—and examining such features has a major significance for our purposes—we are going to use the concepts of necessary and sufficient features used in the study of formal logic, as explained below.

In ordinary speech, the relations of indicative features among themselves and also the phenomena which are characterized by different indicative features are often expressed with the help of the connectives *if. . . .*, *then* (let us recall our examples: *if* smoke is coming out of the chimney, *then* there is a fire lit in the fireplace of the house; *if* there are tracks of a specific form in the snow, *then* an animal of a certain species passed by here; and so on). The general form for expressing the implication is: if a, then b. In the symbolic form we write $a \Rightarrow b$, where a is an expression of the presence (or absence) of a certain property, quality, or state of an object, and b is an expression of the presence (or absence) in the same or another object of some other property, quality, or state.

In the presence of such an implicational connective, i.e., in the case of the truthfulness of the complex expression $a \Rightarrow b$, one says that what is stated in a (i.e., the presence or absence of specific properties or relations in a certain object or objects) is a *sufficient* indicative feature of what is stated in b, or a sufficient indicative feature for reaching a conclusion about the presence of what is stated in b. The presence of smoke coming out of a chimney is a sufficient indicative feature of the fact (or for the conclusion) that a fire is lit in the fireplace of the house. The presence of tracks of a specific form in the snow is a sufficient indicative feature that an animal of a certain species passed the given place, and so on. In the absence of what is stated in a, there is necessarily an absence of what is stated in b, i.e., if $\bar{a} \Rightarrow \bar{b}*$ is true, then what is stated in a is a *necessary* indicative feature of what is stated in b. For example, the divisibility of the sum of digits in some number by three is a necessary indicative feature of the divisibility of that number by three. For it is true that if the sum of digits in some number is not divisible by three, then the number itself is not divisible by three. It is known that the divisibility of the sum of digits in a number by three is also a sufficient indicative feature of the divisibility of the number itself by three.

In order to convey the fact that a is necessary and sufficient for b, the expression "if and only if a, then b" or in symbolic form: $a \Leftrightarrow b$ is used. This expression is equivalent to a conjunction of the two statements:

"If a, then b," and "if b, then a" $((a \Rightarrow b) \, \& \, (b \Rightarrow a))**$

$*\bar{a}$ signifies a negation (i.e., the statement: it is untrue that it is a); it is analogous for $\bar{b}$.

$**a$ and b in the structure of the complex expression $a \Leftrightarrow b$ are also expressions. This complex expression conveys the fact that what is stated in a is a necessary and sufficient indicative feature of that which is stated in b (and vice versa). To simplify the matter, the phenomena we refer to in propositions a and b we also identify with the symbols a and b, and that is why we say that the proposition conveys the circumstance that a is a necessary and sufficient indicative feature of b (and vice versa).

In logic, if the statement $a \Rightarrow b$ is true, then the statement $\overline{b}$ $\Rightarrow \overline{a}$ is also true. Conversely, if the second is true, then so is the first. This signifies that if a is a sufficient indicative feature of b, then $\overline{b}$ is a sufficient indicative feature of $\overline{a}$. And this, in turn, signifies that b is a necessary indicative feature of a. Generally, if b is a necessary indicative feature of a, then a is a sufficient indicative feature of b.

One phenomenon may be an indicative feature of another, because of the presence of some relationship between them in objective reality. These relationships are diverse. Thus, it could be a causal relationship, a functional relationship, simply a temporal connection, and so on. The logical relationship of the type *if . . ., then* is a reflection of what is common in these different concrete relationships. For example, in the implicational propositions "if a body is heated, it expands," "if smoke is coming out of the chimney, then a fire is lit," the basis of the logical relationship is the causal relationship in the corresponding phenomena. In the implicational proposition "if the number two is squared, then one obtains four," such a basis is a functional relationship, whereas in the judgment "if today is Monday, then tomorrow is Tuesday," it is a temporal relationship (order) of phenomena.

From all this, it is evident that an indicative feature of an object can express either its reason (heating is an indicative feature that the body is expanding, and at the same time a reason for that phenomenon), or its consequence (the presence of smoke coming out of the chimney is a sign that a fire is lit and a result of it). Finally, an indicative feature need not express a causal relationship with a phenomenon at all (the divisibility of the sum of digits of some number by three is a sufficient indicative feature of the divisibility of the same number by three, but it is not its reason).

The indicative features of objects and phenomena can be *simple* or *complex* (compound). Complex indicative features may be considered as logical constructions (combinations) built from simple indicative features with the help of logical operations. Thus, mercury has the complex indicative feature of "being liquid and metal." The sufficient indicative feature of some number's belonging to a class of numbers divisible by five is a complex indicative feature of "ending in five or zero," and so on. Later we will examine in more detail the possible structures of complex indicative features.

In speaking of necessary and sufficient indicative features of different phenomena, we must keep in mind not only the simple, but also the complex indicative features. Thus, in definitions of specific objects of a certain class, a specific aggregate of indicative features is usually pointed out. This aggregate of features can be considered as a complex indicative feature representing a conjunction of simple indicative features or their disjunction, or a construction formed by way of using both conjunctions and disjunctions, or perhaps even other logical operations. If the definition is correctly formed, and the simple indicative features are conjunctively joined, then such a complex indicative feature is sufficient for attributing the object to a specific class, and each of the indicative features of this aggregate is necessary for this attribution. In general, if, for example, the expression $(a \ \& \ b) \leftrightarrow c$ is true, then the complex indicative feature $a \ \& \ b$ (or, in other words, the totality of the indicative features a, b) is the necessary and sufficient (since we are dealing with the definition) indicative feature of c, and each of the simple indicative features a and b is the necessary (but insufficient) indicative feature of c. Conversely, if it is true that $a \lor b \leftrightarrow c$ (where $\lor$ signifies *or*), then each of the indicative features a, b is a sufficient indicative feature of c, but neither a nor b are separately the necessary indicative features of c. The necessary indicative feature is only their disjunction.

By knowing necessary and sufficient indicative features, we can identify objects and draw conclusions about phenomena. In doing this, we can, with the help of indicative features, identify objects (phenomena) which we do not perceive (and sometimes cannot perceive at all) by means of the sense organs. Indicative features, therefore, create the possibility of indirect cognition.

2. Selection of Indicative Features

Each object has various indicative features. Owing to this, one can usually define it not only by some one means, but by several. Moreover, indicative features can occur not only in definitions, but in other theoretical statements—in theorems, for example. The presence of various indicative features in an object, in which many of them may also be sufficient, permits identifying it by various methods. Let us call to mind, for example, the three indicative features of the equality of a triangle and, hence, the three methods for establishing this equality; the five indicative features of parallelograms and the corresponding number of methods for identifying them; the various indicative features of the north, south, and other geographical reference frames which provide the possibility for geographical orientation and identification by different methods, and so on. Various indicative features may also be taken as a basis for identification in grammar.

When formulating definitions and other theoretical statements on which algorithms of identification are based, the following questions arise:

(1) Which indicative features out of all the possible ones is it expedient to use as the basis of definitions and other theoretical statements describing objects and phenomena with the purpose of distinguishing them from other objects and phenomena?[3]

(2) What are the criteria for selecting such indicative features and their requirements, or, in other words, what are the characteristics of "good" and "bad" indicative features?

Among the many indicative features characterizing one or another object, some are always present while others are sometimes there in the object and sometimes not. In definitions, one tries to select the necessary indicative features, i.e., those which are most essential. However, it is most important for devising algorithms of identification that the indicative features be sufficient for discriminating the objects interesting us where, in a number of cases, not only essential indicative features could be sufficient, but also inessential ones.[4]

The essentiality of indicative features is a property, one might say, which is ontological, related to nature, and to the existence of the object itself. But indicative features may be considered not only from the standpoint of their essentiality for the existence of the object, but also from the standpoint of their significance for the process of cognition (and identification) of the object. These are the gnoseological and psychological properties of indicative features.

Since it is important, in the process of learning-teaching, to take into account not only the gnoseological and psychological properties of indicative features, but also the properties relevant to the tasks of instruction, it is proper to speak as well of the psycho-didactic properties of indicative features.

Psycho-didactic properties of indicative features do not have any significance for the design of algorithms by which computers work (these properties, therefore, are not examined by mathematics, logic, or cybernetics), but they do have critical significance for designing algorithms for the purpose of instruction. The indicative features on the basis of which algorithms for teaching purposes are designed must satisfy specific psychological and didactic requirements, and it is important to formulate these requirements as precisely as possible.

One of the basic requirements which must be satisfied by the indicative features selected for formulating definitions and other theoretical statements for the purpose of subsequently devising corresponding algorithms of identification is the *possibility of their operational establishment*, i.e., establishment by means of some sufficiently elementary operations. In other words, it is required of indicative features that operations exist for them through which these indicative features may be established. In so doing, the operations must be such that they may be precisely pointed out and taught.

There are indicative features which are identified by means of elementary receptive acts. Thus, in order to identify the color of an object, for a person who knows the color and possesses normal vision, it suffices merely to look at the object. The sensory contact with the object is that action by means of which the color of the object is identified. It is, therefore, natural that the color of the object should be considered as an operationally determinable indicative feature. To establish this feature, one could indicate the operation "look at the object" and the indicative feature will be identified. The possibility of a univocal identification is inherent in the given instance in the very anatomical-physiological structure of the visual analyzer and in the presence of an elementary association between the color and its name. Elementary receptive acts are insufficient for identifying other indicative features, and it is necessary to carry out ideational acts (operations).

The next requirement which indicative features used for establishing definitions and other theoretical statements (and also algorithms) must satisfy is their *being known* by the learners. One must see to it that the indicative features which are made the basis of theoretical statements and algorithms do not need some other indicative features in order to be understood and evaluated. If these indicative features are not known, they cannot serve as a means of identification. If, for example, some object x is described by indicative features a and b, but the person does not know what a and b are (or these indicative features are vague), he will, of course, not be able to use these indicative features. The knowledge of those features will be purely formal.

In the practice of teaching, we see in many instances that students are referred to indicative features which themselves require clarification. Sometimes they are not at all sufficiently specific.

The knowability of indicative features is a characteristic which depends on the experience of the students, the level of their educational development, and their prior preparation. Therefore, when formulating definitions and explaining the indicative features to students, one must not proceed only from the way one or another concept is disclosed in science. One must be sure to take into account to what degree these indicative features by which the content of new concepts is presented are clear and well known to students of a certain age level.

There exists a stage-by-stage process in the assimilation of concepts and in their presentation to students as a specific sequence of levels. At the very first stage indicative feature a (also $b, c,$ etc.) is unknown and becomes known through sensory evidence or is presented by the teacher as indicative features $\alpha, \beta, \gamma,$ with which the student has had experience: a is $\alpha, \beta, \gamma.$ What was previously unknown and has been ascertained through sensory evidence[5] or items of prior experience now becomes known, and the new unknown A (also B, C, etc.) may be elucidated now through the earlier indicative features a, b, c. This is the second stage. The new unknown $\mathfrak{A}$ may now be elucidated via the already known indicative features A, B, C. . . . This is the third stage, and so on.

It is important to take into consideration the fact that the contents of the new concepts may be elucidated through indicative features at various levels and of different degrees of substantiality. One must try to reduce any new concepts to indicative features which are already known to the student at a given level of his development and which are, as far as possible, evident. This is realized by means of chains of substantiations (i.e., rendering more and more concrete, *Ed.*). This process could be represented thus: $\mathfrak{A}$ is A, B, C,. . .; A is a, b, c,. . .; B is a_1, b_1, c_1,. . .; . . .; a is α, β, γ, . . .; b is α_1, β_1, γ_1, . . .; . . .; α, β, γ, . . . α_1, β_1, γ_1, . . . are already known or evident, i.e., they can be established by certain elementary operations and univocally evaluated.

From a didactic viewpoint, providing a correct definition presupposes obligatorily taking into account "the stages" at which the students are, and bringing them to discover the contents of the new concept with the help of indicative features belonging to the corresponding level, i.e., of those which are clear and known to the students of a given age level. If this is impossible, and the contents of the new concepts must be discovered through indicative features unknown to the students, which do not correspond to the "stage" at which they are, it will be necessary to bring them to discover these unknown indicative features by reducing them to known ones.

The indicative features of a specific object can be fixed sometimes by specifying the applicability of specific actions to a given object. The indicative features are specified in the following way: If one may apply some given action to some given object x, then the given object x possesses a specific indicative feature (say a), and it belongs to a particular class. If one cannot apply this action to the given object, then it does not possess the indicative feature a, and it does not belong to the given class. The possibility of applying certain questions to some words is an indicative feature of this type. Posing questions is an action, and when we, for example, say that "words which are called adjectives are those which designate the properties of objects and answer the questions 'what, whose?,' " then one of the indicative features of the fact that the given word (*pink*, for example) is an adjective is the applicability of the question *what* to it. In so doing, it is supposed that a person is able to evaluate the applicability of the question itself, and that the knowledge and experience which permit him to qualify univocally the applicability of the action to the object have been formed in him.

Let us conditionally name those indicative features which consist of the applicability of specific actions to an object, indicative features given in the form of procedures. Indicative features stated in the form of procedures may be considered as a particular instance of features given operationally. Indicative features (and objects in general) may be defined not only by genus and species distinctions (and also by some other means known from traditional logic), but also by means of pointing out certain operations. For example, we may say that *a* is what is obtained when certain actions are taken relative to object *x*. This is an operational definition. At the present time, operational definitions are beginning to play a growing role in science. They must obviously occupy a corresponding place in instruction.

The next requirement for indicative features is their *univocality*.

One of the tasks in formulating definitions and other theoretical statements (as well as, we shall see later, algorithms) consists in choosing, from all the possible indicative features of the phenomenon studied, those which provide the highest probability of identifying the object most reliably.

Indicative features, in that regard, are usually unequal. Thus, for example, if one must recognize an unfamiliar person from a description, this can be done reliably by pointing out such indicative features as height, clothing, and hair color. If only such indicative features as the expression of his face and his way of behaving are given, it is far more difficult to identify the person with certainty.

The indicative features of the first kind are easily measurable (for example, height) and precisely distinguishable (for example, wearing an overcoat or without a coat; blond or brunette). The psychological particularities of that person whom the above indicative features reveal and specify do not influence, or influence only slightly, his recognizability; nor do the character of his views or the level of his intellectual development. With people who can see well, there can be no difference of opinion as to whether the person standing before them is wearing a coat or not, or whether he is blond or brunette. Yet they might have different opinions as to his style of behavior. The recognition (evaluation) of this and similar indicative features depends to a great extent on the subjective particularities of the perceiver: his comprehension, ideas, experience, and so on.

Let us call indicative features which are easily distinguished, precisely differentiated, and basically evaluated in the same way by everyone, *univocal* indicative features. They will signify the very same thing for everyone (or nearly everyone). Their recognition is usually the same in all perceivers, and does not depend, or depends only to an insignificant degree, on psychological particularities of the perceiver. We will call *equivocal* those indicative features whose recognition depends on the subjective particularities of the perceivers.

In so doing, it must be said that the concept of univocality is not absolute. On the one hand, there are no sharp boundaries between many phenomena, and there are often transitional stages and states which could be evaluated differently. On the other hand, equivocal indicative features, owing to instruction and training, can acquire univocality for different people. Incidentally, one of the tasks of instruction consists of evolving ideas and comprehensions that will permit the univocal evaluation of these phenomena. This, of course, does not at all signify that in the process of instruction and training, the individual particularities of a person and his individual uniqueness must be levelled—that similar evaluations must be developed in students with respect to *all* objects and phenomena. We refer here only to those cases for which univocal evaluations are necessary and constitute a necessary condition for the acquisition of certain concepts, habits, skills, and so on.

It is absolutely clear that instruction can be more successful only in the case where the students, whenever possible, are told the available univocal indicative features of objects and phenomena. If this does not happen, indicative features do not fulfill their orienting function. A lack of uniformity appears in the students' actions: some evaluate one or another indicative feature one way, and others another way. From this point of view, if one were to analyze the indicative features which are sometimes specified to students, we would see that some of them do not satisfy the indicated requirement and are therefore imperfect.

Consequently, such indicative features not only do not guarantee the absence of mistakes, but often lead to them.

Finally, still another requirement of indicative features presented to students is that of maximal *ease* of their detection in objects and convenience of operating on them. Thus, for example, the indicative feature whose determination in an object requires a rougher differentiation is easier than an indicative feature which requires a finer differentiation in order to become apparent. Obviously, other things being equal, the kind of indicative features which are psychologically more easily detectable are preferable as the basis of definitions and other theoretical statements. It would seem that this property of indicative features is very vague. However, as we will show later, it is possible to point out a criterion which will permit comparisons of indicative features according to this property. It is the mean time for detecting (or verifying) the indicative feature.

We examined several psycho-didactic requirements for indicative features chosen for formulating definitions and other theoretical statements on the basis of which algorithms of identification must be designed. However, identification rarely takes place through verification of some *one* indicative feature in the object. As a rule, in order to identify an object, it is necessary to verify the presence or absence of *several* indicative features in it—some *complex* of them, a *system*. In other words, one has to deal with complex indicative features. In this connection, the question arises of the possible structures of complex indicative features and the principles of operating with them.

3. Logical Structures of Indicative Features: Concept and Classification

Let us look at two substantially different examples—one from biology and one from arithmetic.* In the classification of animals all members of the class *insect* are characterized by six legs and three clearly defined body regions (i.e., head, thorax, abdomen). In arithmetic all those numbers are divisible by 5 whose last digit is either 0 or 5. Both definitions—of insects and of the set of numbers divisible by 5—involve two indicative features, but the indicative features are not related to each other in the same way. How is this shown?

Let us separate the indicative features from the definitions, number them, and write them one below the other. Then let us examine the logical connectives that join them.

Those animals are called insects which

    (1)   have six legs

             and

    (2)   possess three clearly defined body regions.

The set of numbers divisible by 5 includes all those numbers

    (1)   whose last digit is 0

             or

    (2)   whose last digit is 5.

Obviously, in the first case the indicative features (six legs, three body regions) are joined by the logical connective & (and), while in the second case the indicative features (0, 5 as last digits) are joined by the logical connective V (or).

Each of the indicative features in the first definition is necessary, but not sufficient. Only the combination of necessary indicative features—their aggregate—is sufficient to identify an insect. The indicative features given in the second definition, by

---

*These examples are substituted for examples drawn from Russian grammar for which there is no parallel in English. (*Editor.*)

contrast, are not necessary. However, each one of them is sufficient. In the first case, the presence of six legs alone does not justify the classification of insect (other creatures have six legs); in the second case, if the number ends in a zero, we know it is divisible by 5 and we do not, and indeed cannot, have the 5 present as well.

The analysis of the indicative features of the most diverse phenomena shows that those occurring within complexes are always logically linked among themselves in a special way. These links are realized through various logical *copulae* (operators) including the connectives *and* and *or* which were just encountered in the examples discussed. If the indicative features are joined, for example, by the logical connective *and*, then in order to draw a positive conclusion about the object's belonging to a certain class, based on these indicative features, it is necessary that it have both of the indicative features. It is necessary to verify the presence of each of them in the process of identification. If, however, the indicative features are joined by the logical connective *or*, then the presence of only one of them is sufficient for a positive conclusion.

It often happens that some indicative features are joined by the connective *and*, and their (conjunctively formed) groups by the connective *or*. Conversely, the indicative features may be joined by the connective *or*, and their (disjunctively formed) groups by the connective *and*.

To illustrate the case of conjunctively formed groups joined by the disjunction *or* we turn to a simple example from the field of genetics. The indicative features of the genetic *possibility* of blue-eyedness in unborn human beings are as follows.*

I.   (1)   the father is homozygotic (carries only blue genes)
            and
      (2)   the mother is homozygotic,
            or

---

*The example is substituted for one from Russian grammar. This substitute example deals with the establishment of membership in the class of *potentially* blue-eyed people, but the probability of such membership is being ignored. (*Editor.*)

II. (1)   the father is homozygotic
           and
     (2)   the mother is heterozygotic (carries blue and brown genes),
           or

III. (1)   the father is heterozygotic
           and
     (2)   the mother is homozygotic,
           or

IV. (1)   the father is heterozygotic
           and
     (2)   the mother is heterozygotic.

The indicative features designated by the Roman numerals I, II, III, and IV, are complex (compound); they represent a combination of simpler indicative features. In order to establish whether an unborn child can possibly belong to the class of blue-eyed people, the presence of just one of the combinations of indicative features suffices. In our example, the indicative features entering into the combination are joined by the conjunction *and*. In order to establish the possibility of blue-eyedness on the basis of any one of the combinations, it is necessary to verify the presence in it of both the indicative features.

We cited the example from the field of genetics when combinations (groups) of indicative features are joined by the conjunction *or*, and when indicative features entering into each of the combinations are joined by the conjunction *and*. Let us give an opposite example, which will be drawn from the field of law.* In the United States, an income tax return must be filed by any unmarried person who is:

I. (1)   single
           or
   (2)   an unmarried head of a household
           or
   (3)   a widow(er) with a dependent child
           and

*The example substitutes for one drawn from Russian grammar. (*Editor.*)

II. (1) had an income greater than $1700 (under age 65)
or
(2) had an income greater than $2300 (over age 65).

The complex indicative features, designated by Roman numerals here, are necessary (they are joined by the conjunction *and*), while those indicative features of which these necessary features consist are sufficient for the necessary ones (i.e., sufficient for the features of I and II).

We will call the joining of indicative features by one or another logical connective showing the internal relations of features *the logical structure of indicative features,* and the method of writing which we used (writing each indicative feature on a separate line, indicating its numbers, and also the connective joining it with the other features) *the logic diagram of indicative features.*[6]

It is useful, for the analysis below of phenomena, to designate two basic types of logical structure for indicative features: the first type occurs when indicative features are joined by the connective *and* (conjunctive), and the second occurs when they are joined by the connective *or* (disjunctive).

The grammatical conjunction *or*, as a logical connective, is used in speech and in logic in two basic senses: in a non-strictly divisive sense and in an alternative sense. An expression of the type "*a* or *b*" where *or* is used in the non-strictly divisive sense signifies that one of the two indicated possibilities *a, b* necessarily takes place, but it is not excluded that both possibilities could take place (the question of whether both can take place remains open in the proposition). The connective *or* (in this sense) can join more members than just two. When using *or* in the alternative (strictly divisive) sense, the proposition "*a* or *b*" means that only one of the indicated possibilities takes place. A more precise statement of this case can be given as either "*a* or *b*, and only one of the two but not both.*"

The logical connective *or* in the non-strictly divisive (unifying-dividing) sense is precisely defined in logic by the operation of (weak) disjunction.

A proposition of the sort $a_1 \lor a_2$ is, according to the definition of (weak) disjunction, true even when just one of the propositions a_1, a_2 is true. By analogy, the proposition $a_1 \lor a_2 \lor \ldots \lor a_n$ (the parentheses specifying the order for carrying out the operations of disjunction are omitted in view of the commutativity and associativity of this operation) is true in any case where just one of the propositions a_1, a_2, ..., a_n is true.

The logical connective *or*, in the alternative sense with two members, corresponds to the operation of strict disjunction. A proposition of the sort $a_1 \lor a_2$ is, according to the definition of this operation, true if and only if one of the two propositions a_1, a_2 is true and the other is false. The situation, however, changes if the alternative *or* joins more than two members. If there is the assertion "a_1, or a_2, or a_3, ..., a_k in which *or* has an alternative sense, then it—as numerous linguistic examples show—signifies the following: one and only one of the statements a_1, a_2, ..., a_k is true. Strict disjunction for more than two members does not have this meaning, and therefore cannot serve to transmit numerous divisive assertions with an alternative *or*. The latter are conveyed by formulae of the type:

$$(a_1 \lor a_2 \lor \ldots \lor a_k) \,\&\, \overline{a_1 \,\&\, a_2} \,\&\, \overline{a_1 \,\&\, a_3} \,\&\, \ldots \,\&\, \overline{a_{k-1} \,\&\, a_k}.*$$

A non-strict *or* is more flexible from the standpoint of the possibilities of applying it than is the alternative *or* (strictly divisive).

It is natural to distinguish the questions: what is stated about different events (phenomena), and what do these events represent? In reality, events (or statements) *a*, *b* could exclude each other, but we cannot know this or we may only have to underline the fact that one of them necessarily takes place (is true). In such cases, we use the non-strictly divisive *or*. For example, one may say that (in Russian, *Ed.*) the adverbial modifier of manner signifies the method of acting or the degree of action, keeping in

---

*Comparisons of a strict disjunction with an alternative "or" of normal language and substantiating the statements cited are contained in the notes to Biryukov's brochure [82].

mind the non-strictly divisive *or*. But in the given instance, an even stronger assertion will also be true: "the adverbial modifier of manner signifies methods of acting or the degree of action and only one of the two," since in each particular instance, the indicated possibilities exclude each other. But, in the definition, "a sentence in which the principal or secondary part is omitted is incomplete,"* the disjunction *or* must be understood only in the non-strictly divisive sense, since, in reality, the omission of the principal part of the sentence in an incomplete sentence does not exclude the fact that the secondary part could be omitted at the same time.

Henceforth, we will use *or* in the non-strictly divisive sense everywhere with the exception of instances which will be specially stipulated.

The logical structures of indicative features are economically and clearly expressed in the language of symbolic logic. The description of indicative features and their structures can be realized through two methods: in the language of propositional (or sentential) logic and in the language of predicate logic.

Propositional logic makes up the section of contemporary logic (within the logic of predicates) in which the reaching of conclusions from complex propositions is studied, each of which is composed of several elementary propositions and logical copulae (connectives): *and, or, if . . ., then* and negations (not, it is untrue that). In this sense elementary propositions are those which are taken as a whole without dividing them into subject and predicate (in other words, their internal structure is not taken into account). Thus, in that section of logic, the connection between components within the statements (or propositions, *Ed.*) is not analyzed. The subject-predicate structure of simple statements is not examined. In the predicate logic, not only the structure of complex statements (propositions) is taken into account, but also the subject-predicate structure of simple propositions, and the forms of conclusions in which this internal structure of the statements or sentences plays a role is studied. For example, the statement on

*In Russian grammar. (*Editor.*)

the belonging of indicative feature *a* to a certain object *x* may be expressed in the form of "*a* of *x*" or *a* (*x*) as it is accepted in symbolic logic. In using the symbols &, V, ⇒ corresponding to the logical connectives *and, or, if . . ., then* and taking *x* to be some word, one may write the above-cited definition for a verb in the following form: "*x* is a verb $\underset{Df}{\Leftrightarrow}$ (*x* signifies action V *x* signifies a state) & (*x* answers the question *what has to be done?* V *x* answers the question *what will have to be done?*)."[7] Or, in another way of writing it: The verb (*x*) $\underset{Df}{\Leftrightarrow}$ (signifies action (*x*) V signifies state (*x*)) & (answers the question "what is to be done?" (*x*) V answers the question "what will have to be done?" (*x*)).

Or, in designating indicative features in the order in which they are enumerated, *a*, b_1, b_2, . . ., b_n, we obtain:

$$a(x)\underset{Df}{\Leftrightarrow} (b_1 (x) \text{ V } b_2 (x)) \text{ & } (b_3(x) \text{ V } b_4 (x)).^*$$

Here, we used the language of predicate logic. But in abstracting ourselves from the internal structure of the statements *a* (*x*), b_1 (*x*), and so on, and in designating them through a_1, b_1, and so on, we can express the same thing in the form of a formula of sentential (propositional) logic:

$$a\underset{Df}{\Leftrightarrow} (b_1 \text{ V } b_2) \text{ & } (b_3 \text{ V } b_4).$$

The use of the language of predicate logic ensures the possibility of a more detailed and precise expression of the indicative features and their structures. The application of this language is especially important when the relations of one thing to another are used as the indicative features. For example, the indicative feature consisting of the divisibility of some number *x*

---

*A more precise way of writing these definitions is this:

$$\forall x \ (a(x)) \Leftrightarrow (b_1 (x) \text{ V } b_2 (x)) \text{ & } (b_3 (x) \text{ V } b_4 (x)).$$

The sign ∀ signifies the universal quantifier (∀ *x* means for all *x*," in the given instance: for any word). However, since definitions always relate to all objects of some class, to simplify the writing of this, we can omit the universal quantifier when writing those rules which indicate modes of action relative to all objects of that class.

by 5 may be expressed in the form "divides $(x\ 5)$," which signifies the expression "x is divisible by 5." The indicative feature "is divisible by some number" can be expressed in the form "$\exists\ y$ divides (x, y)," where $\exists$ is the existential quantifier, corresponding to the word "some" in ordinary language. The whole expression represents the statement "there is some number y which is such, that x is divisible by y."

To illustrate the symbolic formulation of definitions, we will return to the example of the indicative features characterizing the class *insect*, i.e., animals having six legs and three clearly defined body regions. To cast this definition into symbolic form we will first rephrase it more precisely. We will say: all animals (x) are members of the class of insects (i), if and only if they possess the characteristics or property of six-leggedness (l) and the characteristic of three distinct body regions (s). In symbolic form we write:

$$(\forall\ x)\ (i\ (x) \underset{\mathrm{Df}}{\Leftrightarrow} l\ (x)\ \&\ s\ (x)$$

which reflects predicate logic. In sentential logic the formula

$$i \underset{\mathrm{Df}}{\Leftrightarrow} l\ \&\ s$$

means the statement "this animal is an insect" is true, if and only if the statements "the animal is six-legged" and "the animal has three body regions" are both true.

The symbolic formulation becomes considerably more complex in the case of the indicative feature for the possibility of blue eyes in human beings (i.e., an as yet unborn child *may* have blue eyes). Let us recall from elementary genetics that blue-eyedness is synonymous with the possession of two recessive genes (bb). Recessive genes do not manifest their characteristic in outward appearance when paired with a dominant gene, such as that for brown eyes (B). Of course, genes are the genetic material passed on to every individual by parents. To put it most succinctly, an offspring (i.e., an individual person) can have the genetic characteristic bb (manifested by blue eyes) only when both parents have passed on a b-gene. To do so, both, father and mother must (1) carry *b*-genes (though they may also carry *B*-genes) and (2) pass a *b*-gene on to the offspring; if one or the other passes a *B*-gene, the offspring will be brown-eyed rather than blue-eyed.

We reformulate: for all individuals (x, y, z), if y is the mother of (M) x, and z is the father of (F) x, then x may have genetic characteristic *bb* (T) if and only if z and y (i.e., father and mother) have property T, or z has property O (i.e., *Bb* or *bB*) and y has property T and z passes a recessive *b*-gene to x (P), or z has property T and y has property O and y also has property P (passes on a *b*-gene), or z has property O and y has property O and z has property P and y has property P.

To recapitulate, first, the definitions of properties and relations to be used in the symbolic formulation.

$$T(x) \quad \Leftrightarrow \quad x \quad \text{has } bb \text{ (two recessive genes)}$$
$$O(x) \quad \Leftrightarrow \quad x \quad \text{has } Bb \text{ (one recessive gene)}$$
$$P(yx) \quad \Leftrightarrow \quad y \quad \text{passes a } b\text{-gene to } x$$
$$M(yx) \quad \Leftrightarrow \quad y \quad \text{is the mother of } x$$
$$F(zx) \quad \Leftrightarrow \quad z \quad \text{is the father of } x$$

Now we are able to write in predicate logic:

$$(\forall x, \forall y, \forall z) \, (M(yx) \, \& \, F(zx) \Rightarrow T(x) \Leftrightarrow ((T(z) \, \& \, T(y)) \lor (O(z) \, \& $$
$$T(y) \, \& \, P(zx)) \lor (T(z) \, \& \, O(y) \, \& \, P(yx)) \lor$$
$$(O(z) \, \& \, O(y) \, \& \, P(zx) \, \& \, P(zy))))$$

The significance of symbolically writing definitions and rules consists of the fact that symbolic formulations help to determine and make clear not only the logical structure of indicative features but also the structure of definitions and rules in general. Moreover, in trying precisely to determine this structure, the impreciseness and simple falsity of some definitions and rules in textbooks are discovered. Starting from verbal definitions, it is often impossible to understand what is essential in some phenomenon, to distinguish phenomena from one another, and to identify them univocally. Also, only after determining the logical structure of indicative features in phenomena and after describing them precisely (in the language of symbolic logic) does it become clear to what degree, at times, these structures are complex and what exceedingly difficult intellectual efforts students must perform in order to master certain aspects of subject-matter (e.g., grammar, arithmetic, genetics, etc.). It also becomes clear why mastery of these aspects of subject-matter causes such great difficulties for some students and why lack of success in studying, for example, even the native language is so great. The logical structures of indicative features are, as we saw, complex in themselves. The difficulty in mastering them, however, is magnified many times over by the fact that, in many instances, these structures are in no way disclosed to the students, and—most important—students are not taught standard and general procedures for independently determining and using these structures.[8]

These deficiencies are provoked not only by the fact that, as a rule, there is no special practice on analyzing the logical structure of indicative features in the classroom or in assigned homework, but, most important, they are provoked by defects in explicating concepts in the textbooks.[9]

Notes

1.   On these and similar processes of identification taking place on the level of perception, see, e.g., [638], [639], [663-665], [842], [843], [846], [847].

2.   The problem of expediency in algorithmizing the processes of identification is solved, in principle, just like the problem of expediency in algorithmizing the processes of transformation (see Chapter Six).

3.   Let us recall the statements of Gorsky and Tavanets [15] on the two functions of ideas in reasoning which were cited above (Chapter Seven).

4.   A person, for example, is easier to identify by some external indicative features than by those deriving from his internal and essential particularities. In biology, some types of plants and animals are more easily distinguished on the basis of descriptions in which the external sensorily perceptible traits are enumerated, and so on.

5.   In contemporary logic, the question of whether or not all

notions can be reduced to concrete sensorily perceptible indicative features (empirical notions) is being considered. This question has not had a final solution. Nonetheless, when we want to devise some algorithm of identification, we strive for such a reduction.

6.   These concepts must be distinguished, because the logical structure of indicative features can be expressed not only in the form of logic diagrams of indicative features, but also by other means, for example, by writing it symbolically in the form of a specific formula.

7.   We say here "signify," and not "designate," since the term "designate" is applied in logic only to names (linguistic expressions naming objects), but not to variables which in the given instance is what the letter x is (see Church [175]). The letters Df under the arrows signify "by definition."

8.   When the manuscript of this book (the Russian edition, *Ed.*) was already at the publishers, we conducted an additional experiment on teaching students of the seventh and eighth grades how to express symbolically definitions in propositional and in predicate logic. The experiment showed that students are entirely able to master the methods of writing down theoretical statements in a generalized symbolic form. Moreover, the mastery of these methods has an enormous general importance for developing in students some general intellectual abilities, since some important notions about the general structure of theoretical knowledge and the methods of scientific reasoning are formed in the students. They are also made acquainted with contemporary scientific language. There is no longer a need to mention the fact that understanding the common structures of theoretical knowledge and the general methods of their construction and description significantly facilitates understanding and mastering each particular theoretical statement.

9.   On this question, see also Farber [336-338] .

Chapter Nine

Deficiencies in Presenting Logical Structures of Indicative Features in Textbooks

The greatest deficiency commonly found in textbooks is that the indicative features mentioned in definitions and other theoretical statements are not emphasized by the authors. This greatly impedes the process of becoming aware of and differentiating the indicative features of phenomena. Not only do textbooks not single out individual indicative features and separate them into parts, they also do not show the characteristics of their logical properties (necessary, but insufficient, indicative features, and sufficient, but unnecessary, and so on), and they do not establish their structures. The essence and the significance of logical connectives forming complex indicative features (even in those cases where these connectives are contained in obvious form in definitions and rules) are not made clear to students. No effort whatever is usually made to create an awareness of the role of these connectives. Indications are completely lacking in textbooks that indicative features can be found in various relationships, that they can have different logical properties, that their internal structures can be different, and that this determines the difference in procedures for dealing with them. It is not surprising, then, that there are also no exercises whatever in textbooks which are directed at forming in students the ability to separate indicative features into parts, to establish their properties and structure, and to determine the systems of operations required by these structures. If one takes into account the fact that teachers also usually do not arrange for such practice, one can understand why the mastery of grammatical (and many other) concepts and the

214

forming of corresponding skills and abilities often entails great difficulties for students and is accompanied by many mistakes.

The authors of textbooks who do not make it their business to establish clearly the logical structure of indicative features for students show small concern for the necessity that the grammatical form of definitions and other theoretical statements correspond to the logical structure of indicative features, and that the grammatical form reveal the logical structure.[1]

Let us examine a number of definitions from textbooks where the logical structure of indicative features is not presented or is incorrectly presented. We will also show, from some examples, how it is possible to translate such definitions from ordinary language into the language of logic.

1. Definitions Where No Logical Operator Is Mentioned

Example 1

In a textbook for the Russian language [533], the following definition is given of the adverbial modifier of place: "Secondary parts of a sentence which answer the questions *where?, where to?, where from?* are called adverbial modifiers of place."

Though it is not shown in the definition which connectives join the indicative features of an adverbial modifier of place, it is clear by the meaning that they are joined by the connective *or* (those secondary parts of a sentence which answer the question *where?, or where to?, or where from?* are adverbial modifiers of place).

In other definitions where the connectives are not shown, it is often difficult to specify from the meaning of the sentence just how the indicative features are linked. The absence of logical connectives often impedes comprehension and leads to misinterpretation and to mistakes. Such a definition will be cited in the next example.

Example 2

In the previously mentioned textbook of the Russian language [533], the following definition for agreement is given: "The relation in which a dependent word is placed in the same number, gender, case or person as the basic word is called agreement."

The structure of the indicative features is not established here. Instead of logical connectives there are commas, and the connective *"or"* in front of the last indicative feature would lead one to think that all of the indicative features are joined by *or* (by analogy, for example, with such a sentence as: *Sunday I am going to the theater, cinema, or the circus*).

However, further examples used in the textbook show that it is not so. We cite two of them:

"1) In the combination *the sun peeps out*, the verb of the present tense, *peeps out*, agrees with the noun *sun* in singular number **and** in third person (the conjunction is stressed by the present author); 2) in the combination *April sun*, the adjective *April* agrees with the noun sun, in singular number, neuter gender, nominative case" (here the meaning also requires the conjunction **and**).

How are indicative features joined: by the disjunction **or**, or the conjunction **and**? Or is it by the disjunction *or* in a non-divisive sense which does not exclude simultaneous linkage by the conjunction **and**? There is no answer to these questions. It is impossible to look into the structure of indicative features. It is not by chance that students encounter considerable difficulties in mastering this section, and make many mistakes in determining the type of relation among the words.

2. Definitions in Which Logical Connectives *"And"* and *"Or"* Are Expressed by Different Grammatical Conjunctions

Not infrequently, instead of the logical connectives **and** and *or*, other grammatical conjunctions are used in definitions. This often obscures and conceals the logical structure of the indicative features, hampers its determination, and leads to mistakes.

It must be noted that, in ordinary speech, one and the same logical connective can be expressed by different grammatical conjunctions. For example, one may say:

I came to your place, *but* you were not there.

I came to your place, *yet* you were not at home.

I came to your place, *however*, you were not at home.

I came to your place, *and* you were not at home.

The grammatical conjunctions, *but, yet, however, and*, which bring certain shades of meaning into the sentence (opposition, temporal sequence, and so on) essentially express one and the same type of logical link—the conjunctive link (expressed in logic through the logical conjunction *and*). The shades of meaning of opposition or time expressed by these conjunctions, that is, as grammatical conjunctions, do not have any relevance for logic. These shades of meaning do not influence the method of operating with the indicative features, i.e., the method of identifying phenomena with their help. (This circumstance, that one and the same type of logical relation can be expressed by different grammatical conjunctions, serves, as we said, as a basis for introducing the concept of "a logical connective" as distinct from the concept of "a grammatical conjunction.")

When we speak of the connectives *and* and *or* as logical, we abstract them from those shades of meanings which the grammatical conjunctions *while, but, however, as well as*, etc. express. In grammatically expressed conjunctions having logical meanings, what is important to us is only their logical meaning on which the method of operating with corresponding indicative features depends.

An important task of the teacher consists of teaching students to determine the logical meaning of grammatical conjunctions and teaching them to replace the grammatical conjunctions by a corresponding logical connective.

Here are some examples of definitions in which the logical connectives *and* and *or* are expressed by different grammatical conjunctions.

Example 1

"A quadrangle which has two opposite sides parallel, *while* the other two are not parallel is called a trapezoid" (italics mine—L.L.) [732].

Here, the grammatical conjunction *while* is used in the sense of the logical conjunction *and*. The logic diagram of this definition is:

A trapezoid is a quadrangle with

1)   two sides parallel

           and

2)   two other sides not parallel.

Example 2

"Words which designate quantity *as well as* the order of objects when counting are called numerals (italics mine—L.L.) [533].

Here the grammatical conjunction *as well as* is used in the meaning of the logical disjunction *or*.

The logic diagram of this definition is:

Numerals are those words which

1)   designate a quantity of objects in counting

           or

2)   designate the order of objects in counting.

Example 3

"Interjections are words which express feelings or motives *but* do not name them" (italics mine—L.L.) [533].

Here, the grammatical conjunction *but* is used in the meaning of the logical conjunction *and*.

The logic diagram of this definition is:

Interjections are words which

1)   express feelings or motives

           and

2)   do not name them.[2]

Special attention should be paid to instances where the grammatical conjunction *and* is used in the sense of the logical disjunction *or*.

Example 4

"The first letter of the first word in every sentence which follows after a period, question mark *and* [sic] exclamation point is written with a capital letter. Furthermore . . . (italics mine—L.L.) [533].

It is clear that one writes the first word of every sentence *either* after a period *or* after a question mark *or* after an exclamation point with a capital letter. In the definition, instead of the disjunction *or*, the conjunction *and* is used. We find an analogous deficiency in the definition of an adverb.

Example 5
"Unvarying words which explain verbs *and* adjectives and designate various circumstances are called adverbs" (italics mine—L.L.) [533].
Clearly, one and the same adverb cannot simultaneously relate to both a verb and an adjective, explaining both at once. Therefore, the correct definition should be: "Unvarying words which explain verbs *or* adjectives and designate various circumstances are called adverbs."[3]
In the latter two examples, the use of the conjunction *and* instead of *or* cannot evoke serious misunderstandings (although it does not foster the cultivation of logical reasoning), since it is clear from the meaning that it is a case of conditions (indicative features) linked by the disjunction *or*. If the grammatical conjunction *and* in the definitions cited were interpreted as the logical conjunction *and*, one would have to begin a sentence with a capital letter only when a period, question mark, and exclamation point are all present together. Adverbs, according to the definition, ought to be a kind of unvarying words which *simultaneously* explain both a verb and an adjective. It is clear that the authors of the definitions had something completely different in mind.
There are, however, instances when the logical sense of grammatical conjunctions is not at all evident. And here, inattention to the logical structure of indicative features and their expression by a haphazardly chosen grammatical conjunction leads to the kind of ambiguity which can engender serious difficulties and mistakes in mastering and applying knowledge.

Example 6
In one science textbook [654], there is the following definition: "Every change of form in a solid *and* its volume is

called a deformation" (italics mine—L.L.).

If this definition were to be taken literally, then deformation takes place only when *both* the form *and* its volume change. Actually, of course, deformation takes place when *either* the form of a solid *or* its volume change, but both can undergo changes together (for a deformation to occur, the change of just one of the properties suffices).[4] Thus, the definition given in the textbook easily can lead students to make mistakes.

It is evident from the examples considered on the preceding pages that the logical connectives *and* and *or*[5] are often expressed by different grammatical conjunctions. There are even instances when the logical disjunction *or* is expressed by the grammatical conjunction *and*. Obviously, one can scarcely consider such a situation as a virtue of textbooks. But since it does take place, one must take this into account and, as mentioned above, students must be specially taught to "translate" grammatical conjunctions into logical connectives and to see the precise logical meaning behind the various grammatical forms for expressing a thought. This must also be taught, because in ordinary speech, people often use grammatical conjunctions instead of logical connectives, and the former do not always correspond with the latter.

3. Definitions with Incorrectly Described Structures of Indicative Features

In the definitions and rules examined above, the fact that the logical structure of indicative features was not revealed, hampered the mastery and application of concepts, and repeatedly led to ambiguity in understanding the phenomena reflected by them. However, ignoring the logical structure of indicative features and failing to make a logical analysis of those systems of indicative features which, in fact, must be the basis of definitions and other theoretical statements, leads to outright mistakes in these definitions and similar theoretical statements. Here is an example of an incorrectly formulated definition.

In a secondary school textbook of the Russian language [533], this definition of parenthetical words is given: "Those words with the help of which the speaker expresses his relation to,

for example, what he is communicating are called parenthetical . . . Parenthetical words are distinguished by intonation." The
defects of this definition are seen easily when we try to apply it.
Let us consider the sentence: *I was late, first, because I was held
up at work, and second, because I waited for the bus for a long
time.* If the definition were applied to the words *first*, and *second*,
one would have to admit that they are not parenthetical, since
they do not reflect the relation of the speaker to the thought
expressed. Rather, they seem to indicate an order of reasons for
the specific event. Nonetheless, these are parenthetical words.
Therefore, not all of the sufficient indicative features are shown in
the definition. This definition does not allow the identification of
all types of parenthetical words and actually provokes mistakes in
many cases.

It is true that in the textbooks, after several paragraphs, it is
said that "with the help of parenthetical words, the speaker may
also indicate the order in which events follow each other and the
links between them," but this does not save the situation; rather,
it confuses it even more.

If the definition were to be followed literally, it would be
impossible to consider these words as parenthetical. If one were to
go by the explanation which follows and consider them as
parenthetical, then one would enter into a contradiction with the
definition.

The reason for this confusion lies in the fact that the
indicative feature given in the definition as necessary is actually
unnecessary. As shown, there are parenthetical words which do
not possess this indicative feature and do not reflect the
relationship of the speaker to the thought expressed.[6]

Notes

1.   For an understanding of the logical structure of indicative
      features and the precise meaning of rules and definitions, it is
      expedient to teach students the analysis of the grammatical
      form of rules and definitions so as to teach them to

"translate" rules and definitions from ordinary, natural language into the language of logic. Generally speaking, the frequent lack of congruence between the grammatical form and the logical structure of an idea is a very important fact which must always be kept in mind in the process of teaching. It is necessary consistently to teach students to see the logical structure of the propositions, which lies beyond the grammatical form of the sentences. It is also necessary to develop in them the ability to make up logic diagrams of indicative features on the basis of definitions expressed in various grammatical forms.

2. It is easy to note that the first indicative feature is complex (compound). It consists of two indicative features joined by the disjunction *or*. A more precise (detailed) logic diagram of that definition would be:

> Interjections are words which
>
> I   1)   express feelings
>
> *or*
>
> 2)   express motives
>
> *and*
>
> II   1)   do not name them.

3. If the letter x refers to unvarying words, and the indicative features of the adverb are designated by the letters a, b, c, then the definition given in the book should be written thus:

$$a \, (x) \, \& \, b \, (x) \, \& \, c \, (x) \underset{Df}{\Leftrightarrow} \text{Adverb} \, (x).$$

As a matter of fact, the definition should be written thus:

$$(a(x) \vee b \, (x)) \, \& \, c \, (x) \underset{Df}{\Leftrightarrow} \text{Adverb} \, (x),$$

or taking the divisive character of the conjunction *or* into account, thus:

$$(a(x) \vee b \, (x)) \, \& \, \overline{a \, (x) \, \& \, b \, (x)} \, \& \, c \, (x) \underset{Df}{\Leftrightarrow} \text{Adverb} \, (x),$$

or in another form:

$$(a(x) \; \dot{V} \; b \; (x)) \; \& \; c \; (x) \underset{\text{Df}}{\Leftrightarrow} \text{Adverb} \; (x).$$

Let us note that the symbolic presentation of definitions requires of the author of a textbook and of the teachers a precise knowledge of the logical structures of indicative features. It also reveals vagueness and ambiguity in understanding the character of the phenomena defined, and, further, can avert the appearance of such vagueness and ambiguity.

4.   Let us formulate the textbook's definition as an implication: "If the change x of a solid is expressed as a change in its form (indicative feature a) and as a change in its volume (indicative feature b), then such a change x is called deformation (D)." In the textbook, therefore, what is said, is:

$$a \; (x) \; \& \; b \; (x) \underset{\text{Df}}{\Leftrightarrow} D \; (x).$$

However, the correct definition is:

$$a \; (x) \; V \; b \; (x) \underset{\text{Df}}{\Leftrightarrow} D \; (x).$$

5.   With different meanings: (a) in the sense of non-strict (weak, non-divisive) disjunction; (b) in the sense of strict disjunction; (c) in the sense of a means of expressing judgments of the type "one and only one of the members of the judgment which are linked by the disjunction '*or*' is true."

6.   Many other examples could be cited of instances when parenthetical words do not express the relation of the speaker to the thought expressed, but they are nevertheless parenthetical.

Chapter Ten

Methods of Identification as a Function of the Logical Structure of Indicative Features

In the preceding chapter, we showed that it is important, when formulating definitions and other theoretical statements, to establish the logical structure of the indicative features of objects (situations, phenomena, *Ed.*) and to express it correctly. Somewhat earlier (see Chapter Seven) we spoke of how a specific algorithm or, more often, several algorithms correspond to each theoretical statement. Thus, for example, a whole series of possible algorithms distinguished by the order in which actions indicated by the formula are carried out corresponds to the formula for the area of a triangle—$A = \frac{1}{2} bh$—on the condition that the expression, written without the use of parentheses, signifies any arrangement of the latter which is permissible according to the concept of a formula in algebra.

An analogous situation takes place when applying definitions, laws, grammatical rules, theorems, and other theoretical statements.

Let us examine, for example, the definition of an adverbial participle given in a textbook for the Russian language [533]: "An unchanging form of the verb which designates additional actions and refers to the predicate is called an adverbial participle." Let us express this definition in the form of a conditional proposition: "If and only if the form of the verb x is unchanging (indicative feature a), and designates additional action (indicative feature b), and refers to the predicate (indicative feature c), then this form of the verb x is called an adverbial participle (AP)." Let us write this definition in symbolic language:

$$a \ (x) \ \& \ b \ (x) \ \& \ c \ (x) \underset{\text{D f}}{\Leftrightarrow} AP \ (x).$$

Now, let us suppose that we encountered the Russian sentence *The friends walked along hugging each other* (this might be better translated *with their arms around each other*, but the adverbial participle cannot be illustrated in English, *Trans.*), and we have to discern whether the word *hugging each other* (all one word in Russian—a reflexive adverbial participle, *Trans.*) is an adverbial participle. We must verify whether that word has the indicative features of an adverbial participle. As is evident from the definition, the indicative features of the adverbial participle are joined by the logical conjunction *and*, and this logical conjunction (like the disjunction *or*) has among its properties— similar to the arithmetical operations of addition and multiplication—the properties of commutativity and associativity.[1] This means that the operations of identification may be carried out in any sequence. For example, one may at first verify indicative feature *a*, then *b*, then *c*, or at first *b*, then *a*, then *c*, or at first *c*, then *a*, then *b*, and so on.[2] Just as the formula $A = \frac{1}{2} \ bh$ does not require one specific (rigid) sequence of operations for calculating the area of a triangle, but has a series of possible algorithms of calculation corresponding to it, so the formula $a \ (x) \ \& \ b \ (x) \ \& \ c \ (x) \underset{\text{D f}}{\Leftrightarrow} AP \ (x)$ does not demand one specific (rigid) sequence of operations for identifying an adverbial participle, but admits several possible algorithms of identification. These algorithms are contained in the definition of an adverbial participle in a hidden form, and solving any particular problem of identification consists of making the transition from the definition to the algorithm, having chosen one of the possible sequences of actions as the means of identifying it.

How can the transition from the definition to the algorithm of identification be realized? What is the relation between the structure of the indicative features given in the definition (and more generally in any theoretical statement) and the structure of actions for identifying the corresponding phenomena?

There is a formal correspondence between the logical structure of the indicative features of a phenomenon and the structure of actions for identifying it on the basis of those indicative features. This correspondence is so important that one may speak of *the relation of the logical structure of operations for identifying a phenomenon to the logical structure of its indicative features.* The logical structure of indicative features by its nature specifies the structure of operations.[3]

This link between the structure of operations and the structure of indicative features is specified by some general principles concerning the truth value of complex expressions which are established in symbolic logic. Thus, the expression having the form a & b is true only in that instance when both a and b which make it up are true; the expression $a \lor b$ is true, when just one of the members of the expression a, b is true. An expression like $a \Leftrightarrow b$ indicates that the expressions a and b are either both true or both false. Therefore, if it is known that $a \Leftrightarrow b$, then in order to establish whether b is true, it suffices to establish the truth value of a (and vice versa).

In general, when A is a complex proposition (for example, the conjunction or disjunction of some elementary propositions), the question of its truth value is solved by starting from a knowledge of the conditions for the truth value of the complex proposition.

The exceptional importance of the above indicated relation between the logical structure of operations and the logical structure of indicative features consists of the fact that when we know the logical structure of the indicative features of some phenomenon and the truth value of the complex proposition reflecting this structure, we may univocally[4] specify the structure of actions for identifying this phenomenon.

The logical structure of indicative features does not specify, as we said, one single algorithm for identifying a phenomenon. Rather, it specifies some finite number of these algorithms, which are distinguished from one another only by the sequence of operations for verifying the indicative features. The common trait of a given number of algorithms is the community of the logical structure of operations entering into these algorithms. What is important is that all of these algorithms when applied to one and the same object (or—what is the same thing—to one and the same initial datum) lead to one and the same result. Therefore, there is often no need to single out one of the algorithms which determine the sequence of operations for identifying the phenomenon. The logical structure of indicative features, without specifying one unique sequence of operations and one single algorithm, determines some logical structure of operations, i.e., some *general method* of actions which is distinguished from the algorithm only by the fact that the sequence of operations for identification is not fixed.

Since the logical structures of indicative features are distinguished by the ways in which the indicative features are joined, namely by the logical connectives through which the structures of the indicative features are formed, a general method of identifying a phenomenon depends on which logical connective(s) joins the indicative features of the phenomenon.

Let us show, through two examples, how the logical structure of indicative features of a phenomenon specifies the general procedure for identifying it.

Suppose we observe two situations in which two living organisms exist in close association—perhaps a sea anemone and a hermit crab in the first case, and some algae and barnacles in the second case. We want to determine in which of these two cases there exists a *symbiotic* relationship.

In order to solve this problem, we must establish the indicative features for a symbiotic relationship and determine their logical structure. *Symbiosis* is commonly defined as a close association of *different* organisms from which both partners benefit.[5]

We shall reformulate this definition into the following conditional statement. If and only if organism x and organism y are different, i.e., not of the same species (indicative feature a) and if they mutually benefit from their association (indicative feature b), then these organisms are symbiotic (S). Written symbolically:

$$a\ (x, y)\ \&\ b\ (x, y) \underset{\mathrm{Df}}{\Leftrightarrow} S\ (x, y).$$

The indicative features of symbiotic biological partners are conjunctively joined by the conjunction *and*. The procedure for identifying the corresponding biological phenomenon follows directly from the way the indicative features are joined (by the logical conjunction *and*). In order for some associated parts of a set of biological organisms $x_1, \ldots, x_n$ to be symbiotic, it is necessary that they possess both indicative features of symbiotic association. The absence of even one indicative feature is the basis for a conclusion to the effect that the given members of the set were not symbiotic. The procedure for identifying the indicative features of symbiotic members follows directly from their logical structure. This can be described in the following manner.

In order to establish whether some associated organisms are symbiotic, we must:

1. Verify whether they have one of the indicative features shown (for example, the first).

    If not, then stop and draw a negative conclusion.

    If yes, then—

2. Verify whether they have the second (and in the given case the last) indicative feature.

    If not, draw a negative conclusion.

    If yes, draw a positive conclusion.

The prescription given is a general procedure, but it is not an algorithm in the precise sense, since there is some uncertainty as to which indicative feature should be verified first and which second. It is easy to understand how such a general procedure can be turned into an algorithm. One must eliminate the uncertainty in the prescription about the sequence of operations for verifying the indicative features and point out exactly which indicative features must be verified first and which second.[6]

In applying the described procedure to the arbitrary examples of the sea anemone and the hermit crab we would, of course, find both indicative features confirmed and may regard the association of these organisms as symbiotic. In the case of the algae and the barnacles there is no mutual benefit and so one indicative feature is absent. Therefore, the association is not symbiotic.

This method is easy to extend to a general case (a case of n indicative features):

$$a_1 (x) \ \& \ a_2 (x) \ \& \ \ldots \ \& \ a_n (x) \underset{Df}{\Leftrightarrow} A (x).^*$$

In order to establish whether some phenomenon x belongs to a certain class A (i.e., whether it is true that A (x), or in another notation, $x \in$ A), we must:

1. Verify whether it has the first indicative feature.[7]

    If not, then stop further verification and draw a negative conclusion.

    If so, then

2. Verify whether it has the second indicative feature.

    If not, then stop further verification and draw a negative conclusion.

    If so, then

    .

    .

    .

n. Verify whether it has the nth indicative feature.

    If not, then draw a negative conclusion.

    If so, then draw a positive conclusion.

We examined the general procedure for analyzing phenomena by indicative features which are conjunctively joined, i.e., by the conjunction **and**. Let us now examine the general procedure for identifying phenomena by indicative features joined disjunctively, i.e., by the disjunction **or**.

Suppose we must make clear to some law students under which circumstances a person who suffers from severe visual impairments may claim the legal benefits of blindness. In order to provide students with the indicative features with which they may

---

*We are examining a case where all indicative features $a_1, a_2, \ldots, a_n$ (and also A itself) are properties (one-place predicates). But these indicative features— some or all—can also be relations. The most general case of the given logical structure of the indicative features is encompassed by the formula:

$$a_1 \ \& \ a_2 \ \& \ \ldots \ \& \ a_n \underset{Df}{\Leftrightarrow} A.$$

determine the existence of legal blindness in any given case we must examine the legal definition of blindness. In the United States[8] it is required that (a) central visual acuity not exceed 20/200 in the better eye with correcting lenses, or (b) that the widest diameter of the visual field subtend an angle no greater than 20 degrees. Reformulated as a conditional statement this definition reads as follows: If and only if the person x has a central visual acuity that does not exceed 20/200 in the better eye with correcting lenses (indicative feature a), or person x has such vision that the widest diameter of the visual field subtends an angle no greater than 20 degrees (indicative feature b), may person x be considered legally blind (B). Written symbolically, we have:

$$a\ (x)\ \vee\ b\ (x) \underset{Df}{\Leftrightarrow} B\ (x).$$

As we see, the indicative features of legal blindness are joined by the disjunction *or*. Each of them, consequently, is sufficient but is not necessary. In order to establish legal blindness in a given case it suffices to establish the existence of at least one of the conditions listed as the indicative features. The procedure for identifying the legal claim to blindness follows directly from the way the indicative features are joined (by the logical disjunction *or*). It can be formulated in the following manner.

In order to establish whether a certain person may legally claim blindness one must:

1. Verify whether one of the indicative features is present (for example, the first).

    If so, stop further verification and draw a positive conclusion.

    If not, then—

2. See whether the second (and in the given example, last) indicative feature applies.

    If so, draw a positive conclusion.

    If not, draw a negative conclusion.

The given prescription is a general procedure for identification, but it is not an algorithm for the very same reasons that were shown for the procedure for identifying phenomena by conjunctively joined indicative features.

The method examined can easily be extended to the case where there are k indicative features joined disjunctively, i.e., when the logical structure of the indicative features is expressed in the form:

$$a_1 \lor a_2 \lor \ldots \lor a_k \underset{\mathrm{Df}}{\Leftrightarrow} A.$$

The method in general form can be formulated in the following manner.

In order to establish whether some phenomenon belongs to class A, one should:

1.   Verify whether it has the first of the indicative features which were pointed out.[9]

        If so, then stop further verification and draw a positive conclusion.

        If not, then

2.   Verify whether it has the second indicative feature.

        If so, then stop further verification and draw a positive conclusion.

        If not, then

      .

      .

      .

k.   Verify whether it has the kth indicative feature.

        If so, then draw a positive conclusion.

        If not, then draw a negative conclusion.

As we see, the method of identification in the case where the indicative features are conjunctively joined by the conjunction *and*, differs fundamentally from the method of identification in the case where the indicative features are disjunctively joined by the disjunction *or*. If, in the first case, the *absence* of one of the indicative features is a signal *to stop* further verification and to formulate a negative conclusion, in the second case the *absence* of one of the indicative features is a signal *to continue* verification. If, in the first instance, the *presence* of the indicative feature is a signal to verify the following indicative feature, then in the second case, this signal is the *absence* of the indicative feature. When indicative features are joined by the conjunction ***and***, one may draw a positive conclusion only in the case where the phenomenon has *all* indicative features; when, however, indicative features are joined by the disjunction ***or***, then the presence of *just one* of the indicative features suffices for a positive conclusion. On the contrary, if in the first case, the absence of *just one* indicative feature leads to a negative conclusion, then in the second case, the absence of *all* indicative features is necessary for a negative conclusion.

The dependence of the structure of operations on the logical structures which were shown has a very general character; it is valid for any subject-matter in any context and with all kinds of indicative features. The structure of operations is specified by the logical structure of indicative features regardless of the nature of these indicative features themselves. The procedure depends not on the content of the indicative features, but only on the ways in which they are joined and on their logical structure. Let us note that this, obviously, makes it possible to form general methods of reasoning and general intellectual abilities in a person—things which were once called formal qualities of the mind. Since the indicative features of *different* phenomena can have *one and the same* logical structure, then also, the method of acting with them in the process of identification of phenomena can be *identical* and general. Formation of general intellectual abilities and general methods of reasoning obviously resides in the fact that children begin to note the common logical structure in the indicative features of different phenomena and to act with respect to these indicative features in conformity with these structures. The dependence of methods of reasoning on the logical structures of the indicative features of phenomena and not on their content is a most important psychological prerequisite for transferring already formed methods of reasoning to new conditions and to a new content.

We examined the general methods of identification for the "pure" structures of indicative features. However, as we said above, the structures of indicative features are often mixed. For example, disjunctively conjunctive (the disjunction of conjunctions) and conjunctively disjunctive (conjunction of disjunctions). In these cases, the method of identification is also "mixed."

Let us suppose that a certain phenomenon has the following logical structure of indicative features:

$$(a\ (x)\ \&\ b\ (x))\ V\ (c\ (x)\ \&\ d\ (x)) \underset{\mathrm{D\,f}}{\Leftrightarrow} A\ (x).$$

It is evident from the formula that the logical structure of indicative features is a disjunction of conjunctions. In order to draw a conclusion about whether phenomenon x belongs to class A, it suffices that it have one of the combinations of indicative features: a & b or c & d. The combination of indicative features is joined here by the disjunction *or*, while the indicative features entering into the combination are joined by the conjunction *and*. Therefore, in order to attribute object x to class A, one must establish the presence in it of not some of the indicative features a, b, c, d, but a specific combination of them. In order to verify the presence of combinations of indicative features in an object, one must act according to the procedure characteristic of the disjunctive structure of indicative features and to verify the indicative features entering into each of the combinations (these indicative features which are conjunctively joined are necessary indicative features within the limits of sufficiency); one must act according to the procedure which is characteristic of the conjunctive structure of indicative features.

In another situation the indicative features might have a conjunctively disjunctive structure, i.e., be a conjunction of disjunctions. For example:

$$(a\ (x)\ V\ b\ (x))\ \&\ (c\ (x)\ V\ d\ (x)) \underset{\mathrm{D\,f}}{\Leftrightarrow} A\ (x).$$

Here the combination of indicative features is conjunctively joined, while the indicative features within the combination are disjunctively joined. Here, consequently, to verify the presence in the object of a combination of indicative features, one must act according to the procedure which is characteristic of a conjunctive structure and to verify the indicative features within the combination (they are disjunctively joined) by the procedure characteristic of the disjunctive structure of indicative features. Procedures for the case of mixed structures of indicative features will be illustrated in more detail later.

We noted above the fact that students, as a rule, are not taught that indicative features are always logically joined in a definite manner nor are they taught the types of these logical structures. Usually, they are also not taught the principles of passing from the logical structure of indicative features to a procedure for identifying phenomena on the basis of one or another structure. They are not shown the dependence of the structure of operations on the logical structure of indicative features (i.e., they are not taught methods of identifying phenomena depending on the types of the logical structures of their indicative features).[10] This, as experimental data show (part of the data will be cited in Part Two of the book), hampers and slows down the mastery of knowledge, skills, and habits, leads to a large number of mistakes in solving problems, and hinders the forming of general methods of reasoning. Later chapters of this book will be devoted to an analysis of these mistakes and to an illustration of a way of preventing them.

Let us summarize briefly.

We saw that to design a general method of identifying phenomena, we must:

1) Single out the indicative features on the basis of which identification can be carried out;

2) Select from these indicative features the ones which satisfy the ontological and psycho-didactic requirements set forth in Chapter Seven (one must verify these indicative features by corresponding criteria, in particular, those shown earlier);

3) Determine and clearly express the logical structure of indicative features and the logical connectives which join them;

4) Formulate theoretical statements on the basis of which identification must be carried out, with a precise indication of the way in which the indicative features are joined;

5) Effect a transition from the theoretical statement to the

identification procedure (on the basis of the relation of the structure of operations for identifying a phenomenon to the logical structure of its indicative features).

We noted that the logical structure of indicative features of a phenomenon specifies only a general identification procedure, i.e., some set of algorithms, but not the specific sequence of operations for verifying indicative features. Meanwhile, each real process of identification can be realized only on the basis of some specific algorithms from among this number, within which a specific sequence of operations is followed.

We emphasized above (note 6) that, in many cases, there is no need to show students ahead of time the sequence of verifying indicative features when identifying phenomena, i.e., to exhibit the algorithm of identification. It suffices to teach them the abilities: a) to establish the logical structure of the indicative features, and b) to pass from the logical structure of the indicative features to a procedure based on that structure.[11]

In doing this, the freedom of choosing a particular algorithm from among a number of algorithms corresponding to the structure of the indicative features is left to the students. They must decide for themselves which one from among the possible sequences of operations to apply in different instances.

Undoubtedly, the two abilities shown are cardinal and fundamental, and their mastery opens up the possibility for the student to solve correctly any problem of identification. It is quite possible to limit oneself to these abilities when students have to solve but single problems of a certain class. The situation, however, changes considerably when students have to solve repeatedly problems of specific classes. Here, not only is it important for students to be able to pass from the logical structure of indicative features to some one course of actions (i.e., algorithm, *Ed.*) out of several possible ones on the basis of that structure, but they must also come to possess specific *efficient skills* for these purposes. It is quite obvious that in many cases (namely when it is necessary to solve repeatedly homogeneous problems of a certain class, e.g., grammatical ones) one should not have to ponder over which procedure to follow of all the possible ones (and these procedures in some respects could be unequally efficient).[12] He should act without having to solve the problem of choosing the procedure itself. He must act *at once* in the most efficient way. But in order to do this, it does not suffice to teach people to pass from the logical structure of indicative features to some algorithm from among a number of algorithms corresponding to that structure. It is necessary to develop in them the automatic *skill* for acting according to the most efficient (in a specific sense) algorithm, i.e., the skill of applying the best (from the standpoint of efficiency) sequence of operations. A person must know that various algorithms for solving the problem exist, and he must be able, when necessary, to apply any one of them. However, in problems which are major, typical, and frequent, he must act automatically and according to the best procedure.

Thus, one of the aims of teaching consists not only of teaching students to solve problems correctly, but also how to solve them efficiently. An efficient solution presupposes a choice of those procedures for solving a problem from among all possible procedures (and, in particular, of all possible algorithms) the one which permits attaining the goal in the least time and with the least effort. It is clear from what has been said that, in teaching students, often it does not suffice to teach them how to determine the logical structure of indicative features and to select one algorithm from among the possible ones. It is important to teach them how to choose the most efficient one from among all possible algorithms and to act according to this most efficient algorithm. Toward this end, one must cultivate methods for quantitative evaluation of the efficiency of algorithms and thereby gain the possibility of comparing the latter according to the degree of their efficiency. We are going to examine such methods.[13]

Notes

1. We are not mentioning other properties, since, for our discussion, only these properties need be taken into account.
2. What was said does not refer to identification carried out through transformation. There, the sequence of operations in

a general case is not inconsequential, since each subsequent operation can be directed at a "material" result of the preceding operation (a specific state of the object being transformed). If, as a result of changing the required sequence of operations, the corresponding transformation has not been carried out beforehand, and the necessary state is not obtained, then the subsequent operation is applied to a state which is not the one needed, and this could lead (and often does lead) to mistakes. Thus, all that has been said above on the possibility of carrying out operations of identification in any sequence refers only to "pure" processes of identification and is not valid for processes of identification where elements of transformation are included. In the latter case, a strict sequence in carrying out operations can be very important, and destroying this sequence can lead to invalid results.

3. Though precise, the specification does not exclude the rearrangement of the sequence(s) of operations, because of the commutativity and associativity of the logical connectives joining these indicative features.

4. See the preceding note.

5. This example replaces one drawn from Russian grammar for which there is no parallel in English. The author points out that "close association" represents an ambiguous indicative feature and requires that criteria for closeness be specified to a student. (*Editor.*)

6. In many cases, it is not necessary to transform such a general procedure into an algorithm. Moreover, there is no need even to formulate the general procedure itself. Students must be taught to pass independently from a definition, a formula, and any other theoretical statement to a general procedure and an algorithm. The basis of this transition is the conformity of the logical structure of operations for identifying a phenomenon to the logical structure of its indicative features. After students have learned to pass from theoretical statements to general procedures and algorithms, the necessity for formulating algorithms of identification may, in a

number of cases (but by no means always), no longer arise. Why it is expedient to teach students not only the ability to determine the logical structure of indicative features and to find the general procedures corresponding to them, but also to teach them algorithms indicating the sequence of actions for identification, will be explained below. Here we will note only that students should not only be taught independently to *pass* from the logical structure of indicative features to the corresponding general procedure and algorithm, but also independently to *determine* the logical structure of the indicative features. This problem is exceptionally important and methods for solving it will be discussed later (see Part Two).

7. It is assumed that an ordinal numbering of indicative features has been established. The assignment of ordinal numbers to indicative features is arbitrary.

8. This example replaces one drawn from Russian grammar that has no parallel in English. While legal definitions of blindness may differ in various legal codes, parallel illustrations in any legal code could be cited. The issue is not the manifest content but the logical structure of the illustration. (*Editor.*)

9. Regarding the numbering of indicative features, see note 7.

10. In the latter case the common deficiency of instruction in schools at the present time becomes apparent. Students are not taught the basics of logic and, in particular, they are not shown how to determine the truth values of complex statements.

11. In other words, it suffices to teach students the ability to determine the logical structure of indicative features and the ability to pass from this logical structure to some algorithm from among a *number of possible algorithms* corresponding to this structure.

12. On this point, see the next chapter.

13. We emphasize that quantitative mathematical approaches to evaluating (and calculating) efficient algorithms do not substitute for or replace psychological and logical methods of analysis and design of algorithms. Mathematical methods are

applied and ought to be applied *not instead of* psychological and logical methods but *as their foundation after* a qualitative psychological and structural-logical analysis of indicative features and algorithms has been made. This condition, of course, is not always observed.

Part One: Section Four

Mathematical Methods in the Design and Evaluation of Algorithms of Identification

Chapter Eleven

The Importance of Quantitative Methods of Analysis and of the Evaluation of the Efficiency of Algorithms

It was shown in the preceding Section that there can be different algorithms of identification for one and the same phenomenon. The difference may apply, on the one hand, to indicative features which constitute the basis of identification (attributing an object to a specific class may often be done on the basis of different indicative features) and, on the other hand, on the basis of the sequence of verifying the indicative features which underlie the identification. In the last segment of the preceding chapter, we spoke of the fact that a particular set of algorithms whose members are distinguished from each other only by the sequence of verifying the indicative features can refer to one and the same logical structure of these indicative features.[1]

We dealt above with the selection of indicative features on the basis of qualitative psycho-didactic criteria. Often, however, different indicative features and their systems comply with psycho-didactic criteria. The question is, which of them should form the basis for identification, and from which indicative features should one proceed when formulating algorithms?[2]

The criterion of choice in these cases can only be quantitative: one must base the identification of an object on those indicative features which permit solving the problem of identification in the most efficient way, i.e., with the least amount of time and effort.

On what does the efficiency of an algorithm of identification depend?

It depends on the following factors:

243

1) On the "easiness" of the indicative features from the standpoint of establishing them and operating with them;
2) on their number;
3) on the distribution of the probabilities of their presence in the objects;
4) on the sequence of their verification.

We shall dwell on each of these factors in more detail.

As we said above, indicative features are usually unequal from the standpoint of the "easiness" of operating with them. Thus, for example, if establishing the presence of indicative feature a requires a more refined differentiation than establishing indicative feature b, then indicative feature a is "more difficult." If operating with indicative feature c requires greater intellectual effort than operating with indicative feature d, then the former is also "more difficult," and so on.

The facility of operating with indicative features is an important parameter by which indicative features differ from each other. It is natural that an algorithm based on easy indicative features, other things being equal, is better and more efficient than an algorithm based on more difficult indicative features. The degree of difficulty or complexity of an indicative feature specifies the degree of difficulty or complexity of the corresponding step of the algorithm. It is very important to be able to evaluate each step of the algorithm from this standpoint. Without this, it is impossible to evaluate the efficiency (degree of difficulty) of the algorithm on the whole. As a first approximation, in evaluating the degree of difficulty of the step of the algorithm, one could take the mean time for verifying the corresponding indicative feature. This index is an integral expression of the various qualities of the indicative features which are linked with the ease (or, correspondingly, the difficulty or complexity) of operating with them. The mean time for verifying an indicative feature may be considered as a specific index of the degree of its difficulty. It is natural to assume that that indicative feature is easier whose establishment, other things being equal, requires less time.[3]

Here we will provisionally consider the mean time for

verifying an indicative feature as a means for the evaluation of its difficulty (or ease) and subsequently as a means for the evaluation of the complexity (difficulty) of the corresponding step of the algorithm.

If one may evaluate the difficulty (or ease) of a separate step of an algorithm by the mean time, then one may evaluate the difficulty of the algorithm as a whole by the mean time. Obviously, that algorithm is simpler and more efficient whose application requires less time on the average.

What is the relationship between the difficulty of the separate steps of an algorithm and the difficulty of the algorithm as a whole?

Obviously, the complexity of the algorithm on the whole depends not only on the complexity of the separate steps, but on their number. Other things being equal, that algorithm which, on the average, is made up of the fewest steps and requires, on the average, the verification of the fewest indicative features is the most efficient one.

But the mean number of steps in identification depends heavily on how often in the process of identification one or another step has to be carried out, i.e., which indicative features have to be verified more or less often in the object (or situation, *Ed.*). We know that indicative features which have to be verified in the object in the process of applying the algorithm can either be present or absent. Algorithms of identification are usually constructed in such a way that the choice of the next indicative feature to be verified depends on the presence or absence in the object of the preceding indicative feature. For example, an algorithm can be such that if the object (situation) has indicative feature a, then indicative feature b is verified after it; if indicative feature a is absent (i.e., there is indicative feature $\bar{a}$) then the verifying of indicative feature c follows it. Thus, in different cases, when applying one and the same algorithm, different indicative features may be verified. Since, generally, the complexity of different indicative features is unequal, in the general case the difficulty of an algorithm depends on how often it is necessary to verify the more difficult indicative features and how often the less difficult ones. But this, in turn, depends on the distribution of the probabilities of the presence in the object of different indicative features and on the sequence of their verification.

Thus, if one algorithm assumes, for example, the verification of three indicative features, while a second assumes four, this does not mean that the second one is more difficult. The operations which make up the second algorithm might be easier than the operations of the first algorithm. It could also be that some operations of the second algorithm are more difficult than the corresponding operations of the first algorithm, but the probability of their application is comparatively small, and so on.

The efficiency of an algorithm (the degree of its difficulty) is the result of the interaction of the factors shown above. It follows that the evaluation of the difficulty of an algorithm must be a function of these factors. A formula for evaluating the efficiency of an algorithm (degree of its difficulty) will be presented below, which expresses this function. Constructing an efficient algorithm of identification means finding those indicative features and their sequence which permit minimizing the function which was mentioned.[4]

Let us suppose that it is possible to identify certain phenomena on the basis of different groups of indicative features. In the general case it is true that, for each group of indicative features having a specific logical structure and constituting a basis for identification, there are several corresponding algorithms which differ only in the sequence in which the indicative features entering into the group are verified. Different sets of algorithms correspond to different groups of indicative features. In order to find the most efficient, from among all the possible algorithms, one must first find the most efficient algorithms within each set of algorithms corresponding to each of the groups of indicative features. After having compared the algorithms that could be found according to a criterion of efficiency (on the basis of a formula for evaluating the algorithm's degree of difficulty) it will be possible to select the most efficient algorithm or algorithms which are equal in efficiency, from each set of algorithms.

Thus, the problem arises of devising and choosing the most

efficient algorithms within a given set of algorithms, i.e., on the basis of the verification of the same indicative features. Solving this problem amounts to the same thing as finding the most efficient sequence of verifying indicative features under the condition that they are already *isolated* and their logical structure is *specified*. In other words, it is the problem of finding that sequence of verification for indicative features which permits, *with given* indicative features, identifying a phenomenon more quickly and easily. The present chapter will be devoted to methods of solving this problem.

The root of the question consists of the fact that algorithms which are equal to each other in a logical and psychological relation[5] might be unequal from the standpoint of the mean number of operations necessary for solving certain problems. The reason for this is the different *sequence* of applying the operations for verifying the indicative features.

At first glance, when identifying the fact that a phenomenon belongs to a specific class because of its indicative features, e.g., *a* and *b*, the sequence of verifying their presence in the object does not play a decisive role—for, in accordance with the laws of the commutativity of conjunction and disjunction, the expression *a* & *b* equals the expression *b* & *a*, and the expression *a* ∨ *b* equals *b* ∨ *a*. From a purely logical point of view, to solve the problem of the truth (or falsity) of the statements *a* & *b* or *a* ∨ *b*, the sequence of verifying the presence or absence of indicative features *a* and *b* is of no importance. It is another matter if we are interested in the fact that the verification take the least possible amount of time. Here, the probabilities of the presence or absence in the object of the indicative features which interest us and the efficient sequence for verifying them contingent on the probabilities begins to play a decisive role. As will be shown further on, the sequence of verifying indicative features has an essential significance for the purely quantitative characteristics of the flow of intellectual processes. One sequence of verification often turns out to be more economical than another and permits solving a problem of one or another type, on the average, in fewer operations.

Searching for more efficient and economical ways of processing information when solving problems is an important means for increasing the productivity of intellectual work and for perfecting methods of intellectual activity. One must aim for having students not only solve problems correctly, but also quickly, so that in a unit of time they process and master the greatest possible amount of information. This depends, in particular, on the efficiency of the algorithms which they are using.

The latter problem has great practical significance, since there are many problems of an algorithmic character, e.g., grammatical, mathematical, and many others which students have to solve by the tens, hundreds, and even thousands. If they do not master the most efficient algorithms and carry out just one or two extra operations in the process of each solution, then tens and hundreds of thousands of these extra operations accumulate in a sum total. It is easy to imagine what an enormous load that places on the shoulders of students, and how much unproductive work they have to perform, because they were not taught the most efficient—from the standpoint of the *number of operations*—algorithms.

Of course, the possibility of finding and devising efficient algorithms for solving different problems does not in the least mean, as we said, that these algorithms must always be taught. But as soon as the question of teaching students algorithms is resolved positively in a given instance, it is necessary not only to be able to devise algorithms, but also to have at hand methods for a strictly objective evaluation of their efficiency.

Up until recently, the efficiency of different ways and methods of reasoning were evaluated in education and psychology intuitively "by eye," for the most part. There were no precise criteria for evaluation. Often, the question of the efficiency of the different approaches to intellectual work was not raised at all. Now, because of the necessity for raising sharply the effectiveness of instruction, this question acquires a great importance. We must develop methods which would make possible, on the one hand, the comparison with each other of different algorithms which are already designed and the precise evaluation of their efficiency quantitatively; on the other hand, the methods would permit devising and synthesizing efficient algorithms. The methods must be strictly objective. It is desirable that they make it possible to solve the problem theoretically, i.e., putting it simply, that they permit the *calculation* of efficient algorithms and submitting them to a precise mathematical evaluation on the basis of the calculation.[6]

We spoke above about the fact that designing efficient algorithms assumes the specification of the best sequence of operations (in identifying phenomena—there are operations for verifying indicative features).

Before starting a systematic exposition of some methods for calculating the optimal sequence of operations in solving specific classes of problems, let us show the importance of such a sequence from a few simple examples.

Let us imagine that someone who is at home cannot find a book he needs. The book could be either in the bookcase (let us designate the proposition that the book is in the bookcase by the letter a), or on the desk among other books (let us designate the proposition about that by the letter b), or on the bookshelf over the desk (let us designate the proposition about that by the letter c). If finding the object in a certain place is designated by the symbol E, then the situation could be symbolically represented thus: $(x \; E \; a) \; V \; (x \; E \; b) \; V \; (x \; E \; c)$.* The problem consists of carrying out a search in the corresponding places to verify whether or not the book is there. What is the most efficient sequence by which to conduct the search, or, in other words, the verification of corresponding "locations"?

Obviously, in order to answer that question, it is necessary, first of all, to know the probability of the book's being in the bookcase, on the desk, or on the bookshelf. It is obvious, from a practical viewpoint, that if the book is most probably in the bookcase (location a), then the search must be begun there; if it is most probably on the bookshelf (location c), then at the

---

*More precisely: $((x \; E \; a) \; V \; (x \; E \; b) \; V \; (x \; E \; c)) \; \& \; \overline{(x \; E \; a)} \; \& \; \overline{(x \; E \; b)} \; \& \; \overline{(x \; E \; a)} \; \&$ $(x \; E \; c) \; \& \; (x \; E \; b) \; \& \; (x \; E \; c)$, since the book cannot be found in two places simultaneously.

bookshelf, and so on. In general, the efficient sequence of a search is such that the most probable location should be verified first and then the next less probable one.

However, this strategy is correct only in the case where the time spent looking for the book in each of the places indicated will be equal. If the time will not be equal, then the given strategy could turn out to be inefficient. In fact, let us suppose that the probability of the book being in location a is equal to 0.7, while the probability that the book is in location b is equal to 0.2. If the time for searching locations a and b is equal (e.g., three minutes), then one must, of course, begin with location a. If, however, it is known that the time for searching location a is much greater than the time for searching location b (e.g., the time for searching location a can be evaluated at eight minutes, while for b, it consists of three minutes), then it is not at all evident that one must absolutely begin the search with a. The probability that the book is at location a is greater, but also the time spent searching at a is considerably greater. It could turn out that it is considerably more efficient to begin with the less probable place and spend less time, than to begin at the more probable and spend more time. It turns out, in this case, to be difficult intuitively to find the best strategy. There is need for mathematical methods and calculations which would permit solving the problem univocally and precisely.

Working out mathematical methods of looking for the lost item in the room, of course, makes no sense whatever. However, problems analogous to this one are often encountered in industrial, scientific, and educational activities. Here the ability to find the optimal method of search is of great importance. For example, industry incurs huge losses because of idle equipment resulting from malfunctions. There can be many different reasons for a stoppage. In order to reduce equipment idleness to a minimum, it is necessary to find the malfunction as quickly as possible. The question arises: in what sequence should the search be carried out? Which order of verifying the proper functioning of the machine's components permits finding the malfunction in the shortest time? Problems such as this are being solved successfully at the present time.

In education, too, many such problems can be found. Such,

for example, is the problem of identifying an object when it is necessary to establish the presence in it of certain indicative features. In what sequence should the object's indicative features be verified? What is the best strategy for its analysis?

We have cited examples related to the analysis of situations and to methods of efficient search for objects. However, what was said also is of importance for devising efficient methods of conveying information.

Let us start once more with an example from everyday life.

A visitor addresses the cashier of a ticket office in the lobby of a hotel.

"I'm here on a business trip. Do you have tickets for some good show?"

"Yes, there are tickets for 'A Passionate Heart' at the Moscow Art Theater. The People's Artists of the USSR (i.e., well known actors of recognized stature, *Ed.*), Stanitsyn, Yanshin, and Gribov are starring. It's a wonderful show and is very successful. I recommend it highly. You won't regret it."

"That suits me fine," interrupts the visitor. "The tickets are for what date?"

"That will be Wednesday the 28th."

"But that's two weeks away," says the visitor with annoyance, "and I'm leaving in three days. What difference does it make to me who's starring in that show!"

Let us suppose that the cashier did not have any tickets for a good show at an earlier date and that he was not so stupid as to offer tickets first for those shows taking place later on. His mistake consisted of stating the specific information in an incorrect sequence.

Obviously, the probability that a ticket to a good show at the Moscow Art Theater would suit the visitor is great. The probability that a ticket to a show taking place two weeks later would suit him is small (for he is on a business trip!).

An efficient means of conveying the information (if one assumes availability of a ticket for that show) consists of beginning with those indicative features of the show which will least probably suit the purchaser, i.e., which would evoke, with the greatest probability, a refusal to buy the ticket.[7] If the cashier, in a similar situation, had begun by informing the visitor of the date, he would not, in most cases, have had to spend the time telling him who was starring in the show, how successful it was, and so on. The visitor would have said at once that a ticket for that date would not suit him. The cashier, however, began with those indicative features which would most probably suit the purchaser, and he spent time needlessly giving information which turned out to be useless.

That example might seem trivial. However, we often encounter mistakes of a similar kind.

Let us suppose that some firm in New York has need of a staff member who has the following characteristics:

1) resides in New York
2) has a higher education
3) is acquainted with the basics of electronics
4) knows the Japanese language

What is the best way to compose an advertisement, for example, in the New York *Times*? In what order should the requirements for the staff member be enumerated?

Often, as practice indicates in analogous cases, the requirements are enumerated in approximately the same order as they are cited here. However, such an order is inefficient. The probabilities of satisfying the requirements are not taken into consideration. The requirements whose probability of being fulfilled are the greatest are cited first, and then the requirements whose probability of being fulfilled are the least are cited last. Actually, the sequence of the statements should be exactly the opposite. This is easy to show by means of a simple calculation.

Let us suppose that the advertisement is put in the newspaper, and 50,000 people who are looking for work read it through. Let us suppose, further, that 49,000 of them are New Yorkers; 15,000 of these New Yorkers are familiar with the basics of electronics; 8,000 of these New Yorkers who are familiar with the basics of electronics have a higher education; and ten of them know Japanese. How much time will be spent reading that advertisement, if we assume that it takes one second to become familiar with each requirement?

Fifty thousand read through the first requirement, spending 50,000 seconds. Since the first requirement is fulfilled by 49,000 people, then 1,000 people will read no further, but 49,000 will read through the second requirement, spending another 49,000 seconds. Since 15,000 people of those 49,000 satisfy the second requirement, then 34,000 will read no further. Fifteen thousand people will read the third requirement spending another 15,000 seconds. Since 8,000 people satisfy the third requirement, the 7,000 who have no higher education will read no further, but 8,000 will read the fourth requirement spending another 8,000 seconds. The over-all amount of time spent on reading that advertisement will equal 122,000 seconds or about 34 man-hours.

How much time will reading the advertisement take, if the requirements are stated in the opposite order, beginning with those whose probability of being fulfilled is the least? The first requirement—in the former enumeration of requirements it was last—will be read by 50,000 people, who spend 50,000 seconds. Since only ten people satisfy the first requirement, 49,990 people who do not know Japanese will read no further, and no more than 10 people will read the last three requirements and spend no more than 3 seconds on them. The total time spent on reading the advertisement will, in this case, be equal to no more than 50,030 seconds or about 14 man-hours.

The sequence of presenting information has significance, obviously, not only for situations analogous to the one cited. The choice of an optimal sequence for presenting information has an enormous significance for the instructional process and particularly so for determining the optimal methods of stating rules in textbooks and for instructing students when presenting them with a different problem type in assignments, and so on.

Let us go on to a more systematic examination of problems connected with optimizing the verification of indicative features in the process of identifying them and with devising efficient algorithms for the analysis and identification of phenomena.

Since the analysis and identification of phenomena is carried out by verifying specific indicative features in them, one may interpret the process of analysis and identification as a *search* during which a person looks for specific indicative features in the phenomena.

The process of analysis may be directed at solving two problems. The first is the problem of identification proper. A person is given some object or phenomenon (e.g., a word, a sentence, a chemical substance, a plant, etc.), and he has to determine to which class, from a given number of classes, this object or phenomenon is related (e.g., which part of speech is the given word, what is the given substance, etc.).

The second problem is that of choice on the basis of identification. Here, the person is given a specific number of objects or phenomena, and he has to select a certain number of objects among them. For example, there could be the problem of selecting people out of a group who are capable of carrying heavy physical loads, or the problem of picking out all the nouns from among a number of words, and so on. Any choice can be carried out only on the basis of identification, and it represents a sequential application of given indicative features to the individual objects—which are elements within an original set.

Let us call the search in the process of solving the first problem, a search for the purpose of identification or an *identifying search*, and the search in the process of solving the second problem, a search for the purpose of choice or a *selective search*. In this book, we will examine only the issues of optimizing the identifying search.

Notes

1. Two algorithms of identification of one and the same object are different, if they are distinguished either by the indicative features underlying the identification or by the sequence of verifying these indicative features.

2. This statement of the issue does not exclude the fact that, in many cases, it is expedient to teach students different indicative features and their systems and to teach them

various algorithms for solving one and the same problem. However, since teaching several different algorithms for solving problems of a given type would take up a great deal of time and prolong the instructional period, it is often expedient to teach students just one algorithm for solving the given problems, but a highly efficient and economical one. Hence, it is important to specify a more efficient (from the standpoint of specific criteria) algorithm from among a number of possible ones.

3. The mean time for verifying an indicative feature is established experimentally. Its value is a statistic. It is possible that, with time, other indices of the difficulty of indicative features (and correspondingly, of separate steps of the algorithm of identification) will be found. In the present chapter, however, we are discussing only the question of how to devise an efficient algorithm and obtain a mathematical evaluation of its efficiency, when the evaluation of the difficulty of each of its steps as measured by the time for carrying it out has already been made.

4. Practically, it is important to consider not only the difficulty (efficiency) of an algorithm once it is mastered, but also the difficulty of the actual process of mastery. However, this is a completely different characteristic of an algorithm which requires consideration of other factors and development of other criteria. It is quite clear, that when choosing strategies of instruction, one must take into consideration not only the degree of difficulty of the algorithm after its mastery, but the very difficulty itself of the process of mastery. Theoretically, one may assume the existence of cases where a more efficient algorithm (efficient from the standpoint of the mean time spent on solving the problems with its help) is more difficult to master than a less efficient one. Which of the algorithms it is more expedient to teach depends on how much of the time and effort spent on teaching are compensated for by the economy which will be obtained *posteriori* from the application of a more efficient algorithm.

5. Two or more algorithms can be called equivalent logically

and psychologically, if they prescribe the identification of phenomena through verification of the same indicative features and the differing sequences of verification lead not only to the same results, but are equally difficult, convenient, and so on.

6.   The importance of such an evaluation for finding optimal methods of solving different kinds of problems led to the emergence of a special branch of mathematics called operations research. The wide circle of problems related to searching for optimal methods of solving practical (and especially industrial) problems of planning and organizing human activities, is studied by such mathematical disciplines as decision theory, linear and dynamic programming, the theory of games, queuing theory, and others. At the present time, operations research and adjacent mathematical disciplines are growing very fast.

7.   Here, we consider also the date when the show takes place as an indicative feature.

Chapter Twelve

Optimization of the Process of Search and Identification [1]

The purpose of optimizing the process of search and identification is to ensure the identification of objects and phenomena by a shorter and more economical way in the fewest operations (if they are of equal complexity) or the least amount of time. In this chapter, we are going to use the concept of strategy along with the concept of algorithm.[2]

The concept of strategy is used in decision theory, in the theory of games, and in some other mathematical disciplines studying the methods for specifying the optimal conduct in specific situations. In the theory of games, in particular, it is assumed that "those playing" make specific moves, where each move represents a choice of one action from among some number of possible actions (such an interpretation of the concept of a game is possible where nature acts as one of the "opponents" whose actions may also be considered as specific moves[3]: in these cases, one sometimes speaks of "games against nature" or "games with nature"[4]). Actions in the course of games are chosen depending on the situation. These actions can be different in different situations. However, often one can imagine all of the possible situations ahead of time and specify one's actions (moves) in each one of them. An exhaustive enumeration of the actions which must be performed in each of the situations which could arise in the course of the game is called strategy (on the concept of strategy, see, for example, [70], [73], [124], [145]).

In comparing the concepts of "strategy" and "algorithm," it is not difficult to see that they coincide in many ways. If one were

257

to use certain analogies with situations examined in decision theory and in the theory of games, namely with situations of the type "games with nature," it is proper to speak about the fact that a specific strategy of search lies at the basis of any algorithm of identification. Before deciding how one must act (an algorithm is always an indication of how one *must* act), it is necessary to specify how it is *possible* to act and why one or another method is more advantageous (from the standpoint of a specific criterion). In other words, at first one must take into account the *possible* means of acting (strategies), evaluate them, and then choose one of the strategies as the *prescription* for action (algorithm). An algorithm may be considered as one of the possible strategies accepted as guide (a prescription) to action.[5]

Research has shown that the optimality of one or another sequence for verifying indicative features depends on the type of problem which must be solved in the process of identification. Problems can be broken down into the following types.

First type. Object x is given, and we must specify whether it does or does not belong to the class which interests us. For example, it is necessary to specify whether the given part of a sentence is the subject. If the class we are interested in (for example, the class "subjects") is designated by the letter A, and the class "non-subject" by the letter $\bar{A}$, then the problem is described in the following manner:

Given: object x. It is necessary to identify: $x \in A$ or $x \in \bar{A}$?

Second type. Object x is given, and we must specify to what class from a finite number of classes it belongs. For example, is the part of the sentence x, a subject, predicate, object, adverb, or attribute? If the classes are designated by capital letters, then the problem is described thus:

Given: object x. It is necessary to identify: $x \in A$, or $x \in B$, or $x \in C$, $x \in D$, or $x \in E$?

When solving problems of the first type, one must choose one answer out of two, where one is a negation of the other. In solving a problem of the second type, one must choose one answer from among some number of answers greater than two. It is expedient to call problems of the first type *problems of alternative choice* and those of the second type, *problems of multiple choice.*[6]

Since the difference of these types of problems has an important significance for the optimal strategy of identification, these problems must be examined separately. Let us begin with problems of alternative choice.

1. Identification Under Alternative Choice

It was shown in Chapter Eight that the indicative features of objects and phenomena may have different logical structures. They may be joined by the logical disjunction "or" in its different meanings—we will call such a connective which is independent from the meaning of "or," disjunctive. They can also be joined conjunctively (by the connective *and*), or have a mixed structure (be joined disjunctively-conjunctively or conjunctively-disjunctively).

If the indicative features are joined disjunctively* (for example, the situation $a(x) \lor b(x) \lor c(x) \Leftrightarrow x \in A$ exists), then in order to draw a positive conclusion about whether the object belongs to class A it suffices to establish the presence of just one of the indicative features, a, b, or c.

If the indicative features are joined conjunctively (for example, the situation $a(x) \& b(x) \& c(x) \Leftrightarrow x \in A$ exists), then in order to draw a positive conclusion as to whether the object belongs to class A it is necessary to establish the presence of all three indicative features, a, b, and c.

If the indicative features are disjunctive-conjunctive or conjunctive-disjunctive, then they must be broken down into groups, and one must act in accordance with the meaning of those conjunctions which first join the groups of indicative features, and then the indicative features within the groups.

It will be evident from further exposition that there is a principal difference in the strategy of the process of search and identification, depending on how the indicative features are

---

*Here, as earlier, if the contrary is not indicated by a disjunctive link, we will mean a link with the aid of the logical connective *or*, corresponding to a weak disjunction.

joined. There are specific functional dependencies between ways in which indicative features are joined and the strategy of identification. We are going to examine them now.

We will call the verification (identification) of indicative features which are disjunctively joined—*disjunctive search and identification* and the verification (identification) of indicative features which are conjunctively joined *conjunctive search and identification*. We will call the remaining types, correspondingly, *disjunctive-conjunctive* and *conjunctive-disjunctive search and identification.*

Optimal Strategy of Disjunctive
Search and Identification

A certain situation is always given with disjunctive search and identification in which one must find (verify the presence of) one of the disjunctively joined objects $a \lor b \lor c \lor \ldots \lor n$.*

These objects could be indicative features whose presence or absence must be established in a certain object. They could also be expressions about the presence of certain objects or about their "location" which are the objects of identification, and so on. Conclusions about the presence or absence of the objects sought can be different in different cases. However, this does not influence the strategy of the search.

Thus, if certain "locations" must be explored (for example, with the purpose of finding out whether the object or fault sought is found there, and so on), the conclusion of the exploration is establishment of the whereabouts of the thing sought after. If it is necessary to explore a certain object (for example, for the purpose of finding out whether it has specific indicative features), then the conclusion deriving from the exploration is the attribution of that object to a specific class. From this conclusion, in turn, other conclusions can emerge, including practical ones.

When verifying certain indicative features in an object, there can be a dependence of the type $a(x) \lor b(x) \lor c(x) \Rightarrow x \in A$. In this formula, the arrow is directed to one side. This means that the

---

*We will not consider here or later on instances where it is not known in advance what objects must be found (this often happens, for example, in scientific research).

presence in the object x of (at least) one of the indicative features a, b, c confers the right to draw a conclusion that the given object belongs to class A. However, the absence of all of the indicative features does not confer the right to draw a conclusion that the given object belongs to class $\overline{A}$.

When verifying certain disjunctively joined indicative features in the object, another dependence can take place of the type, $a(x)$ $\vee$ $b(x)$ $\vee$ $c(x)$ $\Leftrightarrow$ x ϵ A (the biconditional implication "if and only if," Ed.). Here, the arrows are directed to both sides. In this case, the absence of all indicative features does confer the right to draw a conclusion that the given object belongs to class $\overline{A}$.

Indicative features which must be verified in the object during the course of search and identification could be incompatible in pairs, or compatible.[7] Thus, in some situations, the object can have any indicative feature from among some aggregate, but always only one. Here, the indicative features are incompatible. They exclude one another. The sum of the probabilities of these indicative features cannot be greater than one. In other situations, the object can have either one indicative feature from the given aggregate or several at once. Here, the indicative features are compatible. They do not exclude one another and are joined by the conjunction "or" in a non-strictly divisive sense. The sum of the probabilities of such indicative features may be greater than one.

Here, and further on, we will not introduce any limitations on the interrelations of indicative features. We will be examining precisely those cases where disjunctively joined indicative features may be compatible (i.e., the sum of their probabilities may be greater than one). Moreover, it is not necessary that the object have one of the indicative features without fail.

What is the optimal strategy of search and identification in the case where the indicative features are disjunctively joined?

For simplicity, let us take the case where it suffices to establish one of the indicative features a or b or c in the object in order to attribute it to class A. If the object has none of the indicative features, it belongs to class $\overline{A}$. Formally, the situation may be described thus:

$$a(x) \lor b(x) \lor c(x) \Leftrightarrow x \in A.$$

The following could be an example of this type of situation. Suppose we face the familiar problem of the correct case in using the personal pronoun (I or me, he or him). To solve this problem correctly, we must verify the presence in the sentence of specific indicative features. When the pronoun is the subject of a verb *or* when it is the predicate noun the *nominative* case is correct (I, he); when the pronoun is the object of a verb *or* the object of a preposition *or* the object of a gerund, infinitive, or participle, the *objective* case (me, him) is correct.

It is easy to see that in other grammatical situations there may be just one indicative feature, or several indicative features simultaneously. Some, all, or none may be present. If perchance the several features present are disjunctively joined, these features are compatible and the sum of their separate probabilities may be greater than one.

The optimal strategy of search and identification[8] in these cases depends on the correlation of two factors: on the one hand, on the complexity of the indicative features (we agree to express it through the mean time of their verification); on the other hand, on their probabilities.[9]

Here, several possibilities exist.

First case. The probabilities of all indicative features are equal, and the time necessary for verifying them is identical.[10] In this case, the sequence for verifying the indicative features makes no difference.

Second case. The probabilities of the indicative features are identical, but the time necessary for verifying them is unequal. In this case, one must begin with the indicative features whose verification requires less time. Such a strategy increases the chance of discovering some of the sufficient indicative features more rapidly.

In order to express the optimal strategy of identification in a more general form, we will use the following notation:

We will designate the probabilities of indicative features a, b, c by $p(a)$, $p(b)$, $p(c)$;

the time necessary for verifying indicative features a, b, c, we will correspondingly designate by $t(a)$, $t(b)$, $t(c)$;

we will designate the operator for verifying indicative features by the letter T;

the inscription T(a, b, c) will signify that indicative features a, b, c are verified in this order: first a, then b, then c.

One may write the optimal strategy for verifying indicative features in the case where the probabilities of the indicative features are identical, but the time is unequal thus:

$$((p(a) = p(b) = p(c)) \ \& \ (t(a) \leqslant t(b) \leqslant t(c))) \Rightarrow T \ (a, b, c).^* \quad (1)$$

This formula is read in the following manner: if the probabilities of indicative features a, b, and c are identical, but the time necessary for verifying indicative feature a is less (or the same) as the time needed to verify indicative feature b, and the time needed to verify b is less (or the same) as the time for verifying indicative feature c, then the optimal strategy is thus: first, one should verify indicative feature a, then indicative feature b, then indicative feature c.

Third case. The probabilities of the indicative features in the general case, are different, but the time necessary for their verification is identical. In this instance, one must begin the verification with the most probable indicative features. The optimal strategy in the case where all the probabilities are not equal is:

---

*In this—and in analogous—formulae in this chapter, we will apply non-strict inequality. This is done for the purpose of generality, in order to include, as well, those cases where some of the values linked by the signs $\leqslant$ or $\geqslant$ are identical. Also, here and later on, we will introduce formulae for three indicative features, but they are all easy to generalize for the case of n indicative features.

$$((p(a) \geqslant p(b) \geqslant p(c)) \ \& \ (t(a) = t(b) = t(c))) \Rightarrow \mathrm{T}(a, b, c). \quad (2)$$

Fourth case. Both the probabilities of the indicative features and the time needed to verify them are, in the general case, different. In this case, one must specify the ratio of each indicative feature to the time it takes for verifying it, and one must begin with verifying those indicative features for which this ratio is greater. The optimal strategy of search and identification may be described in the following manner:

$$(\frac{p(a)}{t(a)} \geqslant \frac{p(b)}{t(b)} \geqslant \frac{p(c)}{t(c)}) \ \Rightarrow \ \mathrm{T} \ (a, b, c).* \quad (3)$$

In dividing the probability of an indicative feature by the time needed to verify it, we find out what "fraction of the probability" coincides with the unit of time for search and verification. Thus, we can relate this case to the preceding one: the time for verifying indicative features turns out to be the same, but the probabilities are different. But when the probabilities of the indicative features (with equal time) are different, then the verification of indicative features according to formula (2) must be begun with the more probable ones.

It is easy to see that strategy (2) is the particular case of strategy (3) and can be deduced from it.

The question arises as to how to calculate the mean time spent on identifying a phenomenon when its indicative features are disjunctively joined. What, in other words, is the mean time for a disjunctive search and identification? This mean time, as pointed out, is a quantitative evaluation of the quality of the strategy.

Let us return to the personal pronoun situation examined earlier in this chapter and described by the logical expression: $a(x)$

---

*The optimality of the strategy indicated may be proved with the help of standard arguments in the theory of probabilities. However, in view of the special character of such proofs, we will not cite them here and further on in analogous situations.

$\vee\ b(x)\ \vee\ c(x) \Leftrightarrow x \in A$. If it is necessary to emphasize especially that in the case where all indicative features are lacking in the object, it belongs to class $\overline{A}$, then it may be expressed in this way:

$$\overline{a(x)}\ \&\ \overline{b(x)}\ \&\ \overline{c(x)} \Leftrightarrow x \in \overline{A}.$$

This statement follows from the formula just cited (and is equivalent to it).

Let us suppose that as a result of exploring a large number of objects x, it is established that indicative features a, b, c are encountered in them in the following frequencies (probabilities)[11] : $p(a) = 0.65$, $p(b) = 0.55$, $p(c) = 0.45$; $p(a) = 0.65$ signifies that when encountering an object of a given kind, we have the chance of discovering indicative feature a 65 times out of 100; $p(b) = 0.55$ signifies that in encountering an object of that kind, we have the chance of discovering indicative feature b 55 times out of 100, and so on.

Let us now examine what the mean time for identifying x is, if the sequence of verification is a, b, c, and the time of verification $t(a) = 2$ units of time, $t(b) = 3$ units of time, $t(c) = 1$ unit of time.[12]

The process of identification according to the strategy S_v (a, b, c) consists of the following.* At first, indicative feature a is verified. For verifying a, the time $t(a)$ is spent. Since the probability of this indicative feature $p(a) = 0.65$, then in 65 cases out of 100, the object will be identified within the time $t(a)$.** For identifying objects according to feature a, we will spend,

---

*Henceforth, for brevity, we will designate the sequence of verifying indicative features (i.e., the strategy of searching) by the letter S with indices suggesting the character of the strategy of identification. Thus, the strategy $T(a,\ b,\ c)$ with disjunctive identification will be designated by S_v (a, b, c). Strategy T_v (b, c, a) will be designated by S_v (b, c, a), and so on.

**Establishing an indicative feature is a random process. When calculating the mean time for verifying each indicative feature (more precisely, the operations for establishing it) two values are ascribed: its probability and the mean time of its establishment (verification).

therefore, in 100 tests, $65 \cdot t(a)$ units of time = $65 \cdot 2 = 130$ time units. If indicative feature a is discovered in the given object, the verification for this object will be finished.

If indicative feature a will not be in the given object, and this occurs in 35 cases out of 100, then we will verify indicative feature b in it. Indicative feature b is encountered in 55 percent of the cases out of the remaining 35 cases, i.e., $\frac{35 \cdot 55}{100} = 19.25$ cases. For identifying objects according to indicative feature b, one must spend, in addition to the time needed for verifying that indicative feature, still more time for verifying indicative feature a, i.e., the time $t(a) + t(b) = 2$ units of time + 3 units of time = 5 units of time. To identify objects according to indicative feature b in 100 tests, one must spend, therefore, $19.25 \cdot 5 = 96.25$ units of time.

If indicative feature b is discovered in the object examined, then the verification ends with this, for the object.

If indicative feature b is not discovered, and this happens in the remaining $100 - (65 + 19.25) = 15.75$ cases, then we will verify remaining indicative feature c. For identifying objects according to indicative feature c, we must spend, in addition to the time necessary for verifying indicative feature c, also the time for verifying indicative features a and b, i.e., the time $t(a) + t(b) + t(c)$ = 2 units of time + 3 units of time + 1 unit of time = 6 units of time. For identifying an object according to indicative feature c, in 100 tests, we must spend, therefore, $15.75 \cdot 6 = 94.5$ units of time.

Since indicative feature c is the last, then independently of whether we discover it in the object or not, the verification will end with it.

Therefore, the time which is necessary for identifying objects in 100 tests equals:

$$65 \cdot t(a) + \frac{35 \cdot 55}{100}(t(a) + t(b)) + [100 - (65 + \frac{35 \cdot 55}{100})] \cdot (t(a) +$$

$$t(b) + t(c)) = 65 \cdot 2 + 19.25 \cdot 5 + 15.75 \cdot 6 = 320.75. \quad (\dagger)$$

Now we may calculate the mean time τ which must be spent on identifying objects during one test. In order to do that, it is necessary to divide the number obtained by 100. The result we get is that the mean time for identification in our example equals 3.2075 time units. Let us analyze the result. τ is obtained if each of the items of the expression (†) is divided by 100. We have:

$$\tau = \frac{65}{100} t(a) + \frac{35}{100} \cdot \frac{55}{100} (t(a) + t(b)) + [\frac{100}{100} - (\frac{65}{100} + \frac{35}{100} \cdot \frac{55}{100})]$$
$$\cdot (t(a) + t(b) + t(c)).$$

It is easy to see that, in each item, there is a number which represents either the probability of separate indicative features (for example, $0.65 = p(a)$), or the difference between a probability for an indicative feature and unity (for example, $0.35 = 1 - p(a)$).

In a general form, the mean time τ for identifying objects according to our three indicative features can be written thus:

$$\tau = p(a) \cdot t(a) + (1-p(a)) \cdot p(b) \cdot (t(a) + t(b)) + [1 - (p(a) +$$
$$(1 - p(a)) \cdot p(b))] \cdot (t(a) + t(b) + t(c)). \tag{4}$$

In order to shorten the inscription, let us now replace the designations $p(a)$, $p(b)$, ..., $t(a)$, $t(b)$ in our formulae by the designations p_1, p_2, ..., p_n, t_1, t_2, ..., t_n. We will designate $(1-p_i)$ by q_i ($i = 1, 2, ... n$). Then for n indicative features, the formula will be:

$$\tau = p_1 t_1 + q_1 p_2 (t_1 + t_2) + q_1 q_2 p_3 (t_1 + t_2 + t_3) + ... +$$
$$q_1 q_2 ... q_{n-2} p_{n-1} (t_1 + t_2 + ... + t_{n-1}) +$$
$$[1 - (p_1 + q_1 p_2 + ... q_1 q_2 ... q_{n-2} p_{n-1})] (t_1 + t_2 + ... + t_n). \tag{5}$$

This will be the formula of the mean time for a disjunctive search and identification or the evaluation of the difficulty (ease) of the given strategy of search and identification. It is easy to see that this formula evaluates the difficulty (mean time) of the given strategy of search and identification as a function of the difficulty of the separate operations for verifying the indicative features,[13] their quantity, sequence, and also their probabilities. It represents a mathematical expectation of the time that has to be spent on identifying objects according to the given strategy. The formulae of mean time which will be cited below will represent the functions from these same arguments.

The meaning of formula (5) can be explained somewhat differently.

The time t_1 (and only this time) is that which is required in the case where the first indicative feature of the object is present in the object (with the presence of the first indicative feature, it is unnecessary to verify the others). The probability of spending only t_1 time is equal to p_1. The component of the mean time which it takes to verify the first indicative feature, when it is present in the object, is equal to $p_1 t_1$.

The time $t_1 + t_2$ (and only this time) is spent in the case where the first indicative feature (of the object) is lacking and the second one is present. The probability of the event "the first indicative feature is missing and the second one is present" is equal to $q_1 p_2$. The component of the mean time which is spent in verifying the second indicative feature when it is present in the object is equal to $q_1 p_2 (t_1 + t_2)$.

The time $t_1 + t_2 + t_3$ is spent in the case where the first and second indicative features are lacking, but the third is there. The probability of the event "the first and second indicative features are missing, but the third is present" is equal to $q_1 q_2 p_3$. The component of the mean time which verification of the third indicative feature introduces when it is present, is equal to $q_1 q_2 p_3 (t_1 + t_2 + t_3)$.

The time $t_1 + t_2 + t_3 + \ldots + t_n$ is spent in the case where the object or situation has an nth indicative feature and all those up to and including n-1 are missing, or the object has none of the indicative features verified and does not belong to class A. Here, the probability of the last indicative feature has no significance. The probability of an event which we are discussing is equal to unity minus the sum of the probabilities of all the preceding events, i.e., is equal to $1 - (p_1 + q_1 p_2 + \ldots + q_1 q_2 \ldots q_{n-1} p_{n-1})$. The component of the mean time spent verifying the last indicative feature equals $[1 - (p_1 + q_1 p_2 + \ldots + q_1 q_2 \ldots q_{n-2} p_{n-1})]$ $(t_1 + t_2 + t_3 + \ldots + t_n)$. The sum of the components of the mean time spent on identifying the object represents the time spent on identifying the object according to the given strategy. This mean time is expressed by formula (5).[14]

We examined the case where the indicative features of objects are compatible or can be incompatible. If the indicative features are incompatible, the formula becomes simpler:

$$\tau = p_1 t_1 + p_2 (t_1 + t_2) + \ldots + p_{n-1} (t_1 + t_2 + \ldots + t_{n-1}) + [1 - (p_1 + p_2 + \ldots + p_{n-1})] (t_1 + t_2 + \ldots t_n). \qquad (6)$$

The last formula is obtained from the preceding one in the following manner.

Let us examine the second component of formula (5) for an example. When the indicative features are incompatible, then the probability is zero that in verifying the second indicative feature in the object, the first one is present. Consequently, $q_1 = 1 - 0 = 1$. The second component of formula (5), therefore, takes the form $1 \cdot p_2 (t_1 + t_2) = p_2 (t_1 + t_2)$. What has been said is also correct in reference to all the other components.

Optimal Strategy of
Conjunctive Search and Identification

As in the case of disjunctive search and identification, some situation is always given in conjunctive identification in which something must be found, and some system of objects (for example, indicative features) which must be found. However, these indicative features are not joined disjunctively, but conjunctively: a_1 & a_2 & . . . & a_n.

The question arises: in what sequence is it expedient to carry out verification of indicative features when they are conjunctively joined?

As in the preceding instance, when examining the question of an optimal method of search and identification, we enlist the means of logic.

Let us suppose that in order to attribute objects of a certain kind to a certain class, they must possess indicative features a, b, and c.

If the object has all of these indicative features, then it belongs to class A. If just one of the indicative features is lacking, then it does not belong to class A, i.e., it belongs to class $\overline{A}$. In the language of predicate logic, this can be described in the following manner: $a(x)$ & $b(x)$ & $c(x) \Leftrightarrow x \in A$. If it is necessary to emphasize especially that in the case of the absence in that object of one of the indicative features, it belongs to class $\overline{A}$, then this can be expressed by: $a(x) \vee b(x) \vee c(x) = x \in \overline{A}$. This expression derives from the formula just cited (or is equal to this formula).

The general number of possible strategies in conjunctive search and identification is equal to $P_n = n$. In the given case, this is $P_3 = 3 \cdot 2 \cdot 1 = 6$.

As with disjunctive search and identification, the strategy of conjunctive search and identification depends on correlation between two factors: the probability of the indicative features and the time needed for their verification. Here, also, several cases can take place.

First case. The probabilities of the indicative features and the time needed to verify them are identical. In this instance, the sequence of verifying the indicative features is unimportant.

Second case. The probabilities of the indicative features are identical, but the time needed for their verification is unequal. Under these conditions, as in the case of disjunctive identification, we must begin with the indicative features whose verification requires the least time. Such a strategy increases the chance of discovering more quickly the absence of some one of the indicative features.

One may describe the optimal strategy of search and identification for the second case in the following manner:

$$((p(a) = p(b) = p(c)) \,\&\, (t(a) \leqslant t(b) \leqslant t(c))) \Rightarrow T(a, b, c).^* \quad (7)$$

Third case. The probabilities of the indicative features are different, but the time needed to verify them is identical. In this case—as distinct from disjunctive search and identification—verifi-

*See the footnote on page 263.

cation must be started with the least probable indicative features in the course of verification and pass from the less probable to the more probable indicative features during the course of verification.

With conjunctive search and identification, the verification of each subsequent indicative feature takes place only in the instance where the preceding one is present. If the preceding indicative feature is absent, its successor is not verified and a conclusion is drawn that the object belongs to class $\overline{A}$. Therefore, the more probable it is that the preceding indicative feature is missing the fewer are the mean number of indicative features needing to be verified in order to reach a conclusion that the object belongs to class $\overline{A}$. A verification during the course of which one passes from the less probable to more probable indicative features normally will lead to an earlier completion of the procedure.

The optimal strategy of search and identification for the third instance may be written thus:

$$((p(a) \leqslant p(b) \leqslant p(c)) \,\&\, (t(a) = t(b) = t(c))) \Rightarrow \mathrm{T}\,(a, b, c). \quad (8)$$

Fourth case. Both the probabilities of the indicative features as well as the time needed for verifying them are different. In this case, one must specify the ratio of the probability of each indicative feature to the time of its verification and begin identifying the object with those indicative features where this ratio is the least. The optimal strategy of identification in the fourth case is:

$$\left(\frac{1 - p(a)}{t(a)} \ \geqslant\ \frac{1 - p(b)}{t(b)} \ \geqslant\ \frac{1 - p(c)}{t(c)} \right) \Rightarrow\ \ \mathrm{T}(a, b, c). \quad (9)$$

In dividing the probability of the indicative feature by the time needed for its verification, we find out what "probability component" coincides with a unit of time for search and verification. In this way, we can relate this case to the preceding one. The time for verifying the indicative features turns out to be identical, but the probabilities are different. If the probabilities of the indicative features (with identical time) are different, then one must begin verification of indicative features with the least probable, according to formula (8).

What about the mean time of conjunctive search and identification? Let us first deduce a formula for the special case of three indicative features in order to generalize it later to the case of n indicative features. We will consider that indicative feature a is encountered with the probability $p(a)$, indicative feature b with the probability $p(b)$, and indicative feature c with $p(c)$. The time which must be spent on verifying each indicative feature is equal, correspondingly, to $t(a)$, $t(b)$, $t(c)$. We shall consider that we carry out identification according to strategy $S_{\&}$ (a, b, c).

First, indicative feature a is verified in object x. The time spent for verifying indicative feature a is $t(a)$. If indicative feature a is missing in the object (and this occurs with the probability $1 - p(a)$), the conclusion is drawn that x belongs to class $\overline{A}$, and the verification ends there; the time $t(a)$ is spent in this case with the probability $1 - p(a)$.

If the first indicative feature is present in the object (the probability of this is equal to $p(a)$), then the presence of the second indicative feature is verified in the object. The time $t(a) + t(b)$ is spent verifying the second indicative feature. In which case does the process of identification stop with the verification of the second indicative feature? This happens in the case where the first indicative feature is in the object and the second is not. The probability of the event "the first indicative feature is in the object, but the second is not" is equal to $p(a) \cdot (1 - p(b))$.* Therefore, the time, $t(a) + t(b)$ is spent with the probability $p(a) \cdot (1 - p(b))$.

If the objects have both the first and the second indicative features, verification of the third indicative feature takes place. On

---

*Here, and henceforth, in analogous instances, we assume that all of the indicative features considered are independent. In a general case, instead of absolute probabilities, one must consider the corresponding probabilities conditional. (For concepts which are relevant here see, e.g., [72], [89], [92], [152].)

such a verification, the time, $t(a) + t(b) + t(c)$ is spent (the time $t(a) + t(b)$ is spent in order to ascertain the presence of the first two indicative features, and the time $t(c)$ is spent on verifying the third indicative feature). Since the third indicative feature is the last, it is verified with the probability of unity (i.e., 1) minus the sum of the probabilities of all the preceding indicative features (events). Therefore, the time $t(a) + t(b) + t(c)$ is spent with the probability $1 - ((1 - p(a)) + p(a) \cdot (1 - p(b))$.

Hence, the mean time of conjunctive search and identification according to strategy $S_\&(a, b, c)$ equals:

$$\tau = (1 - p(a)) \cdot t(a) + p(a) \cdot (1 - p(b)) \cdot (t(a) + t(b)) +$$
$$[1 - ((1 - p(a)) + p(a) \cdot (1 - p(b)))] \cdot (t(a) + t(b) + t(c)). \quad (10)$$

Now, using the designations cited earlier in this chapter, we can rewrite formula (10) in the following way:

$$\tau = q_1 t_1 + p_1 q_2 (t_1 + t_2) + [1 - (q_1 + p_1 q_2)] (t_1 + t_2 + t_3). \quad (11)$$

In a general form for n indicative features, we obtain:

$$\tau = q_1 t_1 + p_1 q_2 (t_1 + t_2) + \ldots + (p_1 p_2 \ldots p_{n-2} q_{n-1}) \cdot$$
$$(t_1 + t_2 + \ldots + t_{n-1}) + [1 - (q_1 + p_1 q_2 + \ldots + p_1 p_2 \ldots p_{n-2}$$
$$q_{n-1})] \cdot (t_1 + t_2 + \ldots + t_n). \quad (12)$$

In comparing formula (5) and formula (12), it is easy to see that in transferring from one formula to another, q is substituted for p, and vice versa, so that formula (12) can be immediately deduced from formula (5).[15]

Optimal Strategy of Disjunctive-Conjunctive
Search and Identification

We examined instances when the indicative features are joined either disjunctively or conjunctively. However, the structure of indicative features can be disjunctive-conjunctive, i.e., represent a disjunction of conjunctions, and conjunctive-disjunctive, i.e., represent a conjunction of disjunctions. In which sequence is it expedient to verify indicative features in the instance where they represent a disjunctive conjunction?

Let us suppose that object x belongs to class A if and only if it possesses the indicative features: a, or (b and c), or (d and e). If the object has just one of the sufficient indicative features (in the last two instances the sufficient indicative features in most cases are compound, made up of a conjunction of two indicative features), then it belongs to class A. If none of the sufficient indicative features is present, then it does not belong to class A, but to class $\overline{\text{A}}$. In the language of predicate logic, it may be described thus:

$$a(x) \lor (b(x) \ \& \ c(x)) \lor (d(x) \ \& \ e(x)) \Leftrightarrow x \in \text{A}.$$

If it must be specially emphasized that in the case where all of the sufficient indicative features are lacking in the object, it belongs to class $\overline{\text{A}}$, it may be expressed thus:

$$\overline{a}\,(x) \ \& \ \overline{(b(x) \lor c(x))} \ \& \ \overline{(d(x) \lor e(x))} \Leftrightarrow x \in \overline{\text{A}}.$$

A calculation of the number of possible strategies with a disjunctive-conjunctive structure proceeds as follows.

If each disjunctive member were to consist of just one letter, then the number of strategies would be equal to $P_3 = 3!,$* but the second member consists of two letters which yields $P_2 = 2!$ permutations. The third member also consists of two letters which give $P_2 = 2!$ permutations. The number corresponding to the total number of strategies is equal to $P_3 \cdot P_2 \cdot P_2 = 3! \cdot 2! \cdot 2!$ In order to write this down uniformly, one may also include in the formula that number of permutations contributed by the disjunctive member consisting of one letter. It is equal to $P_1 = 1! = 1$. Then the number of strategies for identifying phenomena according to disjunctively joined indicative features $a \lor (b \ \& \ c) \lor (d \ \& \ e)$ may be written: $P_3 \cdot P_1 \cdot P_2 \cdot P_2 = 3! \cdot 1! \cdot 2! \cdot 2! = 24$.

We will deduce a general formula for n indicative features. We will consider that there are n disjunctive members. We will

---

*The factorial of 3, i.e., the product of all positive integers from one to the given number: $3! = 3 \cdot 2 \cdot 1 = 6$. (*Editor.*)

designate the number of indicative features which make up the first disjunctive member by the letter K_1, the number of indicative features which make up the second indicative feature by the letter K_2, and the number of indicative features which make up the kth disjunctive member by the letter K_m. Then the number of possible strategies will be equal to:

$$n! \cdot K_1! \cdot K_2! \ldots \cdot K_m! = n! \prod_{i=1}^{m} K_i!* \qquad (13)$$

The method of specifying the optimal strategy with disjunctive-conjunctive search and identification may be described in the following manner:

1) According to the formula for the optimal strategy of a conjunctive search and identification (for the case of three indicative features, this was formula (9)), the strategy of verifying indicative features within each compound disjunctive member which represents a conjunction of indicative features is specified.

2) According to the formula for the mean time of a conjunctive search and identification (12), the mean time for verifying each compound disjunctive member which is composed of conjunctions of indicative features is specified.

3) According to the probabilities of the conjunctively joined indicative features, the probability of each disjunctive member is specified.[16] The calculations yield probabilities and mean times for verifying each disjunctive member (the sequence for verifying the indicative features within each disjunctive member is the best one chosen). Now each compound disjunctive member may be considered as a simple one.

4) According to the formula of the best strategy for disjunctive search and identification (for the case of three indicative features, this was formula (3)), the optimal sequence for verifying disjunctive members is specified.

The strategy found will be the optimal one. The mean time

---

*The formulae cited above of the number of strategies with "pure" conjunctive and "pure" disjunctive search and identification are particular cases of this formula and can be deduced from it.

for identification according to this strategy may be specified according to formula (5).

Optimal Strategy of Conjunctive-Disjunctive
Search and Identification

In the case examined in the preceding segment, the indicative features were joined disjunctive-conjunctively, i.e., they represent a disjunction of conjunctions. However, they might also be joined conjunctive-disjunctively, i.e., represent a conjunction of disjunctions. In what sequence is it expedient to carry out verification of indicative features in this case?

Let us suppose that object x belongs to class A if and only if it possesses indicative features a and (b or c) and (d or e). In the language of the predicate logic, this may be described in this way:

$$a(x) \ \& \ (b(x) \lor c(x)) \ \& \ (d(x) \lor e(x)) \Leftrightarrow x \in A.$$

If it is necessary to emphasize especially that in the case where one of the necessary indicative features is missing, it belongs to class $\overline{A}$, it may be expressed thus:

$$a(x) \lor (b(x) \ \& \ c(x)) \lor (d(x) \ \& \ e(x)) \Leftrightarrow x \in \overline{A}.$$

The number of strategies with a conjunctive-disjunctive search and identification is calculated according to the same formula (13) as the number of strategies with a disjunctive-conjunctive search and identification.

The method of specifying the optimal strategy with a conjunctive-disjunctive search and identification is the following:

1) According to the formula for the best strategy for disjunctive search and identification [for the case of three indicative features, this was formula (3)], the best strategy for verifying indicative features within each compound conjunctive member which represents a disjunction of indicative features is specified.

2) According to formula (5) for the mean time of a

disjunctive search and identification, the mean time for verification of each compound conjunctive member composed of a disjunction of indicative features is specified.

3) On the basis of the probabilities of the disjunctively joined indicative features, the probability of each conjunctive member is specified.[17]

4) According to the formula for the best strategy of conjunctive search and identification [in the case of three indicative features, this was formula (9)] the optimal sequence for verifying conjunctive members is specified.

The strategy found will be the optimal one. The mean time for search and identification according to this strategy can be specified according to formula (12).

The procedures for calculating the optimal strategy of search and identification and the formula for evaluating the mean time of search which were shown in the earlier segment of this chapter and which make possible the evaluation (by the given criterion) of the optimality of any strategy chosen have a general character. These methods and formulae are applicable to the most varied cases of search under conditions of alternative choice. However, in order that such an application, in particular, to the analysis of the process of search, become practically possible, it is necessary to amass statistical data in psychology and pedagogy which pertain to the frequency of the specific indicative features and the mean time of their verification. Since such statistical investigations of educational material and of the processes of its assimilation have not yet been carried out, it is possible to apply mathematical methods only by making certain assumptions. An example of calculating the optimal strategy of identification (an optimal algorithm) with assumptions will be outlined below.

2. Search in Multiple Choice

Up to this point, we have examined methods for specifying the optimal strategy of search and identification under conditions of alternative choice. Let us now examine methods of specifying the optimal strategy of identification in conditions of multiple choice.[18]

If it is necessary to specify in problems of alternative choice whether a given object x enters into a class A which interests us (i.e., whether it enters into A or $\overline{A}$), then in problems of multiple choice, the problem consists of deciding into which class the object enters from among a certain finite number of classes which interest us.

First of all, it must be said that problems of multiple choice can be reduced to problems of alternative choice. Let us suppose that a certain object x is given, about which it is known that it belongs to one of the classes, A, B, C and that it must be specified to which of them it belongs. Such a problem may be solved in the following manner. At first, one may specify whether the given object belongs to class A or $\overline{A}$, where it is understood that class $\overline{A}$ is a combination of classes B and C. If it happens that the given object belongs to class $\overline{A}$, then it must be specified whether it belongs to B or $\overline{B}$ where $\overline{B}$ means class C. Such a method, however, is inefficient in many cases. It is much more efficient to verify the object for the presence of indicative features not for every class separately, but for indicative features common to the different classes.

Up until now, we have used the concepts of the theory of probability for specifying the strategy of optimal search and identification in conditions of alternative choice. For specifying the strategy of optimal search and identification under conditions of multiple choice, however, the concepts and methods of information theory are of major significance.

Some Concepts of Information Theory[19]
Since, in the course of subsequent statements, we will have to produce separate calculations and use a number of concepts from information theory, we will briefly characterize these concepts.

One of the basic notions with which the theory of information is linked is that of *uncertainty*. When some phenomenon or process has several possible outcomes, and it is not known which of them will occur, there is uncertainty. Thus, when tossing a coin, we do not know whether heads or tails will come up; when an average student is given a surprise quiz, we do not know whether his grade will be A or B or C or even D; if there are eight balls of different colors in an urn, and we draw one of them at random, we do not know which ball we will pick out: red, white, black, or some other color. If one considers that all the results in each of the situations are equally probable, then these events and their probabilities may be portrayed in the following manner:

Outcomes	Heads	Tails
Probabilities	1/2	1/2

Outcomes	A	B	C	D
Probabilities	1/4	1/4	1/4	1/4

Outcomes	Red	Yellow	Green	Blue	Violet	White	Gray	Black
Probabilities	1/8	1/8	1/8	1/8	1/8	1/8	1/8	1/8

It is evident from the diagrams that in the cases where the number of possible outcomes is greater, the probability of each separate outcome is less. Thus, if the situation has 4 possible outcomes, then each probability is equal to 1/4, if it has 8, then 1/8, and so on. The sum of the probabilities of all outcomes with a complete system of events is equal to 1.

The greater the number of possible outcomes that a situation has, the more uncertain it is, in the sense that there are fewer chances of correctly foretelling which of the outcomes will in fact occur. The uncertainty could, therefore, be measured by the number of possible outcomes (under the condition that they are equally probable). However, in information theory, the uncertainty is measured not by the number of outcomes, but by the logarithm (to the base 2, as a rule) of that number.* Thus, the uncertainty of the situation with the tossed coin is not considered equal to 2, but the $\log_2$ of 2; the uncertainty of the grade which the student will get is considered to have that of $\log_2$ 4; the uncertainty of the situation with drawing balls amounts to $\log_2$ 8. Generally, the uncertainty of situations where there are n equally probable outcomes is equal to $\log_2 n$.**

From this link which exists between the number of possible outcomes and the probability of the separate results, it is clear that the uncertainty of a situation can be expressed not only by the logarithm of the number of outcomes, but also by the logarithm of the probability of any of the separate results. Thus, if some situation has 8 equally probable outcomes, then its uncertainty is equal to $\log_2$ 8. But $\log_2 8 = -\log_2 1/8$; and 1/8 is the probability (p) of any one of the outcomes. In a general form, $\log_2 n = -\log_2 1/n$, but since $1/n = p$, then $\log_2 1/n = -\log_2 n$. Subsequently, $\log_2 n = -\log_2 p$. If the left part of the equation expresses uncertainty through the logarithm of the number of outcomes, then the right part expresses it through the logarithm of the probability of the separate outcomes.

Through the concept of uncertainty, it is possible to elucidate the concept of information. Information is that which removes or limits uncertainty.[20] If one, for example, does not know what the weather will be like tomorrow: rain, snow, cloudy or clear, and the weather bureau reports that it will rain, the prognosis removes the uncertainty, and provides information. A message which does not remove any uncertainty provides no information. Since one knows for certain that the sun will rise tomorrow (there is no uncertainty, since the sun cannot but rise), a message about the sun's rising will contain no information.

The link of information with uncertainty makes possible a quantitative measurement of information. It is completely obvious

---

*Henceforth, we will use logarithms with a base of two everywhere.

**On the reasons for selecting logarithmic measures for measuring uncertainty, see point three of note 21.

that the greater the information contained in a message, the more uncertainty it removes. It is natural, therefore, to measure information by the uncertainty it removes.[21] Thus, a message to the effect that, when answering a question, the student received the grade B removes the uncertainty equal to $\log_2 4$, and therefore has information equal to $\log_2 4 = 2$. If we start from the fact that the teacher also grades F (i.e., the student can receive one out of five grades), the message on the grade that the student got removes the uncertainty equal to $\log_2 5$ and has the quantity of information equal to $\log_2 5 = 2.32$. As we can see, the information (or more precisely, the quantity of information) is measured by specific numbers. In our examples, these numbers were greater than one.

What should be considered a unit of the quantity of information? What is that uncertainty with whose elimination it is natural to link the concept of the unit of the quantity of information? It is natural to consider in this connection the uncertainty of a situation having two equally probable results (e.g., the uncertainty linked with tossing a coin). A report on how the coin fell removes uncertainty equal to $\log_2 2$ and has the quantity of information equal to $\log_2 2 = 1$. Information removing uncertainty linked with two equally probable events and accepted as the unit of the quantity of information received the name of "bit" (from *bi*nary uni*t*).

What has been said can be restated somewhat differently.

Let us suppose that we are uncertain concerning the occurrence of some events. One of the ways of reducing or eliminating uncertainty is to ask a competent person a question about it and get his answer. The essence of any question consists of the fact that a person makes known his ignorance about something and wants to receive information which would reduce or even eliminate his ignorance.

Questions may be compared among themselves from various points of view. It is especially interesting to compare questions from the point of view of the degree of ignorance they reflect. How should this degree be measured and evaluated?

Obviously, this is possible by way of evaluating that

uncertainty which is reflected in a certain question and which is removed by obtaining the corresponding information. Let us take two questions as examples: 1) Are you going out of town on Sunday? and 2) On what day next year will there be a solar eclipse?

The uncertainty contained within the first question is equal to $\log_2 2$. Here, only two answers (results) are possible: either the person is going out of town on Sunday, or he is not going (the results are supposed to be equally probable). The uncertainty contained within the second question is equal to $\log_2 365$, since a person who does not know astronomy might suppose that an eclipse could take place on any of the 365 days of the year. A different uncertainty—or the same thing, a different degree of ignorance—stands behind these questions. The answer to each of the questions which removes a different uncertainty also contains a different quantity of information. It is easy to calculate. The quantity of information which is contained in the answer to the second question is $\dfrac{\log_2 365}{\log_2 2} = 8.51$ times greater than the quantity of information contained in the first question.

Following this method, we have the possibility of comparing the degree of ignorance included in those or other questions independently of their content and abstracting ourselves from that content. Thus, the questions: "Are you going out of town on Sunday?" and "Is the natural number n wholly divisible by 2?" express the same degree of our ignorance (in both instances, two equally probable results are possible). The information contained in the answers to these questions is identical.

According to how much uncertainty the answer to the given question must remove, all questions may be classified as binary, trinary, quadrinary, and generally n-ary. The unit of information, the "bit," may also be thought of as a quantity of information which is contained in the answer to one binary question.

Any n-ary question may be expressed in the form of a set of binary questions.* For example, a quadrinary question about the

*This circumstance, by the way, casts light on the selection for the basic unit of the quantity of information of a binary unit (any selection may be represented in the form of the sequence of choices with two outcomes).

grade which the student received can be broken down into two binary questions. To do this, all the grades must be divided into two groups, and in the first question, one must ask whether the student got a grade belonging to the first group (whether "high" or "low"). After the answer to that question ("yes" or "no"), one can find out precisely what grade the student did get. The uncertainty contained in one quadrinary question is equal to $\log_2$ 4. Equal to this value is the quantity of information which the answer to that question has. The uncertainty which is contained in two binary questions is also equal to $\log_2$ 4. Actually, $2 \log_2 2 =$ $\log_2 2^2 = \log_2$ 4. Hence, the quantity of information which the answers to both binary questions have, is equal to the same two binary units: $2 \log_2 2 = 2$.

In the examples which we examined above, the information received by a person removed all of the uncertainty at once, and therefore, its quantity was precisely equal to the value of that uncertainty. If, for example, the uncertainty was equal to $\log_2$ 4, the quantity of the information was also equal to $\log_2$ 4 = 2 binary units. Often, however, the information received does not remove all of the uncertainty, but only a part of it. How can one measure the quantity of information in that case? Obviously, in this case also, one must measure the information by the value of that uncertainty which it removes. One may calculate this uncertainty in the following manner: subtract the uncertainty which remains after the information is received from the original uncertainty (the uncertainty before receiving information). Let us consider an example. Suppose that during the first ten days of a specific month, a store is going to receive a certain book which interests someone. Suppose now that this person went to the store on the first of the month, and the salesgirl told him that they would not get the book before the third. To what is the quantity of information which the salesgirl gave to the purchaser equal?

The original uncertainty was equal to $\log_2$ 10 (one could expect receipt of the book on one of ten days). After the information of the salesgirl, the uncertainty became equal to 7 (the receipt of the book can now be expected in the course of one of seven days). The uncertainty which was removed by the information of the salesgirl is equal to $\log_2 10 - \log_2 3 = 1.74$. This figure also indicates the quantity of information in binary units which is contained in the salesgirl's report.

Let us designate the uncertainty contained in the original situation β by the symbol H (β) and the uncertainty of situation β which remains after receiving the report by the symbol H_α (β). Let us designate the quantity of information contained in report α concerning situation β by I (α, β). It is obvious that I (α, β) = H (β) - H_α (β). One may read the formula thus: the quantity of information contained in report α concerning situation β is equal to the original uncertainty minus that uncertainty which remains after receiving report α.

It is easy to see that cases where a report eliminates all uncertainty at once are also embraced by this formula. But, here, H_α (β) = 0 (after receiving the report, no uncertainty remains). The formula in this case becomes: I (α, β) = H (β) - O = H (β).

Up until now, we have considered situations where all the events or results were equally probable (or we supposed them to be equally probable). Very often, however, one has to deal with situations where the events and results are not equally probable. Thus, if we supposed that an average student could receive with equal probability A, B, C or D, then with a good student, the probability of receiving A or B is higher than that of receiving C or D, and vice versa for the poor student. It is easier to foretell the grade for a good or poor student than it is for an average student. This is related to the fact that the uncertainty of a situation where all the results are equally probable is greater than the uncertainty of a situation where all the results are not equally probable, i.e., some results can be expected more often than others.

It is clear from what has been said that the uncertainty of a situation depends not only on the number of events (as considered up until now), but also on their probabilities. In other words, the uncertainty of a situation is the function of both these factors. The following formula permits the evaluation of the uncertainty of a situation even in those cases where the results are not equally probable:

$$H = -p_1 \log_2 p_1 - p_2 \log_2 p_2 - \ldots - p_n \log_2 p_n = \sum_{i=1}^{n} (-p_i \log_2 p_i).$$

The value H designating an average value of uncertainty is called entropy in information theory. The formula cited above for uncertainty of n equally probable events is a special case of this general formula for entropy which can be obtained from it and applied when all events are equally likely to occur, i.e., one may consider that $p_1 = p_2 = \ldots = p_n = p.$

A Method of Calculating the Optimal Strategy of Identification in Multiple Choice

We discussed the fact that identifying some event or phenomenon means specifying to which class it belongs. For example, identifying which part of speech a certain word is, means establishing whether it is a noun, adjective, numeral, pronoun, verb, participle, adverb, preposition, conjunction, article, or exclamation.

It is not difficult to see that a situation in which a person has to identify some sort of phenomenon is a situation of uncertainty. As a matter of fact, the problem of identification consists of removing this uncertainty by choosing one of the possibilities.

In the examples cited above, the uncertainty was removed by reports issuing from other people (from the weather bureau, the salesgirl, etc.). In the situation to be examined by us now, the person *himself* must remove the uncertainty by means of *his own cognitive activities*. This is possible because the cognitive activity is directed at obtaining specific information. An approach to cognitive activity, in particular, reasoning activity, from the point of view of which information the specific cognitive acts extract, makes possible, within specific limits, the application of precise methods of information theory to the analysis of cognitive activity. For example, the statement of the following problem becomes justifiable: how should one organize the cognitive activity of a person (e.g., a student) so that he will obtain the greatest quantity of information in the least time? Under what conditions will the information extracted be the greatest?[22]

Let us examine an example from the type of problem related to identifying forms of simple sentences. Although the example from which the procedure of specifying the optimal strategy of identification will be developed is particular, the procedure itself has a sufficiently general character and can be applied for solving other problems and not just grammatical ones.

In USSR schools, four kinds of Russian simple sentences are studied: definite-personal, indefinite-personal, impersonal, and elliptical (also called "nominal"). The definite-personal sentences have two forms or types, each of which has its own particular indicative features. Let us call them "definite-personal sentences type I" and "definite-personal sentences type II." If each of the types of personal sentences is considered as an independent variety of simple sentences, then, in practice, a student has to deal with five forms of simple sentences.[23] Let us designate them by capital letters of the Latin alphabet: "definite-personal type I"—A, "definite-personal type II"—B, "indefinite-personal"—C, "impersonal"—D, "elliptical" (or "nominal")—E.

When a student encounters some sentence x, whose form he must specify, he finds himself in a situation of uncertainty with respect to five possible outcomes: x is either, A, or B, or C, or D, or E $(x \in A \lor x \in B \lor x \in C \lor x \in D \lor x \in E)$. The "outcomes" in the given case are five possible hypotheses about the x's belonging to one of the kinds of sentences, and the problem consists of specifying which of them is true.

If we know with what frequency each of the kinds of sentence is encountered in the language, i.e., know the probabilities of each of the outcomes (this knowledge may be obtained by a statistical study of the language), it is easy to evaluate the uncertainty of the problem with which the student has to deal. Since, however, at the present time, we do not have information available on the comparative frequency of the different kinds of sentences, here we will have to proceed from the simplification of accepting all results as equally probable. This is how one acts at times when one does not know the probabilities associated with some phenomenon.[24] In this condition the uncertainty in the situation which the student encounters when he has to identify the kind of simple sentence is equal to $\log_2 5$.[25]

The efficient algorithm for solving this problem must remove the given uncertainty in the least number of operations.[26] In which case is this possible? Obviously, in the case where one can find operations where each one will extract (in the given situation) the *maximum possible* information. The property of intellectual operations for removing uncertainty and extracting information we will call *informativity* of intellectual operations. Therefore, the problem of making up an efficient algorithm amounts to finding those operations which—in the framework of the given problem— remove the most uncertainty and extract the most information, i.e., possess the greatest informativity.

On what then, in its turn, does the informativity of operations depend? By what is it specified? The answer to this question is linked to uncovering the function of intellectual operations in the process of identifying objects and phenomena.

As we have often said, the process of identification consists of establishing through intellectual operations specific indicative features in the phenomena which are reflected in the concepts about phenomena. Depending on whether the phenomenon has these indicative features or whether they are lacking, a conclusion is drawn as to which class this phenomenon belongs.

The indicative features of simple sentences can be described in the form of the following structural diagram, and we will proceed from them as given data.[27]

Definite-personal type I	Definite-personal type II	Indefinite-personal	Impersonal	Elliptical
1) Subject present	1) Subject absent	1) Subject absent	1) Subject absent	1) Subject present
and	and	and	and	and
2) Predicate present	2) The predicate is expressed by a verb of the first or second person	2) The predicate is expressed by a verb of the third person plural (in the past tense simply by a verb in the plural)	2) The predicate is expressed by any part of speech except verbs of the first and second person singular and also by third person plural	2) Predicate absent[28]

We will designate each indicative feature (more precisely, an expression about the indicative feature) by a specific letter:

The subject is present—a

The predicate is present—b

The predicate is expressed by a verb of the first or second person—c

The predicate is expressed by a verb of the third person plural (in the past tense, simply by a verb in the plural)—d

The subject is absent—$\bar{a}$

The predicate is absent—$\bar{b}$

The predicate is expressed neither by a verb of the first nor second person—$\bar{c}$

The predicate is not expressed by a verb of the third person plural (in the past tense simply by a verb in the plural)—$\bar{d}$

As we see, the number of indicative features by which the kinds of simple sentences are described which are studied in school equal 4 (a, b, c, d). But the kinds of sentences are defined not by isolated indicative features but by their specific combinations. It is important to note that the combinations of indicative features cited above are not the only possible ones.

The methods and symbols of mathematical logic make it possible to describe all the possible combinations of indicative features of forms or types of simple sentences easily, clearly, and economically. This is how the indicative features of simple sentences which were shown above will look in symbolic notation:

definite-personal type I — a & b

definite-personal type II[29] — $\bar{a}$ & b & c

indefinite-personal—$\bar{a}$ & b & d

impersonal — $\bar{a}$ & b & $\bar{c}$ & $\bar{d}$

elliptical (nominal) — a & $\bar{b}$

Now it is possible to describe more precisely the situation of that problem which must be solved in order to identify the form of one or another sentence x. The question of what kind of sentence is the one given can be made concrete in the following manner. It must be ascertained whether or not it possesses indicative features *a* & *b*, or indicative features $\bar{a}$ & *b* & *c*, or $\bar{a}$ & *b* & *d*, or $\bar{a}$ & *b* & $\bar{c}$ & $\bar{d}$, or *a* & $\bar{b}$.

In what order should the operations for verifying the indicative features be carried out, and what is the most efficient sequence of these verifications? The different sequences for verifying indicative features will be the different algorithms for solving the given problem (with those initial indicative features).

If the indicative features which must be verified in a certain object in order to attribute it (in the process of multiple choice) to one of the classes are independent, and all the indicative features in the object must be verified in each case in the process of identification, then the formula specifying the number of possible strategies of identification is:

$$n \cdot (n-1)^2 \cdot (n-2)^4 \cdot (n-3)^8 \cdot \ldots \cdot (n-2)^{2^{n-2}} = \prod_{i=0}^{n-2} (n-i)^{2^i}.$$

(One of the *n* indicative features can be verified first. Since each of the indicative features first verified can either be present in the sentence or missing, there remain *n* - 1 possibilities for the second operation. There are only $(n-1)^2$ operations possible for the second case. For the third operation, there are correspondingly $(n-2)^4$ possibilities, and so on. The last operation is univocally specified by the preceding ones.)

This formula differs from the formula for the number of strategies of identification in conditions of alternative choice. When solving problems of alternative choice, the absence of one of the conjunctively joined indicative features leads to halting the process of identification and drawing a negative conclusion. In solving problems of multiple choice, the absence of the indicative features verified does not, in general, lead to halting the process of identification.

If the indicative features of simple sentences were independent, and all features would need to be verified so as to ascribe a sentence to one or another type, the number of possible strategies, according to the formula just cited, would be equal to 576.

However, since these indicative features (at least partly) are not independent, and since there is no need to verify all indicative features in identifying a simple sentence, the number of possible strategies will be significantly less.

Obviously, some of these strategies are inefficient and some efficient, but not the most efficient. Finally, some of the strategies (and perhaps, just a single one) are most efficient. It is necessary to find strategies belonging to the third group in order to accept them as algorithms.

Above we discussed that function of intellectual operations through which one verifies the presence of specific indicative features in the object or phenomenon. But what is the function of indicative features in the process of identification?

Reflecting on an example will give the answer. Let us suppose that we have to specify what kind of sentence x is. Before we begin analyzing it and verifying the presence in it of different indicative features, we have reason to suppose that it could belong to any one of five possible types. The initial uncertainty (with the simplified assumption equal probability for the different results) is equal to $\log_2 5$. But here, we begin to verify this sentence for the presence of indicative feature a (is there a subject in the sentence?). If there is a subject in the sentence, we can at once draw the conclusion that this sentence is neither a definite-personal type II sentence nor an indefinite-personal sentence, nor an impersonal sentence. Of the five possibilities (hypotheses) only two remain: either it is a definite-personal sentence type I or it is an elliptical sentence. If the subject is missing in the sentence, this means that it is neither definite-personal type I nor elliptical, and of the five hypotheses, three remain (it is either definite-personal type II, or indefinite-personal, or impersonal).

Thus, independently of whether the indicative feature verified is in the sentence or not, in any case, a limiting of the original uncertainty takes place, as well as a narrowing of the field of subsequent searches. This is what constitutes one of the most important functions of indicative features in the process of identification. *By partially or wholly removing the original uncertainty, indicative features provide the necessary information on the character of those phenomena in which these indicative features appear present or absent during the process of identification.* Since we are able, in the light of the assumptions we have made, to measure both the original uncertainty and that which remains after verifying indicative features, we can precisely evaluate the information which the verification of each indicative feature provides, i.e., precisely evaluate its informativity. It is easy to see that the informativity of the intellectual operations discussed earlier is directly linked with the informativity of the indicative features which a person encounters in the process of identification and is determined by it. Thus, the informativity of an intellectual operation can be evaluated by the informativity of the corresponding indicative feature, and the informativity of the indicative feature by that uncertainty which its verification removes.

From what has been said, the approach to solving problems by determining the optimal strategy of identification and the most efficient algorithm becomes clear. Obviously, of all the possible sequences of verification of indicative features, one must choose the one which ensures obtaining the most information from each indicative feature and, because of this, makes it possible to solve the problem with the least number of operations.

Indicative features are not alike with respect to their informativity. Moreover, one and the same indicative feature may yield a different quantity of information depending on what indicative features were verified before the given present one and, in particular, the last previous one. For example, some indicative feature may yield much information if it is verified *after* indicative feature γ and little information if it is verified *before* it.

The strict method of solving the stated problem consists, therefore, of calculating the quantity of information for different indicative features, taking into account all of the possible ordinal positions which they could occupy among other indicative features, and choosing that sequence of verification which permits the removal of the original uncertainty (on the average) in the least number of operations. However, in a number of cases the way to a solution can be considerably simplified. It is completely obvious that if, at every stage of constructing an algorithm, one were to select those indicative features whose verification yielded the greatest quantity of information, the algorithm made up of such a sequence for verifying indicative features would be the most efficient.

Determining the quantity of information yielded by different indicative features which have been verified at one and the same "place" may be done by way of the following calculation. Suppose we are interested in the question of the indicative feature with which it would be advisable to start an analysis of a sentence in order to determine its type? To do this, one must calculate the quantity of information that each indicative feature would yield, if the process of verification were begun with it. Let us calculate, for example, the quantity for indicative feature a.

In the sentence which we are verifying for the presence of indicative feature a, this indicative feature can either be present or not. If indicative feature a is clearly present, there remain only two possible results, and $H_a(\beta) = \log_2 2$.* The quantity of information which the establishment of this fact yields—that the sentence has indicative feature a—is equal to $H(\beta) - H_a(\beta) = \log_2 5 - \log_2 2 = 1.32$ bits. If indicative feature a is not present in the sentence (i.e., there is indicative feature $\bar{a}$), then there are three results which remain, and the fact that indicative feature $\bar{a}$ is present in the sentence equals $H(\beta) - H_{\bar{a}}(\beta) = \log_2 5 - \log_2 3 = 0.74$ bits.

As we see, in some cases, verifying indicative feature a (when it is present in the sentence) yields 1.32 bits of information and in others (when it is lacking), 0.74 bits. What is the *average* information which verifying indicative feature a will yield, if

---

*As before, we will designate the uncertainty of the original situation by the symbol $H(\beta)$ (in the given case, the uncertainty derives from the fact that the sentence identified can belong to one of five types; this uncertainty, according to our assumption, equals $\log_2 5$). The symbol $H_a(\beta)$ designates the uncertainty which remains after verifying indicative feature a in the sentence and after establishing the fact that the sentence has indicative feature a.

analysis of the sentence is begun with it? It can be calculated by the formula for the mean value of a random magnitude.

The probability of receiving 1.32 bits of information is equal to 2/5 (since the subject is found in two kinds of sentences out of five); the probability of obtaining 0.74 bits is equal to 3/5 (since there is no subject in three of the types of sentences out of five). Hence the mean information which verifying indicative feature *a*, regardless of whether it is present or absent in the sentence is equal to 2/5 · 1.32 + 3/5 · 0.74 = 0.97 bits.

An analogous calculation for the remaining indicative features shows that indicative feature *b*—if one begins the analysis with it—yields 0.72 bits of information, and indicative features *c* and *d*, 0.52 bits each. These figures show that the first operation in an algorithm for identifying types or forms of sentence must be the operation for verifying indicative feature *a*, since this indicative feature removes more uncertainty and yields more information in comparison with all others.[30]

In a general form, the mean amount of information which is yielded by the verification of indicative feature α in a situation having β outcomes may be expressed by the formula:

$$I(\alpha, \beta) = p(\alpha) \cdot (H(\beta) - H_\alpha(\beta)) + p(\overline{\alpha}) \cdot (H(\beta) - H_{\overline{\alpha}}(\beta)) = p(\alpha) \cdot I(\alpha, \beta) + p(\overline{\alpha}) \cdot I(\overline{\alpha}, \beta).$$

Thus, we determined which of the indicative features, if verified first, yields the greatest quantity of information and from which indicative feature, therefore, one must begin the analysis of the sentence in order to identify its type in the most efficient way. Now, we have to specify which indicative feature (of the remaining three) it is expedient to verify as the second one.

By acting precisely as we have done up until now, it is possible to calculate the quantity of information yielded by the verification of each of the subsequent indicative features under conditions *a* and $\overline{a}$. In comparing the informativity of indicative features *b*, *c*, and *d* under conditions *a* and $\overline{a}$ (this may be called *conditional informativity*) we specify which of the indicative features it is most expedient to verify second.[31] Which of the indicative features it is most expedient to verify third is specified analogously (and subsequently which indicative feature will be verified fourth and last). At the end of all calculations we will find that sequence for verifying indicative features which will be optimal.

As is evident, constructing an efficient algorithm represents a multi-step process. At first, it is specified which indicative feature must be verified first. Then, having solved this problem, another must be solved: specify which indicative feature must be verified second (separately for the presence and absence of the preceding indicative feature), and so on. The general problem of constructing an efficient algorithm is thus solved step-by-step.

What has been said with respect to a criterion for choosing indicative features at each step (choosing the most informative indicative feature) refers only to the case where the time for verifying the different indicative features is identical. If the time is not the same, one must take into consideration not only the informativity of the indicative features, but also the time for their verification. In this case, what is important is not the informativity of the indicative feature in itself, but the quantity of information which can be obtained in verifying the given indicative feature within a unit of time. This is, so to speak, the *specific informativity* of the indicative feature. If we were to designate the informativity of a certain indicative feature k by the symbol Inf (k), then the specific informativity would equal $\frac{\text{Inf}(k)}{t(k)}$, where $t(k)$ is the time for verifying indicative feature k. For the case where the times for verifying the different indicative features are not the same, the rule by which we must specify which indicative feature should be verified at the given step, may be formulated thus: one must verify that indicative feature whose specific informativity is greatest. Thus, for example, if, at the given step, one may verify one of the three indicative features, k_1, k_2, k_3, and $\frac{\text{Inf}(k_1)}{t(k_1)} \geqslant \frac{\text{Inf}(k_2)}{t(k_2)} \geqslant \frac{\text{Inf}(k_3)}{t(k_3)}$, one must begin by verifying k_1.

In a general form, the rule for choosing an indicative feature at each step when there are n indicative features may be formulated thus:

$$\frac{\mathrm{Inf}\,(k_1)}{t\,(k_1)} \geqslant \frac{\mathrm{Inf}\,(k_2)}{t\,(k_2)} \geqslant \ldots \geqslant \frac{\mathrm{Inf}\,(k_n)}{t\,(k_n)} \Rightarrow \mathrm{T}\,(k_1).$$

The method described for constructing an efficient algorithm (or algorithms) is sufficiently general and reliable.[32] However, it is very cumbersome and requires a large number of calculations.

Would it not be possible to find another more simple and convenient procedure for solving the problem of designing efficient algorithms? So as to devise such a procedure we have resorted not only to information theory, but also to mathematical logic. It was only necessary to combine them in a specific way in order to devise a joint method.

The assertion proved in the theory of information that the most economic method of removing uncertainty consists of breaking down the number of possible results into parts equal in probability and determining to which of these parts the unknown phenomenon belongs, was the point of departure for us.[33] The closer to each other the probabilities are of those parts into which the number of possible results are broken down, the more uncertainty is removed, and the greater is the information received.*

If we examine the role of indicative features in the process of identification from the standpoint shown, and we raise the question of why verifying different indicative features yields different amounts of information, we find that it is because different indicative features (their presence and absence) differentially divide the possible outcomes. The ratios of the probabilities of the parts which are obtained as results of such a breaking down are unequal in different instances. This is easy to show by an example.

a		$\bar{a}$		
A	B	C	D	E
2/5		3/5		

b				$\bar{b}$
A	C	D	E	B
4/5				1/5

*We mean here and elsewhere the *average* information which one obtains in repeatedly solving a problem.

It is evident from a comparison of these tables that, for example, indicative feature a and its negation (this indicative feature is found in two types of sentences and is absent in three) break down a number of possible results (types of sentences) into parts whose probabilities are equal, respectively, to 2/5 and 3/5. Indicative feature b and its negation (this indicative feature is present in four types of sentence and is missing in one) breaks down the number of possible indicative features into parts whose probabilities are equal to 4/5 and 1/5. Indicative feature a and its negation break down the number of possible results into parts whose probabilities are closer than the probabilities of the parts into which indicative feature b and its negation break them down. It is precisely for this reason that verifying indicative feature a (the first) yields more information than verifying indicative feature b.

What has been said suggests a simple and convenient way of finding the most efficient algorithm without making calculations as to the quantity of information which verifying each indicative feature at different "locations" yields. In constructing an efficient algorithm one must at every step of the process determine the probabilities of all the indicative features and of their negations, compare them among themselves, and select those indicative features whose probabilities of presence and absence are closest to each other.[34] To do this, one must find the way of describing the logical structure of indicative features so that it becomes possible through application of a formula to determine the probabilities of the indicative features and of their negations. The usual symbolic description of the logical structure of indicative features (a simple sentence may be characterized by one of the following combinations of indicative features: $ab \vee \overline{a}bc \vee \overline{a}bd \vee \overline{a}b\overline{c}\overline{d} \vee a\overline{b}$) does not make it possible to solve the stated problem, since, with such a description, it is impossible to determine the probability of each indicative feature and of its negation. In the disjunctive members, as it is easy to see, not all of the indicative features are represented.

There is, however, a method which makes possible the

introduction of the missing indicative features into each disjunctive member—this is reducing the formulae of propositional logic to the perfect disjunctive normal form. Reduction to the perfect disjunctive normal form is carried out by a sequential, conjunctive addition of the missing letters and their negations to each member of the disjunction with the following opening of parentheses (in our case, the letters and their negations signify the presence or absence of indicative features). The formula obtained is logically equivalent to the original.[35]

Let us sequentially describe all of the "steps" which must be performed in order to construct the most efficient algorithm for the identification of a phenomenon according to a logico-mathematical description of the structure of indicative features and the probabilities of their combinations, i.e., according to the probabilities of the different outcomes of that phenomenon. We will illustrate the method of constructing an algorithm by the example of the problem discussed above on identifying the types of simple sentence.

1. After the indicative features which specify this or that phenomenon (in our case—types of simple sentences) are found and selected, the logical structure of the indicative features is established. This structure is described in the language of logic as a formula. The logical structure of the indicative features of simple sentences was shown above, i.e., $ab \lor \overline{a}bc \lor \overline{a}bd \lor \overline{a}bcd \lor a\overline{b}$.

2. On the basis of statistical calculations or of certain assumptions, the probabilities of each of the results (in our case—each of the types of a simple sentence) are established. We arbitrarily assumed (for lack of empirical data) that these probabilities equal 1/5.* Since each disjunctive part describes the indicative features of one of the types of sentences, the probability of each type of sentence may be ascribed to a

---

*Since each sentence can belong to one and only one class, the sum of the probabilities of belonging to each of the five classes equals 1. This must be taken into consideration when reading further in the text of this chapter.

corresponding disjunctive part:

$$ab \quad \lor \quad \overline{a}bc \quad \lor \quad \overline{a}b\overline{d} \quad \lor \quad \overline{a}\overline{b}\overline{c}\overline{d} \quad \lor \quad a\overline{b}.*$$

$$1/5 \qquad 1/5 \qquad\quad 1/5 \qquad\quad\; 1/5 \qquad\qquad 1/5$$

(Here, under each disjunctive member the accepted probability of the given disjunctive member is indicated.)

3. In order to determine with which indicative feature it is most expedient to start the process of identification, one must determine which of them is the most informative, i.e., breaks down, along with its negation, a number of possible outcomes into parts whose probabilities are closest to each other. But for this, one must know the probabilities of each indicative feature and of its negation. From our formula, however, it is not possible to know this, since in the different disjunctive members, all of the indicative features are not represented (certain indicative features are contained in them in "hidden" form).

In order to reveal the hidden indicative features and represent all of the indicative features and their negations in the disjunctive member, we reduce the original formula to the perfect disjunctive normal form. Toward this end, we "multiply" each disjunctive member by the missing letters:

$$ab\,(c \lor \overline{c})\,(d \lor \overline{d}) \lor \overline{a}bc\,(d \lor \overline{d}) \lor \overline{a}b\overline{d}\,(c \lor \overline{c}) \lor \overline{a}\overline{b}\overline{c}\overline{d} \lor a\overline{b}$$
$$(c \lor \overline{c})\,(d \lor \overline{d}) \Leftrightarrow abcd \lor abc\overline{d} \lor ab\overline{c}\overline{d} \lor ab\overline{c}\overline{d} \lor \overline{a}bdc \lor$$
$$\overline{a}b\overline{d}c \lor \overline{a}b\overline{c}\overline{d} \lor \overline{a}b\overline{c}d \lor \overline{a}\overline{b}\overline{c}\overline{d} \lor a\overline{b}cd \lor a\overline{b}\overline{c}\overline{d}.**$$

---

*From what has been said, it is clear that all the disjunctive members entering into this formula are incompatible in pairs.

**In order to understand this formula, one must keep in mind the following: Omitting a sign from the formula signifies an operation of conjunction; the latter is viewed as joining the members more closely than an operation of disjunction. The expression cited is analogous to formulae of ordinary algebra where multiplication joins the members of a formula more closely than addition. In this sense, the operation of conjunction (in the notation used here) is analogous to the multiplication in algebra—similar to the way in which ordinary algebra $a(b + c) = ab + ac$, in the propositional logic $a\,(c \lor \overline{c})$ $\Leftrightarrow ac \lor a\overline{c}$. In the former case the letters are variables for the numbers (they replace the numbers); in the latter case, they are variables for propositions. The symbol $\Leftrightarrow$ also has a somewhat different sense than the equal sign. As distinct from the symbol =, the symbol $\Leftrightarrow$ signifies logical equivalence.

4. Since indicative features are often not independent, not every indicative feature can be combined in an object with any of the others. Thus, in our case, indicative features c and d cannot be combined in a sentence (the predicate cannot simultaneously be expressed by a verb of the first or second person and the third person plural), and likewise, indicative features $\bar{b}$ and c and $\bar{b}$ and d (the predicate cannot be missing and at the same time be expressed by verbs of the first or second or third person). In complete sentences of the types examined, the absence of all principal parts is also impossible (i.e., indicative features $\bar{a}$ & $\bar{b}$ & $\bar{c}$ & $\bar{d}$ present).

Therefore, after reducing the original formula to a perfect disjunctive normal form, we must verify whether there occur among the disjunctive members, members which correspond to combinations of indicative features which cannot exist together in the object examined. If there are such disjunctive members in the formula, then they must be excluded from it. In our formula, members 1, 5, 7, 10, 11, and 12 are such disjunctive members. After crossing them out, the formula shrinks to:

$$ abc\bar{d} \ \lor \ ab\bar{c}d \ \lor \ ab\bar{c}\bar{d} \ \lor \ \bar{a}b\bar{c}d \ \lor \ \bar{a}b\bar{c}\bar{d} \ \lor \ \bar{a}\bar{b}\bar{c}d \ \lor \ \bar{a}\bar{b}\bar{c}\bar{d} $$

Definite-personal type I	Definite-personal type II	Indefinite-personal	Impersonal	Elliptical (Nominal)

5. Since after reducing the original formula to a perfect disjunctive normal form, there appeared new members in the formula (expressing specific varieties of types of sentences), a corresponding redistribution of probabilities is required.

We see that a definite-personal sentence of type or form I which had the probability of 1/5 "broke down" into three varieties. Let us assume that the probabilities of these varieties are identical. Then each of the disjunctive members reflecting varieties of definite-personal sentence type I takes on the probability of 1/15. The remaining disjunctive members retain the same probabilities. The new distribution of probabilities is:

$$abc\bar{d} \lor ab\bar{c}d \lor ab\bar{c}\bar{d} \lor \bar{a}bcd \lor \bar{a}\bar{b}cd \lor \bar{a}\bar{b}\bar{c}d \lor \bar{a}\bar{b}\bar{c}\bar{d}.$$

1/15   1/15   1/15   1/5   1/5   1/5   1/5

6. Now we may begin calculating the probabilities for each separate indicative feature and for its negation.

If the probability, for example, of the first disjunctive member equals 1/15, then this means that in 1/15 cases we encounter each of the indicative features in the sentences of the given variety. Therefore, the probability of each of the indicative features when encountering a sentence of the given variety is equal to the probability of the given disjunctive member. But we will also encounter these indicative features (or their negations) when we have sentences of other types before us. In order to calculate the probability of an indicative feature and the probability of its negation independently of the type of sentence in which it is found (overall probability), obviously one needs only to add the probabilities of occurrence and the probabilities of non-occurrence (negation) of each indicative feature in each disjunctive member. Numerically, the probability of each entry is equal to the probability of that disjunctive member into which the given indicative feature enters.

Thus, indicative feature a enters into the first, second, third, and seventh disjunctive members of the latter formula. Their probabilities are 1/15, 1/15, 1/15, and 1/5. Therefore, the "general" overall probability of indicative feature a (i.e., the probability of the fact that in encountering some simple sentence, we discover a subject) equals the sum of these probabilities: $p\,(a)$ = 1/15 + 1/15 + 1/15 + 1/5 = 2/5.

Indicative feature $\bar{a}$ enters into the fourth, fifth, and sixth disjunctive members. Their probabilities correspondingly equal 1/5, 1/5 and 1/5. Therefore, $p\,(\bar{a})$ = 1/5 + 1/5 + 1/5 = 3/5.[36]

The probabilities of all the other indicative features and their negations are calculated analogously. A summary table of all the other indicative features has the following form:

$p\,(a) = 2/5$, $p\,(\bar{a}) = 3/5$
$p\,(b) = 4/5$, $p\,(\bar{b}) = 1/5$

$$p\,(c) = 4/15, \qquad p\,(\overline{c}) = 11/15$$
$$p\,(d) = 4/15, \qquad p\,(\overline{d}) = 11/15$$

7. After the probabilities of all the indicative features and of their negations are calculated, one must see in which of the indicative features the value of the probability of the presence of the indicative feature and of its absence are closest to each other.[37] In our table, this is indicative feature a.* Therefore, this indicative feature divides the number of possible outcomes into the parts that differ least in their probabilities and provides on the average, the most information. The process of identification must be begun with it (on the condition that the time for verifying the indicative features is the same).[38]

Having performed the indicated cycle of operations, we may determine in an analogous manner which indicative feature must be verified second, third, and so on. The only difference lies in the fact that now we will have to proceed not from the absolute probabilities of the indicative features and their negations, but from conditional ones. But the value of the conditional probabilities is easy to determine when we know the absolute probabilities.

The calculation shows (we will not demonstrate it, since the principle of the calculation is clear from what has been said) that the most efficient of the possible algorithms are two algorithms whose sequence of operations corresponds to the order of indicative features cited in the tables on page 302.

These two algorithms are equal to each other in efficiency.[39] They may be expressed not only as in the tables, but also in the form of a strict verbal prescription.[40] One may also use Lyapunov's [120] notation to describe these algorithms. (We taught students one of these algorithms in the course of a pedagogical experiment which will be described in the second Part of the book.)

These algorithms are the most efficient with the assumptions regarding probabilities which were already mentioned above. If calculations were made on the basis of precise statistical data on

---

*When indicative features are linked together in a specific way (as in the given case), one should take into account what has been said in note 30.

the frequency of different types of simple sentences and the degree of complexity of the different indicative features (or of the operations for verifying them), the efficient sequence for verifying the indicative features could turn out to be somewhat different. However, since the statistical data are lacking at the present time, the algorithms proposed may, for the time being, be accepted as the most efficient.

a		ā		
b	b̄	c	c̄	
			d	d̄
Definite-personal Type I	Elliptical (Nominal)	Definite-personal Type II	Indefin-ite-per-sonal	Imperson-al

a		ā		
b	b̄	d	d̄	
			c	c̄
Definite-personal Type I	Elliptical (Nominal)	Indefin-ite-per-sonal	Definite-personal Type II	Imperson-al

It follows from what has been said, incidentally, that still more precise methodological recommendations will be possible only when specific statistical data have been obtained in psychology and linguistics (and also in other disciplines taught to students). The significance of a precise approach to designing procedures of instruction consists, in particular, of the fact that this approach presents psychology and the other substantive scientific disciplines with hitherto unencountered problems. Thus the methodology of instruction begins to influence to some degree content and orientation of investigations in psychology and in various subject-matter areas.

It was shown above how, in proceeding from the ideas of information theory and by means of mathematical logic, it is possible to construct efficient algorithms for solving specific problems. However, we have not yet given a precise mathematical evaluation of the degree of efficiency of the algorithms cited.

In order to make such an evaluation, it is necessary to specify the average number of operations required for solving a problem according to the given algorithms.[41] It is not difficult to do this. It is evident that in order to identify definite-personal sentences types I and II and elliptical sentences, it is necessary to verify two indicative features each and, therefore, to carry out two operations each. In order to identify indefinite-personal and impersonal sentences, it is necessary to verify three indicative features each[42] and carry out three operations. To what is the average number of operations which have to be performed in order to identify any sentence according to this algorithm equal? This number depends on how often we have to apply two operations and how often three. But the frequency of applying one or another number of operations depends on the frequency of "encountering" the different types of sentences. The latter frequency is determined by the probability of each type of sentence which is given, known or assumed.

Thus, we encounter definite-personal sentences type I (according to our assumption) in one case out of five (the probability of definite-personal sentences type I equals 1/5). Therefore, in 1/5 cases, we engage in two operations. The same number of operations and with the same frequency we apply to elliptical sentences and to definite-personal sentences type II. In the remaining two of the five cases, we engage in three operations each. The average number of operations in which we engage when acting according to the given algorithm equals, therefore: $1/5 \cdot 2 + 1/5 \cdot 2 + 1/5 \cdot 2 + 1/5 \cdot 3 + 1/5 - 3 = 3/5 \cdot 2 + 2/5 \cdot 3 = 12/5 = 2 \ 2/5 = 2.4$. We engage in precisely the same number of operations in acting according to the second algorithm.

The result shown can be obtained by an even simpler way, if the indicative features which must be verified in acting according to this or another algorithm, are described by the symbols of mathematical logic. Thus, the indicative features which we must verify according to the first algorithm are described in the following manner: $ab \lor \overline{ab} \lor \overline{ac} \lor \overline{acd} \lor \overline{acd}$.

Since we know the probability of encountering each of the types of sentences, we also know the probabilities of each indicative feature and its negation. Having added the probabilities of all the entries of letters and letters with a line over them into the formula, we obtain at once the value for the average number of operations which must be carried out in identifying any type of sentence by a given algorithm. It is easy to see that there are 12 letters in the formula given above. The probability of each letter is 1/5. The mean number of operations equals, therefore, $1/5 \cdot 12 = 12/5 = 2.4$. We obtained the very same value as with the preceding method of calculation.

The method for designing the efficient algorithms described above consisted of a sequential selection of the most informative indicative features in "lining them up" in a specific sequence.[43] As we have shown, these algorithms give the solution to the problem examined, on an average of 2.4 operations. The algorithms devised by us are the most efficient (with standing assumptions regarding probabilities), which is a direct result of the method of their design.

What about a general formula for evaluating the efficiency of one or another strategy (algorithm) of identification in a condition of multiple choice?

In examining the method for designing efficient algorithms as illustrated in the example of the algorithm for identifying types of simple sentences, we proceeded, for the simplicity of the examination, from the assumption that all operations for verifying the indicative features of simple sentences have an identical complexity and therefore demand an identical mean time for carrying them out. Hence, it was possible to accept a mean number of operations as a measure for evaluating the efficiency of the algorithm. However, as it was mentioned at the beginning of this Section of the book, different operations for verifying indicative features may have different degrees of difficulty and therefore require a different amount of time to carry them out. That is why the general formula for evaluating the efficiency of an algorithm when solving problems of multiple choice should express this evaluation not by a mean number of operations, but by the mean time of identification according to the given algorithm.

We introduce the following designations. Let A_1, A_2, A_3, . . . ,A_n be the classes to which objects x subject to identification might belong;[44] let p ($x \in A_i$) be the probability that object x belongs to class A $_i$ (i = 1, 2, . . . n); let t (A_i) be the time needed to identify the object's belonging to class A_i. Since the object's belonging to one or another class is a random event (in one case, the object belongs to class A_1, in another class to A_2, and so on), the time which must be spent in different cases for identifying the object is a random magnitude taking the values t (A_1), t (A_2), . . . ,t (A_n). If time t (A_1) is spent with the probability p ($x \in A_1$), time t (A_2) with probability p ($x \in A_2$), . . ., time t (A_n) with probability p ($x \in A_n$), then the mean time spent on identifying the object according to a specific strategy (algorithm) may be calculated by the formula for the mean value of a random magnitude:

$$\tau = p \ (x \in A_1) \cdot t \ (A_1) + p \ (x \in A_2) \cdot t \ (A_2) + \ldots + p \ (x \in A_n)$$
$$\cdot t \ (A_n) = \sum_{i = 1}^{n} p \ (x \in A_i) \cdot t \ (A_i).$$

If p ($x \in A_i$) and t (A_i) are designated respectively as p_i and t_i, then the formula may be written thus:

$$\tau = p_1 \ t_1 + p_2 \ t_2 + \ldots + p_n \ t_n = \sum_{i = 1}^{n} p_i \ t_i.$$

This formula expresses the mathematical expectation of the time needed for identifying the object according to the corresponding algorithm.

The formula cited provides the evaluation of the efficiency of the algorithm depending on the mean time needed to identify whether the object belongs to a specific class and the probability of its belonging to that class. In a number of cases, however, it could be useful to express the functions shown in a different way, taking the probabilities of separate indicative features and the mean time of their verification as arguments. In order to do that, the values p $(x \in A_i)$ and t (A_i) must be expressed through the probabilities of the corresponding indicative features and the mean time of their verification. Once we have the definite statistical material, it is easy to substitute some values for others.

Algorithms are created for the repeated solution of problems of one and the same type, and are applied many times by a person. An objective criterion of the efficiency of an algorithm, such as the mean number of operations, or the mean time for identification according to the given algorithm, has a great practical significance. The possibility emerges of comparing different algorithms through this criterion, and therefore certain aspects of the methods of instruction. For it is perfectly clear that, other things being equal, those methods are to be preferred which are based on more efficient algorithms of solution for specific problems and which teach students a more economical and perfect method of intellectual activity.

In the present chapter, the methods of specifying the optimal strategy for verifying indicative features under the condition that these indicative features are given, have been examined. These methods do not show and cannot show the method for selecting the indicative features themselves.[45] However, the formulae for evaluating the mean time of searching during identification (or what is the same thing—the degree of difficulty or efficiency of the corresponding algorithms) permits a comparison among alternative designs of the algorithms on the basis of *different* indicative features and yields a precise comparative evaluation of their efficiency. Due to this, the prerequisites for surmounting the empirical and intuitive evaluations of the efficiency of this or that algorithm for solving specific problems are created. Empirical and intuitive evaluations are inevitable prior to the development of precise mathematical criteria.

Is there any sense in applying the resources of mathematical logic and information theory so as to solve an uncomplicated problem such as the design of an algorithm for identifying the types of simple sentences? The question may be answered thus: the problem of identifying the types of simple sentences was used here only as an illustrative example of a general method. It is obvious that this problem is by no means unique. When teaching, it is necessary to solve hundreds and thousands of similar problems. Therefore, working out a general method for their solution has an important significance.

The method described for designing efficient algorithms and also the manner of evaluating their efficiency emanates from an analysis of the objective logical structures of instructional material and of the informativity of its separate elements. Such an analysis is important because the thinking activity of students in the process of learning is directed at the processing (assimilation) and mastery of this material. Without knowing its logical structure and without being able to analyze it, it is impossible to investigate profoundly the psychology of mastering knowledge, and it is also impossible scientifically to construct a method of teaching. One of the most important conditions for studying and forming the thinking activity of students in the process of instruction must be the exposure of those objective requirements which the educational material presents for reasoning processes and of those methods of operating with it which emanate from its logical structure.

We saw that these methods—the most efficient ones—may be precisely calculated. They represent those ideal forms (models) of processes which must then be projected in the consciousness of the students (on the condition that they satisfy the psychologico-pedagogical requirements).

Such models of the processes have significance not only for efficient instructional design, but also for an analysis of the real mechanism of the students' reasoning activity during studies. Knowing how the process of reasoning or thinking should proceed in its perfect form opens up the possibility of a precise evaluation of the degree of development of the corresponding reasoning process in each given student. It becomes possible to specify—of course, with varying degrees of approximation—to what degree this process corresponds to the required one and which of its components need special treatment toward their formation and mastery.

Devising effective teaching procedures requires an analysis

and evaluation of different factors. It is perfectly clear that from knowing *what* to teach students and which is the best structure of knowledge and operations, it does not univocally follow *how* to teach. Although *how* to teach directly depends on *what* must be taught, the methods of teaching are not just conditional on the instructional content and the structure of those processes which have to be formed.

In the present chapter, we dealt only with those aspects of the instructional methods which are specified by the logical structure of instructional material and by its informative properties.

A number of psychologico-pedagogical problems connected with this problem were already partially examined above and will be further elucidated in following chapters.

Notes

1. The basic content of this chapter was stated in Landa [272], [275]. On this question, see also Freidman [341], [342].
2. In speaking of the strategy of searching in the process of identification, we will mean the sequence of verifying some system of indicative features of the object or situation with the purpose of attributing it to some class. In this sense, the sequences *ab* and *ba* will be different strategies.
3. A "move" can be the turning up of "heads" or "tails" when tossing a coin, the ensuing of cold or warm weather, and so on. In the problem of identifying phenomena, the presence or absence in the given object of the indicative feature which interests us may be considered a "move." Various "moves of nature" (by analogy with a person's actions) can have different probabilities.
4. These concepts are widely used by different authors (see, for example: Lyapunov's introduction to [124] and that of Poltayev to [73]).
5. It does not follow from what has been said that an algorithm is always a good strategy by several criteria. One could

choose a poor strategy as a prescription for action, and then the chosen algorithm will be inefficient.

6. The choice of one possibility from several is often called alternative choice in the literature on mathematics and cybernetics. The term "alternative *or*" applied by us in Chapter Eight corresponds to this. But in this chapter, we are going to call the choice of one possibility from two (where they are such that one possibility is a negation of the other) alternative choice.

7. It could also be that some of the indicative features of the object are compatible in pairs, while others are incompatible in pairs. Indicative features may also be dependent and independent, but we will only examine independent indicative features in this section of the chapter.

8. The general number of possible strategies for n indicative features is equal to $P_n = n!$ In the given case, the general number of possible strategies will be $P_3 = 3 \cdot 2 \cdot 1 = 6$. Actually, one may first verify one of the three indicative features, second, one of the remaining two, and third, the one last indicative feature.

9. Compare the problem on searching for the lost object or searching for a malfunction in some equipment (machine) where the strategy of verifying locations depends on the probability of finding the object (malfunction) in the given location (assembly) and on the time necessary for verifying each location (assembly).

10. The time for verifying indicative features which reflects their complexity is established experimentally.

11. Let us suppose, for simplicity, that indicative features a, b, c, are independent, i.e., the probability of the presence or absence of one of them, e.g., c, does not depend on what we already know about the presence or absence of the remaining indicative features.

12. "Unit of time"—this may be a second, a microsecond, and so on.

13. According to the assumption accepted above that the mean time for verifying an indicative feature may be considered as

the index of the difficulty (ease) of this indicative feature (as a distinctive coefficient of complexity).

14. Formula (5) is equal to the formula:

$$t_1 + q_1 t_2 + q_1 q_2 t_3 + \ldots + q_1 q_2 \ldots q_{n-1} t_n$$

which may be obtained from formula (12) by transforming the latter (opening parentheses). But one may arrive directly at this formula by the following argument. The first indicative feature is always verified in the process of identification (and consequently time t_1 is always spent). If the first indicative feature is not in the object (and this occurs with the probability of q_1), then the second indicative feature is verified. In that case, in addition to time t_1 with probability q_1, time t_2 is spent (the additional expenditure of time is thus equal to $q_1 t_2$). If there is neither the first nor the second indicative feature (and this occurs with the probability of $q_1 q_2$), then time t_3 is spent on this probability (in this instance, the additional expenditures of time are equal to $q_1 q_2 t_3$), and so on. The last indicative feature n is verified in the object in the case where there are none of the preceding indicative features (the probability of this is equal to $q_1 q_2 \ldots q_{n-1}$). Therefore, the additional expenditures of time will be $q_1 q_2 \ldots q_{n-1} t_n$. The summation of all these times provides the cited formula.

15. Equal to formula (12) is the formula:

$$t_1 + p_1 t_2 + p_1 p_2 t_3 + \ldots + p_1 p_2 \ldots p_{n-1} t_n$$

which may be obtained from formula (12) by transforming the latter. But this formula may be obtained directly by the argument analogous to that cited in note 14 above (or through the substitution of q by p in the formula cited on page 269).

16. Calculation is carried out according to the rule of multiplying probabilities. Thus, the probability $p(ab) = p(a) \cdot p_a(b) = p$

$p\,(b) \cdot p_b\,(a)$.

17. The calculation is carried out according to the rule of adding probabilities. Thus, the probability $p\,(a + b) = p\,(a) + p_a\,(b) = p\,(b) + p_b\,(a)$.

18. Let us note that search and identification under conditions of alternative choice may be considered as a special case of search within conditions of multiple choice, specifically where the number of classes consist of just two and where the second class is a complement of the first to an aggregate of all the objects from which the choice is made.

19. A more strict and complete statement of the question examined in this paragraph may be found in works on information theory (see, e.g., [93], [181], [182]).

20. In the present work, we will speak for the most part about the uncertainty in our *knowledge* with respect to some events and the information which removes this uncertainty. But the concepts of uncertainty and information are applicable not only to knowledge. Thus, if the electric current from some source can pass through one of eight circuits, then the given uncertainty in this case equals $\log_2 8$. But in turning on the specific switch, we make the current pass through one specific circuit and thereby remove the original uncertainty, having "reported" specific information to the system.

21. The agreement according to which a logarithm of the probability of an event which is taken with the inverse sign is chosen as a measure of information (and uncertainty) is based on the following natural requirements:
   1) the information must grow with the lessening of the probability of the event carrying this information;
   2) with $p = 1$, the information must equal zero;
   3) the information supplied to us in the simultaneous occurrence of two events must equal the sum of information which each of them carries. The sole function $f\,(p)$ which possesses the properties
   1) $f\,(p_1) > f\,(p_2)$ when $p_1 < p_2$,
   2) $f\,(1) = 0$,
   3) $f\,(p_1 \cdot p_2) = f\,(p_1) + f\,(p_2)$

is the function—log p; the choice of a logarithmic measure of the uncertainty and of information is specified by this. For more about this in detail, see [182].

22. Since the time for identification depends on the number of operations, their probabilities, sequences, and the time spent on each operation, in a particular case when the time spent on different operations is identical—and we will be examining precisely such a case—the problem indicated may be formulated thus: how should the cognitive activity of a person be organized in order that he may obtain the greatest quantity of information in the least number of "steps" or "moves" (cognitive acts)? Before examining this problem in a general form, we will examine a particular variant.

23. With reference to generalized-personal sentences, explanations will be given below, in Chapter Sixteen.

24. Let us note that, independently of whether we consider the results equally probable or not equally probable, the principal course of reasoning and the methods of calculation remain the same. And that is exactly what is important for us now. At the same time, such a statement of the problem of finding the optimal strategy for identification shows what data are lacking in contemporary linguistics and methods of teaching languages, and what data need to be collected.

25. The uncertainty of a situation is, evidently, one of the objective indices of the degree of difficulty of the problem (although it is not the only one). It is completely obvious that the greater the uncertainty of a situation, the more possible ways of searching for the solution of the problem exist and the more difficult is the problem, all other things being equal. An information theoretic approach, especially if one calculates the entropies of the problem situations, makes it possible to formulate some objective criteria for evaluating the problem difficulty.

26. Since there are no statistical data for the degree of difficulty (ease, complexity) of separate operations (nor are there, so far as we know, data on the frequency of occurrence of the

different types of simple sentences), we will start from the assumption that these operations have an identical difficulty (or—what amounts to the same thing—the time for verifying the different indicative features of simple sentences is identical). Accepting this assumption as well as that of the equal probabilities of the different types of simple sentences does not change the course of reasoning, because the task of the present chapter consists merely of showing the possible ways of an approximate mathematical model of the process of identification. In so doing, the particular values of the different parameters entering into the model do not play an important part.

27. It will be shown below (see Chapter Sixteen) precisely why these indicative features were taken as the bases for identifying the types of simple sentences.

28. A sufficient indicative feature of an elliptical (nominal) sentence is "the predicate is absent." Here, however, we cited the indicative feature "the subject is present" since we taught students these two indicative features. Identification of an elliptical sentence when acting according to an efficient algorithm is carried out, as we will see later, on the basis of the two indicative features pointed out. It is important to note also, that, everywhere in this chapter, only complete sentences are examined. This corresponds to the way in which the theme "types of simple sentences" is studied in school. The study of simple sentences begins with complete sentences. The indicative features of incomplete sentences are introduced only after students learn how to identify the types of complete sentences. In order to teach students how to distinguish incomplete sentences from complete sentences where one of the principal members is missing, one must add certain supplementary indicative features. They will be examined below, in Chapter Sixteen.

29. When describing the indicative features of definite-personal sentences type II, indefinite-personal sentences, and impersonal sentences, we show in the obvious form, the presence of predicates in them, although it is possible not to do this by

limiting oneself to the indication of the form of the verb which expresses the predicate (if it is said that the predicate in the sentence is expressed by a verb in the third person plural, this automatically implies that there is a predicate in the sentence). Let us also note the following. Since the letters only describe the composition of the indicative features of corresponding types of simple sentences but do not show the sequence of their verification, they may generally be written in any order.

30. Having established that it is more expedient to begin the analysis of the sentence by verifying indicative feature a, since it yields the most information, we may draw the conclusion that the efficient algorithm or algorithms are found among those which "begin" by verifying indicative feature a. The remaining algorithms are inefficient, and must be excluded from further examination.

31. In this case, the question is resolved more simply in view of the relation of the indicative features c and d with indicative feature b (indicative features c and d cannot be verified without indicative feature b having been first verified, i.e., without first having found the predicate). In view of this, calculating the information which verifying indicative features c and d could yield before verification of indicative feature b makes no sense.

32. While examining the manuscript of this book and articles [272], [275] in which this method was described, B.V. Gnedenko brought our attention to the fact that an analogous method (true, for solving another problem—searching for faults in electronic equipment and determining the most efficient search with different sequences for verifying possible faults) was described by several American authors and particularly Kletsky [190]. In the period when we were formulating this method, Kletsky's article had not yet been published. Thus the same method was developed simultaneously, though for different reasons, in different countries by different authors independently of each other (the corresponding American publication did appear somewhat earlier than ours).

33. Suppose we wish to find out of which number between 1 and 8 someone is thinking, but may ask only questions that can be answered by "yes" and "no." In order to establish this number with the fewest questions, one must successively partition the set of numbers into equiprobable parts (or, if this is impossible, into as nearly equiprobable partitions as possible) and ask which partition contains the number in question. If the probabilities of all outcomes (numbers being thought of) are equal, the set of outcomes must be broken down into equal partitions. It is easy to see that it is possible to determine the number being thought of in three questions.

34. It is not difficult to see that if, when solving a problem of alternative choice, the optimal strategy consisted of beginning the verification of indicative features either with the most probable (disjunctive search and identification) or with the least probable (conjunctive search and identification), then in solving problems of multiple choice, one must begin the verification with those indicative features which break down the number of results into equal parts whose probabilities are equal or as close as possible to each other (on the stipulated condition for simplifying calculations, that the time for verifying the indicative features is identical).

35. One may read in more detail about the concept of the perfect disjunctive normal form in books on mathematical logic (see, e.g., [83], [104], [139]; also, [62], [105], [159]).

36. $p(\bar{a})$ may be calculated even more simply. It equals $1 - p(a)$, i.e., $1 - 2/5 = 3/5$.

37. This can be determined in the following way. Those probability values are closest to each other whose difference is the least absolute magnitude. Therefore, in order to find out for which of the indicative features the probabilities of its presence and absence are closest, one must subtract one probability from another. The absolute magnitude of the difference will be the index of how close to each other the two probability values are.

38. If the time necessary for verifying the different indicative features is unequal, then the specific informativity of each

indicative feature is defined, and the indicative feature having the most specific informativity is selected.

39. From the logical structure of the indicative features of simple sentences (complete), it follows that if there is a subject in a certain sentence whose type must be identified, one should verify whether it has a predicate (the presence or absence of the predicate is, in this case, the differentiating informative indicative feature). If there is no subject in the sentence, then there is always a predicate, and this is why there is no sense in verifying that indicative feature by itself (it is not the differentiating and informative one). One must verify the sentence only by the indicative feature: what kind of verb expresses the predicate. Although one cannot verify what expresses the predicate without finding the predicate itself, these two operations may be combined into one in the prescription. The problem solver can be told: verify whether the predicate is expressed by a verb of such a form (for example, is the predicate expressed by a verb of the first or second person?). Anyone carrying out this instruction will either first find the predicate and then determine how it is expressed, or he will solve the two problems simultaneously. In the diagram on page 302, features c or d follow right after indicative feature $\bar{a}$ (and not indicative feature b) which is conditional on the combination of the two operations.

40. See Chapter Sixteen.

41. If the different operations require different times, the calculation may be carried out, as we did above, in units of time.

42. On the condition that we united the operations of verifying the presence of the predicate and specifying how it was expressed into one operation.

43. After this method was described in [272], a number of authors (e.g., Chentsov [344], Gentilhomme [421], and Pohl [484]) applied it to solving other problems. In particular, Chentsov applied it to the problem of calculating the efficient sequence of operations in conducting certain laboratory procedures in physics and electrical engineering. The

data published by these authors show that this method was completely validated and led to the theoretical prediction of the most efficient sequence of operations when solving problems from different fields.

44. Above, we also called the object's belonging to one or another class "outcomes" and considered the possible results as "hypotheses."

45. The problem of selecting indicative features according to qualitative psycho-didactic criteria and certain methods of solving specific problems arising in this connection were examined in Part One, Section Three of this book.

Chapter Thirteen

The Stochastic Mechanisms of Intellectual Activity in the Process of Perception and Identification of Phenomena

We examined some mathematical methods of calculating efficient algorithms and the ways of specifying the optimal sequence for verifying indicative features in the process of identification.

What significance does the choice of one or another sequence of verifying indicative features have if, as a result of mastering the algorithm, a person learns to "see" the necessary indicative features directly and immediately without following that chain of sequential verifications which he was taught and which he practiced? Is it at all necessary to teach a sequential verification of the indicative features of phenomena if the person, once the skill of analyzing and identifying phenomena on the basis of an algorithm has been formed, will not follow such an elaborate, consecutive verification of the indicative features?

Let us start with the latter question.

Although a person with an established habit of analysis and identification isolates the necessary indicative features in a phenomenon directly and "sees" them at once, the most important condition for the speedy and correct formation of such a skill is precisely such training in sequential verification of indicative features according to a specific program. Why?

In the process of perceiving each object or phenomenon, an enormous number of stimuli *qua* indicative features act upon our sense organs, but we attend selectively only to certain ones. In so doing, it often happens that those stimuli *qua* indicative features which by their physical nature are weaker are noticed, while those which are stronger are ignored.

It is possible only to explain the selective attending to some indicative features and not others by the fact that our brain is unequally "tuned" to perceive different indicative features, and the excitability of its different systems is different.[1] Owing to this, the receptivity (sensitivity) for one indicative feature is higher in it than for another. The fact of unequal sensitivity for different stimuli has long been described in physiology and psychology.

Especially important in connection with this, is the fact that a highly selective sensitivity is formed in the course of activity and is linked to the fact that specific properties of things take on a special significance for people.

The physiological mechanism of the "tuning in" of the brain when selectively attending to specific stimuli was studied by Pavlov and his colleagues. As is known, during the establishment of conditioned reflexes in the brain, a link is formed between stimuli which until a specific moment are non-significant (for example, specific sounds, lights, etc.) and stimuli which are vitally important to the organism (food, etc.). When entering into a relation with the vital stimuli, the non-significant stimuli become their *signals* and therefore take on a specific significance for the organism.[2] Herein lies the most important biological significance of conditioned reflexes.

The signalling function of conditioned reflexes consists of the fact that they notify the organism about forthcoming influences, preparing it for them in a specific way ("anticipatory activity," according to Pavlov). One may suppose that to the physiological mechanisms of this preparation belongs the selective anticipatory excitation of specific points in cortical and subcortical regions which according to Anokhin [525] underlie the precursory reflection of reality.

Among the conditioned stimuli acting on the organism, Anokhin distinguishes between those fulfilling the role of a starting signal and the system of situational stimuli that prepare the brain to perceive the conditioned stimulus. Because of them, "the precursory heightening of excitability takes place in the corresponding regions of the cortex and subcortex" [42] .

Apparently this explains the heightened selective sensitivity of the brain to specific stimuli. Raising the sensitivity of specific points (centers)* creates a *preparedness* for responding more quickly and easily to some stimuli and their properties (for example, the indicative features of some object) as compared to others, and the possibility of "cognizing them directly." Obviously, one of the most important tasks of methodology and didactics consists of designing instruction in such a way that students would first isolate from among all possible indicative features those significant for solving problems of a specific class. In different cases, there will, undoubtedly, be different indicative features. The result of instruction should be the creation in the brain of specific "mosaics of excitations" (Pavlov's expression) and the formation of centers and systems in which the necessary "precursory excitation" originates.

It has been proven in psychology and physiology that the most important condition for raising the level of arousal in the brain is the occurrence of physical and intellectual activity. It was established by Ananyev [522], [523], Leontyev [693], [697-699], Rubenstein [772], [773], Smirnov [798], [799], and other psychologists that of all the properties of an object, it is easiest to realize and recall those at which a person's actions are directed. Because of the arousing property of actions, including intellectual ones, the possibility is created for isolating at will the necessary properties of the objects, putting the objects in the best position for perception, operating with them in different ways, and including them in different systems of relations. Active processing of objects and their properties requires the presence of heightened excitations in specific centers (systems) in the brain. In the course of action, this increased excitation is most successfully created. That "mosaic of excitations" develops which is easily actualized in response to specific stimuli.

It follows from what has been said, that the best method of creating the necessary mosaic of increased excitation in the brain

---

*In saying "center" or "point," we mean the system of afferent cells integrated into a single entity (see [524]).

is that of actively operating with certain objects and, in particular, of verifying specific indicative features in them. By consciously and purposefully verifying specific indicative features in an object (for the purpose of identifying it), we create in the brain a system of increased excitations which then ensures the immediate and direct isolation of those indicative features in the process of perceiving and thinking. The internal mechanisms of this process can be conceived of as follows.

Verifying certain indicative features in an object in the course of identification leads to the formation of a specific system of increased excitations in the brain and prepares the way for isolating each of these indicative features in the future. Once the centers of arousal (increased excitation) "have been formed" and the corresponding links between them have been established, the conditions are created for the image of the object, in the presence of a given problem, to evoke, by conditioned reflex, the heightened excitation of those centers. The latter intensifies the perceptual sensitivity for those indicative features which up until then were verified in a specific system, and permits singling them out quickly and easily in the object, often by means of "direct cognition." Such an ability is the highest stage of mastering an algorithm, and it is the result toward which one should strive.[3]

Although often the goal of instruction is the development of the ability to perceive "directly" the necessary indicative features, the path to it leads through exercises involving active operation of indicative features and conscious, sequential verification of them in the object according to the elaborately detailed steps of algorithms.

We attempted to answer the question as to why it is necessary to be trained in conscious, energetic, and detailed verification of indicative features according to algorithms, so as to form the ability to distinguish directly the necessary indicative features in the object. Let us go on to the second question, concerning the value of exercises involving a specific *sequence* of verifying indicative features in the order of their calculated frequency (probability).

The research of Pavlov and his co-workers on systems in brain activity and on the role of reinforcement in the formation of conditioned reflexes (see, for example: [611], [635], [645], [653], [740-742], [744], [788]) showed that in systems of temporary neural connections, the *sequence* of acting stimuli is reflected on the one hand and their *frequency* or repetition on the other. Thus, if a system of stimuli acts on the organism in the sequence of A, B, C, D, the excitations in the corresponding centers in the brain also take place in this sequence. In so doing, after establishing a dynamic stereotype, the excitation of the first center evokes, by conditioned reflex, the excitation in the second center; excitation in the second center the third; and so on. The frequency of the influence of the different stimuli affects the level of the anticipatory arousal of the corresponding centers.

In order to form a fixed stereotype it is necessary that the stimuli act upon the brain in a strict sequence. However, in the life of both animals and man, it happens very often that the sequence of the stimuli is not rigid (strictly determined), but rather stochastic. This is reflected by the fact that stimulus B does not always follow stimulus A, but only in a certain percentage of cases (i.e., with a specific probability). In other cases A is followed by C or D and so on (each of them also with a specific probability). In Anokhin's discussion of the problem of the precursory reflection of reality and of the physiological mechanisms of such a reflection [525], two schemes are presented, which the author explains in the following manner:

"Let us suppose that a consecutive series of phenomena, A, B, C, D, E, unfolds in the external world. These phenomena act on the organism at significant time intervals in the course of half a day. Suppose that each of these phenomena evokes in the protoplasm of a living organism a consecutive series of chemical transformations which we will designate, correspondingly, by the symbols *a, b, c, d, e.* (See the diagram on page 323.)

"Let us suppose further that the series of external influences on the organism (A, B, C, D, E) are systematically repeated during the course of many years and thereby establish more perfect chains of chemical reactions. As a result of prolonged and frequent repetition in the living organism's protoplasm of this specific series

A, B, C, D, E—consecutively developing external events at different intervals;

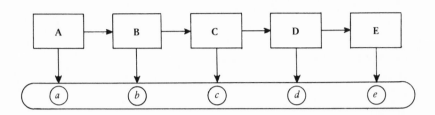

a, b, c, d, e—protoplasmic reactions arising from individual energetic properties of each separate external action.

of chemical reactions, the bond between them is so strengthened that the whole series of transformations, *a—b—c—d—e*, becomes an uninterrupted and quickly developing chain of chemical reactions.

"In the protoplasm of a living organism, a single unitary chain of chemical reactions is established which was earlier evoked by the consecutive action of external factors A, B, C, D, E. Now even when these factors are separated by significant intervals of time, *the action of just the first factor A* [the italics are mine—L.L.] "is capable of activating the whole chain of consecutive chemical reactions." (See the diagram on page 324.)

Later on, Anokhin shows that analogous processes develop in the nervous system, and because of them, the precursory reflection of reality and the preparation of the organisms for future influences become more perfect.[4] Actually, the lower case letters need not be viewed as designating the chemical reaction in the protoplasm, but as designating specific centers of excitation in the central nervous system.

The schemes above represent simplified models of processes unfolding in the organism, including the brain, in response to the influence of stimuli which follow each other in strict sequence.

How might one conceive of a model of processes evolving in the brain in response to stimuli having a stochastic nature and succeeding each other with a specific probability? We shall try to devise such a model.

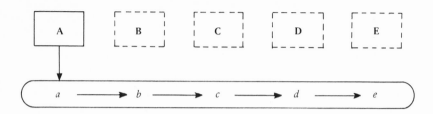

After many iterations of a consecutive series of phenomena, A, B, C, D, E, an uninterrupted chain of chemical transformations is formed that is elicited by the first event, A, the external world alone. The process of reflection in the protoplasm outstrips the course of sequential events in the external world. The reaction *e* of the protoplasm takes place at the time when event E would be expected to take place. (Or it might not ensue at all.)

Let us suppose that there are eight stimuli A, B, C, D, E, F, G, H which repeatedly affect someone in the following sequence:

The influence always begins with stimulus A.

After A in five out of 10 cases (with a frequency of 0.5) follows stimulus B; stimulus C follows with a frequency of 0.3 and stimulus D with a frequency of 0.2, i.e., P_A (B) = 0.5; P_A (C) = 0.3; P_A (D) = 0.2.

Stimulus E follows stimulus B with a frequency of 0.8, and stimulus F follows B with a frequency of 0.2, i.e., P_{AB} (E) = 0.8; P_{AB} (F) = 0.2.

Stimulus G follows C with a frequency of 0.3, and stimulus H follows C with a frequency of 0.7, i.e., P_{AC} (G) = 0.3; P_{AC} (H) = 0.7.

Nothing follows stimulus D.[5]

The given sequence of stimuli is reflected in the brain in a specific system of excitations, each of which occurs with a specific frequency. Let us depict this situation in a form analogous to Anokhin's schema. Under each letter there is given the probability of the corresponding event (the stimulus and the corresponding excitation) conditional upon the occurrence of the preceding event.

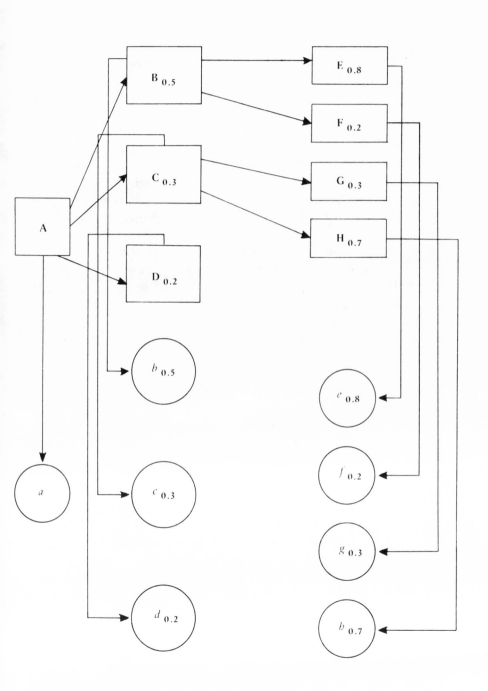

A, B, C, D, E, F, G, H are external events following one another with a specific probability; *a, b, c, d, e, f, g, h* are the excitations in the brain arising in response to the action of corresponding stimuli.

As a result of prolonged and repeated influence of the indicated system of stimuli on the brain, specific linkages are established among all the centers of excitation. Eventually the action of stimulus A alone leads to the reproduction (actualization) of these linkages, evoking a precursory excitation of centers *b, c, d, e, f, g, h* and creating a preparedness in them for perceiving the corresponding stimuli.

From the schema it is obvious that, after many iterations of phenomena A, B, . . . H, which followed each other with a specific probability, a system of centers with intensified excitability is formed in the brain. Excitation in these centers now occurs in response to just the first stimulus, A. If one considers that the levels of excitation (reflecting the conditional probability, *Ed.*) in each of the centers corresponds to the frequency of the corresponding stimuli, then the numerical subscripts to the lower case letters may be considered as an expression of the different levels of anticipatory excitation in the corresponding centers (measured in some conditional probability units).

Will the level of excitation in all of the centers be the same? Obviously not, since the corresponding stimuli act upon them with unequal frequency. If stimulus E follows stimulus B with a probability of 0.8 and stimulus F with a probability of 0.2, this means that following B, the stimulus affected center *e* on an average of eight times out of 10 and at center *f* two times out of 10. It is natural that the level of anticipatory excitation of center *e* will be higher than the level of excitation of center *f*. The difference in the frequency of conditional reinforcements—and the difference related to it in the level of excitation—creates in given situations a greater readiness in the brain for receiving stimulus E and a lesser readiness to perceive stimulus F.[6]

One may suppose that the proportion of anticipatory excitations of different centers (and consequently of "expectations" and "readinesses") expresses the probabilities of those events (stimuli) which evoke excitation in these centers. In other words, the proportion of anticipatory excitations reflects the stochastic picture of the world and creates in the organism an unequal readiness for perceiving different stimuli.[7] Since every situation in which stimuli follow one another with a specific frequency represents a situation of uncertainty having some entropy, one might also say that the proportions of anticipatory excitations reflect the degree of the uncertainty in the situation, i.e., its entropy.

This proposition may be corroborated by facts. We will cite just one of them.

In the last few years, the study (especially with information-theoretic treatment of experimental data) of human reaction time to stimuli has been much pursued (see, e.g., [550], [671], [700-702], [867]).

One of the facts established in research on this problem is illustrated in the following experiment. In front of the subject is a panel on which there are a number of lamps and a corresponding number of switches. The subject is given instructions to throw a specific switch in response to the lighting up of a specific lamp. The illumination sequence of the lamps is programmed in advance. In the course of the experiment, the reaction times of the subject and his mistakes are registered.

If the different lights go on with the same average frequency (i.e., the appearance of different stimuli has an identical probability) the average reaction times to different stimuli will be identical. It is proportional to the logarithm of the number of lamps to which one must react. The wider the field of choice, the greater the reaction time (with specific limitations which we will not discuss here). If the different lights go on with unequal frequency (i.e., the appearance of some stimuli is more probable than that of others), the reaction time for the more frequent stimuli is less than the reaction time for the less frequent ones. In other words, the reaction time for more probable stimuli is shorter.* As Yaglom and Yaglom [181] rightly point out, the

*We will not pause for a mathematical description of the patterns that were established, referring the reader instead to the indicated references.

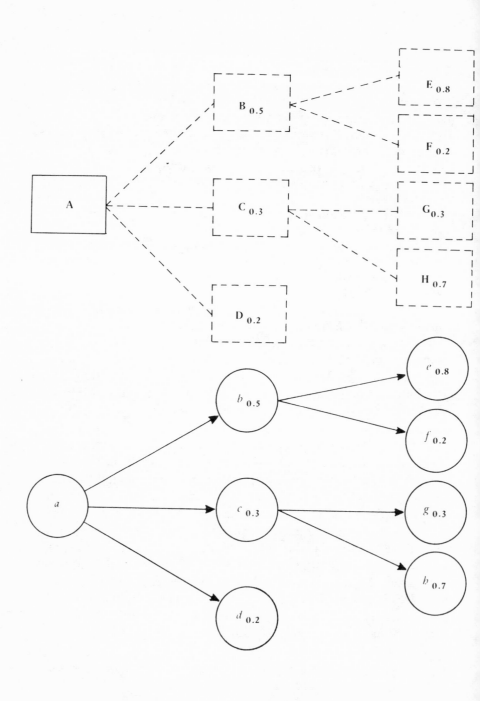

reason for such a difference in time lies in the fact that the rarer signals (the subject reacts to them more slowly) are more unexpected for them. In this connection, Leontyev and Krinchik [700] note that "when perceiving such a type of signal, a specific orienting activity appears which is evoked not by the 'newness' of the object, since the stimulus appears many times in the test, but by a special property—its unexpectedness."

One may assume that the reaction time for specific stimuli is directly related to the readiness to perceive these stimuli and to the expectation of them, while the readiness itself is related to the greater ease of occurrence of the process of anticipatory excitation in the corresponding centers of the brain. Since the level of anticipatory excitation of the brain centers depends to a great extent on the number of influences on them, the distribution of anticipatory excitations reflects the distribution of the probabilities of the eliciting stimuli. The "unexpectedness" of a stimulus is the consequence of a weak readiness for its perception; a weak readiness is the direct result of the low probability of a stimulus and the "rarity" with which its influence has been felt.[8]

Hence, it follows that one of the important tasks of instruction is to create in human beings the correct stochastic representation of the world and to develop a distribution of "readiness" for perceiving different stimuli such that it reflects their objective frequency.* Since a specific level of excitation in some center underlies each "readiness," it is a question of establishing the correct distribution of anticipatory excitations in the different centers.

At present, the question of how the readiness for perceiving and reacting to future stimuli is created in the brain is attracting a great deal of attention from both psychologists and physiologists. The efforts of Georgian psychologists of the school of Uznadze on the theory of set are essentially devoted to this problem (see, e.g., [851]). But, if in the theory of implantation, the problem of readiness has been examined without including probability con-

---

*Where the significance of the stimuli differs, the "readiness" must also reflect this significance.

siderations (although many facts obtained by the Georgian psychologists lend themselves to a probabilistic interpretation), Bernstein [535-538], Grashchenkov, Latash, Feigenberg [594], [828], [829], and others, have dealt explicitly with the problem of probabilistic foresight and probabilistic prognostication in the brain.

Of great interest in connection with this problem are the attempts to construct probabilistic models of perception (see, e.g., [528], [801], [802]), and also certain approaches to a probabilistic analysis of the processes of reasoning (see, e.g., [818], [819], [821]). Serious research has begun which is directed at the construction of stochastic models of learning (a detailed bibliography on stochastic and mathematical models in general can be found in Itelson [258], [259]).

We spoke above of the fact that, for a more rapid formation of the necessary centers of intensified excitation (i.e., sensitivity, *Ed.*), active operation with objects and their indicative features is of great importance. In order for the levels of excitation of the various centers to correspond to the objective frequency of the corresponding indicative features, a more frequent operation with the more probable indicative features is necessary. A strategy of verifying indicative features in problems of identification deriving from the probabilities of the indicative features leads to precisely this outcome.

We will clarify what has been said by an example. Let us suppose that in order to attribute some object x to a certain class A, it is necessary to establish the presence in it of one of the indicative features a, b, c, when it is known that $p(a) = 0.6$, $p(b) = 0.2$, and $p(c) = 0.1$. In order that the distribution of excitations in the brain and, therefore, the degree of readiness to perceive the given indicative features, correspond to their objective frequency, it is necessary that, of the 10 contacts with object x, a person encounter indicative feature a in it on the average of six times, indicative feature b two times, and indicative feature c once.

Two ways of developing a correct evaluation of the probabilities of the indicative features and a correct distribution of the "expectations" and "readinesses" to perceive them are possible.

The first consists of verifying the indicative features in the course of identifying objects in a random, arbitrary sequence (sometimes the verification begins with indicative feature a, sometimes with indicative feature b, and sometimes with indicative feature c). The second way consists of verifying the indicative features in a strict, optimal (from the standpoint of the principle outlined above) sequence.

Whatever way a person might take, in the final analysis, he would arrive at a more or less correct evaluation of the probabilities of the indicative features. For, out of each 10 contacts with objects of a given type, a person discovers indicative feature a in it on an average of six times, indicative feature b two times, and c once. Naturally, a person ultimately will expect indicative feature a more often than indicative features b and c.

However, if one follows the first way, a very large number of contacts with the object and, consequently, a considerable amount of time are required in order to develop in someone the correct evaluation of the probabilities of the indicative features and to form a correct distribution of "expectations" and "readinesses" to perceive the corresponding indicative features.[9] Moreover, each separate act of identification, as shown above, also requires a considerable amount of time. If, from the very start, the objective frequency of the indicative features is determined, and proceeding from that, the optimal sequence for their verification is devised, a correct evaluation of the probabilities of the indicative features can be established much more quickly. The optimal strategy for verifying indicative features given to the student in the form of some prescription (algorithm), from the very beginning may develop in him the most efficient sequence for perceiving indicative features and the necessary distribution of excitations in the corresponding centers. It creates a readiness for a more rapid reaction to the more probable indicative features and a less rapid reaction to the less probable ones. All this, having created a correct stochastic image of the world, facilitates and speeds up perception and processing of information in a person, as well as the choice of actions. It lessens the number of mistakes and generally raises the productivity of intellectual efforts. The absence of a correct stochastic image of the world and the lack of correspondence between the distribution of "expectations" of the appearance of specific indicative features and their objective frequency hinders intellectual activity and leads to an excessive amount of time and effort being spent.[10] Herein lies the importance of devising optimal strategies for the analysis and identification of objects by proceeding from the estimated probabilities of different indicative features. Therein lies the importance of teaching specific sequences for verifying indicative features (algorithms) and not just teaching the indicative features and their logical structures.

In discussing the dependence of the sequence of verifying indicative features on their frequency, we have until now meant cases where the significance of the different indicative features (and in general of phenomena with which a person deals and to which he must react) are equal. Very often, however, the significance of the different indicative features and phenomena in general with which he deals and to which he must react is unequal. In that case, the number of necessary "operations" (i.e., encounters, *Ed.*) with the different indicative features and phenomena in the process of training may not correspond to their frequency. Thus, it can turn out that unlikely but significant indicative features must be operated with (encountered, *Ed.*) more often than the more probable but less significant ones. What has been said can be clarified with a simple example.

Let us suppose that one must teach an operator to control a certain process of production. On the control panel are switches which must be turned off in case of a breakdown. Breakdowns happen rarely, but since the correct reaction to these situations is very important (an inopportune or faulty reaction could lead to a dangerous accident), the amount of training for reacting to a breakdown situation must be considerably greater than that which follows from a calculation of just the frequency (probabilities) of such situations. How much greater it is, is specified by the consequence of an inopportune or faulty reaction of the operator, and perhaps, in particular, it is evaluated on the basis of the ratio of the probability of the situation and "the cost of the mistake" (or the cost of the "delay") in terms of a faulty or delayed reaction to this situation. The lack of correspondence between the number of "encounters" with indicative features and their objective frequency can take place not only in industrial instruction, but also in the exercises for mastering school subjects. Here, the divergence in the number of "encounters" from the objective frequency of indicative features may be specified, obviously, not only by the significance of the indicative features, but also by some other factors. However, this question requires additional research.

The development of mathematical methods for evaluating the necessary number of exercises for each of the subjects studied in school not only in terms of the frequency of specific phenomena and their indicative features, but particularly in terms of their significance, has an enormous importance for pedagogy. The creation of such methods permits the specification in a strictly scientific way not only of what students must be taught, but also in *what sequence* the formation of corresponding operations should be carried out. Further, it will specify which and how many exercises must be executed to develop the different operations for forming the required abilities and skills in the best way and for obtaining the greatest progress toward the instructional goals. The existing empiricism in specifying the number and character of exercises to be presented to students will thereby be overcome.

Notes

1. We are speaking here of excitability in that broad sense in which Pavlov and his disciples used it (see, e.g., [743]).
2. Dobrynin and his associates have been studying the problem of significance in the last few years from the point of view of psychology [608] , [609] , [734] .
3. Our associate, L.V. Shenshev, aptly compared algorithmic prescriptions with the scaffolding that is raised while building a house. A scaffolding is necessary so as to build the house,

but once the house is built the scaffolding is unnecessary.

4. The comparison of the image of "expected" reality with the image of "actual" reality is carried out by a special apparatus which Anokhin called an "action acceptor" [524], and Bernstein a "collating apparatus" [534].

5. A situation of a similar kind takes place, for example, in the process of reading, when different letters follow one another with a specific probability. The same can be said about specific words following one after another. At present, these probabilities, which are specific for each language, are the object of much study. The circumstance that some elements of a language follow others with specific probability, permits the prediction (also with a specific probability) of which element will appear after which other one. This plays an important role in the origin of guesses about the meaning of unfamiliar words when reading a foreign text.

6. Differences in the probability of the succession of the various stimuli to each other is reflected psychologically in the different degrees of their expectation. Thus, if stimulus E follows stimulus B more often than stimulus F, the influence of stimulus B evokes a stronger expectation of stimulus E. The person is more ready to perceive E than F.

7. The issue of the perception of the stochastic structure of the sequence of signals shown and of the forming of specific "expectations" for specific signals was examined by Leontyev and Krinchik [701].

8. As Leontyev and Krinchik [700], [701] rightly note, the speed of the reaction depends not only on the frequency of the stimulus' influence—one might say its "expectedness"— but also on its significance for the subject (for the organism). The significance of the stimulus does not influence the speed of a person's reaction if this significance is identical in all of the stimuli. However, it influences the reaction time considerably if the various stimuli have a different significance. Investigations particularly devoted to the influence on reaction time of the significance of stimuli and to the analysis of patterns arising from this have been conducted by

Leontyev and Krinchik recently [701], [702]. In a number of investigations (see, e.g., [545], [820]) the dependence of the reaction time on some other factors is also shown.

9. If a person does not have a great number of contacts with an object, he will not be able to perceive the distribution of the probabilities of the indicative features and evaluate them correctly. Very often one encounters in life inadequate evaluations of the probabilities of indicative features and incorrect distributions of "expectations."

10. Special instruction in "expectations" has a great importance for forming the ability to solve problems and, in particular, for forming the ability to see orthographic difficulties. Students make many mistakes, not because they do not know the grammatical rules, but because they do not see the orthographic difficulties and do not distinguish those spots in words and in sentences which they have to "think over" and with which they have to "be careful." This stems from the fact that they do not know where to expect "dirty tricks," and *they do not expect them*. That "system of expectations" which creates a heightened sensitivity to specific stimuli and, therefore, leads to their isolation has not been cultivated in them. Exercises in verifying specific indicative features in objects give students a knowledge of which indicative features to expect in objects of specific classes and *train* them to expect these indicative features when perceiving these objects. That "system of expectations" is thereby developed which creates a readiness to distinguish the necessary stimuli and permits seeing orthographic difficulties.

PART TWO

Research Findings

Part Two: Section One

Organization of Experimental Instruction and Research

Chapter Fourteen

Objectives and Methods
of Experimental Instruction

1. Objectives of Experiment

In the preceding Part, the results of theoretical research into the problem of teaching students general methods of reasoning, specifically algorithms, were stated, and ways of designing algorithms were shown. In this Part, the results of experimental research into the problem of teaching students general methods of reasoning and algorithms will be set forth, and certain problems connected with these instructional methods will be discussed.

The syntax of the Russian language was chosen as material for experimental teaching. The correctness or incorrectness of punctuation serves as a good indication of whether or not the student has mastered the correct methods of reasoning; and students have been observed to experience great difficulties in learning the Russian language. The Russian language is one of the school subjects which is taught (and learned) least successfully.

It is doubtful whether one could find an adult even with an incomplete secondary education who would not be able to divide 243 by 3, whereas many people with secondary or even higher education cannot correctly punctuate a more or less difficult sentence (especially with conjunctions like *and* and *as*, and in a number of other cases). Without knowing the algorithms for solving grammatical problems, they often act by intuition and guesswork, which leads to a great number of mistakes.[1]

Since, according to our assumption, the important reason for mistakes and for difficulties in mastering the Russian language consists of the fact that students are not taught (or are poorly taught, or not always taught) the algorithms for solving grammatical problems, and most importantly algorithms of identification, one of the general goals of the experiment consisted of beginning to teach these algorithms to the students—once the algorithms of analysis (identification) for syntactical phenomena were devised, just as, for example, they are taught algorithms for multiplying or dividing in arithmetic. Such instruction by concept ought to facilitate significantly the mastery of a course in syntax, raise the quality of this mastery, lessen the number of mistakes, provide the students with some general methods of reasoning, and cultivate in them the ability to make correct judgments.

It is entirely clear that if students were to master grammatical algorithms the way they master arithmetical algorithms, the process of solving the problems of correct punctuation would not, in principle, be different from solving problems, for example, of dividing two numbers. It would become just as algorithmic. Actions with words and sentences would be just as precise and sure as actions with numbers. Of course, the number of operations necessary for solving grammatical and for solving arithmetical problems might be different, and the operations themselves might differ in terms of composition and difficulty (or ease). It is important, however, that the application of an algorithm to the solving of a syntactical problem would lead to the correct punctuation with the same necessity as the application of an algorithm for dividing two numbers would lead to the correct solution.* It is clear how much this could facilitate the mastery of syntax and raise the effectiveness of instruction.

What role is to be assigned to the teaching of algorithms in the study of the native language?

It is difficult to give a definitive answer to that question at this time. There is, no doubt, however, that teaching a language must not and cannot amount to the same thing as teaching algorithms. Forming the ability to speak well and express one's

---

*By contrast with English, rules of punctuation in Russian are strict and quite precise. Ability to punctuate sentences correctly is an absolute requirement for any educated person with cultural pretensions. (*Editor.*)

thoughts correctly and precisely, i.e., developing the culture of speech in the broad sense of the word, should have a major role in instruction. As for the role to be assigned to the teaching of algorithms, one may believe that algorithms are applicable for the most part to the study of the structural-grammatical and logical aspects of language phenomena, and they are less applicable to the study of the expressive-semantic aspects of the language and to the solution of expressive-semantic problems.[2]

In this book, we concentrate our attention on problems of a structural-grammatical and logical character. A great many of these problems must be solved in mastering the language. It suffices to say that the overwhelming majority of problems related to the ability of writing grammatically are these problems of structural-grammatical and logical character, and that it is possible to devise sufficiently general methods of solution for them and in algorithmic form (algorithmic prescriptions). Such problems will be the object of our examination in subsequent chapters.

The fundamental issues of experimentation on teaching students general methods of solving problems, particularly algorithms, may be formulated thus:

(1) Ascertain what significance the deliberate teaching of general methods of reasoning and particularly of algorithms has for raising the quality of knowledge, abilities, and skills in the Russian language. Ascertain further how, with the help of such teaching, one can raise the punctuation skills of students and facilitate the assimilation and mastery of grammar in general. Finally, what is the effectiveness of the deliberate teaching of algorithms?

(2) Devise procedures for teaching students general methods of thinking and algorithms, and perfect these procedures. In particular, find the means for teaching students not only how to apply pre-designed algorithms, but also how to develop algorithms independently.

(3) Design some special training devices for teaching algorithms; and test, evaluate, and perfect these devices.

(4) Determine the psychological particularities that enable students independently to devise and master algorithms. In

particular, determine the influence of instruction in algorithms on the development of motivation for learning and interest in the native language as a school subject.

(5) Establish how deliberate teaching of methods of reasoning and in particular of algorithms influences the development of certain general characteristics of intellectual activity and of certain general mental abilities in the students.

2. Materials and Methods of Experiment

Four topics in the syntax of the Russian language were chosen for the actual research: "the subject," "forms of simple sentences," "homogeneous parts of a sentence," and "the compound sentence." The first three are part of the seventh and eighth grade curriculum.[3]

These topics were chosen, first of all, because while studying them (and even after they have already been learned), students make many mistakes,[4] and second, because these topics are closely related to each other. (Thus, in not being able to find the subject in a sentence, it is also impossible to specify the type of simple sentence, or to distinguish a complex sentence from a simple one. The operations which are applied when identifying a simple sentence with homogeneous parts are also applied when identifying a compound sentence, and so on. Mastering all of these topics is also important for developing the culture of speech in students.)

The teaching experiment was conducted as part of a lesson under normal classroom conditions, first in the eighth grade of School No. 312 in Moscow and later (for the first three topics) in the seventh grades of five other schools. The experimenter conducted part of the lessons in the seventh grades, and the teacher conducted the remaining part.

Each topic was developed methodically and in detail. Consideration was given not only to the sequencing in the treatment of individual problems, but also to all aspects of independent study, including the types of exercises. After studying each topic, the students were assigned control study problems.

The experimental instruction in the lesson was always

combined with an individual diagnostic experiment. The latter made it possible to deduce those internal processes and changes, hidden from external observation, which took place in the students while mastering the material. Moreover, studying the dynamics and the course of mastery (and not just its results) gave the experimenter the necessary information for making corrections in the methods of instruction and thereby perfecting them.

The results of instruction in the experimental classes (for brevity, we will thus designate the classes in which experimental instruction was conducted) were compared with the results of instruction in control classes where the teachers conducted the class according to their usual methods. There were 22 such classes (in seven different schools). In some of the control classes, the experimenter observed the lessons. In all of the control classes, after studying the corresponding topics, students were assigned the same control study problems as the experimental classes. With some of the students in the control classes, the same diagnostic experiments were conducted as with the students in the experimental classes. In all, there were 60 students from control and experimental classes in the individual diagnostic experiment.

Two of the topics out of the four, namely, "types of simple sentences" and "compound sentences," will be described below, with the aim of illustrating the course of experimental instruction.

Before directly describing the study of each topic, the rationale for the methods underlying the instruction will be presented. It includes the following issues:

(a) analysis of the concept of the related grammatical phenomenon (i.e., its indicative features and their structure) as it is explicated in the textbook; the traditional method of presenting the topic in question is characterized;

(b) analysis of students' mistakes and difficulties connected with learning the corresponding concept and its application when solving grammatical problems;

(c) substantiation of the necessity for another definition of the concept to that given in the textbook, if the indicative features pointed out in the textbook are unsatisfactory; proposed indicative features and their logical structure are characterized;

(d) a description of the most efficient possible algorithms (in which the proposed indicative features are used) for solving the corresponding grammatical problems; and

(e) a description of certain principles and, in some cases, of algorithms of teaching.

One of the reasons for difficulties in mastering grammar and solving grammatical problems derives from the fact that the majority of students do not know the general form of an indicative feature, nor wherein lie its functions. Also, they do not know what kind of logical structures of indicative features exist and how the approach to identifying grammatical phenomena depends on the structure of their indicative features.

In this connection, we considered it expedient to conduct a special lesson devoted to exploring these important logical concepts and to forming the corresponding logical abilities before beginning the experimental instruction on specific grammatical topics. We conveniently designated this lesson as the "logic lesson."

Before proceeding to a description of the lesson in logic, we will pause for a short characterization of mistakes provoked by students' lack of the necessary knowledge of logic and absence of the corresponding logical abilities.

Notes

1. Teachers and instructional methodologists have pointed out many times the difficulties and mistakes in mastering the Russian language. Thus, Dudnikov [613] in his dissertation devoted to the problems of teaching punctuation, reaches the following conclusion from results of a mass investigation conducted by him in the schools of the Russian Federation: "From an analysis of punctuations in students' work in the eighth to eleventh grades, we came to the conclusion that punctuation skills in the upper grades, especially in the ninth grade, are extremely unsteady and precarious and that the level of punctuational competence of the students in these

grades is very low." Abakumov [518], Ustritsky [823], and others called attention to this earlier.

2. The science of language does not amount to the same thing as grammar and is not absorbed by logic, although the grammatical and logical aspects of language occupy an important place in this science and are very closely linked.

3. At present, the topics "types of simple sentences" and "homogeneous parts of a sentence" are studied in the eighth grade, but in the period when the experiment was carried out (1957-1959), they were studied in the seventh grade.

4. Quantitative data will be cited in the following chapter.

Chapter Fifteen

Forming in Students the Concepts of Indicative Features, Their Logical Structure, and Corresponding Reasoning Methods

1. Students' Mistakes Due to the Lack of Definite
Logical Knowledge and of Corresponding Abilities

In a paper [686] concerning the forming of methods of mathematical reasoning in students, we have already shown on the basis of experimental data how many students of the eighth and ninth grades, when using the words "indicative feature," actually did not know what they are, why indicative features are necessary, what their functions are, and what must be done with them in the process of identification and problem solving. The data we obtained when studying how students solve grammatical problems were analogous.

Here is an excerpt from a report of a conversation with the student Viktor Ye. The student parses the sentence *The translucent forest grows dark in solitude, and the fir tree shows green through the frost* (verse). In distinguishing the subject and the predicate, he says that the predicates here are homogeneous.*

Experimenter. Are you sure that they are homogeneous?

Student. Yes.

Ex. Remember what we call homogeneous parts.

---

*While Russian grammar is likely to be an unfamiliar subject-matter for most readers, and many Russian grammatical features have no parallels in English, the point here will be clear. Most of the task content in the accounts of experimentation hereafter is taken from Russian grammar. *A knowledge of Russian, however, will not be necessary* to appreciate the didactic rather than the grammatical aspects of the examples given. (*Editor.*)

Stu. Homogeneous parts are those words which answer one and the same question and refer to one and the same word.

Ex. Now apply that definition to the words *grows dark* and *shows green*. Are they homogeneous parts or not?

The student does not reply.

Ex. Do they answer one and the same question?

Stu. They do.

Ex. Can one say they are homogeneous?

Silence.

Ex. Do these words refer to one and the same word?

Stu. No.

Ex. Well, all the same, are they homogeneous parts of the sentence?

Stu. (After a pause, confusedly) I don't know.

As we can see, the student's knowledge of the indicative features of homogeneous parts is purely verbal. He names them, mechanically repeating the definition, but he does not know for what they are necessary and what to do with them when solving grammatical problems. He does not understand that the indicative features given in the definition must be isolated, that the character of their relation, i.e., their structure, must be established, that each of them must be "applied" to the object to be identified, and that, depending on the presence or absence of these indicative features in the object, a specific conclusion must be drawn. The student has no idea (concept) of all this, and he does not operate with indicative features as they really are. His knowledge of the definition does not help him in the least to solve the problem confronting him.

A large number of mistakes are related to the fact that students do not know what relations exist among indicative features and that some relations exist at all. They do not set themselves the task of establishing, nor do they establish, the logical structure of the indicative features of the phenomena analyzed, and their actions do not correspond to this structure. Thus, for example, in the course of an individual experiment, many students were given definitions in which the connectives linking the indicative features were omitted (examples of such definitions were cited earlier in this book), and the question was raised as to which connective could link them. This question provoked bewilderment in many students. They did not even understand what they were being questioned about.

Another experiment consisted of giving students definitions in which not only the indicative features were precisely distinguished, but also the connectives which linked them. Then objects were presented which had to be identified by means of these indicative features. The students were confronted with the question of how to act in order to determine what the object would be if the indicative features given in the definition were linked by the conjunction *and* or if they were linked by the disjunction *or*.

The most successful students correctly answered this question. However, the majority of average and poor students could not. Many of them did not even understand the question.

Let us cite several examples of students' mistakes which are related to the fact that they do not even try to establish the logical structure of the indicative features of objects (or phenomena, situations, etc., *Ed.*) and that they do not know how to act with the object when identifying it according to the structure of its indicative features.

It is known that students of the sixth grade, while correctly defining an adjective and naming the connective which links its indicative features, often consider words of the type *whiteness, hardness,* and *roughness* adjectives. The question as to why they think these words are adjectives usually elicits the answer: because they designate the indicative features of an object.

It is clear from what has been said what the nature of this mistake is. In order that a word be an adjective, it does not suffice that it designate an indicative feature of an object. It is also necessary that it answer the question "*which?*" or "*whose?*," i.e., that an adjective have a second indicative feature. When students do not recognize the logical structure of an adjective's indicative features and do not know that, for a positive conclusion about a word's belonging to the class of adjectives, it is necessary to establish the presence in it of both indicative features of an adjective, they draw such a conclusion after having verified merely the first indicative feature. This conclusion is untrue and incorrect.

Many of the students' mistakes are evoked by shortcomings in the exposition of the logical structure of indicative features in textbooks. We have discussed these shortcomings earlier in this book. We will cite examples of students' mistakes which are engendered by such shortcomings.

In the preceding Part of this book, work on the theoretical proposition connected with the role of recognizing the logical structures of indicative features when identifying phenomena was developed, and the basic tasks and the design of a "logic lesson" were specified.

These tasks are:

(1) Give students an understanding of what an indicative feature is, and why it is necessary to know the indicative features of phenomena.

(2) Form in students a concept of the ways of linking indicative features by logical connectives, i.e., of the logical structures of indicative features and, in connection with this, a concept of necessary and sufficient indicative features.

(3) Show students the dependence of actions for identifying phenomena on the logical structure of their indicative features and develop general methods for identifying phenomena according to the types of the logical structures of indicative features.

For the sake of space, we will not break down the tasks indicated into more elementary ones and describe those teaching operations which were applied for solving those problems. These operations are easily seen from the description of the course of the lesson. Let us just note that not one of the propositions was given to students in prepared ("ready-made") form. The method for conducting the course was such that the teacher always placed the students in such situations that they had to solve specific problems, and in solving these problems independently discover the necessary propositions and carry out the necessary intellectual operations.

2. Design and Implementation of a
"Logic Lesson"[1]

At the beginning of the lesson, the teacher said that, in order to learn to write grammatically, it is very important to know how to identify grammatical phenomena—parts of speech and parts of a sentence—and to know what must be done to distinguish correctly one grammatical phenomenon from another. He showed in certain examples how grammatical phenomena are confused by students and how the inability to distinguish them leads to mistakes.

Teacher. Here is why before going directly to the topic "the subject," we will pause at the question of on what the correct identification of one or another phenomenon depends. Let us imagine the following situation: A friend phones me at work and the following conversation takes place: "I promised to return a book to you today which I took from you" my friend says to me, "but I'm busy right now, and I asked my colleague Ivan Ivanovich to bring the book to you (he's passing your way). He'll be there in an hour."—"Unfortunately I have to leave in a half an hour and can't wait for him."—"Then meet him at the entrance of the subway station 'Kirovskaya,' and he'll give you the book."—"But I don't know him." What must my friend tell me about Ivan Ivanovich so that I (or any one of you, if you were in the same situation) could recognize him among other people?

Student.[2] He has to have certain distinctive marks.

Teacher. Right. Now think, what word could be used here instead of the words "distinctive marks"?

Stu. One could say, "indicative features."*

Tea. Good. Actually, indicative features are the same as distinctive marks. Indicative features are marks by which we recognize, i.e., identify, something. Henceforth, we will say "indicative feature" instead of the words "distinctive marks." Now, what indicative features of Ivan Ivanovich could my friend mention to me so that I could recognize him at the subway?

Stu. He could describe how he was dressed; for example,

---

*If the students are unable to answer this question, then it is the teacher who introduces the term "indicative features."

that he was wearing a brown overcoat.

Tea. Good. Let's write down that indicative feature of Ivan Ivanovich. (The teacher writes on the board: "in a brown overcoat.") *Does it suffice* to know that indicative feature of Ivan Ivanovich in order to recognize him among the other people? (The teacher emphasizes the word suffice by intonation.)

Stu. No.

Tea. Why do you think so?

Stu. Because there are many men in brown overcoats.

Tea. Right. We won't be able to distinguish Ivan Ivanovich among other people by the one indicative feature, "in a brown overcoat." What other indicative features could you name which Ivan Ivanovich might have?

Stu. For example, he could be tall.

Tea. What do you think—is that a good indicative feature for us?

Stu. No, it isn't.

Tea. Why?

Stu. Because it is difficult to measure a person's height by eye.

Tea. It really is difficult to measure height by eye. But that's not the only problem. Let us suppose that, having the corresponding instruments for measuring, we could measure the height of the people passing by us. Why would the indicative feature "tall" be bad in that case?

Stu. Because what is meant by tall has not been previously indicated.

Tea. Right. This indicative feature is bad in such a formulation because it is imprecise (it is not specified or not precisely indicated what is meant by tall and which height can be considered tall) and, therefore, this indicative feature is not univocal: some people could consider a person tall, and others will think he is of medium height.* What other more appropriate indicative feature could you propose?

---

*It is very important to draw the students' attention to the requirements for indicative features. An analysis of indicative features from the standpoint of their specificity, univocality, and other qualities is of great importance for the development of students' general intellectual abilities.

Stu. Ivan Ivanovich in a gray hat.

Tea. Well, I think that indicative feature is considerably better. It is more specific. Here, there can be no divergence of opinions, as with an evaluation of height. Let's write down this indicative feature, also.

The second indicative feature is written under the first, and they are numbered.

Tea. Does it suffice to know both of these indicative features in order to recognize Ivan Ivanovich?

Stu. No. There are also many men in brown overcoats and gray hats.

Tea. All the same, how are we going to distinguish Ivan Ivanovich?

Stu. We still have to show some more of his indicative features, for example, that he has a book in his hand.

Tea. Let's also write down that indicative feature.

The indicative feature is written under the preceding ones.

Tea. What do you think now? Are there enough indicative features to recognize Ivan Ivanovich?

Stu. Not yet. Suppose two men suddenly appeared at the subway entrance in brown overcoats and gray hats, each one with a book. In order to be sure, it is necessary to have one more indicative feature.

The students propose several indicative features: "a scar on the right cheek," "wearing horn-rimmed glasses," "wearing a gray and black striped scarf."

Tea. Let us take one of the proposed indicative features, for example, "scar on the right cheek."

The indicative feature is written under the preceding ones.

Then the teacher asks whether there are now enough indicative features in order to distinguish Ivan Ivanovich from other people, and the students say that now, perhaps, there are enough.[3]

Tea. Really? Could it possibly happen that two men wearing brown overcoats, gray hats, carrying a book in their hand, and both having scars on their right cheek turn up at the same place? In practice, these indicative features suffice in order to recognize

Ivan Ivanovich in a large number of other people.[4] Now, think. What connective could link these indicative features?

The students answer without difficulty that the conjunction *and*[5] could link the indicative features. This conjunction is written between the indicative features and the logic diagram looks like this:

Ivan Ivanovich's Indicative Features

(1) wearing a brown overcoat
   and
(2) wearing a gray hat
   and
(3) with a book in his hand
   and
(4) having a scar on his right cheek.

After a logic diagram of the indicative features is drawn up, the teacher must bring the students to an understanding of how one must act according to these indicative features, which system of operations follows from the logic diagram of the indicative features, and how the given system stipulates the method of acting with respect to the phenomenon which must be identified. This is the second and most important stage of the task.

Tea. Suppose now that we went together to meet Ivan Ivanovich. We get there a bit earlier than the appointed time, and we stand in the vestibule of the subway. Different people go past us. What must we do in order to recognize Ivan Ivanovich among them?

Stu. We have to see which of the people have Ivan Ivanovich's indicative features.

Tea. Right. Here we are standing at the entrance to the subway and watching the people going by us. Suppose I look at a certain person. What must I do to establish that he is Ivan Ivanovich?

Stu. In order to do that, it is necessary to see what color overcoat he is wearing.

Tea. Of course, I will verify whether he has the indicative feature "wearing a brown overcoat." Suppose he does not have the first indicative feature, i.e., he is not wearing a brown overcoat?

Stu. Then it is clear that he is not Ivan Ivanovich.

Tea. Do you consider it necessary to examine that man any further? Is it necessary to verify the presence of the remaining indicative features in him? (Instead of the words "see whether he has the remaining indicative features," let's say "verify the remaining indicative features.")

The students say that it is unnecessary.

Tea. Actually, if the first indicative feature is missing, there is no point to verifying the remaining indicative features. We can immediately draw a negative conclusion: that man is not Ivan Ivanovich. Now, let us suppose that the person going past us does have the first indicative feature, i.e., he is wearing a brown overcoat. Can we draw the conclusion from this that it is Ivan Ivanovich? Does the presence of one indicative feature in the person *suffice* in order to conclude that this is Ivan Ivanovich?

Stu. No. It is necessary to verify whether he has the other indicative features of Ivan Ivanovich.

Tea. And why do you think the other indicative features of Ivan Ivanovich must be verified?

Stu. Because the person going past us will be Ivan Ivanovich only if he is wearing a brown overcoat and gray hat, is carrying a book in his hand, and has a scar on his right cheek.

Tea. Right. When the indicative features are linked by the conjunction *and*, the presence of one indicative feature does not suffice for a positive conclusion (i.e., in order to say "this is Ivan Ivanovich"). The presence of the other indicative features must be verified. Which indicative feature will you verify now?

Stu. We have to verify whether the person going by has the second indicative feature of Ivan Ivanovich—to see what color hat he is wearing.

Tea. Suppose he does not have the second indicative feature. The person is not wearing a gray hat. What do we do then? Is it necessary to verify the remaining indicative features?

Stu. No, if there is no second indicative feature, it is not Ivan

Ivanovich.

Tea. And if he does have it? Does the presence of two indicative features *suffice* in order to draw a positive conclusion that this person is Ivan Ivanovich?

Stu. No. We have to verify the presence in him of the third indicative feature.

The remaining indicative features are examined in an analogous fashion. Then the teacher summarizes, drawing the following conclusion from the conversation with the students.

Tea. Now, we have established that the indicative features of Ivan Ivanovich are linked by the conjunction *and*. This means that, for some person to be Ivan Ivanovich, he must have all of the indicative features of Ivan Ivanovich: the first, and the second, and the third, and the fourth. How must we act in order to find out whether the person going by us is Ivan Ivanovich? We have to verify the presence in that person of *all* the indicative features of Ivan Ivanovich. If even one indicative feature is missing, he is not Ivan Ivanovich.

Tea. (continues) In which order are we going to act?

(1) First of all we are going to verify the presence of the first indicative feature in the person going past us.

If the first indicative feature is not there, then we will not verify any more indicative features and at once draw a negative conclusion: he is not Ivan Ivanovich.

If the first indicative feature is there, then:

(2) We verify the presence of the second indicative feature in him.

If the second indicative feature is not there, then we won't verify any more indicative features and at once draw the negative conclusion: he is not Ivan Ivanovich.

If the second indicative feature is there, too, then:

(3) We verify the third indicative feature . . . . We act in this way until we have arrived at the last indicative feature.

If the last indicative feature is not there, then we won't verify any more indicative features and at once draw the negative conclusion: he is not Ivan Ivanovich.

If the last indicative feature is there (i.e., all the indicative features are there), then—and only then—do we draw a positive conclusion that he is Ivan Ivanovich.

Tea. (continues) Since in order to draw a positive conclusion it is necessary that *all* indicative features be present, then each indicative feature is *necessary*. For if one of the indicative features is missing, we cannot draw a positive conclusion. But each indicative feature in being necessary is not *sufficient*, as we saw. If the person going past me has some *one* indicative feature of Ivan Ivanovich, this is still *not sufficient* in order to come to the conclusion that he is Ivan Ivanovich. Only the *sum total of indicative features*, their *combination*, is *sufficient* for a positive conclusion. Now, let us examine, in a diagram, everything we have talked about so far. We will designate each indicative feature by a separate small circle. We will designate the presence of an indicative feature by a plus sign (+) and the lack of one by a minus sign (-).

The teacher draws the diagram.[6]

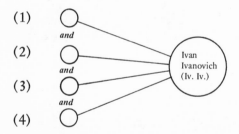

The teacher conducts the following exercise with reference to the diagram. He writes pluses and minuses in the small circles, and the students have to say what is obtained in the large circle. We show several completed diagrams on page 359 as an illustration.* After training, the teacher draws the conclusion.

Tea. A plus in the circle on the right can only be there when there are pluses in all of the small circles to the left. If there is a

---

*The absence of a plus or minus sign in a circle signifies that the corresponding indicative feature was not verified (a conclusion was drawn on the basis of verifying the preceding indicative features).

minus in just one of them, then there will be a minus in the circle on the right.

The students memorize this rule quickly and easily.[7]

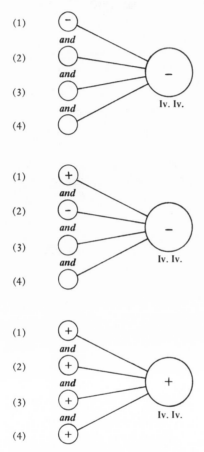

After the exercise with diagrams is finished, the teacher goes on to the forming of an even more general method of reasoning. He states a thesis to the effect that all that has been said about indicative features, their structures, and actions with them has significance not only for recognizing some individual, but also for recognizing, discerning, identifying any object or phenomenon.

Tea. We saw with you that, in order to identify Ivan

Ivanovich and to distinguish him from other people, we have to know his indicative features. What do you think? Is it only Ivan Ivanovich who has indicative features, and is it only Ivan Ivanovich who can be recognized by his indicative features?

The students say that all people, animals, things, etc., have indicative features.

Tea. Actually, every object and every phenomenon has its indicative features, and in order to identify objects, i.e., in order to isolate them from among other objects and distinguish them from other phenomena, we must know their indicative features. Remember the definition of an angle. What is an angle?

Stu. An angle is a figure formed by two lines emanating from the same point.

Tea. Which indicative features are pointed out in this definition? Which indicative features must a geometrical figure possess in order that one can say that it is an angle?

The students name the indicative features of an angle; they are written down on the board.

Tea. Which connective joins these indicative features?

Stu. They are joined by the conjunction **and**.

The conjunction **and** is written between the indicative features and the diagram of the indicative features is as follows.

Indicative Features of an Angle

A figure is called an angle which
(1)   is formed by two lines
     and
(2)   these lines originate from the same point.

A more abstract diagram of the indicative features of an angle is sketched:

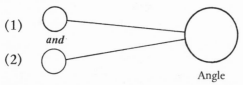

Tea. Now, let us suppose that we have many geometric

figures, and we have to find an angle from among them.
The teacher sketches several figures on the board.

Tea. These figures are like the people who pass by us in the
subway, and the angle is our "Ivan Ivanovich"—the object which
we have to find and identify. How must we act in order to
recognize which of the figures is an angle, i.e., as with respect to
Ivan Ivanovich?

Stu. We have to verify whether it has the indicative features
of an angle.

Tea. Using the indicative features of an angle, let's verify
whether the figure at the left is an angle. What must we do?

Stu. We have to verify whether it has the first indicative
feature of an angle.

The students establish the fact that the figure being verified
has the first indicative feature.

Tea. Does the presence of just this one indicative feature
suffice in order to draw a conclusion that this given figure is an
angle?

Stu. No, it is still necessary that the lines originate from a
single point.

Tea. Right. We have to verify whether the figure has the
second indicative feature.

It is established that there is no second indicative feature.

Tea. What conclusion may we draw from this?

Stu. Since the second indicative feature is not there, the
figure is not an angle.

The other figures are analyzed in an analogous fashion.[8]

Thereafter, the students are given a task which requires that
they analyze quite independently the indicative features, determine their logical structure, construct a logic diagram of the
indicative features, and on this basis specify the operations
necessary for identifying the phenomenon.

Tea. With the help of indicative features, we can recognize
not only people and geometrical figures, but also grammatical

phenomena. Open the textbook of the Russian language, read the definition of an adjective, isolate the indicative features, number them, and put down the connective which joins them. In short, act as we did in the case of the indicative features of Ivan Ivanovich and in the case of the indicative features of an angle.[9]

After the students accomplish the task in their notebooks, a structural diagram of the indicative features of an adjective is written on the board.

The Indicative Features of an Adjective

Words are called adjectives which
(1)   designate the characteristics of objects
        and
(2)   answer the questions *which? whose?*

Then the students are given a series of words (*walking, wooden, wood, redness, red*), and it is proposed that they determine which of these words are adjectives and which are not, and why, using the indicative features shown.

The students, having pointed out the necessary order of actions, solve this problem without difficulty. Not only do they find the adjectives *wooden* and *red* from among the words, but they also prove why the words *walking* and *redness* are not adjectives (because one of the necessary indicative features or both are missing).

After completing this stage of the "logic lesson," it is helpful to draw the students' attention to the fact that for a correct identification of phenomena, the position of the indicative features in the diagram and also the sequence of their verification (which indicative feature is verified first, which second, and so on) is of no importance. The correct answer may be obtained through any sequence in the positions of indicative features and of their verification. But if the indicative features are joined by the conjunction *and*, it is more efficient to begin the verification with the rarest indicative features. This narrows down the field of search and obtains an answer (e.g., whether the person going by is

Ivan Ivanovich) more quickly. To substantiate the reason why it is more efficient to start verification with the rarest indicative features when they are joined by the conjunction *and*, one can conduct reasoning exercises and calculations analogous to those cited by us in Chapter Eleven. In the example of identifying Ivan Ivanovich, the rarest indicative feature is a scar on the cheek. One must begin the verification with it. Practically speaking, with the necessity of recognizing Ivan Ivanovich in a crowd of people going by, one must look at their faces, more precisely at their right cheek, not their coat, hat, or what they are carrying in their hands. If one were to imagine that a huge stream of people is going by, and it is difficult in practice to look at each person carefully (notice what color overcoat he is wearing, what color hat, what he is carrying in his hands, and whether he has a scar on his cheek) then the advantage of the efficient strategy of examination (if there is no scar on his cheek, then it is not Ivan Ivanovich, and there is no further need to look at the person) is obvious. Since the indicative feature "a scar on the cheek" is very rare, then in the overwhelming majority of cases, for a conclusion that the person going past is not Ivan Ivanovich, it suffices to limit oneself to verifying just one indicative feature. With another strategy, it will be necessary to verify each person by a greater number of indicative features and examine him more "minutely" which, in a large stream of people and with the necessity of quickly examining each person, is very difficult and tiring.[10] From this living example, the students can understand very easily how important it is to think about the sequence in which the indicative features should be verified and how even such a "simple" action as recognizing a person must be done "with intelligence." In our time, problems of optimization are the most important category of problems which people in the different spheres of industry, community life, etc., have to solve. That is why it is important to train children from an early age to reflect about the efficiency of their actions and to train them even in the least important things they do to seek an efficient solution.

After the students have familiarized themselves with the structure of indicative features joined by the conjunction *and* and have mastered the procedure corresponding to this structure, the teacher goes on to familiarize them with the structure of indicative features linked by the disjunction *or*.

Tea. We saw that indicative features can be linked by the conjunction *and*, and we know what to do to identify an object or phenomenon in this case. But it so happens that linking indicative features by the conjunction *and* is not the only method of joining them. Let us suppose that we go to a certain small town (our touring trip is going to start from there), and my friends ask me to give a package to their relative, Maria Ivanovna. We arrive in this town, and I telephone Maria Ivanovna to tell her about the package. She says that unfortunately repairs are being made in her apartment and that she cannot invite me over.—"Well, then let's meet at three o'clock by the post office."—"Fine," answers Maria Ivanovna.—"But how are we going to recognize each other?"—"I'll be wearing a big white hat with a black feather (no one else has a hat like it—I ordered it specially), or I'll be carrying a dark blue suede bag which has 'Moscow-Prague' written on it. My brother, who recently returned from Czechoslovakia on a business trip, gave it to me. I also haven't seen any other such bags in our town."

The teacher asks the students:

Tea. What are Maria Ivanovna's indicative features? Let's write them down. (The indicative features are written down.) By what connective can they be joined?

Stu. The indicative features may be joined by the connective *or*.

The disjunction is written down and the diagram is this:

Maria Ivanovna's Indicative Features

(1)   Wearing a white hat with a black feather
  or
(2)   Carrying a dark blue suede bag with the inscription "Moscow-Prague."

Tea. Let us suppose that we are standing at the post office and waiting for Maria Ivanovna. A woman walks by. What must we do to find out whether it is Maria Ivanovna or not?

Stu. Verify whether or not she has Maria Ivanovna's indicative features.

Tea. Which one shall we start with?

Stu. Let's start with the first one.

Tea. Let's assume that the woman going by us has the first indicative feature: she's wearing a white hat with a black feather.

Stu. Then it is Maria Ivanovna.

Tea. Well, what about the second indicative feature? Do we have to verify it?

The students say it is unnecessary to verify the second indicative feature.

Tea. Why do you think that the second indicative feature does not have to be verified?

Stu. Because the woman going past is Maria Ivanovna if she is wearing a white hat with a black feather or carrying a dark blue suede bag with the inscription "Moscow-Prague." One indicative feature suffices.

Tea. Correct. If the indicative features are linked by the conjunction *or*, then the presence of just one of the indicative features suffices for a positive conclusion. The second indicative feature does not even have to be verified. But if the woman passing by is not wearing a white hat with a black feather, i.e., there is no first indicative feature, how must we act?

Stu. We must verify whether she has the second indicative feature: whether she is carrying a dark blue suede bag with the inscription "Moscow-Prague."

It is established through the conversation that if the woman has such a bag, then she is Maria Ivanovna. If not, then she is not Maria Ivanovna.

Tea. Now, think: Is each of Maria Ivanovna's indicative features necessary? Does she have to be wearing a white hat with a black feather and carrying a dark blue suede bag with the inscription "Moscow-Prague?"

Stu. No, it is not necessary. She can be without the hat or without the bag.

Tea. Actually, if one or the other is said, then it is not at all necessary that there be both, although the case where both indicative features might be present is not excluded.[11] And is each of the indicative features sufficient? Does it suffice for us to see a woman wearing a white hat with a black feather or carrying a dark blue suede bag with the inscription "Moscow-Prague" in order to say that she is Maria Ivanovna?

The students say that it suffices.

Here, again, it is expedient to raise the question of the efficient sequence for verifying the indicative features. If, with indicative features linked by the conjunction *and*, it is more efficient to begin with the rarer indicative features, when verifying indicative features linked by the disjunction *or*, it is more efficient to do the opposite—to begin with the least rare (i.e., the most probable) indicative features. If it is known that Maria Ivanovna is *most likely* to be wearing a white hat with a black feather, then one must first of all see whether the woman going by is wearing a hat and if so, what kind. If it is known that Maria Ivanovna is most likely to be carrying the bag, then one must first of all verify what the woman going by is carrying. With such a "strategy of examination" when meeting Maria Ivanovna, we will probably be able to "economize" our operation.

It must be said that in spite of the somewhat contrived character of the situation, the students easily see the main idea behind such problems. They react to this problem as they react to many interesting and highly useful problems which are contained, for example, in published collections of interesting problems. They accept them with humor, but at the same time seriously. The students accept right away as analogous to real-life situations invented tales about the sequence in which one ought to take a cabbage, a goat, and a wolf across the river so as to keep all of them. The real meaning of the problem on determining the efficient sequence for verifying indicative features linked by the disjunction *or* can also be shown to students through examples on finding lost objects, looking for a malfunction in a machine, establishing a diagnosis of illness, etc.

Later, based on comparisons of how the indicative features of Ivan Ivanovich and Maria Ivanovna are linked, they come to the conclusion that each of Ivan Ivanovich's indicative features is necessary, but not sufficient. Each of Maria Ivanovna's indicative features is, on the contrary, sufficient, but not necessary.

Then the teacher goes on to the generalized procedure for identifying phenomena in the case where their indicative features are joined by the logical disjunction *or*. In order to do this, the indicative features are again depicted as circles, and this diagram is sketched on the board.[1][2]

The teacher puts pluses and minuses in the circles designating the indicative features, and the students have to say what the answer will be (i.e., in the circle designating Maria Ivanovna).

Here are the variants which were analyzed together with the students:

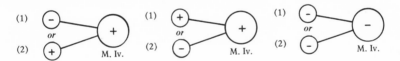

Then the teacher shows the students the way of formulating the procedural order in a general form.

Tea. How are we going to act in those cases where the indicative features are joined by the disjunction *or*, i.e., when each of them is sufficient but not necessary?

Stu. First we must verify whether the object has the first indicative feature. If it is in the object, then we draw a conclusion right away to put a plus sign in the circle to the right.

Tea. So, this signifies that we have found the object sought. Well, what if the first indicative feature is not there?

Stu. Then we verify the second indicative feature; and if it is there, we conclude that this is what we are looking for.

Tea. And if the second indicative feature does not turn up?

Stu. Then this means that it is not what we are looking for, and we have to put a minus in the circle to the right.

Tea. So in which case should we put a minus in the circle to the right?

Stu. When the object has none of the indicative features that we are verifying.

Tea. Well, how many indicative features suffice for a positive conclusion?

Stu. The presence of just one indicative feature.

Then the teacher goes on to comparing the procedural order in cases where the indicative features are linked by the conjunction **and** and when they are linked by the disjunction **or**. The students are guided toward recognizing those differences in the character of procedures which follow from the differences in the logical structures of indicative features (what this difference consists of is described above: see Chapter Ten).

Thereafter, it is proposed that the students recall the indicative features of a number divisible by five, that they write down these indicative features, and establish which connective joins them. The following structural diagram of the indicative features is constructed.

Indicative Features of a Number Divisible by Five

A number is divisible by five if
(1)   it ends in the number five
  or
(2)   it ends in a zero.

Then the teacher offers a series of numbers (150, 83, 15, 1066) and the students have to determine whether these numbers are divisible by five. Using these indicative features and reasoning similarly to the ways illustrated above, they have to specify whether these numbers are divisible by five.

The teacher asks how the indicative features for the divisibility of a number by five are joined: like those for Ivan

Ivanovich, or like those for Maria Ivanovna. Then they draw the conclusion that in order to find numbers which are divisible by five, one must reason precisely as one reasoned when looking for Maria Ivanovna.

We quote the reasoning of one of the students. He was asked to establish whether the number 150 was divisible by five.

Tea. What must be done to find out whether 150 is divisible by five?

Stu. We have to see whether the figure 150 has the first indicative feature: does it end with 5? This figure does not end with five.

Tea. Is it possible or not to draw a negative conclusion right away?

Stu. No, we have to verify the second indicative feature: does the figure end with a zero? It does. This means that 150 is divisible by five.

Tea. Is the presence of one of our two indicative features enough to draw a positive conclusion that it is divisible by five?

Stu. Yes, it is enough.

Tea. But now tell me, in which case would the number not be divisible by five?

Stu. If it did not end in either five or zero.

Tea. Correct. If both indicative features were missing—both the indicative feature "ends in the number five" and the indicative feature "ends in the number zero"—it could not be divisible by five.

Then the students are asked to open the textbook of the Russian language and write down the indicative features for writing words in capital letters and to determine the conjunction which joins these words. The students arrive at the following diagram of the indicative features.

Indicative Features for Writing Words with Capital Letters

Words are written with capital letters which
(1)   begin a sentence
      or

(2) designate people's names, patronymics, surnames, and nicknames
or

(3) designate the names of animals (not generic names, but names like Fido, Puss, etc.)
or

(4) designate geographical names
or

(5) designate the honorary titles of the USSR
or

(6) designate the names of holidays of the Revolution or commemorative days
or

(7) designate the names of historical events
or

(8) designate the names of newspapers, magazines, and literary and musical works.[13]

Thereafter, the students are given a number of words, and they have to specify whether these words should be written with a capital letter or not. They use this diagram and reason according to it.

Finally, the teacher goes on to the concluding part of the "logic lesson"—to cases where indicative features have a mixed structure.

Tea. Up until now we have examined indicative features which are either joined by the conjunction *and* or by the disjunction *or*. But it happens very often that some indicative features (or groups of indicative features) are joined by the conjunction *and* and others (or their groups) are linked by the disjunction *or*. Let us imagine such a situation. My friend who was supposed to give me back my book said the following: One of two of my colleagues, Ivan Ivanovich or Pyotr Petrovich has to come into your area. I still do not know who is going and who will give you the book, but it will be either Ivan Ivanovich or Pyotr Petrovich. Then my friend named the indicative features of each of his colleagues.

Indicative Features of Iv. Iv.	Indicative Features of P. Pet.

Indicative Features of Iv. Iv.

(1)   in a brown overcoat
      and
(2)   in a gray hat
      and
(3)   carrying a book
      and
(4)   with a scar on his right cheek

Indicative Features of P. Pet.

(1)   in a leather jacket
      and
(2)   in a dark blue cap
      and
(3)   bearded
      and
(4)   carrying a black portfolio

Tea. Let us call the person who has to give me the book the "messenger." This messenger can be either Ivan Ivanovich or Pyotr Petrovich. Let us write down the indicative features of the messenger, one under the other. (The indicative features are written down.)

Tea. We have two groups of indicative features before us. How are the indicative features joined to each other within each group?

Stu. By the conjunction *and*.

Tea. And how are the groups of indicative features joined to each other?

Stu. By the disjunction *or*.

The disjunction *or* is put into the diagram, and it takes this form:

Indicative Features of the Messenger

I   (1)   in a brown overcoat
            and
       (2)   in a gray hat
            and
       (3)   carrying a book
            and
       (4)   with a scar on his right cheek
            OR

II   (1)   in a leather jacket
            and
      (2)   in a dark blue cap
            and
      (3)   bearded
            and
      (4)   carrying a black portfolio

Tea. How will we act now in order to find out whether the man going by us is the messenger?

Stu. First, we have to see whether the person going by us is Ivan Ivanovich. If yes, then this means that he is the messenger.

Tea. And what if it is not Ivan Ivanovich—can we conclude that this person is not the messenger?

Stu. No, because the messenger could be Pyotr Petrovich.

Tea. So what do we have to do then?

Stu. We have to see whether the person going by us has the indicative features of Pyotr Petrovich. If yes, then this means that he is the messenger, and if not, then he is not the messenger.

Tea. Now, when we have indicative features united in a group, we have to determine first of all which connective joins the group of indicative features, and then we see which connective joins the indicative features within each group. Then we act according to these connectives. Let us represent the indicative features by circles and do some exercising on the diagram. (The diagram is sketched.)[14]

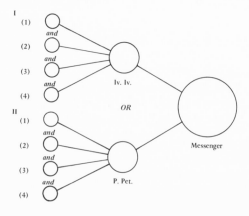

Tea. Now, we are standing in the subway and waiting for the messenger. What do we do in order to find out whether he is the messenger?

Stu. We look to see whether he has Ivan Ivanovich's first indicative feature.

Tea. Let us suppose that the first indicative feature is there (a plus is put in the first circle). Can we conclude from this that it is Ivan Ivanovich?

Stu. No, we have to verify whether the man going by has the second indicative feature.

Tea. Let us suppose that he has the second indicative feature (a plus sign is put in the second circle). What do we do then?

Stu. We verify whether the third indicative feature is there.

Tea. Suppose the third indicative feature is not there (a minus sign is put in the third circle).

Stu. Then this means that it is not Ivan Ivanovich (a minus is put in the circle over the words "Ivan Ivanovich").

Tea. Is it or is it not necessary to verify the fourth indicative feature of Ivan Ivanovich in the same way?

Stu. No, it is not necessary.

Tea. So, we have established that the man going by us is not Ivan Ivanovich. Can we conclude that he is not the messenger?

Stu. No, because he could be Pyotr Petrovich.

Tea. What do we have to do then?

Stu. We have to verify whether he has the first indicative feature of Pyotr Petrovich.*

Tea. Let us suppose that the man passing by has the first indicative feature of Pyotr Petrovich (a plus sign is put in the circle). Then what?

---

*Attention must be drawn to a difficulty in this illustration. Above it was assumed that the person in question (whose indicative features are being verified) possesses Ivan Ivanovich's first indicative feature (brown overcoat). Clearly, this rules out the possibility of his possessing Pyotr Petrovich's first indicative feature (leather jacket) as asserted below. A parallel problem exists with respect to the two second indicative features. Therefore, some different stipulation of Pyotr Petrovich's first two indicative features—drawn from categories that are not mutually exclusive with those stipulated for Ivan Ivanovich (e.g., eyeglasses, scarf)—would seem necessary. (*Editor.*)

Stu. Then we verify Pyotr Petrovich's second indicative feature.

Tea. Let us suppose that he has the second indicative feature (a plus sign is put in the circle). Then what?

Stu. Then we verify the third indicative feature.

Tea. Let us suppose that the third indicative feature is there (a plus sign is put in the circle). Then what?

Stu. We verify the fourth indicative feature.

Tea. But the fourth one is not there (a minus sign is put in the circle). What conclusion can be drawn from this?

Stu. This means that it is not Pyotr Petrovich (a minus sign is placed in the circle over the words "Pyotr Petrovich").

Tea. So the man going by us is neither Ivan Ivanovich nor Pyotr Petrovich. What conclusion can be drawn from this: is he the messenger or not?

Stu. It is clear that he is not the messenger (a minus sign is put in the circle over the word "messenger").

Tea. But now let us suppose that the man passing by also has the fourth indicative feature (instead of a minus, a plus sign is put in the circle).

Stu. This means that it is Pyotr Petrovich, and he is the messenger.

The minuses in the circles over the words "Pyotr Petrovich" and "messenger" are replaced by pluses. The diagrams have this form:

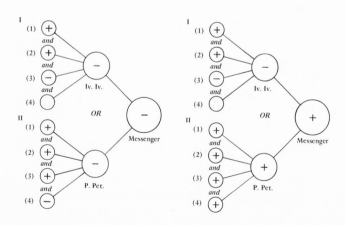

The students can be given still other data, and they have to reach the corresponding conclusions.

The logical structure of the indicative features of the messenger represent a disjunction of conjunctions. It is also very useful to present phenomena to the students for analysis, the structure of whose indicative features represents a conjunction of disjunctions as well as a disjunction from parts, some of which are compound (joined by *and*) and others which are simple; and finally the same thing can be said for the disjunction of conjunctions.

Exercises according to diagrams expressing the mixed structure of indicative features have a special significance, since the indicative features of many grammatical and mathematical phenomena have precisely such a mixed structure of indicative features.

After training by diagram has finished, and the students have obtained a general idea of the structure of operations with a mixed structure of indicative features, the teacher summarizes.

Tea. We have seen from a number of examples that one can correctly recognize objects or phenomena only when one knows their indicative features and can establish which connectives join them. The procedure for identifying these objects and phenomena depends on which connectives join the indicative features. Later on, we are going to study different grammatical phenomena, and in order to learn how to identify them without making mistakes, how to distinguish one from another, and how to write correctly, we shall always:

(1)   determine the indicative features of these phenomena;
(2)   establish which conjunctions join these indicative features;
(3)   construct a diagram of the indicative features; and
(4)   depending on the character of the diagram (the way the indicative features depicted in it are joined), specify how one must proceed when identifying the corresponding phenomena.[15]

Such an order of selecting and analyzing phenomena does not depend on the phenomenon with which we are dealing. We kept this order when we had to specify the sequence of actions for identifying Ivan Ivanovich and Maria Ivanovna, when we had to specify whether a geometrical figure which was examined was an angle, whether a given word was an adjective, etc. Thus, this order is *general*. It is applicable to the most *diverse* objects and phenomena.

    Tea. (continues) When in the course of identifying different phenomena, we drew diagrams of their indicative features, we noted that the most diverse phenomena could have identical diagrams of indicative features. It follows that the character of the procedures for identifying them can also be the same. For example, the indicative features of Ivan Ivanovich, of an angle, and of an adjective were joined by the conjunction *and*, and the procedures for identifying them were one and the same. This also applies to the indicative features of Maria Ivanovna, the indicative features of a number divisible by five, and the indicative features for writing words with capital letters, where the joining by the disjunction *or* specifies another procedure for identification, but one that is also identical for all the phenomena examined. What has been said also applies to those cases where the indicative features of different phenomena are linked by the conjunction *and* and their groups by the disjunction *or* and vice versa. This was precisely how the indicative features of the messenger were joined.

    Tea. (continues) One important conclusion follows from all this, which many of you have already drawn: the phenomena which we discuss and the problems which confront us can be different, but the procedure and the way (method) of acting are the same. And this does not depend on the content of the indicative features of phenomena (they could be the indicative features of both people and things, the indicative features of mathematical figures or numbers, the indicative features of words, etc.), but on how the indicative features are joined and what their structure is like. We will be convinced of all of this many times during the study of the different topics of our course. (This is the end of the lesson.)

    We reproduced the basic fragments of the "logic lesson." Let us examine several problems related to the content and to the way in which it was conducted.[16]

Teaching students methods of identifying phenomena is carried out, as we recall, via an example of identifying the conventional characters—Ivan Ivanovich and Maria Ivanovna. It was only after the students assimilated the procedures from these examples that they were presented with mathematical and grammatical problems requiring the same procedures for their solution.

Is it necessary to teach students a method of reasoning based on an example of conventional characters? Is it not better to start immediately with some grammatical or other school subject-matter and carry out the corresponding work with it?

These questions must be answered negatively. Here is why.

One of the basic positions of didactics has to do with the fact that it is extremely important to rely on the life experiences of the students in the process of teaching. The more new knowledge is related to the life experience of the students, the more quickly it is assimilated.

It has been shown that if students do not have any personal experience in one or another field on which one may rely when introducing new knowledge, then it is often expedient to especially create such experience before introducing this knowledge. In order to do this, one must go through some types of practical exercises (see, e.g., [562], [601], [784], [860]).

However, in speaking of the necessity for relying upon the life experience of students, in didactics, psychology, and methodology, one usually means just the *knowledge* which the students have. For example, when introducing the notion of adjacent angles, it is recommended that one use the students' knowledge about adjacent rooms [641]; when introducing the concept of inertia, make use of the knowledge of what happens when a car in motion suddenly brakes abruptly, etc. Using concepts which have been formed in the students is very important, but we think it does not suffice.

There is still another sphere of experience which could be used successfully in teaching—this is, in particular, *logical experience*, the habit of reasoning, of carrying out specific intellectual operations which are formed in the process of solving the problems of life and other practical problems.

We propose in connection with this that the requirement of relying on the life experience of students should be understood more broadly. When introducing new knowledge and when acquainting students with specific methods of reasoning, one must rely not only on the experience contained in the students' knowledge, but also on the experience which they have with logical reasoning, i.e., on those intellectual operations and their systems which were formed in them in the course of life and studies, but which they often do not realize they possess.

In our experiment, the appeal to the students' life experience was also an appeal to their logical experience (where this was possible, we relied on experience contained in knowledge). This took place in the course of selecting problems of identifying the conventional characters Ivan Ivanovich and Maria Ivanovna. Students constantly meet with problems of identification in life, and undoubtedly in the course of solving them, habits of carrying out specific logical actions are formed which it is important to make precise and generalize. They must be made the object of recognition and raised to the level of an entirely general method.

It must be said here that it is not absolutely necessary to proceed in the instruction from the example of identifying people. One may place before the students the problem of identifying some kind of object (a particular make of car, etc.). The teacher G., in conducting experimental lessons in one of the experimental classes, gave the students a problem in this form: during World War II, two Soviet agents were supposed to meet behind the German lines. What must each one know about the other in order that they will be able to recognize one another?[17] One may think up a number of other situations, too.

The use of problems from the various fields of study (problems from life, mathematics, grammar) in the course of the "logic lesson" had as its purpose *maximally generalizing methods of reasoning* which students mastered in the process of discussing the problems with Ivan Ivanovich and Maria Ivanovna. The more varied the material which the teacher uses to show the applicability of one or another method, the more this method will be recognized as being general.[18] We even believe that the use of not just varied material, but particularly material from the subject fields which are most distant from one another and, at first glance, not related to each other, has an especially great significance for forming broad and deep generalizations. The phenomena from life, mathematics, and grammar whose relationships the students did not see before instruction were just such material as was used in the experimental instruction.

The consistent comparing of phenomena from life, mathematics, and grammar when conducting the "logic lesson" must show students that, in the most diverse phenomena, there can be general and similar structures of indicative features. It is precisely for this reason that the method of approaching *different* phenomena could be *identical* and *general*. It is precisely for this reason that methods of reasoning can be transferred from one context to another. Realizing the fact that problems which are different at first glance (for example, the problem of recognizing Ivan Ivanovich, the problem of specifying which part of speech a given word is, etc.) require one and the same mode of reasoning was a great and unexpected discovery for the students, and we tried to guide them to this discovery.

Thus, the work done during the "logic lesson" made it possible not only to teach students some general principles for approaching phenomena which differed in content through a general method of analysis, but it also confronted them with several problems of the methodology of reasoning. It seems to us that these problems must be given in forms accessible to the students—and the earlier the better. This is important not only for raising the general culture of the students and their general development, but also immediately for the practice of their reasoning.

Let us go on now to analyze the contents and methods of teaching separate grammatical topics and to presenting the course of lessons on these topics.

We shall not describe the whole course of studying each topic, limiting ourselves to a description of just those parts of the lessons where the students familiarized themselves with the indicative features of grammatical phenomena, constructed logic diagrams of indicative features, specified the sequence of actions for identifying phenomena-algorithms, and then trained in applying these algorithms. Our task consisted of attaining the maximum independence of students at each of the stages of work and creating such conditions of teaching that the students themselves reached all the necessary conclusions.

If applying some algorithm to solving a problem is not a creative process, then the analysis of grammatical phenomena, the establishment of their indicative features, the establishment of their structure, and the structure of operations depending on it are, in a psychological sense, processes which are creative to a considerable extent. In experimental teaching, the great attention paid to forming in students the ability to independently construct algorithms for solving different problems had precisely the goal of training them for creative abilities.

Notes

1. For the presentation of the material below, it suffices to allow one-and-a half or two lessons, but it can be studied in parts during several lessons by setting apart 10-15 minutes during each for this purpose. It is not necessary to hold a "logic lesson" immediately before the topic of "the subject." We conducted such a lesson before this topic only because the experimental teaching began with it. We are convinced that work on the students' mastery of the notions of an indicative feature and of the logical structure of indicative features and work on forming the ability to determine the logical structures of indicative features (as well as establishing the dependency of the procedure with a phenomenon on the character of the structure of its indicative features) must be conducted considerably earlier, even in the beginning grades of elementary school. Several special preliminary observations and experiments show that this is entirely possible. Of

course, there is no need to conduct special "logic lessons" like the one described above in the beginning grades. Work on the concept of the logical structure of indicative features and the formation of corresponding logical abilities must be carried out gradually but systematically within the material of those subjects which the students are studying. If such work were established in the beginning grades, then in the seventh grade, there would be no necessity for conducting a special "logic lesson."

2. Since, in summarizing the course of the lesson, it is not important which students answered the questions and whether it was one student or different students who answered, we will simply say "student" in the fragments of the lessons cited.

3. A teacher of the Russian language in School No. 34 in Moscow, S.F. Ivanova, when looking through the manuscript of this book, observed that another variant for conducting the work described was possible. The main point of this variant is the following: The teacher hangs a picture of Ivan Ivanovich at the board and asks the students to indicate, according to the picture, the distinctive features of the person depicted. In this case, the students will not have to "think up" the indicative features by themselves (although in certain respects this is quite beneficial). They will be able to refer to the drawing. It seems to us that this variant deserves attention and is interesting. Experience ought to show which variant is more expedient. It is quite possible that it is the second.

4. One can tell Ivan Ivanovich's indicative features to the students, but we deliberately preferred that method where the students independently had to point out Ivan Ivanovich's possible indicative features. Work on searching for and selecting indicative features had to show the students that indicative features have to be distinctive. Such work, conducted systematically, has to teach the students to independently *construct* definitions (and not just *remember* them, as is often the case now in schools). Searching for and

formulating the indicative features which distinguish Ivan Ivanovich from other people is none other than the process of constructing definitions.

5. An analogous question posed to students in connection with indicative features which are not derived from a definition and which have to do with concepts remote from their life experience provoked difficulties for the students. Here, however, the question was understood at once and provoked no difficulties.

6. If Ivan Ivanovich's indicative features are designated by the letters a, b, c, d, and x signifies the person whose indicative features are being verified, then the logical structure of Ivan Ivanovich's (Iv. Iv.) could be described as follows in the language of predicate logic:

$$a(x) \ \& \ b \ (x) \ \& \ c \ (x) \ \& \ d \ (x) \Leftrightarrow \text{Iv. Iv. } (x).$$

In the language of sentential (propositional) logic, the proposition expressed would have the form:

$$a \ \& \ b \ \& \ c \ \& \ d \Leftrightarrow \text{Iv. Iv.}$$

The expression Iv. Iv. (x) is an abridgement for the expression "x is the person having the name Ivan Ivanovich."

7. The importance of using diagrams consists of the fact that, first of all, it reveals the logical structure of the indicative features in a general form; secondly, it shows (also in a general form) how this structure conditions the structure of operations on these indicative features. Revealing the structures of indicative features and operations on them in a general form, as well as the dependence of the method of reasoning on the structure of the indicative features, ensures a quicker and more profound formation of the necessary generalization. This, in turn, is of great importance for the students' development, since general methods of thinking are cultivated.

8. The cited methods for forming concepts are somewhat

analogous to those described by Reshetova and Kaloshina [312], and by Talyzina [328]. However, there is a significant difference between the methods described by them and the methods applied in our experimental teaching. In our methods, students are provided with special knowledge about the different logical connectives joining the indicative features. A general understanding of the logical structure of indicative features and a knowledge of the types of logical structures (although the word "structure" is not used) are formed. Students are brought to an understanding of the important condition that the method of identification and the procedure with respect to a phenomenon depend on the type of logical structure of its indicative features. In our approach to students, special exercises are conducted which require choosing the procedure depending upon the type of logical structure. During the course of instruction, they are also given some knowledge of the properties of indicative features from the standpoint of specificity, univocality, and so on. Moreover, the importance of the different sequences for verifying indicative features in connection with the problem of making efficient the process of identification (in particular, some criteria for selecting the optimal sequence for verifying indicative features depending on the logical connectives which join them and the frequency of these indicative features are shown) is established. All this knowledge is of a rather high level of generality, and serves as a basis for very general methods of reasoning.

9. As we see, the order of actions shown represents a particular general procedure for definitions and a general method for mastering a concept and subsequently applying it. Students are taught the general system of operations which will make possible mastering different concepts (geometrical, grammatical, etc.) and correctly identifying the most varied phenomena. Therein lies the developmental significance of such instruction. In learning how to correctly operate with one subject-matter, the students acquire the ability to operate with *different* subjects, and they master several general

methods of thinking.

10. The assumption that the time for verifying all the indicative features is the same underlies all reasoning here. From this or other examples, one may also show the students how the time for verifying each indicative feature influences the selection of the sequence for verifying the indicative features. The students also understand this dependency very easily.

11. Here, the teacher digresses briefly and analyzes with the students the meanings (strictly divisive and non-strictly divisive) that the disjunction *or* may have. After several examples, students independently draw a conclusion about the difference between a "strict" and "non-strict" *or*. However, for the process of identifying objects according to indicative features—and the teacher especially underlines this—it is often not important in which sense the disjunction *or* is used. The procedure for identification will be identical in both cases.

12. If the letters *m* and *n* designate Maria Ivanovna's indicative features, and *x* signifies the woman whose indicative features are being verified, then the logical structure of Maria Ivanovna's (M. Iv.) indicative features may be described in the language of predicate logic thus:

$$m\ (x) \lor n\ (x) \Leftrightarrow \text{M. Iv. } (x).$$

In the language of propositional (sentential) logic, the expression will have this form:

$$m \lor n \Leftrightarrow \text{M. Iv.}$$

13. It is easy to see that the second, sixth, and eighth indicative features are complex (compound). Indicative features entering into the composition of these indicative features in their turn are joined by the disjunction *or*. Indicative features entering into the composition of the sixth indicative feature are joined in the textbook by the conjunction *and*, whereas in reality, they should be joined by the disjunction *or*.

14. If the letters a, b, c, d, as formerly, designate Ivan Ivanovich's indicative features and e, f, g, h, Pyotr Petrovich's indicative features, then the logical structure of the messenger's indicative features may be described thus in the language of predicate logic:

$$(a\ (x)\ \&\ b\ (x)\ \&\ c\ (x)\ \&\ d\ (x))\ \lor\ (e\ (x)\ \&$$
$$f\ (x)\ \&\ g\ (x)\ \&\ h\ (x))_{\overset{\leftrightarrow}{\text{Df}}}\ \text{Messenger}\ (x).$$

In the language of sentential logic, this expression has the form:

$$(a\ \&\ b\ \&\ c\ \&\ d)\ \&\ (e\ \&\ f\ \&\ g\ \&\ h)_{\overset{\leftrightarrow}{\text{Df}}}\ \text{Messenger}.$$

15. In clearly formulating the given sequence of actions for the students, the teacher disclosed the *general method* for approaching different phenomena, including grammatical ones, and a general plan for studying different topics. One must note that the presence of general methods of approaching different phenomena and of different aspects in studying different topics ensures the generality of the methods of teaching these topics and makes possible the development of a *particular general method of teaching* general methods of thinking. Henceforth, as an example of studying different topics, we will show these general traits in teaching methods and this generality of pedagogical actions and pedagogical operations when teaching different topics.

16. We said above that, when methods of logical thinking are properly presented and started in the lower grades, there may be no need to conduct a special "logic lesson," and such work can be handled as part of regular lessons.

17. It must be admitted that such a form of problem presentation provoked much animation in the class and so many collateral associations that this overshadowed the logical content of the problem to some degree. In using diversion, it is very important to observe moderation. The scenarios involving Ivan Ivanovich and Maria Ivanovna seem more

successful to us. Experience has shown that a "logic lesson" based on the example of identifying Ivan Ivanovich and Maria Ivanovna provokes a high degree of interest in the students, but this interest is not linked to the external diversion of the situation discussed (there is nothing interesting in the situation itself) but only with the problem itself and with its logical content.

18. Variations in visual material are very important for forming concepts correctly (see, e.g., Menchinskaya and her colleagues: [543], [643], [648], [717], [737], and others). If such variations are not introduced, then the concepts, as these authors showed, do not possess a sufficient degree of generality, and unessential indicative features are often included in them. There is no doubt, however, that variation is important, not only for forming concepts but also for forming intellectual operations and thinking procedures in general. Thus, the significance of variation is evidently wider. Another point of view on this question is that of Galperin [228]. According to this viewpoint, if concepts are formed in a stage-by-stage procedure, variation of unessential indicative features is not required; in this case, students are merely oriented toward essential indicative features. Galperin's assertion does not raise any doubts. However, one may believe that variation of unessential indicative features is nevertheless very beneficial, since it ensures the establishment of *direct* links between forms of objects and actions with them and the formation of *direct* generalizations, which is very important for automatizing intellectual activities and the development of habits. Undoubtedly, one may also form correct concepts without varying unessential indicative features, but undoubtedly also, formation of these concepts proceeds more quickly and easily, if actions are mastered step-by-step from *varied* visual material.

Chapter Sixteen

Studying the Topic
"Types of Simple Sentences"

1. Substantiation of Instructional Methods

The definitions of the four types of sentences given in earlier editions of the textbook for the Russian language for high school [532] suffered (and partially suffer now) from a number of defects.[1]

Let us examine these definitions in sequence.

"Personal sentences," it says in the textbook, "are those sentences which consist of a subject and a predicate or of just the verb which clearly indicates a subject by its endings (the pronouns *I, we, you*)."

First, let us discuss the name for this type of sentence. One may say that personal sentences subdivide into two types—definite-personal and indefinite-personal. This is correct logically, and it at once creates the correct notion of the interrelation of the types of sentences. In the textbook, the generic words "personal sentence" designated just one of the types of personal sentence. In the program constructed for experimental instruction, this deficiency was corrected, and the sentences designated in the textbook as personal were called definite-personal.

Further, there are definite-personal sentences *with* subjects and *without* subjects.* Without a doubt, these are two types (or

---

*The indicative feature "there is no subject" is used extensively by the author in the chapters which describe his experimental work with students, and it must be clarified for English-speaking readers. There are several instances in Russian syntax where the subject in the sentence is understood. Here are the ones used most frequently by the author: (1) in present tense verbs where the conjugation ending clearly indicates the subject (I, you, he, we, you, they); (2) reflexive verbs in the third person singular with the noun or personal

two varieties) of definite-personal sentences (for each of them has its own specific indicative features). It is expedient to indicate this to the students after having named the corresponding form of these types (or varieties). We called them definite-personal type I and definite-personal type II.

In the textbook, it was said about definite-personal sentences type I (our terminology) that "personal sentences are those which consist of a subject and a predicate." This is expressed unsuccessfully and ambiguously: one could understand it in such a way that definite-personal sentences consist of just a subject and a predicate. It is better to say that "there is both a subject and a predicate" in definite-personal sentences. From now on, we shall keep to this formulation.

In the definition of definite-personal sentences type II, the indicative feature "the subject is absent" is not clearly shown (in the definition it says: "consists of just the predicate"). In order to create a perfectly clear-cut understanding of the logical structure of the indicative features for definite-personal sentences type II, this indicative feature must be shown plainly. Then it will be clear immediately that there is a subject in definite-personal sentences type I and that there is no subject in definite-personal sentences type II.

The second indicative feature of a definite-personal sentence type II is unsuccessfully put as "consists of a verb whose endings clearly indicate the subject (the pronouns *I, we, you*)." The indicative feature is formulated in such a way that, in order to establish it in a sentence, one must mentally introduce one of the

---

pronoun in the dative case; (3) verbs in the third person singular, neuter form of the past tense with the noun or personal pronoun in the dative case; (4) indefinite verbs in the third person plural; (5) nouns or personal pronouns in the accusative case with an impersonal verb in the third person and a noun [the agent] in the instrumental case—this is translated by the passive voice in English. Since, in most cases, sentences with understood subjects cannot be translated literally from the Russian, from now on, when the author gives sample sentences of this type, the understood subjects will be enclosed in brackets [ ]. (*Translator.*)

pronouns *I, we, you* into the sentence. Meanwhile, this operation is superfluous. One may say that the immediate indicative feature of predicates characteristic of definite-personal sentences type II is that they are expressed by verbs of the first or second person (indicative mood). We took this feature as an indicative feature of definite-personal sentences type II.

Thus, the indicative features of definite-personal sentences can be formulated in this way:[2]

Indicative Features of Definite-Personal Sentences Type I	Indicative Features of Definite-Personal Sentences Type II
(1) The subject is present *and*	(1) The subject is absent *and*
(2) the predicate is present.	(2) the predicate is expressed by a verb of the first or second person.[3]

If the indicative feature* "the subject is present" is designated by the letter a, the indicative feature "the predicate is present" by the letter b, "the predicate is expressed by a verb of the first or second person" by letter c, the class of definite-personal sentences by the letters $D\text{-}p$, and x means a certain sentence, then the logical structure of the indicative features of a definite-personal sentence may be symbolically described thus:

$$(a\ (x)\ \&\ b\ (x))\ \lor\ (\overline{a}\ (x)\ \&\ c\ (x))\underset{\mathrm{Df}}{\Leftrightarrow} D\text{-}p\ (x).$$

In the language of sentential (propositional) logic, the same thing may be written thus:

$$(a\ \&\ b)\ \lor\ (\overline{a}\ \&\ c)\underset{\mathrm{Df}}{\Leftrightarrow} D\text{-}p$$

Let us go on now to analyze the indicative features of an

---

*More precisely, the statement about the indicative feature, but here, as above, we shall simply say "indicative feature" for brevity.

indefinite-personal sentence. The latter was defined in the text-book in the following manner: "A sentence whose predicate refers to all persons in general, or to an unspecified number of persons." Orlova quite correctly pointed out the basic defects of this definition [736]. She noted that the most important indicative feature of an indefinite-personal sentence was not shown—namely that it has no subject. When the indicative feature "the subject is absent" is missing, definite-personal sentences of the type *Every-one read the fables of Krylov* (the predicate refers to all persons) and *[They] made the bookshelf themselves* (the predicate refers to an unspecified number of persons) completely fit the indicative feature "the predicate refers to all persons or to an unspecified number of persons." In applying the definition given in the textbook to these sentences, we inevitably arrive at a mistake.

Let us indicate still two more essential defects in this definition.

Let us analyze the sentence *[They] called on me today in arithmetic, and [they] gave me the mark "very good"* (a sentence which a boy says when coming home from school). If one were to go according to the indicative feature given in the definition "the predicate refers to all persons in general or to an unspecified number of persons," then this sentence could not be called indefinite-personal (for a completely specified person called on the boy—the arithmetic teacher), whereas this sentence is indefinite-personal.

Another defect consists of the following. In the textbook, there appear sentences of the type *You can't get a fish out of a pond without work.** This, however, is incorrect also. These sentences are generalized-personal (see *Grammar of the Russian Language*, Vol. II, *Syntax*, Moscow, 1954). The attribution of these sentences to indefinite-personal sentences (this was done, evidently, to reduce the number of types of sentences which students have to study) can only provoke confusion in the students' minds, since ignorance of both the objective logic of grammatical phenomena as well as of the psychology of assimila-

---

*You can't get something for nothing. (*Translator.*)

tion underlies the inclusion of one type of sentence with another. It is impossible with such an approach to formulate precise indicative features which would make possible the distinction of one kind of sentence from another.

In experimental instruction, we refused to make such a mixture of grammatical phenomena, and we excluded sentences of the type *You can't get a fish out of a pond without work* (i.e., generalized-personal) from the examples of indefinite-personal sentences.

The question arises as to whether one should teach generalized-personal sentences in school. Its solution requires special discussion, but since sentences of this type are encountered comparatively rarely and knowledge of the indicative features of this type of sentence does not have a great theoretical or practical significance, we did not study generalized-personal sentences with the students. To act otherwise would have meant broadening the existing program in which there was no topic "generalized-personal sentences."

Defects in the textbook, when the concept of an indefinite-personal sentence is presented, lead to great difficulties for the students when they assimilate this type of sentence—and to a great many mistakes. In Orlova's work, which has already been mentioned, many of these mistakes are cited. Our data corroborate those of Orlova.

Thus, for example, mistakes where students considered definite-personal sentences of the type *Someone knocked at the door* and *Some people can't swim*, indefinite-personal sentences were typical. They based their decision on the fact that "the predicate refers to an unspecified number of persons." On the other hand, many students consider indefinite-personal sentences impersonal. Since the indicative feature "the subject is absent" was indicated in a clear form only in the definition of an impersonal sentence, the students attributed sentences having no subjects to impersonal sentences.

The following indicative features of indefinite-personal sentences which eliminate the defects and difficulties mentioned can be formulated.

Indicative Features of Indefinite-personal Sentences

(1)   The subject is absent
      and
(2)   the predicate is expressed by a verb of the third person
      plural (in the past tense, simply by a verb in the
      plural).[4]

If, as earlier, the indicative feature "the subject is present" is
designated by the letter a, the indicative feature "the predicate is
expressed by a verb of the third person plural (in the past tense,
simply by a verb in the plural)" by the letter d, and the class of
indefinite-personal by the letters $In\text{-}p$, then the logical structure of
the indicative features of indefinite-personal sentences may be
symbolically described thus:

$$\overline{a\ (x)}\ \&\ d\ (x) \Leftrightarrow In\text{-}p\ (x).^*$$

When we formulated the indicative features of indefinite-
personal sentences, we began by not including generalized-personal
sentences in the class of indefinite-personal sentences. If we admit
that generalized-personal sentences must be studied in school, then
we must include an additional indicative feature in the diagram
shown above. The characteristic of generalized-personal sentences
consists of the fact that, in their formal-grammatical properties,
they can also resemble definite-personal sentences type I (*We
willingly give what we ourselves do not need*) and definite-personal
sentences type II ([*You*] *do not put a handkerchief up to another
man's mouth*) and indefinite-personal ([*People*] *count their chicks
in the fall*).** The specific indicative feature of generalized-per-

---

*Or, in the language of sentential (propositional) logic:

$$\overline{a}\ \&\ d \Leftrightarrow In\text{-}p\ (x).$$

**You cannot prevent another person from speaking his mind. Don't count
your chickens before they hatch. (*Translator.*)

sonal sentences consists of the fact that the verb designates an action which "in equal measure refers to or could be attributed to every or any person." (*Grammar of the Russian Language*, Vol. II, Part 2, Moscow, 1954).

It must be said that the indicative feature proposed in *Grammar of the Russian Language* is not wholly successful, since it does not permit one to distinguish generalized-personal sentences from certain definite-personal sentences where the subject designates general classes of subjects.

Let us go on to the analysis of the indicative features of impersonal sentences. Let us recall how it is defined in the school textbook: "a sentence whose predicate does not have a subject and cannot have a subject is called impersonal." The second of the indicative features is equivocal. A number of students consider sentences of the type [*It*] *blinds the eyes,* [*It*] *jounces the truck on the bumps* (The truck jounces on the bumps, *Trans.*), etc., as personal sentences, since one can put in the subject. (*The light blinds the eyes. Some sort of force jounces the truck on the bumps.*) Moreover, as we shall show later, the indicative feature "there cannot be a subject" is not at all distinctive in relation to impersonal sentences, since there can be no subject in indefinite-personal sentences. Inserting some subject into an indefinite-personal sentence changes its meaning and transforms it into a sentence of another type. Therefore, the indicative features "there is no subject and there cannot be one" are characteristic not only for impersonal, but also for indefinite-personal sentences. The difference between these types of sentences must be sought not in whether or not there can or cannot be a subject, but in how the predicate is expressed in them (in the absence of the subject).

Thus, the second indicative feature of an impersonal sentence may be made more univocal if one begins with the way in which the predicate is expressed. Impersonal sentences are the third type in which there are no subjects (after definite-personal sentences type II and indefinite-personal sentences). If the predicate is expressed by verbs of the first or second person in definite-personal sentences type II and in indefinite-personal sentences by verbs of the third person plural, then in impersonal sentences, it may be expressed by other differing forms with the exception of the persons and numbers just indicated.

Indicative Features of Impersonal Sentences

(1)   The subject is absent
      and
(2)   the predicate is expressed by any part of speech except verbs of the first and second person, as well as verbs of the third person plural (in the past tense, simply by verbs in the plural).[5]

If the second indicative feature of an impersonal sentence is expressed by negating indicative features c and d, and the class of impersonal sentences is designated by *Imp.*, then the logical structure of the indicative features of an impersonal sentence may be expressed thus:

$$\overline{a\ (x)} \ \& \ \overline{c\ (x)} \ \& \ \overline{d\ (x)} \Leftrightarrow Imp\ (x).^6$$

Let us turn to the indicative features of an elliptical (nominal) sentence. The following definition is given in the textbook [533] for an elliptical sentence: "Elliptical sentences are those which have one principal part—the subject." In this definition, the second indicative feature of an elliptical sentence, "there is no predicate," is not shown in a clear form. In order to give students a precise understanding of the logical structure of the indicative features of an elliptical sentence, they may be formulated thus:

The Indicative Features of an Elliptical Sentence

(1)   The subject is present
      and
(2)   the predicate is absent.[7]

If the indicative feature "the predicate is present" is

designated as before by the letter b, and the class of elliptical sentences by El, then the logical structure of the indicative features on an elliptical sentence may be symbolically described thus:

$$a \ (x) \ \& \ \overline{b \ (x)} \Leftrightarrow El \ (x).^*$$

The algorithm for the identification of types of simple sentences was presented in Chapter Twelve.

Beginning with the task of the formation of an efficient algorithm for the identification of types of simple sentences, we took the following order for their study: definite-personal sentences type I, elliptical (nominal) sentences, definite-personal sentences type II, indefinite-personal sentences, and impersonal sentences, formulated thus:

(1) to guide students independently to uncover the "good" indicative features of the types of simple sentences and to establish their logical structure (i.e., form a concept of the types of simple sentences);

(2) to teach them to identify the distinct types of simple sentences, starting from the logical structure of the indicative features of each type;

(3) to guide the students independently to formulate a general method (algorithm) for the identification of the types of simple sentences and to form an efficient habit of identification based on this algorithm.

Each of these didactic problems breaks down in the process of instruction into more elementary tasks. Each of the latter represents an indication of the goal which must be attained in consequence of carrying out the given stage of instruction, i.e., of those psychological processes (knowledge, skills, and habits) which must be formed as a result of instruction at that stage where one proceeds from specific conditions of instruction. Before specifying

---

*Or, in the language of the sentential (propositional) logic:

$$a \ \& \ \overline{b} \underset{\mathrm{Df}}{\Leftrightarrow} El.$$

a method for solving a didactic problem, i.e., the pedagogical actions which must be executed to attain the stated goal and to obtain the necessary result, it is necessary to specify those operations which the student must carry out so that the necessary process is formed. Only thereafter is it possible to specify the method for solving the didactic problem stated and those actions which the instructor must carry out in order to "provoke" the necessary operations in the student.

It is expedient to prepare a description of the course of instruction in the following manner. The entire description must be broken down into separate states (or steps). Each step must include an indication of which didactic problem was stated at the given stage and how it was solved. Before presenting the corresponding fragment of the lesson, it is necessary, therefore, to:

(a) precisely formulate the didactic problem (i.e., say precisely what must be obtained as a result of instruction at the given stage, what knowledge, skills, and habits must be formed in the students);

(b) indicate those intellectual and other operations (we shall conditionally give them the general term "psychological operations") which the student must carry out in order that the necessary knowledge, skills, and habits be formed in him;

(c) indicate those actions—pedagogical operations—which the teacher must carry out in order to "provoke" the necessary psychological operations in the student and thereby solve the didactic problem stated at the given stage. If the operations carried out by the teacher are considered elementary, and if it is possible to indicate the specific sequence of these operations, then it is possible to speak of an algorithm for solving the didactic problem (i.e., of an algorithm of instruction at a given stage).

Since, however, such a detailed characterization of the didactic problems of each stage and of the algorithms for their solution would occupy too much space and would hamper an integral perception of the course of the lessons, we shall give only a detailed substantiation of each stage of instruction for the first subtopic: "Definite-personal Sentences Type I and Elliptical Sentences."[8]

2. Description of the Procedures and Lessons (Basic Fragments)

1. Definite-personal Sentences Type I and Elliptical (Nominal) Sentences *

Usually, the study of these types of sentences is separated in time. They are gone through without directly comparing one with another, and the elliptical sentences are the last to be studied. The analysis of types of sentences from the standpoint of common features and of the distinction of their indicative features suggests another logic for studying this topic. First of all, it is expedient to study elliptical sentences not at the end of this period, but at its beginning. Second, it is expedient to study the types of sentences not separately, but by introducing several of them at once and in contrast to each other. This not only significantly economizes time (in 7-10 minutes, the students are already familiar with the indicative features of two out of four kinds of sentences), but it also makes it possible to recognize more easily and in depth the common and distinctive traits of the different types of sentences.

First Stage

Didactic Task: To attain an independent discovery by the students of the common and the distinctive indicative features of two objects (in the given case, of two types of sentences: definite-personal type I and elliptical).[9]

The Psychological Operations which the students must carry out: **

(1) Comparison of specific objects (in the given case, sentences of the two types indicated).

---

*The Russian term is "nominal sentence." This expression is considered a sentence, since one usually omits the verb *to be* for present tense predicate nominative constructions. However, "nominal sentence" in English terminology refers to complete sentences. Therefore, the designation "elliptical sentence," or "elliptical (nominal)," seemed more appropriate. (*Translator.*)

**For the sake of brevity, we shall omit in the future the words: "which the students must carry out." The operations which we call "psychological operations" are considered here and further on as elementary. It is assumed that the students are able to carry them out.

(2) Separation of the common and distinctive indicative features of these objects (in this case, sentences).

The Pedagogical Operations which the teacher must carry out:*

(1) *Operation A.* An exposition of two objects to the students in order to compare them (in the given case, two sentences having the indicative features *a* & *b* and *a* & $\bar{b}$).

(2) *Operation B.* The question: "What do the given objects have in common?" is asked.[10]

(3) *Operation C.* The question: "How do the given objects (in our case, two sentences) differ from one another?" is asked.

(4) *Operation D.* The students are told which are the common indicative features of these objects (two sentences of the two types of sentences mentioned above).

(5) *Operation E.* The students are told which are the distinctive indicative features of these objects (sentences of the types examined).

The pedagogical operations given may be considered as operators of an algorithm of instruction. But in order to construct an algorithm of instruction, it is still necessary to establish the conditions (the so-called logical conditions) on which the application of different operations (operators) depend. In the given case, the logical conditions are the characteristics of intellectual activities of the students which are evoked by the pedagogical operations—more precisely, by their outcomes.

In order to construct an algorithm for solving the didactic problem shown, two logical conditions suffice:

(1) The students independently isolated the common indicative features of the objects examined—*logical condition a*.

(2) The students independently isolated the indicative features distinguishing one object from another (distinctive indicative features)—*logical condition b*.

---

*In the future, we shall also omit the words "which the teacher must carry out" for the sake of brevity. The pedagogical operations are considered here and further on as elementary, and it is assumed that the teacher is able to carry them out.

Let us formulate the algorithm applied by us for solving the stated didactic problem:[11]

Algorithm

In order that the students independently isolate the common and the differing indicative features of the two objects (in the given case, two types of sentences: definite-personal and elliptical), it is necessary to:

(1) *Operation A.* Present the two objects (sentences having the indicative features a & b and a & $\bar{b}$).

(2) *Operation B.* Ask the question: "What do they have in common?"

If the students isolate the common indicative features (i.e., *logical condition a* takes place), then go on to instruction 3.

If the students do not pick out the common indicative features (i.e., *logical condition $\bar{a}$* takes place), then go on to instruction 4.

(3) *Operation C.* Ask the question: "How do the objects differ from each other?"

If the students isolate the differing indicative features (i.e., *logical condition b* takes place), then go on to instruction 6.

If the students do not pick out the differing indicative features (i.e., *logical condition $\bar{b}$* takes place), then go on to instruction 5.

(4) *Operation D.*[12] Tell the students which are the common indicative features of the objects examined and go on to instruction 3.

(5) *Operation E.* Tell the students which are the differing indicative features of the objects examined.

(6) Conclude the given stage of instruction.

We cite an example from the course of instruction at this stage. The teacher writes two sentences on the board:

*The night [is] dark.**

Night.

Tea. Establish what is common among these sentences.
Stu. There is a subject in both of them.
Tea. How do these sentences differ from one another?
Stu. In the first there is a predicate and in the second there is none.

Diagram of the Algorithm**

$$B \rightarrow A \rightarrow B \rightarrow a \underset{-}{\overset{+}{\underset{D}{\overset{C \rightarrow b}{\diagdown}}}} \overset{+}{\underset{E}{\diagup}} \mathbf{8}$$

Logic Diagram of the Algorithm

$$\begin{array}{cccccc} 1 & 2 & 3 & 3 & 1 & 2 \\ ABa \uparrow & C \downarrow & b \uparrow & . \downarrow & E. \downarrow & DC \uparrow \end{array}$$

Second Stage
Didactic Task: Have the students bring together the indicative features of the objects examined (types of sentences).
Psychological Operations: Systematization of the indicative features picked out.
Pedagogical Operations:
(1) *Operation A.*[13] Ask the question: "What are the indicative features of the first object (sentence) and what are the indicative features of the second?"

---

*Since the verb is not necessary in the present tense in Russian to express being, this sentence is made up of a noun in the nominative case and a short-form adjective. (*Translator.*)

The symbols **B and **8**, as before, designate objects (i.e., their states, *Ed.*) which, are given as the input of the algorithm and are obtained, correspondingly, as its output.

(2) *Operation B.* Tell the students the object's system of indicative features.

(3) *Operation C.* Write down the object's indicative features.

The logical conditions in the algorithm for solving the given problem will be the following:

(1) A student points out the system of indicative features (*logical condition c*).

Since it is very cumbersome to present the algorithm in writing, we shall (and from now on) just cite the diagram of the algorithm (the teacher can also construct the logic diagram of the algorithm which corresponds to this diagram without difficulty).

Diagram of the Algorithm

We show here the course of instruction at this stage.

Tea. What are the indicative features of the first sentence and what are those of the second?

Stu. In the first sentence there is a subject and a predicate. In the second there is a subject, but there is no predicate.

Tea. Let's write down these indicative features.

The indicative features are written down and numbered:

(1) the subject is present	(1) the subject is present
(2) the predicate is present	(2) the predicate is absent

Third Stage

Didactic Task: Have the students determine the logical structure of the indicative features which have been isolated.

Psychological Operations:

(1) Join the indicative features by the conjunction *and*.

(2) Join the indicative features by the disjunction *or*.[14]

(3) Compare the results of operations 1 and 2 with the indicative features of the objects analyzed (the sentences *The night* [*is*] *dark* and *Night*).

(4) Choose one of the connectives.

Pedagogical Operations:

(1) *Operation A.* Ask the question "Which connective links the indicative features?"

(2) *Operation B.* Insert the logical connective.

(3) *Operation C.* Tell the students which is the necessary connective.

Here, the logical condition is the following:

(1) Have the students point out the necessary connective in *logical condition a.*

Diagram of the Algorithm

The progress of instruction at this stage is the following.

Tea. Which connective links the indicative features of the sentences?

Stu. They are linked by the conjunction *and.*

Tea. Let us insert this conjunction into the diagram of the indicative features.

The conjunction is inserted. The diagram is as follows:

(1) the subject is present	(1) the subject is present
and	*and*
(2) the predicate is present.	(2) the predicate is absent.

As we see, the students, guided by the teacher's questions, determined the indicative features of the two types of simple sentences (still not knowing their names) independently and specified their logical structure.

Fourth Stage

Didactic Task: Acquaint the students with the names of each of the objects analyzed (the sentences of two types); make sure the reason why these objects (sentences) are given these names and not others is made clear.

Since the students cannot discover the names of the given types of sentences independently (the names are often a matter of convention, as in this case), the given didactic problem is solved by the teacher's explanation.

We cite this explanation:

Tea. The first sentence having both a subject and a predicate is called a definite-personal sentence. It is given this name because there is a completely specified subject about which the predicate states something definite. This subject is the word *night*. What is stated about it is that it is *dark*.

In the second sentence, the subject is also *night*. But nothing is stated about it. It is only named. Such a sentence where the object or phenomenon is just named, but nothing is said about it (since the predicate is missing) is called elliptical.

Thus, the first two indicative features are those of a definite-personal sentence, and the second two are those of an elliptical sentence.

Let us write down the names of the types of sentences over the indicative features of these two types of sentences which we have picked out.

Ind. Fea. of Def.-pers. Sent.	Ind. Fea. of Ell. Sent.
(1) the subject is present	(1) the subject is present
and	*and*
(2) the predicate is present.	(2) the predicate is absent.

As we see, the name of the grammatical phenomenon is introduced after the students already have become acquainted with it in practice and after they know its indicative features— indicative features which they themselves have discovered. The knowledge independently acquired by the students takes shape through the name. It is deduced directly from the indicative features and is therefore comprehensible to the students.

Fifth Stage

Before going into the training for the identification of types of sentences on the basis of the established indicative features, it is necessary that the students repeat the general procedure of reasoning which must be applied in order to recognize an object in a situation where it might belong to one of several known classes.

This specifies the didactic task of the given stage.

Didactic Task: Get the students to repeat the basic steps of reasoning which must be applied in order to identify an object in a situation where it might belong to one of several known classes.

Psychological Operations: Recall the basic steps of reasoning.[15]

Pedagogical Operations:

(1) *Operation A*. Present the object (sentence) to the students.

(2) *Operation B*. Ask the question: "To what class could the given object belong (in the given case, what could be the type of sentence examined)?"

(3) *Operation C*. Ask the question: "What has to be done in order to know to which class the given object belongs (in the given case, in order to establish the type of sentence examined)?"

(4) *Operation D*. Ask the question: "How should such a verification be carried out?"

(5) *Operation E*. Ask the question: "What has to be done if the object does not belong to the first of the classes examined (in the given case: if the sentence is not definite-personal)?"

(6-10) *Operation P_1-P_5*.[16] Tell the students the correct answers to questions, 2, 3, 4, and 5.

The following are taken as *logical conditions*:

(1) *Logical condition a*. The students' answer is: "The object may belong to A or B or C ... (the given sentence may be definite-personal or elliptical)."

(2) *Logical condition b_1*. The students' answer is: "Verify whether the object belongs to the first of the classes examined (is the given sentence definite-personal)?"

(3) *Logical condition b_2*. The students' answer is: "Verify whether the object belongs to the second (and in the given case, last) of the classes examined (is this sentence elliptical?)."

(4) *Logical condition c_1*. The students' answer is: "Verify whether the object has the indicative features of the first of the classes examined (does the given sentence have the indicative features of a definite-personal sentence?)."

(5) *Logical condition c_2*. The students' answer is: "Verify whether the object has the indicative features of the second class examined (and, in the given case, the last)."

Diagram of the Algorithm

$$B \to A \to B \to a \overset{+}{\underset{-}{\nearrow}} \begin{matrix} C \\ \downarrow \\ P_1 \end{matrix} \to b_1 \overset{+}{\underset{-}{\nearrow}} \begin{matrix} D_1 \\ \downarrow \\ P_2 \end{matrix} \to c_1 \overset{+}{\underset{-}{\nearrow}} \begin{matrix} E \\ \downarrow \\ P_3 \end{matrix} \to b_2 \overset{+}{\underset{-}{\nearrow}} \begin{matrix} D_2 \\ \downarrow \\ P_4 \end{matrix} \to c_2 \overset{+}{\underset{-}{\nearrow}} \begin{matrix} B \\ \downarrow \\ P_5 \end{matrix}$$

We shall describe the course of instruction at the given stage. The teacher writes the sentence on the board:

No one came.

Tea. What type of sentence (of the types you know) could this sentence be?

Stu. It could be definite-personal or elliptical.

Tea. What must be done to identify its type?

Stu. First we must verify whether or not it is definite-personal.

Tea. How should this verification be carried out? What has to be done?

Stu. We must verify whether the sentence examined has the indicative features of a definite-personal sentence.

Tea. Let us suppose that it is not a definite-personal sentence. What do we have to do then?

Stu. Then we have to verify whether the sentence is elliptical.

Tea. What has to be done for that?

Stu. We have to verify whether the sentence has the indicative features of an elliptical sentence.

As we see, the method of reasoning here is the same as it was with identifying the "messenger": first we have to verify whether the sentence is definite-personal (compare: is the person going by Ivan Ivanovich?); if not then we have to verify whether it is elliptical (compare: is the person going by Pyotr Petrovich?). At first, the sentence is verified according to the indicative features of the first type of sentence (definite-personal). Then if the sentence is not definite-personal, it is verified according to the indicative features of the second type of sentence (elliptical).

It is easy to see that not only is the procedure of reasoning general, but also the instructional procedure which the teacher applies. He asks the students questions of the same type requiring the same kind of reasoning (from the point of view of logical structure) from them. The generality of the instructional procedure is determined by the generality of the didactic problems which have to be solved in both cases.

Sixth Stage

The students recall the basic stages of reasoning when they identify a phenomenon which could belong to one of several classes, but they still have not recalled (reproduced) the reasoning procedure which depends on the way the indicative features of the phenomenon are joined by the logical connective. This is where the didactic task of the sixth stage originates.

Didactic Task: Get the students to recall the method for the identification of phenomena whose indicative features are linked conjunctively (by the conjunction *and*) and to apply this method to the identification of types of sentences.

Psychological Operations:

(1) Recall the procedure which corresponds to the conjunctive structure of the indicative features.[17]

(2) Apply this method to solving specific problems of identification.

Pedagogical Operations:

(1) *Operation A*. Ask the question: "What has to be done in order to discern whether the object belongs to a specific class (is the given sentence definite-personal)?"

(2) *Operation B*. Ask the question: "What must be done if the first indicative feature is present in the object (sentence)?"

(3) *Operation C*. Ask the question: "Why does the second indicative feature have to be verified?"

(4) *Operation D*. Ask the question: "What conclusion must be drawn if the object has all of its indicative features joined by

the conjunction *and* (in the given case: if the sentence has all the indicative features of a definite-personal sentence)?"

(5-8) *Operations* P_1-P_4. Tell the students the correct answers to questions 1, 2, 3, 4.

Here, as *logical conditions*, the following are distinguished:

(1) *Logical condition* a_1. The students' answer is "Verify whether the object (sentence) has the first indicative feature."

(2) *Logical condition* a_2. The students' answer is: "Verify whether the object (sentence) has the second indicative feature."

(3) *Logical condition b.* The students' answer is: "Because the indicative features are linked by the conjunction *and*."

(4) *Logical condition c.* The students' answer is: "If the object has all of the indicative features (in the given case, a definite-personal sentence), then we conclude that this object belongs to this class (i.e., that the sentence examined is a definite-personal sentence)."

Diagram of the Algorithm

We will now describe the progress of instruction at this stage.

As before, the sentence: *No one came* is written on the board.

Tea. What must we do to identify whether or not this sentence is definite-personal?

Stu. We must verify whether it has the first indicative feature of definite-personal sentences—a subject.

When the students apply the indicative features of a subject, they find that there is a subject in the sentence—*no one* (the process of looking for a subject is already sufficiently algorithmized, and the students "see" the subject right away without going through extensive reasoning).

Tea. Let us suppose that the first indicative feature is present: there is a subject in the sentence. What else must we do then?

Stu. We have to verify whether the sentence has the second indicative feature—a predicate.

Tea. Why do we have to verify the second indicative feature?

Stu. Because the indicative features are linked by the conjunction *and*.

The students find the predicate in the sentence.

Tea. The two indicative features of a definite-personal sentence are in the sentence. What conclusion can we draw from this?

Stu. Since this sentence has both the indicative features of a definite-personal sentence, this means that it is a definite-personal sentence.

As we see, the consolidation of the knowledge of the indicative features of grammatical phenomena and the formation of operations for their application take place immediately after the indicative features are established.

Seventh Stage

After the students have recalled the method for the identification of phenomena which have a conjunctive structure of indicative features, it is necessary to conduct training in the application of this method to the identification of the types of sentences already studied and in the pursuit of the goal of cultivating the habit of identification. The didactic task of the seventh stage emerges from this.

Didactic Task: Cultivate in the students the habit of identifying definite-personal and elliptical sentences on the basis of the identification procedure which they have assimilated.

The students are given a number of sentences for training in identification. The operations which must be carried out during identification are the same as those shown in the description of the preceding stage.

Therefore, we shall not indicate them.

The sentences used in training were chosen in such a way as to prevent and exclude the emergence of mistakes (and even hesitations) in specifying types of sentences in the future. From the very beginning, the students are given the kinds of definite-personal sentences which the experiment in the control classes showed[18] to have been taken often by them as indefinite-personal (*On the other side of the wall, someone could be heard talking loudly*) and even impersonal (*Nothing distressed him*). Teaching the students how to identify the sentences shown according to indicative features ensures that the knowledge of these sentences is correctly formed at once. This prevents the cropping up of faulty notions and generalizations. When examples are selected, the principle of the necessity for varying unessential indicative features is also taken into consideration Thus, for example, among the examples are extended and unextended elliptical sentences. Subjects in definite-personal sentences are in some cases at the beginning of the sentence and in other cases at the end, etc.

Eighth Stage

In the preceding stage, the students encountered in practice the fact that the subject in definite-personal sentences may be expressed by different parts of speech. In order to identify a definite-personal sentence, there is no need to establish how the subject is expressed in it, but it is very beneficial to be aware of the fact that it can be expressed by different parts of speech.

Didactic Task: Make the students aware of the fact that the subject in the definite-personal sentences may be expressed by different parts of speech.

Psychological Operations:

(1) Establish the parts of speech which express the subject in the examples chosen.

(2) Induce conclusion that the subject in definite-personal sentences may be expressed by different parts of speech.

Pedagogical Operations:

Operation A_1. Ask the question: "By what part of speech is the subject expressed in the first sentence?"

Operation A_2. Ask the question: "By what part of speech is the subject expressed in the second sentence?"

Operation B. Ask the question: "By what part of speech may the subject be expressed in definite-personal sentences?"

Operations P_1-P_n. Tell the students the correct answer.

The following are distinguished as *logical conditions*:

Logical condition a_1. The students' answer is: "By a noun."

Logical condition a_2. The students' answer is: "By a negative pronoun."

Logical condition b. The students' answer is: "By different parts of speech."

Diagram of the Algorithm

$$\text{B} \to A_1 \to a_1 \overset{+ \to A_2 \to \ldots \to B \to b \overset{+ \to \text{Я}}{\searrow P_x}}{\searrow P_1}$$

We describe here the progress of instruction at the given stage.

Tea. By what part of speech is the subject in the first sentence expressed?

Stu. It is expressed by a noun.

Tea. By what part of speech is the subject in the second sentence expressed?

Stu. It is expressed by a negative pronoun.

Tea. By what parts of speech may the subject in definite-personal sentences be expressed?

Stu. By different parts of speech.

Ninth Stage

In the preceding stage, the students became convinced that the fact that a sentence belongs to the class of definite-personal sentences does not depend on which part of speech expresses the subject. Now it is necessary that they realize that, when a sentence belongs to the class of definite-personal sentences this does not depend on the semantic meaning of the subject. The didactic task of the ninth stage emerges from this.

Didactic Task: Get the students to have a precise notion of the fact that, if a sentence belongs to the class of definite-personal sentences, this is not specified by semantic indicative features (the semantic meaning of the subject), but by the formal grammatical indicative features (the presence of a subject and a predicate in the sentence no matter what meaning the subject has).

In the interest of economy, the problem is solved by way of an explanation. We cite this explanation.

Tea. So, we have seen that the subject in a definite-personal sentence may be expressed by any part of speech including an indefinite pronoun. And although in the sentence, *On the other side of the wall, someone could be heard talking loudly*, the subject is expressed by an indefinite pronoun, and it is not known who is speaking, nevertheless, the sentence is definite-personal, since there is a subject in the sentence, and it is expressed by a specific word. What designates this subject and what meaning the subject has is unimportant when specifying the type of sentence. It is only important that the sentence has (along with a predicate) a completely specific subject.

In the following two stages, the students are given certain additional information on elliptical sentences. In particular, the teacher draws attention to the fact that, if an elliptical sentence consists of just the subject without any words explaining it, then it is unextended (for example, *Moscow*). If there are explanatory words with the subject (for example, *A hot July noon*), then it is extended. Then the teacher pauses at several functions of elliptical sentences and shows the purposes which they serve in the language.

The study of the two types of simple sentences, definite-personal type I and elliptical, ends with this. Although the description of this section occupied many pages, the study of these two types in fact takes 25-30 minutes (one can convince oneself of this by reading the description of the course of the lessons and omitting the substantiations and the commentary).

In order to shorten the statement and create a more complete picture of the course of instruction for the reader, from now on, we shall not break down the description of the progress of the lessons into stages and pause at the didactic tasks of each stage and at the algorithms of solution for these problems (all the more since many of these problems and algorithms are analogous to those described above). We shall limit ourselves (wherever it is necessary) to just several commentaries. If the reader so desires, it is easy to break down the course of instruction into stages, to fragment the didactic tasks of each stage, and to determine the algorithms for their solution.

2. *Definite-personal Sentences Type II*

After the students have mastered the indicative features of definite-personal sentences type I, which have a subject and a predicate, the teacher acquaints the students with definite-personal sentences type II, which have no subject.

Tea. We talked about the fact that a definite-personal sentence has both a subject and a predicate. But it so happens that this is just one of the types of definite-personal sentences. Now we are going to examine the other type.

Let us now take two sentences:

I want to take a walk.

Want to take a walk.

Tea. What type is the first sentence?

Stu. Definite-personal. It has both a subject and a predicate.

Tea. What type is the second sentence? Although there is no subject in the sentence—it is omitted—this sentence is also definite-personal, since the subject here is conceived of as completely specific and as one entity. If desired, it can be inserted into the sentence: *(I) want to take a walk*. Apropos of this point, the subject may be omitted in similar sentences precisely because it is conceived as completely specific and unique. It is indicated by the personal form of the verb. Try to put another subject in this sentence.

The students try, but with no result.*

(An assignment of this type has the goal of convincing the students of the correctness of different propositions through their personal experience by way of an original language experiment.)

Then the teacher points out that the predicate in sentences of this type may be expressed not only by verbs of the first person, but also by verbs of the second person.

Tea. Cite examples of definite-personal sentences with missing subjects where the predicate will be expressed first by verbs of the first person and then by verbs of the second person.

The students cite examples.

Tea. What do you think, can the subject in a definite-personal sentence be omitted if the predicate is expressed by a verb in the third person singular? Let us take some sentence, for example.**

Eventually, the indicative features are inserted in a diagram. It assumes this form:

Ind. Fea. Def.-per. **Type I**	**Ind. Fea. Def.-per.** **Type II**	**Ind. Fea. Ellip.**
(1) the subject is present *and*	(1) the subject is absent *and*	(1) the subject is present *and*
(2) the predicate is present.	(2) the predicate is expressed by verbs of the first or second person.	(2) the predicate is absent.

*In Russian each person of the verb has a distinctive ending which identifies the person speaking, and it is permissible to omit the pronouns in the first and second person singular and plural. (*Translator.*)

**Since the Russian example has no parallels in English and is not particularly illuminating, it is omitted. Students analyze these examples and also perform some independent experiments with the sentences. Thus they come to discover the indicative features distinguishing definite-personal sentences of type II from definite-personal sentences type I. (*Editor.*)

Then a short exercise is conducted.

Several definite-personal sentences type I are presented, and the students, guided by the person in which the verb is expressed, have to specify which of the sentences can be turned into definite-personal sentences type II and which cannot (i.e., in which of the sentences may the subject be "thrown out"). In those sentences which can be used without a subject, the students have to put the subject in parentheses.

The goal of this exercise is to have the students master the indicative features of definite-personal sentences type II by practice. By transforming definite-personal sentences type I into definite-personal sentences type II, they become more deeply aware of the essence of definite-personal sentences type II, and they understand better the conditions in which the "omission" of the subject is possible. This has a specific value for the development of speech culture. One of the common mistakes of students consists of the fact that they often use sentences without a subject which cannot exist without a subject; and, on the contrary, they often use a subject where it would be better to avoid using a subject. The latter makes speech too "weighty" and deprives it of the necessary brevity and "energy."

After this, the teacher gives the students another exercise the purpose of which is to train them in the identification of types of definite-personal sentences through the use of a logic diagram of indicative features.

As before, when the preceding exercises were conducted, fundamental attention is paid to the correct procedure and the correct course of reasoning. The students establish the fact that the indicative features of definite-personal sentences are "organized" just as the indicative features of the "messenger." Hence, it follows that the reasoning procedure must be the same: at first one must verify whether the given sentence is definite-personal type I or not; if not, then is it definite-personal type II? If the given sentence has the indicative features of neither a definite-personal sentence type I nor type II, then this means that it is not a definite-personal sentence, and it must be verified for the presence of indicative features for sentences of other types.

Among the sentences included in the exercises, there are not only definite-personal sentences but also indefinite-personal and impersonal sentences which the students do not yet know. Why is this done?

It already has been said that the presence in the object of specific indicative features is the basis for a positive conclusion that the given object belongs to a specific class. If some indicative features are missing in the object (any necessary one or all sufficient ones), then this is the basis for a negative conclusion. In order to develop thinking, the ability to come to a negative conclusion (something is not such and such) is no less important than the ability to come to a positive conclusion (something is such and such). Only by teaching the students to draw a negative conclusion can we teach them to verify their hypotheses and assumptions and teach them to use the method of elimination. One may reach a conclusion as to what the given object or phenomenon is not only by a direct route, but also by an indirect one. If the given object is not this or that, then this means that it is something else (of course, with the condition that all possibilities have been studied). For example, if the given sentence could only be definite-personal or indefinite-personal, or impersonal, or elliptical, and it is established that it is neither definite-personal, nor indefinite-personal, nor elliptical, this means that it is impersonal.

In a number of cases, it is much simpler to specify what some phenomenon is with the aid of the method of elimination than by a direct route. This is connected with the fact that sometimes it is much easier to establish the absence of some indicative feature than to establish the presence of other indicative features. The students can choose the most efficient and economical way of reasoning only when they have mastered different methods of reasoning. One must always prepare them for this.

The students have already become acquainted with problems whose solution consists of obtaining negative conclusions when they studied "the subject." There, the students were given sentences where the subject was absent, and they had to draw a conclusion that the word was not a subject after establishing the fact that one or another word did not have the indicative features of a subject. An analogous problem is put before the students in this case.

We illustrate here the analysis of two sentences. The first sentence is:

Go and carry out the instruction.

Tea. What must we do first of all in order to specify the type of this sentence?

Stu. We have to see whether or not this sentence is a definite-personal sentence type I. Let us verify the first indicative feature: is there a subject in the sentence? There is no subject. This means that it is not a definite-personal sentence type I.

Tea. What must we do after this?

Stu. We have to see whether it is a definite-personal sentence type II. Let us verify the first indicative feature: "there is no subject." The first indicative feature is there—there is no subject.

Tea. How must we act then?

Stu. We have to see whether the second indicative feature is there: is the predicate expressed by verbs of the first or second person? *Go* and *carry out* are verbs in the second person. This means that the second indicative feature is there.

Tea. What conclusions may we draw from this?

Stu. If the given sentence has the two indicative features, then this means that it is a definite-personal sentence type II.

Tea. What subject could be inserted in the given sentence? What subject is understood in it?

Stu. The subject *you. (You) go and carry out the instruction.*

When answering the latter question, students often make the following mistake. They say that one may insert different subjects in the sentence, for example, *friends, children*, etc. (*Friends, go and carry out the instruction. Children, go and carry out the instruction,* etc.). The teacher must show that the words *friends, children*, etc., cannot be subjects in the given case. He proposes that they replace *friends* with a pronoun and read the sentence with the pronoun. What is obtained is: *They, go and carry out the instruction.** Therefore, friends is not the subject. The teacher says that the word friends is a noun of direct address, and points out its characteristic traits. The subject which can be inserted in the given sentence may only be the pronoun *you*. Here it is beneficial to show the students that this pronoun may appear both in the function of the subject (*You go and carry out the instruction*) as well as in the function of a direct address (*You, go and carry out the instruction*). Everything depends on the purpose and sense of the statement.

The second sentence is: [*They*] *say good things about him*.

As in the preceding case, the students verify the sentence for indicative features of a definite-personal sentence type I. Since the first indicative feature of a definite-personal sentence type I is missing (there is no subject in the given sentence), the students conclude that the given sentence is not a definite-personal sentence type I.

Then the sentence is verified for the indicative feature of a definite-personal sentence type II. The first indicative feature is there. There is no subject.

Tea. What do we do next?

Stu. We verify the presence in the sentence of the second indicative feature: the predicate in a definite-personal sentence type II must be expressed by a verb in the first or second person. *Say—they* (*they say*). The predicate is expressed by a verb of the third person.** The necessary indicative feature is missing.

Tea. What conclusion should be drawn from this?

Stu. This means that this sentence is not definite-personal: it does not have the indicative features of either type I or type II of definite-personal sentences.

The verification of the sentence for the indicative features of

---

*This is quite clear in Russian, where the verb endings of the second and third persons plural are distinct. (*Translator*.)

**In Russian the ending of the verb is unique to the third person plural. (*Translator*.)

an elliptical sentence shows that it does not have them either. This means that it is not an elliptical sentence. As a result of analyzing the given sentence, three negative conclusions are obtained.

As we see, the teacher is always training the students in the application of a corresponding method of reasoning which is general for solving all of the problems of identification.

The remaining sentences are discussed in an analogous fashion.

3. Indefinite-personal Sentences

After the students have mastered the indicative features of definite-personal and elliptical sentences and have learned how to identify them, the teacher goes on to the study of indefinite-personal sentences.

Tea. Up to now, you and I have analyzed sentences without subjects, where one may establish what the subject is, since it is suggested by the form of the verb and is the only one possible. In the sentence *Want to take a walk* no other subject besides I is imaginable.* Sentences where there can be only one *specific, unique* subject possible are called definite-personal. Whether a person uses a subject or omits it in such a sentence, it is still clear who is doing the acting (I, you, we). We saw that the predicate in such sentences is expressed by verbs of only the first or second person. But what if we use a sentence without a subject whose predicate is expressed by a verb of the third person plural (for example, the sentence: [*They*] *like him in the group*). Can one say that?

Stu. One can.

When the experimenter begins a study of a new type of sentence, as in the preceding cases, he does not furnish the students with the indicative features. At first, he creates conditions for a language experiment so that the students independently discover these indicative features in the course of practical

---

*As mentioned before, the endings of each person in the conjugation of a verb in Russian are different, and there is no question of confusing them. (*Translator.*)

linguistic activities.

Tea. Does the form of the verb here suggest some one specific, unique, possible subject?

Stu. No, it does not.

Tea. Think, what kind of subject could be imagined here?

Stu. For example, *friends, colleagues, children*, etc.

Tea. If the sentence in which the predicate suggests one, specific, unique, possible subject is called definite-personal, then how, in your opinion, should one designate a sentence where the predicate does not suggest a specific subject?

Stu. Indefinite-personal.

Tea. Correct. In grammar such sentences are actually called indefinite-personal.

As we see in this case, the teacher himself does not even have to introduce the name of sentences of this type. When the students have grasped the grammatical meaning of the sentence of the given type, they easily guess their names. The name in the given case stems from the character of the sentences of this type.

Tea. Now, let us solve this kind of problem. Let us suppose that your friend (we'll call him Petya) did something wrong and the teacher (we'll call him Sergei Petrovich) said some unfavorable things about Petya's behavior (Petya was not present at this conversation). Let us suppose, further, that you want to tell Petya of Sergei Petrovich's opinion of his behavior. How would you say that to him?

Stu. Sergei Petrovich says unfavorable things about your behavior.

Tea. And now suppose that you want to tell Petya the opinion of his behavior, but for some reason, you do not want to tell him who it is who has bad things to say. What would you say in this case?

Stu. [They] say unfavorable things about you.

Tea. By means of what type of sentence did you express your thought in the first case?

Stu. By means of a definite-personal sentence type I: there was both a subject and a predicate in the sentence.

Tea. And in the second sentence?

Stu. By means of an indefinite-personal sentence.

Tea. How was the predicate expressed in that sentence?

Stu. It was expressed by a verb in the third person plural.

Tea. And now tell me, if the predicate in the sentence is expressed by a verb in the third person plural, does this signify or not that it is indefinite-personal?

Stu. No, it does not signify this. In order that it be indefinite-personal, there cannot be a subject.

As we see, the grammatical indicative features of indefinite-personal sentences were not introduced directly, but through the solution of a particular kind of problem which can be called communicative-verbal or situational-verbal. The importance of problems like this consists of the fact that the student is placed in an imaginary situation of interlocution (communication) where he has to tell his interlocutor some information. The student is given some information and its meaning (in this case, the purpose consisted of telling Petya the opinion about his behavior without telling who expressed the opinion), and the student independently has to find the verbal form which permits attaining this purpose. In solving similar problems, the students begin to understand more clearly the dependence of the formal grammatical indicative features of a sentence on what idea has to be expressed through it and what information must be transmitted. They master well the fact that the grammatical form is called upon to serve the specific requirements of communication, and it must serve as a means of solving specific problems which arise in the process of communicating. They begin to realize the living practical value of a grammatical form.

The solution of problems similar to the one cited also has great importance for the development of the culture of speech. For the culture of speech consists precisely of finding the appropriate speech form that will make it possible to solve this kind of problem in the best way and to express precisely the idea that is to be transmitted when one has a specific communication problem.

Other functions and conditions for the use of indefinite-personal sentences are shown to the students in an analogous way.

For example, sometimes indefinite-personal sentences are used because it is simply not known who is carrying out the action (*On the other side of the wall, [they] were quietly talking*). Sometimes they are used when it is not important who is doing the acting (*[They] brought the milk*). Indefinite-personal sentences are also used in some other instances.

After the students have established the functions and conditions for the use of indefinite-personal sentences and have in fact isolated their indicative features, these indicative features must be formulated precisely and written down and the connective which joins them indicated.

Tea. What are the indicative features of indefinite-personal sentences?

The students name them, the indicative features are written down, and the connective which joins them is specified. The diagram of the indicative features appears thus:

Indicative Features of Indefinite-personal Sentences

(1)   the subject is absent
       and
(2)   the predicate is expressed by a verb of the third person plural (in the past tense, simply by a verb in the plural).

A comparison of the indicative features of indefinite-personal sentences with those of definite-personal sentences type II helps the students to uncover the indicative features of the former. Both types do not have a subject (this is a common indicative feature). It is clear that one must look for the difference in how the predicate is expressed. In definite-personal sentences type II, the predicate is expressed by verbs of the first or second person. A sentence without a subject, where the predicate would be expressed by a verb of the third person singular, is incomprehensible (the students established this through experience). Therefore, the predicate in indefinite-personal sentences must be expressed by a verb in the third person plural (in the past tense, simply by a verb in the plural). It becomes clear from what has been said why the types of sentences must be studied in precisely the order we chose. The logic underlying the study of the types of sentences in that sequence is the following. At first those types of sentences are studied which have a subject: definite-personal and elliptical; then those sentences which have no subject: definite-personal type II and indefinite-personal. Finally, the sentences are studied where, if one uses the definition of the textbook, "there is no subject, and there cannot be a subject."

Since the indicative features of all the types of sentences must be shown visually to the students, the indicative features of each new type studied by the students are simply added to those indicative features which have already been written down.

Ind. Fea. Def.-per. Type I

(1) the subject is present
       and
(2) the predicate is present.

Ind. Fea. Def.-per. Type II

(1) the subject is absent
       and
(2) predicate expressed by verbs in first or second person.

Ind. Fea. In.-per. Type I

(1) the subject is absent
       and
(2) the predicate is expressed by verbs in third person plural (in past simply by verbs in plural).

Ind. Fea. Ellip.

(1) the subject is present
       and
(2) the predicate is absent.

We already noted the fact that the indicative features of indefinite-personal sentences used by us differ essentially from those given in the textbook. The first indicative feature clearly shows that there is no subject in indefinite-personal sentences. The second shows the form of the verb which expresses the predicate in indefinite-personal sentences and which is characteristic for this type of sentence (when the first indicative feature is present).

Both the first and second indicative features are completely univocal. The application of these indicative features cannot create any uncertainty in the interpretation of this type of sentence. The given indicative features also ensure a relative facility in the identification of this type of sentence.

Let us continue the description of the course of study of the topic: "Indefinite-personal sentences."

As usual, exercises followed the establishment of the logical structure for the indicative features of indefinite-personal sentences. We will not describe how these exercises were conducted because they do not differ, in principle, from those exercises described above. Let us just note that the exercises which were applied when this topic was studied were not limited to exercises on the identification of the types of sentences. Among the exercises were those which required the transformation of one kind of sentence into another and the analysis of the changes of meaning which were produced by this. There were also exercises which required the ability to express one and the same thought in different ways (i.e., exercises directed at the development of the culture of speech and cogitation) and some others. However, the description of these types of exercises does not enter into the task of the present book. In the future, when we cite fragments characterizing the course of instruction, we shall omit everything which is not immediately directed at the formation of specific logical operations and efficient intellectual procedures. To act differently would have meant digressing from the limits of the present book.

Having studied the indicative features of indefinite-personal sentences, the students became acquainted with three (or dividing definite-personal sentences into two types, four) types of simple sentences. If we follow the method which we used up until now, then in order to establish to what type some sentence belongs, we have to sequentially apply to it the indicative features of each of the types distinguished by us: first see whether it is a definite-personal sentence type I, if not then verify whether it is a definite-personal type II, if not, then see whether it is an elliptical sentence, if not, then we should verify, finally, whether it is indefinite-personal. As we see, this is a long way. In order, for example, to "reach" the third type of sentence in the order in which they were studied and to be sure that the sentence is indefinite-personal, one must "try out" a large number of indicative features and carry out many operations.

The method of identification can, however, be considerably simplified and made efficient, if one takes into consideration the fact that the different types of sentences have identical indicative features, though with different signs (i.e., positive or negative), also (for example, "the subject is present"—"the subject is absent," "the predicate is present"—"the predicate is absent"). If the verification of indicative features is not conducted in the order in which the types of sentences are written, but begins with the indicative features which are common to some types of sentences and which at the same time distinguish them from other types of sentences, then the number of operations required for the solution of the problems can be shortened greatly.

Actually, the indicative feature "the subject is present" appears in two kinds of sentences (definite-personal type I and elliptical) and is missing in sentences of the other two types (definite-personal type II and indefinite-personal).[19] This indicative feature is present in sentences of different types with opposite signs: the first two with plus signs ("the subject is present") and the second two with minus signs ("the subject is absent").

Let us assume that we have before us some sentence x whose type must be specified. If, as the first operation, we choose to verify the sentence for the presence of a subject, then an affirmative or negative answer ("yes," "no") at once limits the field of further search. If there is a subject in the sentence (answer "yes"), then it can be either definite-personal type I or elliptical. If there is no subject, then the sentence is either definite-personal or indefinite-personal. The field of search is limited by half in such a way. Let us suppose that there is a subject in the sentence whose type we must specify and that the sentence, therefore, is either definite-personal or elliptical. Now we may specify what kind it is with the aid of just one operation—verify the sentence for the presence of a predicate. If there is a predicate in the sentence (the answer "yes"), then this is a definite-personal sentence type I: if not (the answer "no"), then it is an elliptical sentence. In the case where the subject is missing in the sentence, the problem of the type of sentence is solved in an analogous way; only here, different indicative features are verified—the way the predicate is expressed. Thus, with the method described for identifying the type of sentence, in every case the problem is solved in just two operations. It is clear that this procedure is considerably simpler.

Up to this point, the sequence of the students' operations was specified by the sequence of how the indicative features were written down in the logic diagram of the indicative features. Related to the fact that, with the method just examined, the indicative features must be verified not in the order in which they were written in the diagram, but in another, the problem arose of teaching students the efficient procedure for the verification of indicative features. It was necessary to guide them to an independent discovery of an efficient algorithm for the identification of types of simple sentences and to supply them with some principles for independently specifying the most efficient system of operations. The students must clearly realize that, in general, different ways of solving one and the same problem are possible, including problems of identification, but they must always look for the economical and efficient way.

This stage of instruction begins when the teacher calls the students' attention to the large number of operations which must be carried out in order to specify the type of one or another sentence when they use diagrams of indicative features.

Tea. Let's think whether it would be possible to simplify the work of identifying the type of a sentence. Compare the indicative features of definite-personal sentences type I and the indicative features of elliptical sentences. What do they both have in common?

Revelation of the common and differing indicative features takes place in the process of an independent comparison of the logical diagrams of the indicative features.

Stu. There is a subject in both types of sentence.

Tea. And now compare the indicative features of definite-personal sentences type II with those of indefinite-personal sentences. What do these two types have in common?

Stu. Neither one has a subject.

Tea. Now, let us suppose that we have some kind of sentence before us. We have to specify its type. For what indicative feature is it most expedient to look in this sentence?

Stu. For the presence of a subject in the sentence.

Having affirmed the correctness of the student's answer, the teacher writes the arabic number 1 on the board and alongside it the name of the operation for verifying the indicative feature shown: "Is there a subject?"

Tea. What are the possible results of verifying this indicative feature in a sentence? What instances could we have here?

Stu. Either there is a subject or there is not.

Tea Let's show them on the diagram. As we see, the teacher just asks the students questions which guide them, and the students formulate the sequence of operations independently. Then the teacher draws arrows over which he writes "yes" and "no." The diagram begins to look like this:

1.   Is there a subject in the sentence?

Tea. Let us suppose that there is a subject in the sentence. In this case, to which type can the sentence be ascribed?

Stu. It can be ascribed to definite-personal type I or to elliptical.

The teacher writes the corresponding inscription under the left-hand arrow.

Tea. And if there is no subject, then what kind of sentence could it be?

Stu. It could be either definite-personal type II or indefinite-personal.

The teacher writes the corresponding inscription under the right-hand arrow. The diagram looks like this:

1.   Is there a subject in the sentence?

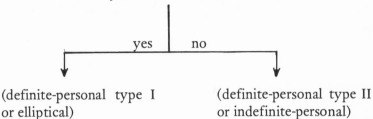

(definite-personal type I   (definite-personal type II
or elliptical)   or indefinite-personal)

Tea. Let us examine the instance where there is a subject in the sentence. We have already established that, in this instance, a sentence may be either definite-personal or elliptical. The method of reasoning by means of analyzing instances which is applied here is widespread in mathematics ("let us examine cases where $a = b$, when $a > b$ and when $a < b$"). What must be done now in order to find out to which type (of the two indicated) our sentence belongs?

Stu. We must verify whether there is a predicate in the sentence.

Tea. Correct. A comparison of the indicative features of definite-personal sentences type I and elliptical sentences shows that these sentences differ only in the presence or absence of the predicate (the presence of the subject is their common indicative feature). Therefore, when we have verified whether the predicate is in the sentence or not, we at once specify the type of this sentence. Let us write this down.

The second operation is written down on the diagram. The diagram looks like this:

1.   Is there a subject in the sentence?

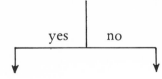

2.   Is there a predicate?

Tea. What are the possible results of verifying the second indicative feature?

Stu. Either there is a predicate in the sentence or there is not.

Tea. Let us indicate this on the diagram.

Two arrows are drawn with the replies "yes" and "no."

Tea. If there are both a subject and a predicate, what type will the sentence be?

Stu. The sentence will be definite-personal type I.

This conclusion is written on the diagram.

Tea. Now suppose there is a subject in the sentence, but no predicate. To what type must the sentence be ascribed then?

Stu. To elliptical sentences.

This conclusion is also written on the diagram, and it looks like this:

1.   Is there a subject in the sentence?

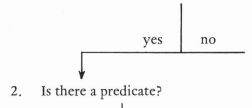

2.   Is there a predicate?

Conclusion: Definite-   *Conclusion*: elliptical.
personal type I.

The identification of types of sentences recalls the game of guessing the name of a famous person of which someone is thinking. The game, let us remember, consists of the following. One of the players thinks of a famous person (writer, artist, politician, etc). The other player asks questions which the first player may answer only by "yes" or "no." What do the questions which the person guessing asks represent from a logical viewpoint? These questions are none other than the verification of specific indicative features. The difference between the game of guessing an important person's name which has been thought of and the process of identifying a type of sentence consists only of the fact that here the person asks *himself* the questions and receives the answers, as it were, from the sentence. The presence or absence in the sentence of the indicative features verified is the basis for the answer "yes" or "no." The analogy between the process of identifying the type of sentence and the game of guessing an important person's name which has been thought of can easily be pointed out to the students.

Acquainting the students with the essence of the given method has a great general educational value, since this method is very general. As we saw above (see Chapter Twelve), when one uses ideas from the theory of information, one can arrive at a more efficient application of the given method.

Tea. Now let us examine the case where there is no subject in the sentence. We established that it could be a sentence of definite-personal type II or indefinite-personal. What do we have to do to find out to which (of these two types) our sentence belongs?

Stu. To do that, we have to see how the predicate is expressed in the sentence.

Tea. True. In other words, we have to verify the indicative features which distinguish the sentences of these two types from one another.

Stu. If the predicate is expressed by the verbs of the first or second person, then it is a definite-personal sentence type II; if by a verb of the third person plural (in the past tense by a verb in the plural), then it is an indefinite-personal sentence.

Tea. Let us write that on the diagram. We will write down the actions in the following order. Our first action is already written down: it consists of verifying whether there is a subject in the sentence. There is no subject. Then, the second action will consist of verifying whether the predicate is expressed by a verb of the first or second person. If it is, then we draw the conclusion that the given sentence is definite-personal type II. If not, then—and the third action consists of this—we verify whether the predicate is expressed by a verb of the third person plural (in the past tense—simply by a verb in the plural). If it is expressed by a verb in the third person plural, then we reach the conclusion: this is an indefinite-personal sentence.

The teacher conducts all these arguments while drawing the diagram. He explains it by these arguments. The diagrams look like this:

1. Is there a subject in the sentence?

yes   no

2. Is there a predi-   2. Is the predicate expressed by a verb
    cate?   in the first or second person?

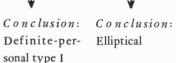

yes   no   yes   no

Conclusion: *Conclusion*:   *Conclusion*:   3. Is the verb ex-
Definite-per-   Elliptical   Definite-per-   pressed by a
sonal type I   sonal type II   verb in the
                                                    third person
                                                    plural (in the
                                                    past tense
                                                    simply by plur-
                                                    al)?

yes   no

Conclusion: Indefinite-
personal

Thereafter, training begins in the identification of types of sentences for the purpose of cultivating the habit of identification.

The students are given a number of sentences whose type they must specify, as they use that sequence of operations which is shown on the diagram.

4. Impersonal Sentences

The last type of sentences—impersonal—remains to be studied.

The teacher writes two sentences on the board:

[*I*] *want to sleep.*

[*There*] *is a need* [*in me*] *to sleep.* *

Once again, he proposes that the students compare these sentences and independently determine those indicative features which they have in common, as well as those which distinguish them from each other. The common indicative feature is easily established. There is no subject in either sentence. Some students propose the following as a second indicative feature: one may not insert a subject (this indicative feature is given in the textbook). By the example of the sentence [*It*] *blinds the eyes*, the teacher shows the students that this indicative feature sometimes leads to mistakes. The sentence [*It*] *blinds the eyes* is impersonal, but many students think that one can insert a subject (The light *blinds the eyes*). On this basis, the given sentence is not recognized as impersonal. Besides that, one may not insert a subject in indefinite-personal sentences. When one inserts the subject (there is a formal possibility of doing this), this changes its meaning and transforms it into a sentence of another type (this is illustrated by examples).

The teacher asks whether it is possible to find some other, more reliable indicative feature. He calls the students' attention to

* Rendered here by a reflexive verb third person singular plus the personal pronoun *me* in the dative case—a common way of expressing different desires. There is no equivalent construction in English. (*Translator.*)

the right-hand part of the diagram, where the indicative features of definite-personal type II and indefinite-personal sentences are shown. "Don't these indicative features suggest the second indicative feature of an impersonal sentence?"

And, actually, the diagram does suggest such an indicative feature. There are three types of sentences in all where there is no "subject" definite-personal type II, indefinite-personal, and impersonal. If the indicative features of definite-personal type II and indefinite-personal sentences consist of the fact that the predicate is expressed by verbs of the first or second person and also by verbs of the third person plural (in the past tense, simply by verbs in the plural), then it is natural that the second indicative feature of an impersonal sentence will be the following: the verb is expressed by any part of speech except the ones mentioned.

The indicative features are written down, and their logical structure is specified.

Indicative Features of Impersonal Sentences

(1)   the subject is absent
      and
(2)   the predicate is expressed by any part of speech except verbs of the first or second person and also by verbs of the third person plural (in the past tense simply by verbs in the plural).

When impersonal sentences are specified by the method of elimination (and, as we said, it is very important to teach the students the different methods and to develop their general logical culture), this is the easiest way of all, since it is not necessary to introduce any additional indicative features. This indicative feature was accepted by us as the second indicative feature of an impersonal sentence.

The general diagram of indicative features takes on the following form:

Indicative Features of Definite-personal Sentences Type I	Indicative Features of Definite-personal Sentences Type II	Indicative Features of Indefinite-personal Sentences	Indicative Features of Impersonal Sentences	Indicative Features of Elliptical Sentences
(1) subject present	(1) subject absent	(1) subject absent	(1) subject absent	(1) subject present
and	*and*	*and*	*and*	*and*
(2) predicate present	(2) the predicate is expressed by a verb in the first or second person	(2) the predicate is expressed by a verb in the third person plural (in the past tense, simply by a verb in the plural)	(2) the predicate is expressed by any part of speech except verbs of the first or second person and also the third person plural (in the past tense, simply by a verb in the plural)	(2) predicate absent

The indicative features shown may also be expressed in the form of the following diagram:*

Types of sentences	Indicative features			
	subject	predicate	the predicate is expressed by verbs of the first or second person	the predicate is expressed by a verb of the third person plural (in the past tense, simply by a verb in the plural)
Definite-personal type I	+	+		
Definite-personal type II	-	+	+	
Elliptical (Nominal)	+	-		
Indefinite-personal	-	+		+
Impersonal	-	+	-	-

Close to the above diagram is the diagram on the following page (we shall give two of its variants). The merit of the three diagrams shown lies in the fact that here the indicative features of all the types of sentences are brought together. It is easy to look them over and compare them with each other. On their basis, one can correctly identify the type of any sentence (except generalized-personal sentences). However, these diagrams do not show directly in which order the verification of indicative features must be carried out. In other words, these diagrams do not show the

*Ths plus in the box signifies the presence of an indicative feature, a minus, its absence. If there is neither a plus nor a minus in the box, then this shows that the given indicative feature for the corresponding type of sentence either is meaningless or that one does not have to verify it.

Variant 1:

The subject is present		The subject is absent		
predicate present	predicate absent	the predicate is expressed by a verb in the first or second person	the predicate is expressed neither by a verb in the first person nor by a verb in the second person	
			the predicate is expressed by a verb in the third person plural (in the past tense, simply by a verb in the plural)	the predicate is not expressed by a verb of the third person plural (in the past tense, simply by a verb in the plural)
Definite-personal type I	Elliptical (Nominal)	Definite-personal type II	Indefinite-personal	Impersonal

Variant 2:

The subject is present		The subject is absent		
predicate present	predicate absent	The predicate is present		
		the predicate is expressed by a verb of the first or second person	the predicate is expressed by a verb of the third person plural (in the past tense, simply by a verb in the plural)	the predicate is expressed neither by a verb of the first or second persons nor by a verb of the third person plural (in the past tense, simply by a verb in the plural)
Definite-personal type I	Elliptical (Nominal)	Definite-personal type II	Indefinite-personal	Impersonal

algorithm and therefore cannot fully replace the diagram-tree which we had begun to construct. It is expedient to apply such diagrams at those stages of the study of types of sentences when work with indicative features, acquaintance with them, and their comparison take place. But as shown earlier, it does not suffice to know the indicative features. One also has to master the method of their operation, i.e., one has to know how one must act with them and what is the most efficient sequence of actions. But this is precisely what the given diagrams do not explicitly show.

We could have limited ourselves to the diagram of indicative features which were cited when the students were taught the types of simple sentences. Since, however, we undertook the task of forming in the students not only a concept of the types of simple sentences, but also the efficient algorithm for identifying types of simple sentences, it was necessary to finish the diagram-tree which we had begun to construct.

As we recall, the part of our diagram on the right-hand side was not finished. At the third question: "Is the predicate expressed by a verb of the third person plural (in the past tense, simply by a verb in the plural)?," there was just the conclusion for the answer "yes" (if yes, then it is an indefinite-personal sentence). We can now draw a conclusion for the answer "no" (if no, then the sentence is impersonal). Thus, the construction of the diagram is entirely completed, and the general algorithm is devised which specifies how to solve the problem of the identification of the type of a simple sentence.

Let us show this algorithm in its entirety. See the following page.

The given diagram contains, in fact, one general algorithm for the identification of the type of a simple sentence, and this algorithm makes it possible to identify types of sentences in three operations at the most.

In an article devoted to the distinction between scientific and artistic thinking, the literary critic K. Zelinsky [637] wrote: "In logic, the movement of thoughts along the rails of mathematics (and, we add, of logic—L.L.) is that compulsion, that of inevitability of conclusions, which enthralls each scientist and paralyzes opposition . . . . All this is railway transport of thoughts which conveys you to the truth as though it were the final station."

Diagram of Actions for Identifying the Types of a Simple Sentence

In order to specify the type of a simple sentence, one must verify:

(1) Is there a subject in it?

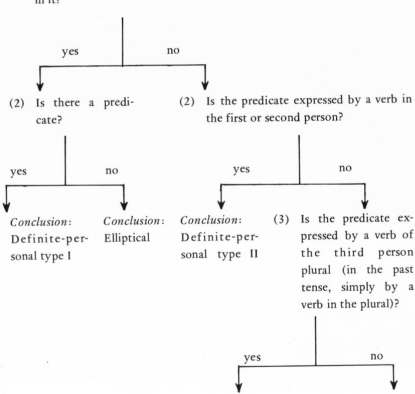

yes   no

(2) Is there a predi- cate?

(2) Is the predicate expressed by a verb in the first or second person?

yes   no   yes   no

Conclusion: Definite-per- sonal type I

Conclusion: Elliptical

Conclusion: Definite-per- sonal type II

(3) Is the predicate ex- pressed by a verb of the third person plural (in the past tense, simply by a verb in the plural)?

yes   no

Conclusion: Indefinite-per- sonal

Conclusion: Impersonal

The characteristic trait of each algorithm—and it is quite possible to apply this to those systems of operations which we taught the students—is the fact that they lead to the truth precisely as though "on rails." This comparison is especially successful if one keeps in mind the systems of operations portrayed in the form of such diagrams where the arrows indicate precisely when and how to advance, depending on the presence of different conditions or indicative features. If a person correctly executes all of the operations in the diagram and keeps to the correct sequence, then he inevitably arrives at the correct answer, and he does not make a mistake.

After the students with the help of the teacher have formulated the indicative features which make it possible to distinguish impersonal sentences from other types of sentences and have formulated the general algorithm for identifying the types of sentences, the teacher tells them in detail what may express impersonal sentences, what is their origin in the language, their function in verbal communication, etc.[20] Then a number of sentences are analyzed. The procedure is the following. The teacher divides the board into three columns. In the left-hand column, the teacher writes the sentence analyzed. Then he asks the students the question: what does this sentence express? And what part of speech expresses the predicate in it? The students answer the questions, and their answers are written in the second and third columns.

As a result of this instructional procedure, the students mastered not only the formal grammatical indicative features of impersonal sentences, but they also grasped the essential characteristics of this type of sentence. Now, one may begin training in identification of the type of sentences by applying the whole system of operations. The training is conducted much as described above.

After the system of operations for identifying the types of sentences is consolidated, the students learn how to specify the type of sentence without going through extensive reasoning. One student said that he "sees right away" how the predicate is expressed and what kind of sentence it is. However, the degree of automatization of operations was different in different students.

The above formulated system of operations for the identification of the types of simple sentences makes it possible to specify the type of a given simple sentence independently of its concrete contents—it really is a general method for the identification of types of simple sentences.

Notes

1. Defects found in the definitions of the earlier editions of the standard Russian textbook used in USSR schools have an independent theoretical interest in this connection. We must note that some of the mistakes were not even eliminated in the latest edition. The basic task of this and analogous sections is not the analysis of one or another specific textbook or of different specific defects in methods of instruction. The aim is to provide an indication of some uniform and general links between methods of instruction and ways of assimilation as well as a plan for a different approach to the design of some aspects of instruction which differ from those most widespread in practice today. The specific defects in the definitions and in the methods of instruction are only the starting point and, in some sense, an occasion for raising the problems which interest us.

2. Here, and further on, when we cite the indicative features of simple sentences, we have not undertaken the task of differentiating between incomplete and complete sentences. This task arises later on. Just those indicative features needed for the identification of the types of simple sentences will be shown in this book.

3. Since the indicative feature "the predicate is expressed by a verb of the first or second person" includes the indicative feature "there is a predicate," we shall for the sake of brevity not include this latter indicative feature in the structural diagram although—as was done previously—it could be shown

explicitly in the diagram. We shall proceed in this way in the future if the indicative feature "there is a predicate" will be included in some other indicative feature.

4. These indicative features differ from the indicative features used by A.M. Orlova in the teaching experiment. Orlova [736] writes: "For indefinite-personal sentences, we considered the mastery of two factors essential: the uncertainty about the person understood to be acting and the absence of his name in the text of the sentence examined. The first of these factors is noted in the definition given in the textbook. Indications of the second are missing in it. In order not to make the definitions bulky and therefore difficult to remember, we did not make any changes in it, but we set ourselves the task of constantly drawing the children's attention to the second of the factors indicated above."

5. If it is necessary (due to difficulties experienced by some students with this formulation), one may break down the second indicative feature into its more elementary indicative features: there is a predicate, and it is not expressed by a verb of the first or second person, and it is not expressed by a verb of the third person plural (in the past tense, simply by a verb in the plural). It is clear from what was said in Chapter Two about the relativity of the concept of an elementary operation, that the level of formulation (breaking down) for indicative features can be different in different cases, and it is specified in the final analysis by the level of development of the students' knowledges and skills.

6. In order that not all of the indicative features be negative, it is expedient in this case to include in explicit form the indicative feature "there is a predicate," which is contained in the words: "the predicate is expressed by any part of speech except . . ." Then the expression becomes:

$$\overline{a\ (x)}\ \&\ b\ (x)\ \&\ \overline{c\ (x)}\ \&\ \overline{d\ (x)}\ \underset{\text{Df}}{\Leftrightarrow}\ \text{Imp}\ (x)$$

Or, in the language of sentential logic:

$$\overline{a}\ \&\ b\ \&\ \overline{c}\ \&\ \overline{d}\ \underset{\text{Df}}{\Leftrightarrow}\ \text{Imp}\ (x)$$

7.   We do not deal with the arguments which occur in linguistics about whether the principal part of an elliptical sentence is the subject or the predicate. Rather, when we formulate indicative features, we start from the definition of the school textbook.

8.   The lessons which will be described below were conducted by the normal unprogrammed method (the method of programmed instruction at the time the experiment was conducted was not yet widespread in the USSR), but the whole course of instruction was strictly calculated and programmed in the broad sense of the word. The study of each topic was broken into a number of stages ("steps"); their sequence was established; and those operations which had to be formed in students so that they would attain the goal set for the given stage were established.

9.   Here, and later on, the didactic task will be formulated as a *general* didactic problem which could arise (and does arise) not only when the given specific topic is studied, but also when other topics and subjects are studied. Such, for example, is the problem of getting students to distinguish the common and distinctive features of two objects independently. This problem arises not only in the Russian language, but also when mathematics, physics, chemistry, biology, and other academic subjects are studied. Therefore, it can have common methods of solution and algorithms, independent of the problem's subject-matter content or in what specific form it appears in different instances (whether one has to separate the common indicative features and the distinctive indicative features of two sentences or of two chemical substances or of two plants, etc.). Since the general didactic problem and the general operations by which it is solved always appear in one or another specific form, we shall show this form in parentheses. Thus, when the didactic problem of the first stage has been formulated—getting the students to separate the common and the distinctive indicative features of two objects independently (this is a general didactic problem)—then we shall show in parentheses that in the given case

two sentences are these objects. We shall proceed in an analogous way in relation to the psychological and pedagogical operations. We shall isolate its general meaning in each operation and show the specific form in which it appears in the given case and with the given subject-matter.

10. The statement of a question is a pedagogical action (operation) which evokes (actualizes) specific thought operations of the students.

11. As mentioned before, different algorithms for the solution of the same pedagogical problem can exist. In the present chapter, algorithms which were used in experimental instruction will be cited. Of the different possible algorithms, we chose those which lead to greater activity and independence of students while they assimilate knowledge and solve problems. When we selected pedagogical actions (methods of instruction), we were guided by propositions established in psychology and didactics regarding the effectiveness of specific pedagogical influences. (Let us recall that one must not confuse algorithms *by means of which* one teaches with algorithms *which* one learns, i.e., teaching algorithms and learning algorithms. Algorithms of the first type are algorithms of the teacher's actions. Algorithms of the second type are algorithms of the students' actions.)

12. Telling the students the correct answer to a question if they cannot answer it is not the only way, and in many cases, not the best way for the teacher to react to the students' difficulties or lack of knowledge. However, we can apply this method here, since the description and inclusion of other methods in the algorithm would occupy a great deal of space, would complicate the algorithm, make it very lengthy, and lead the teacher away from the basic methods for the solution of didactic problems at each stage of instruction. It is all the more justifiable to include such a way to react in the given situation, because during the course of the instruction which will be described below, there were almost no cases where the students could not answer the questions raised independently and where the teacher would have had to give

them the correct answer himself. Let us note that if one were to describe an extensively branched algorithm on paper and one were to try to read it, it would be very difficult to understand. But on the other hand, very lengthy and branched algorithms may be used successfully in programmed instruction with the help of machines. If the machine has a large memory, it can successfully realize very lengthily branched algorithms.

13. One and the same letter at different stages may designate different operations. The value of the letters is separately indicated at each stage. If one were to introduce indices for the letters (since the letters of the Latin alphabet do not suffice for designating all of the necessary operations), then one could isolate the operations common to different stages of operations and designate them by the same symbol. This makes it possible to create a single alphabet of operations. Thus, for example, questions of one and the same type, which however deal with different subject-matter may be designated by the same letters, though with different indices. For example, let us assume any question about an object belonging to a class may be designated by the letter *K*. Then the question of whether a given sentence is elliptical could be designated by *K sent · є elliptical* and the question of whether a certain word is a noun could be designated by *K word · є noun*. One could also use other designations for indices. The only thing that is important is that one and the same letter designate questions of one and the same type, i.e., questions which provoke the same psychological operations (although carried out on different subject-matter). What has been said about the use of the same designations for pedagogical operations of the same type may also be attributed to the designation of psychological operations and their outcomes which appear as logical conditions in an algorithm of instruction.

14. The psychological operations cited which essentially represent trials—attempts to unite indicative features with different conjunctions—may be broken down, if necessary, into

more elementary operations. However, the necessity for this arises only when the students are unable to carry out the operations indicated and when these operations are not elementary for them (for example, when they do not know how to carry out a choice of connectives on the basis of trials and attempt to unite indicative features by connectives and to compare the results of trials with the indicative features of the sentences). But the students of the seventh and eighth grades usually know how to do this after the "logic lesson" and do it as a rule without mistakes.

15. All these stages were characterized when the progress of the "logic lesson" was described.

16. In order to shorten the inscription, we have not written out each operation on a separate line. It is assumed that the teacher knows the correct answer in each specific case and can tell it to the students.

17. The method is described in Chapter Eight.

18. See, also, Orlova [736].

19. We have in mind just those types of sentences which the students have already studied up to the stage described.

20. This is elucidated completely and in detail, for example, in reference [570], and in the lesson the teacher may use a great deal of information from it.

Chapter Seventeen

Studying the Topic
"Compound-Coordinate Sentences"

1. **Substantiation of the Methods of Instruction**

The following definition was given for compound-coordinate sentences in the textbook for secondary schools: "Compound-coordinate sentences consist of equivalent sentences which are independent of each other." However, the main difficulty for the identification of compound sentences lies not in the specification of whether some given sentence is compound-coordinate, but whether it is compound or simple. In other words, before one specifies to what type of compound sentences some given sentence belongs (to compound-coordinate or compound-subordinate), one must specify whether it is compound. What is basically difficult for students is not the distinction between compound-coordinate and compound-subordinate sentences, but the distinction between compound and simple sentences, especially compound and simple sentences with homogeneous parts. It is considerably less difficult to identify compound-subordinate sentences, since in the majority of cases there is a characteristic indicative feature—a subordinate conjunction or a conjunctive word.

How should students be taught to distinguish compound and simple sentences?[1]

There is the following definition in the textbook [532]: "A compound sentence is one which consists of two or more simple sentences."[2]

However, the question arises: how can one find out whether some sentence consists of one or several sentences? What are the indicative features by which one can find out? The textbook says nothing precise about this. A compound sentence is characterized by the indicative feature "consists of two or more simple sentences." This in itself requires that the indicative features for its recognition be explained and pointed out.[3] Without this, it is impossible to operate with the given indicative features. They are "dead," "not functioning." The difficulties students have when they distinguish compound sentences and simple sentences with homogeneous parts are due to the fact that they do not know how to establish the indicative feature "consists of two or more simple sentences" (in a given sentence) or else they use the indicative features unsatisfactorily. Students are usually taught these indicative features stepwise in the formulation of procedures, i.e., of instructions as to which actions must be carried out with a sentence in order to specify its type (compound or simple).[4]

These stepwise procedures are:

(1) Divide the sentence into parts and verify whether these separate parts can be an independent sentence (the basis for the procedure is the indicative feature: "if the given sentence consists of two or more independent sentences, then it is compound; if not, then it is simple").

(2) Specify the "number of thoughts" contained in the sentence (the basis for the procedure is the indicative feature: "if the given sentence expresses two or more thoughts, then it is compound; if not, then it is simple").

(3) Divide the sentence into parts and verify* each part for the indicative features of simple sentences of different types (the basis for this procedure is the indicative feature: "if, after the sentence is divided into parts, each part will be a separate sentence of one of the known types, then the sentence on the whole will be compound; if not, then it is simple").

(4) Isolate pairs of principal parts (the basis for the procedure

---

*The expressions "verification for indicative features" or "verification of indicative features" here and further on are used for brevity. It is a question, of course, of verifying the presence or absence of the indicative features in the phenomenon.

is the indicative feature: "if there are two or more pairs of principal parts related to one another in the sentence, i.e., the sentence consists of two or more principal parts, then such a sentence is compound").

Let us examine each of these procedures in detail.

First Procedure: Divide the sentence into parts and verify whether each of these parts can make up an independent sentence.

Let us suppose that we have to write the sentence: *The sun came out, and [it] became warm.** We do not know what kind it is: a compound sentence consisting of simple sentences, or a simple sentence with homogeneous parts (depending on this, one does or does not place a comma before *and*). In order to specify the type of sentence, some teachers recommend mentally putting a period in place of *and* and reading the parts obtained separately: *The sun came out. [It] became warm.*[5] If each of the parts can exist as an independent sentence, the whole sentence is compound and one must place a comma before *and*; if each of the parts cannot exist independently, the sentence is simple with homogeneous parts, and one should not place a comma before *and*.

But how can one find out whether each of the parts can exist as an independent sentence? On which indicative features, in their turn, should one depend in order to solve this problem? These indicative features are not usually shown. Obviously, it is supposed that this can be specified intuitively. In fact they can only be the indicative features of the sentence—completeness of thought (the presence of information or a question) and intonation, i.e., intonational-semantic indicative features.

It would seem that the use of this procedure would make it possible to easily identify a compound sentence as distinct from a simple sentence with homogeneous parts. But actually, this is not so. Students using this procedure make a large number of mistakes, as the experiment showed.

First of all, there are sentences which can be artificially broken down into word combinations which resemble (each one)

---

*Here and further on, sentences will be cited with the correct punctuation, but they were presented to the students without punctuation marks.

separate sentences in form, but which in reality are not simple sentences within the compound sentence. The most important defect of this method is the fact that, if one evaluates the word combination as to whether or not it can exist as an independent sentence, the outcome is not necessarily univocal. What one person would consider an independent sentence, another person would not, and vice versa.

We cite an example:

Student Vladimir M. incorrectly placed a comma before the first *and* in the sentence *They went across the bridge into the village of Borodino, from there [they] turned left and past a large number of troops and cannons and came out by a high kurgan [burial mound]*.

Experimenter. Why did you place a comma before *and*?

Student. This is a compound sentence; it consists of simple sentences.

Exp. Why do you think that?

Stu. This is a compound sentence because one may put a period in place of the comma.

Exp. And what sentences do you get after that?—Complete sentences which can be read separately?

Stu. Yes, you get complete sentences (he reads both parts of the sentence separately).

Exp. But you see, the second sentence is incomplete. It is not clear who is being spoken about.

Stu. It is clear. We are speaking about *them*. [*They*] turned left and came out by a high kurgan.

What was said above about the given procedure does not, of course, mean that intonational-semantic indicative features do not have any value for the identification of grammatical phenomena and for punctuation. There are cases where punctuation is stipulated only by intonational semantic indicative features. The word combination *To punish impossible pardon* is widely known. Depending on what the speaker wants to say, one should put a dash (or another corresponding punctuation mark) either after the word to *punish* or after the word *impossible: Punish—* [*it is*] *impossible to pardon, Pardon—*[*it is*] *impossible to punish*. There are many similar examples. Actually, they do not indicate an instability or uncertainty of grammar rules as such. Such examples tell us that the use of some formal grammatical characteristics of sentences often does not suffice, if we want the punctuation marks to serve as a means of adequately expressing our thoughts.

As Firsov [830], [831] pointed out convincingly, there are grammatical phenomena where punctuation coincides with intonation. There are phenomena where it does not coincide and even contradicts the intonation. Finally, there are grammatical phenomena where the punctuation sometimes coincides with the intonation and sometimes not.

It is quite obvious that only in the first case may one orient oneself and proceed from the intonation. It is only here that it is an indicative feature for identification. In the second case, one may not proceed from the intonation. In the third case, also, one may not proceed from it, but here, the intonation can serve as a signal for an hypothesis about what kind of phenomenon it is with which we are dealing. But this hypothesis may be verified only with the help of other "non-intonational" grammatical indicative features.

Compound sentences as well as simple sentences with homogeneous parts belong to the third category of grammatical phenomena. When there is a pause in the sentence, in particular before a coordinating conjunction, then one may not univocally conclude that the given sentence is compound. This could be a simple sentence with homogeneous parts. Therefore, in the given case, intonation is not an indicative feature for identification when one solves the problem of distinguishing a compound sentence from a simple sentence with homogeneous parts.

The second procedure, close to the preceding one, consists of specifying the "number of thoughts" contained in the sentence. If one thought is expressed in the sentence, it is simple; if two or more, it is compound.

The experiment showed that the students who use this method commit the greatest number of mistakes in punctuation. This is explained by the fact that it is impossible to "count the thoughts" without conducting a formal grammatical analysis of the sentence. The given indicative feature ("the number of thoughts") by which students often orient themselves is not univocal. The type of a sentence is not specified by its semantic properties, but by its formal grammatical properties.

Let us take two sentences:

The birch was the first to bud and also the first to turn yellow.

The birch was the first to bud, and it was the first to turn yellow.

There is no difference in meaning between these sentences. Nevertheless, the first is a simple sentence with homogeneous parts, and a comma is not necessary before *and*. The second sentence, however, is compound, and a comma before *and* is necessary.

Let us examine another two sentences:

[He is] not well, and he is planning to go to the doctor.

He feels sick and is planning to go to the doctor.

Again, there is no difference in meaning between these sentences, but the first is compound and a comma must be placed before *and*.

It is possible to cite many similar examples. They convince us of the inadequacy and subsequently the incorrectness of such instruction where the students are oriented toward the establishment of just the semantic indicative features. These examples show the inadequacy of the procedure for the identification of types of sentences (whether they are compound or simple with homogeneous parts) by specifying the number of thoughts contained in it.

Experimental data completely confirm this.

Student Natasha Sh. mistakenly placed a comma before the second *and* in the sentence: *The teacher asked the student to hang up the charts and maps on the board and give an explanation.* She said that this was a compound sentence. The experimenter questions her and asks her to substantiate her opinion.

Stu. Here there are two thoughts and a pause. One is to hang up the charts and maps. The other is to give an explanation.

Student Galya O. mistakenly placed a comma before both *ands* in the sentence: *Nekrasov selflessly loved his country, was proud of his people and believed in its mighty strength and its wonderful future.* She said that this was a compound sentence. To the question why it was compound, in her opinion, she answered:

Stu. There are several thoughts here. We separate each by a comma.

The same student incorrectly placed a comma before the verb [*they*] *were* in the sentence: *Large glass jars with coffee, cinnamon and vanilla, crystal and porcelain tea-caddies, and cruets with oil and vinegar were in the kitchen.*

Exp. Why did you place a comma before *were?*

Stu. There we are talking about *oil* and *vinegar* and here *were* [referring to all the things mentioned]—that's different. They must be separated so as not to confuse them.

The student Viktor B. placed a comma before *and* in the sentence: *The river gleams dimly and babbles along the pebbles on the bank.*

Stu. There are two thoughts here: first—the river gleams dimly; second—it babbles along the pebbles on the bank.

The student Tamara S. incorrectly placed a comma before the second *and* in the sentence: *The children ran through the woods and through the field and swam in the lake and in the river.* To the experimenter's question she replied:

Stu. Running through the woods and through the field was one act. Swimming in the lake and in the river was another. They are different things. Therefore a comma is necessary.

The student Sergei M. placed a comma after the word *boy* in the sentence: *This was a well built boy with handsome and fine features.* He said that it was a compound sentence.

Exp. Why do you think this is a compound sentence?

Stu. There are two thoughts here. One is that he is well built.

The other is that he has handsome and fine features.

Many similar examples can be cited. They show how subjective is such a criterion of a compound sentence as the number of thoughts expressed in it. Students are inclined to consider combinations of words as expressing independent thoughts (e.g., two independent sentences) when they designate two different actions (for the students it is one thought to hang up the charts and maps and another to give an explanation), two different subjects or two different groups of characteristics (for the students there is one thought—a well built boy—and another—with handsome, fine features), etc. When they find several "thoughts" in a simple sentence, they take it to be compound and commit errors.

We cited examples where the students saw several thoughts in simple sentences and therefore considered them compound. Now we shall cite examples where, on the contrary, students evaluated compound sentences as simple, considering that just one thought is expressed in them.

The student Tamara Z. did not place a comma before *and* in the sentence: [*It is*] *Night, fog, and it is pitch dark.*

Exp. Why do you think a comma is not necessary before *and?*

Stu. Fog and it is pitch dark is one sentence.

Exp. How do you know?

Stu. It is pitch dark because there is a fog. The latter seems to be the reason for the former. The comma is unnecessary.

Student Vera M. did not put a comma in the sentence: *The bookcase was large, and there were many books in it.* She considered this sentence simple.

Exp. Why do you think this is a simple sentence?

Stu. It is impossible to say independently: *there were many books in it.* This cannot be a sentence because it is not clear what the many books were in, but with the bookcase it is clear. There is one single thought here.

Another student came to an analogous conclusion when he did not put a comma in the sentence: *They are prominent engineers, and* [*it was*] *precisely they who were entrusted with designing the new electric power plant.*

Stu. They were entrusted. Who are they? It is not clear without the beginning.

Student Sasha K. did not place a comma before *and* in the sentence: *The train left, and its lights soon disappeared.*

Stu. The comma is unnecessary. One cannot say separately *its lights soon disappeared.* It is not clear whose lights. There is no independent thought here.

One must admit that the students were right in these cases. If one considers that the criterion of a compound sentence lies in the fact that it consists of parts which express complete independent thoughts, then the sentences cited here actually cannot be considered compound. The second halves of the compound sentences cited (and these are independent simple sentences!) taken by themselves do not express any complete independent thought and are incomprehensible without the first half.

The use of just semantic indicative features engenders yet another type of mistake. Let us examine them.

The student Tamara Z. analyzes the sentence: *The birch was the first to bud, and it was also the first to turn yellow.*

Stu. It is not necessary to put in a comma. Here, there is only one person.

Exp. Is this sentence simple or compound?

Stu. Simple.

Exp. Why do you think so?

Stu. The birch budded, and it also turned yellow.

Exp. But here there are two subjects; *the birch* and *it.*

Stu. Yes. Nevertheless it is one and the same thing. It is the same birch.

As we see, the student is not operating with the language phenomena, and not with the sentences, but with the content expressed by them. Such a formal grammatical indicative feature as the presence of two subjects with a predicate referring to each one of them has no significance for her. What is important for her is that in both cases it is a question of one and the same *object*—the birch.

Into the sentence: *I don't feel well, and I am planning to go to the doctor*, many students also did not put a comma. They considered this to be a simple sentence, since everything said in it refers to one and the same person.

Here is a record of a conversation with one of the students.

Exp. Why do you think that this is a simple sentence?

Stu. Because *I* is talked about: *I* don't feel well and *I* am planning to go to the doctor (the student accentuates the words *I* and *I*).

Exp. But they are in different forms and different cases!*

Stu. That's not important. We are talking about *me*.

The student, guided only by the meaning, completely ignores the formal grammatical distinction. The object of his analysis is not the language but just the content expressed by means of the language, and this is what leads to the mistake. He completely substitutes for the grammatical analysis a semantic analysis. Meanwhile, in the given case, the formal grammatical properties of the sentence entirely specify the correct grammatical way of writing it down.

We see, therefore, that the use of just semantic indicative features while specifying the type of sentence is insufficient and leads to a great number of mistakes. In solving problems of the identification of the type of sentence, a completely different approach is required (this will be discussed below).

Let us examine the *third procedure* applied in practice when teaching the identification of compound sentences in school. This is to divide the sentence into parts and verify each part for the presence of the indicative features of simple sentences. This procedure, as distinct from the preceding ones, creates conditions for the correct specification of the type of a sentence, but it is not efficient, since it solves the problem indirectly through the instrumentality of something else—by the use of the indicative features of simple sentences of different types. Even if the students were able to identify the types of simple sentences without mistakes (which do not exist in practice, particularly

---

*The first part of the sentence is with the pronoun *I* in the dative case, and the second part is with the pronoun *I* in the nominative case. (*Translator.*)

because of the unsatisfactory character of a number of indicative features given in the textbook), nevertheless, this method requires a significant number of superfluous operations. Moreover, this method does not show into which parts the sentence must be broken down in order then to verify each of the parts for the indicative features of simple sentences of various kinds. It is assumed that the students will solve this problem some way or other.

The *fourth procedure* (applied to the identification of compound sentences) which we shall examine is to isolate the pairs of principal parts in the sentence. This way is the closest to the correct method for identifying the type of sentence.[6] However, in the form in which it is usually applied, it ensures the identification of just one of the types of a compound sentence. It is not possible to solve the problem with it in a number of other cases. This procedure, as we said, is guided by the following indicative feature of a compound sentence: a sentence is compound if it has two or more pairs of principal parts which refer to one another, i.e., if it consists of two or more component sentences.

If this indicative feature is expressed diagrammatically, we can say that a compound sentence is one which has the following structure.*

The students who learn this indicative feature qualify such compound sentences as: *The transparent forest grows dark alone, and the fir tree shows green through the frost* without mistakes. In sentences of a similar type, they look for subjects and predicates. They underline them, isolate the pairs of principal parts, and draw the conclusion that this is a compound sentence. Then they put in the corresponding punctuation marks. But they have difficulties

*Here, and below, one line in the diagrams will designate the subject, two lines the predicate, and an arc the circumstance that the given predicate refers to the given subject.

when they encounter sentences where there are no pairs of principal parts and which have one subject and several predicates or just predicates, for example:

‗‗‗   ‾‾‾   ═══ ; ═══   ‾‾‾   ‾‾‾

═══   ═══   and the like.*

    Sentences that have such combinations of elements can be either compound or simple. How does one solve the problem in that case? By what indicative features should one be guided? Not one of the students of the control classes who went through the individual experiment and began the analysis of a sentence by isolating the principal parts could answer this question. Since they do not know the indicative features which must be used in those cases (teachers usually do not mention these indicative features), they are forced to "discover" their own indicative features, which are often incorrect and lead to mistakes.

    Thus, the student Viktor Ye. correctly punctuated compound sentences which had the structure ‿⌢═ ‿⌢═ . But he was given the sentence: *We came up to the house and knocked on the door, and [they] invited us to come in.*

    Stu. There are three predicates and one subject. It is a simple sentence.

    Exp. Why do you think so? Can't it be compound?

    Stu. No. There have to be two subjects and two predicates in a compound sentence, but this is not so here. This means that the sentence is simple.

    As we see, for the student questioned, just the structure of the sentence—the presence of a pair of principal parts where, in each, the predicate refers to a corresponding subject—is an indicative feature of a compound sentence. If the sentence has another structure, he mistakenly considers it simple. For the student, one of the possible and sufficient indicative features of a compound sentence appears as the *unique* and *necessary* indicative feature.

---

*If one of the parts is not connected with another by an arc, this means that it either does not refer to it or that their relationships have not yet been established. In the given example, it is the latter case.

Here is still another example. The student Tamara Z., just like Viktor Ye., acted correctly when confronted with a sentence containing pairs of principal parts which agree with each other.* But here she encounters the compound sentence: *The lake became very shallow, and [it] was not at all difficult to get across it.*

Stu. *Lake* is the subject, *grow shallow* and *get across* are the predicates (she emphasizes them). I think it is not necessary to put in a comma.

Exp. Why?

Stu. I don't know. It just seems that way to me.

Exp. Think it over: is this sentence simple or compound?

Stu. (After a long pause). I think it is simple.

Exp. Why?

Stu. If there had been another subject, then it would be compound, but this way . . .

Thus, the absence of the only indicative feature of a compound sentence the student knows—a pair of principal parts in agreement—puts her in a difficult position. She does not know what to do, hesitates and, guided by an incorrect indicative feature, accepts an incorrect solution.

When certain students encounter a sentence which has no pairs of principal parts ("one of them isn't there") they try to complete the sentence by introducing the "missing" principal part.

Thus, one student incorrectly placed a comma before the *and* in the sentence: *They went across the bridge into the village of Borodino, from there [they] turned left and past a large number of troops and cannons, and came out by a high kurgan [burial mound].*

Exp. Why did you put a comma before *and*?

Stu. This is a compound sentence.

Exp. What makes you think that?

*Here and further on, instead of the expression "a sentence which consists of two or several two-part simple sentences" we shall talk about sentences which include pairs of principal parts referring to one another (or are in agreement), since the students are guided precisely by this indicative feature when they identify the type of a sentence; and, in a number of cases, they themselves talk about pairs of principal parts joined in a specific way.

Stu. It is possible to insert *they* into the sentence: ...*and they came out by the high kurgan*.

When he does not encounter the indicative feature of a compound sentence—pairs of principal parts in agreement—and when he does not know how to specify the type of a sentence and what indicative features must be used in this case, the student thinks up his own indicative feature: if the sentence can be restored to a "normal" structure, then this means that it is compound.

Thus, we see that the procedure for the identification of the type of a sentence by means of separating out the principal parts is essentially correct, since it supposes an analysis of the sentence's structure. However, the indicative feature of compound sentences used in the present case—the presence of pairs of principal parts in agreement—is neither necessary nor sufficient. It leads to the correct solution of the problem in only one case—when the sentence has pairs of principal parts in agreement. This procedure does not show how one should specify the type of a sentence when it has a different structure and which indicative features must be used in this case.

It is clear from what has been said that it does not suffice to know *one* indicative feature in order to solve a problem in a general form and for identifying *any* compound sentence. One has to master a *system* of sufficient indicative features as well as a particular *general method* for operating with them.

Thus, the question naturally arises about searches for such univocal indicative features and about the formulation of such a general method which would make possible the correct identification of any compound sentence with the least number of operations. This should be done directly without using the indicative features of the types of simple sentences.

The comparison of simple sentences having homogeneous parts with compound sentences made the discovery of such indicative features possible.

Let us take two sentences—one simple with homogeneous parts and the other compound:

> *The polar explorers flew off to their wintering place and in eight hours were already in the region of the research station for snow drifts.*

> *The polar explorers flew off to their wintering place, and [they] sent off airplanes with supplies after them.*

How do these sentences differ from one another from a grammatical standpoint? By the fact that in the first sentence, both predicates refer to the subject (*the polar explorers flew off* and *were*). In the second sentence, one of the predicates does not refer to the first subject (the polar explorers flew off, but other people sent the airplanes with the equipment). Obviously, the presence of at least one predicate which does not refer to one of the subjects found in the sentence is therefore (in the presence of a coordinating conjunction or of a coordinating intonation) an indicative feature for a compound sentence. This indicative feature is more general than the indicative feature by which students often orient themselves—the presence in the sentence of pairs of principal parts in agreement. The latter indicative feature is included in the indicative feature shown by us, as a particular case which makes possible identification of a wider circle of sentences by means of the indicative feature which we have shown.

Actually, in order to recognize, for example, the sentence: *The transparent forest grows dark alone, and the fir tree shows green through the frost* as compound, it suffices to establish the fact that the verb *shows green* does not refer to the subject *the forest* (or, on the contrary, that the verb *grows dark* does not refer to the subject *the fir tree*). However, if we started out differently, if we tried to consider the specific indicative feature of a given type of compound sentences—the presence of pairs of principal parts in agreement—as a general indicative feature for compound sentences of a broader group, we would make mistakes in many cases. Thus, we would not be able to recognize the sentence: *The polar explorers flew off to their wintering place, and [they] sent off airplanes with supplies after them* as being compound (it does not have "complete" pairs of principal parts).*

It is often possible to establish the fact that the predicate does not refer to the subject by a purely formal indicative feature—it does not agree with the subject.** But it is not possible to draw the conclusion from this that "if the predicate agrees with the subject, then it refers to it."[7] Such a conclusion would be incorrect. We can see this from the example of the sentence: *The polar explorers flew off to their wintering place, and [they] sent off airplanes with supplies after them* (the second predicate is an indefinite personal verb in the third person plural, *Trans.*), but nevertheless, the second predicate does not refer to the first subject.[8] Subsequently, if the predicate does not agree with the subject, this is a true indicative feature of the fact that it does not refer to it. But if it does agree with it, then on the basis of this, one cannot draw the conclusion that it refers to the subject.

The indicative feature which we pointed out (the presence in the sentence of at least one predicate which does not refer to one of the subjects in the sentence) makes possible the identification of any compound sentence which has at least one subject. However, this type of compound sentence is not the only one. There are compound sentences where there is no subject at all. It follows from this, therefore, that in order to formulate a general method for identifying compound sentences, one must also find indicative features for another type of compound sentences. When we join these indicative features with the disjunction *or* (i.e., each of them is only sufficient but not necessary) we obtain the complete system of indicative features which will make possible the identification of any compound sentence. As for sentences where there are two or more subjects and no predicates, these

*There is no subject in the second part of the Russian sentence. The verb, through its meaning, refers to some indefinite person(s) who is understood, but is not the subject of the first part, i.e., the simple sentence entering into the compound sentence. The verb does not refer to *The polar explorers.* (*Translator.*)

**The Russian example given here has no parallel in English and has been omitted. (*Editor.*)

sentences are always compound, and they are made up of elliptical sentences. Sentences where there are two or more predicates and no subjects, as analysis shows, are compound in all cases except two. The first exception takes place when all the predicates are expressed by a verb of the first and second persons only, for example, [*I*] *walk and sing*. [*You*] *lie down and think*. These are simple sentences. The second exception is sentences where all the predicates are expressed by verbs of the third person plural (in the past tense, simply by verbs in the plural). A sentence with such predicates can be either compound or simple with homogeneous parts. In order to specify what kind it is, one must use a supplementary indicative feature—verify who does the acting. If the person doing the acting expressed is one and the same person, then such a sentence is simple with homogeneous parts. If the persons performing the actions are different, then such a sentence is compound.

We shall clarify what has been said by examples. Let us take the sentence: [*It*] *was already dark, and* [*they*] *set off different colored rockets from the river bank*. We have to specify whether it is compound or simple with homogeneous parts. Since there is no subject (in Russian, both subjects are contained in the verbs and not expressed by pronouns, *Trans.*), it is impossible to use the first sufficient indicative feature. Therefore, we establish how the predicates in the given sentence are expressed. We said that with the exclusion of cases where the predicates are expressed by verbs of the first or second person or by the third person plural (in the past tense, simply by verbs in the plural) a sentence with two or several predicates is compound.[9] Therefore, in order to establish what kind of sentence this is, we have to make one verification: establish whether all of the predicates of the sentence are expressed solely by verbs of the first or second person or solely by third person plural. We verify. This is not so. This means that this sentence is compound. It is made up of simple sentences, and they must be separated from each other by a comma (the comma is necessary before the conjunction *and*).

When the indicative features shown are applied, it is also easy to establish the type of the following sentences: [*It*] *grew dark,*

and [*it*] *turned cold; The boat was carried away by the current,
and then* [*it*] *was difficult to dock at the necessary place*; etc.
There is no subject in these sentences, and not one of the
predicates is expressed by a verb of the first or second person or
by a verb of the third person plural. This means that this is a
compound sentence. (In general, as we know, in order that a
sentence be compound, it suffices that just one of the predicates
not be expressed by the forms of the verbs indicated.)

We have examined compound sentences of three types: (1)
sentences where there is at least one subject and where at least one
of the predicates does not refer to the subject in the sentence; (2)
sentences where there are just subjects and no predicates; and (3)
sentences where there are just predicates and no subjects and
where there is at least one predicate which is not expressed either
by a verb of the first or second person or by a verb of the third
person plural (in the past tense, simply by verbs in the plural).

Let us take two sentences:

At the excavation, [*they*] *found a cache of old coins
and gave it to the museum.*

[*They*] *brought the broken record player to the
workshop, and there* [*they*] *repaired it.*

There are no subjects in either sentence (they are contained
in the verbs, though not in a definite form, *Trans.*), and the
predicates are expressed by verbs in the plural. In these cases, as it
was said, the indicative feature for identification is the following:
if the doers of the actions are one and the same people, then it is a
simple sentence with homogeneous parts. If the doers of the
actions are different, then it is a compound sentence. Let us verify
the sentences by this indicative feature. In the first sentence, the
doers of the actions are one and the same persons (they both
found the cache and gave it to the museum). In the second
sentence, they are different (the owners brought the record player
to the workshop, and the craftsmen repaired it). Therefore, the
first sentence is simple, and the second is compound, and a comma

is necessary before *and* in the second sentence.

It may seem that the indicative feature "the doers of the actions are different people" is equivocal: one person who reads the sentence thinks that there are different people and another that they are the same. However, this is not so. The person dealing with the text can be either in the position of the person writing it or the person reading it. When a person writes, he knows precisely what thought he wishes to express and who is the doer of the actions. If the doers of the actions are different people, then he expresses this with the aid of a comma, placing it before the corresponding conjunction. If they are the same people, then he does not insert a comma. Therefore, the presence or absence of a comma univocally expresses his thought. On the other hand, for the reader, the presence or absence of a comma is a completely univocal indication of how the sentence must be understood. If there is a comma, this means that the doers of the actions are different persons (the comma is inserted to convey this information). If there is no comma, then they are the same people, and no ambiguities in understanding the sentence can arise here. Some difficulties arise only in the process of dictation. But a dictation is an artificial situation which is especially created for academic purposes. Actually, in hearing the sentence: *At the excavation, [they] found a cache of old coins and gave it to the museum*, one person might think that the doers of the actions were the same people, while another might think they are different. But this only speaks for the fact that, in the spoken word, if information about whether the doers of the actions are the same or different people has significance (and the context does not indicate this), one should avoid sentences of this kind of construction. Such sentences could be ambiguous. In the written word, this ambiguity is eliminated with the help of punctuation marks: the addition or omission of the comma. If such sentences are used in a dictation, it is necessary to indicate to the students what thought is contained in the sentence and who is the doer of the actions. The purpose of a dictation is not the verification of how the student guesses what thought is expressed in the ambiguously constructed sentence, but of how the student can precisely express in writing the thought given.

Let us bring all the indicative features together and set up a logic diagram of the indicative features of a compound sentence.

The Indicative Features of a Compound Sentence

A sentence is compound if:

I   (1)   it has a subject (or several subjects)
           and
     (2)   it has a predicate (or several predicates)
           and
     (3)   at least one of the predicates does not refer to any subject in the sentence.
           OR
II   (1)   it has at least two subjects
           and
     (2)   there are no predicates.
           OR
III (1)   there are no subjects
           and
     (2)   not all of the predicates are expressed solely by verbs of the first or second person
           and
     (3)   not all of the predicates are expressed by verbs of the third person plural (in the past tense, simply by verbs in the plural).
           OR
IV   (1)   there are no subjects
           and
     (2)   all the predicates are expressed by verbs of the third person plural (and, in the past tense, simply by verbs in the plural)
           and
     (3)   the doers of the actions are different people.

The simple sentences within a compound sentence can be united either by coordinating conjunctions or by intonation.

Let us describe symbolically the logical structure of the indicative features of a compound sentence. For this we shall designate the indicative features of a compound sentence in the following manner:

there is a subject in the sentence (one or more)—*a*

there are at least two subjects in the sentence—*b*

there is a predicate in the sentence (one or more)—*c*

at least one of the predicates does not refer to the subject in the sentence—*d*

all the predicates are expressed by verbs of the first or second person—*e*

all the predicates are expressed by verbs of the third person plural (in the past tense, simply by verbs in the plural)—*f*

the doers of the actions are the same people—*g*

the simple sentences within the compound sentence are joined by coordinating conjunctions—*h*

the simple sentences within the compound sentence are joined by intonation—*p*

We designate compound sentences by *comp-d*. Then the structure of the indicative features of a compound sentence may be described thus in the language of predicate logic:*

$$[(a\,(x)\,\&\,c\,(x)\,\&\,d\,(x))\vee(b\,(x)\,\&\,\overline{c\,(x)}\vee\overline{(a\,(x)}\,\&\,\overline{e\,(x)}\,\&\,\overline{f(x)}))\vee$$

$$\overline{(a\,(x)}\,\&\,f\,(x)\,\&\,\overline{g\,(x)})]\,\&\,(h\,(x)\vee p\,(x))\underset{\mathrm{Df}}{\Leftrightarrow}\mathit{Comp\text{-}d}.\,(x).$$

In the language of sentential logic, the same thing may be expressed thus:

$$[(a\,\&\,c\,\&\,d)\vee(b\,\&\,\overline{c})\vee(\overline{a}\,\&\,\overline{e}\,\&\,\overline{f})\vee(\overline{a}\,\&\,f\,\&\,\overline{g})]\,\&$$
$$(h\vee p)\underset{\mathrm{Df}}{\Leftrightarrow}\mathit{Comp\text{-}d}.$$

We show the indicative features of a compound-coordinate sentence in a table (see page 467).

---

*By *x* we mean some sentences.

Since compound sentences may be specified by different complexes of indicative features, we shall conditionally divide compound sentences into types (or varieties) as we did with definite-personal sentences. We shall designate each type (or variety) by a corresponding roman numeral.

As we see, the structure of the indicative features for a compound sentence is the same as the structure of indicative features for "the messenger": the indicative features are joined in groups (combinations); each of the combinations is sufficient, but not necessary. The indicative features within each of the combinations are necessary for the sufficiency of this combination. The logic for the identification of compound sentences according to the indicative features which we have proposed is different, in principle, from that which derives from the indicative features shown in the textbook and which is taught in most cases.

The usual reasoning is: if a certain sentence x consists of two or more simple sentences, this means that it is compound. With this logic, the question of how to find out whether the given sentence is made up of two or more simple sentences remains unclear. That which is the premise and is accepted as known is, in fact, unknown, and a satisfactory method for the specification and establishment of this unknown (whether or not it consists of two or more simple sentences) is not given.

The logic which we propose is different. The analysis does not start by establishing whether the given sentence does or does not consist of simple sentences, but by specifying whether or not it is compound. At the basis of our approach lies a reverse reasoning: if the given sentence x is compound (and the precise indicative features of a compound sentence are given as well as the method for its identification), this means that it is made up of two or more simple sentences.

Therefore, what was the starting point of reasoning with the generally accepted logic of analysis ("the sentence consists of two or more simple sentences"), here is the conclusion or the result.

With the present method of instruction, in order to specify whether a given sentence is compound, one must first specify

Indicative Features of Compound-Coordinate Sentences

Type of sentence	There is a subject in the sentence (or more subjects).	There are at least two subjects in the sentence.	There are at least two (or more) predicates in the sentence.	At least one of the predicates does not refer to the subject in the sentence.	All of the predicates are expressed only by verbs of the first or second person.	All of the predicates are expressed by verbs of the third person plural (in the past tense, simply by verbs in the plural).	The doers of the actions are different persons.	The simple sentences within the compound sentence are joined by a coordinating conjunction or by intonation.
Compound (I)	+		+	+				+
Compound (II)	-	+	-					+
Compound (III)	-		+		-	-	-	+
Compound (IV)	-		+			+		+

whether it consists of two or more simple sentences (this is shown as an indicative feature for compound sentences). According to our method, it is not necessary to do this. The problem of whether or not the given sentence is compound is solved in another way, on the basis of a specific system of univocal indicative features which were shown above. Whether or not the sentence does or does not consist of two or more simple sentences is the direct result of this.

The method proposed by us makes it possible not only to identify the type of a sentence (whether it is compound or simple) with a high degree of reliability, but it is also more efficient than the usual method. Actually, if one follows the usual method, one has to look for simple sentences in all sentences, including simple sentences: after all, a person does not know what kind the sentence he encounters will turn out to be—simple or compound. Any sentence can be compound. If our method is followed, then there is no need to look for simple sentences within each sentence. The search for simple sentences within the given one begins only when it is known for sure that they are there. It is known because it was established that it is a compound sentence. With our method, one must look for simple sentences within a given sentence only when they really are there. In the first case, one had to look for them even when they were not there. One can establish the fact that there are no simple sentences in the given sentence and that there was no need to look for them only after trying and failing to find them.

The diagrams above which show the indicative features sufficient for the identification of compound sentences, as distinct from simple sentences with homogeneous parts, nevertheless do not show the expedient sequence of actions for the verification of these indicative features, and they do not represent the efficient algorithm of identification.

On the basis of considerations analogous to those which were cited when the algorithm for the identification of types of simple sentences was established (see Chapter Sixteen), it is possible to propose the following algorithm of identification of the types of sentences (an algorithm for establishing whether the given sentence is compound or simple with homogeneous parts).

Procedural Diagram for the Identification of the Type of Sentence[10]
(compound or simple with homogeneous parts)

In order to specify the type of sentence one must:

(1) Verify whether there are subjects in the sentence

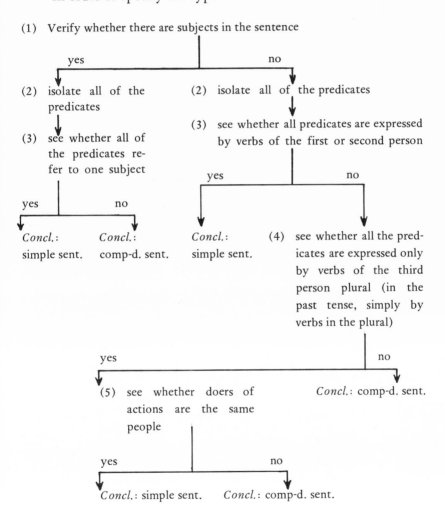

yes   no

(2) isolate all of the predicates   (2) isolate all of the predicates

(3) see whether all predicates are expressed by verbs of the first or second person

(3) see whether all of the predicates refer to one subject

yes   no

yes   no

Concl.:   *Concl.*:   *Concl.*:   (4) see whether all the predicates are expressed only by verbs of the third person plural (in the past tense, simply by verbs in the plural)
simple sent.   comp-d. sent.   simple sent.

yes   no

(5) see whether doers of actions are the same people   *Concl.*: comp-d. sent.

yes   no

Concl.: simple sent.   *Concl.*: comp-d. sent.

We note that, in practice, one does not usually encounter cases where the students take sentences of the type [*I*] *walk and sing* for compound sentences. Starting from this point, the algorithm for the identification of types of sentences may be simplified by excluding one operation from it. Then the algorithm will have this form:

Procedural Diagram for Identifying the Type of a Sentence[11]
(whether compound or simple with homogeneous parts)

In order to specify the type of a sentence, one must:

(1) verify whether there is a subject in the sentence

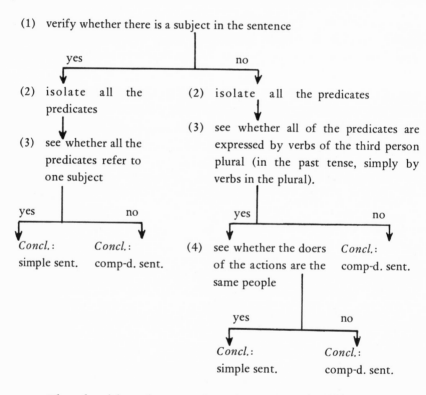

yes   no

(2) isolate all the predicates

(3) see whether all the predicates refer to one subject

(2) isolate all the predicates

(3) see whether all of the predicates are expressed by verbs of the third person plural (in the past tense, simply by verbs in the plural).

yes   no   yes   no

Concl.: simple sent.

Concl.: comp-d. sent.

(4) see whether the doers of the actions are the same people

Concl.: comp-d. sent.

yes   no

Concl.: simple sent.

Concl.: comp-d. sent.

The algorithm shown solves the overwhelming majority of problems of the identification of types of sentences in an efficient way and therefore may be considered as an adequate algorithm in practice.

Can this algorithm be made even simpler? Can it be made

more efficient by reducing the number of operations in it? As the analysis shows, this is not possible with the given indicative features. It is true that here, as in the case of the algorithm for identifying the types of simple sentences, one way of simplification remains—that is to reduce the number of indicative features which describe the types of sentences and their varieties. If one succeeds in doing this (up until now it is not clear how this can be done), one will be able to simplify the algorithm even more. If this cannot be done successfully, and other more "economical" indicative features for compound sentences as distinct from simple sentences with homogeneous parts cannot be found, then the algorithm shown must be considered the most efficient.

The special characteristic in the study of this topic was the fact that the separate indicative features of compound sentences were formulated not only in the natural language but were also graphically depicted by special symbols. In so doing, the dicussion dealt for the most part with indicative features which were graphically depicted.

We shall show how the indicative features specified above may be graphically depicted. As before, we shall designate the subject by one line; the predicate by two lines; the circumstance that the predicate refers to the subject by an arc; and the fact that it does not refer to the subject by two vertical lines. Then the first indicative feature of a compound sentence is visually depicted thus:[1][2]

Here, it is clear at once that there is a subject and that one of the predicates does not refer to it. If we take for an example sufficient indicative feature IV (i.e., the indicative feature of the IV type of compound sentence), it may be depicted thus:

If the lines symbolizing the subject are put in parentheses, then this means that there is no subject in the sentence, but that it could be conceived and formally "inserted."[13] The arcs joining the predicates with the conceivable subjects show that these predicates refer to different conceivable subjects, i.e., that the doers of the actions are different persons. Incidentally, the diagrammatical representation of the indicative features showed that indicative feature IV may be presented as one of the modifications of indicative feature I, namely:

$$\overset{\frown}{\underline{}} \; \overset{\frown}{\equiv} \; , \; \Big\| \; and \; \overset{\frown}{\underline{}} \; \overset{\frown}{\equiv} \; .$$

The only difference consists of the fact that, in the sentences which "fall under" indicative feature I, there really are subjects, while here they are "conceived of." It is important, however, that after these subjects are inserted mentally, the structures of the sentences coincide completely. This is difficult to see without a diagrammatical representation of the indicative features. Bringing this circumstance to light makes the assimilation of the system of indicative features considerably easier, since the number of sentence structures to be remembered is lessened.

The indicative features of a compound sentence are, in essence, structural indicative features. The sentence is compound or simple depending on its grammatical structure. In other words, the grammatical structures of sentences are also their indicative features. The advantage of graphically depicting such indicative features consists, in particular, of the fact that it makes it possible to *show the structure of the sentences visually* and to depict the indicative features.

We shall describe the basic steps in the study of the topic "Compound Sentences," and cite extracts from the record of the lesson.

2. Description of the Course of Lessons (Basic Fragments)

1. Compound Sentences Type I

One of the first assignments given to students when they

studied this subject consisted of making one compound sentence out of two or more simple sentences. This assignment is usually recommended in methodology and is very beneficial, since it shows students "the mechanism" of forming compound sentences out of simple sentences. After they complete this assignment, the teacher describes the types of compound sentences (compound-coordinate and compound-subordinate), and he tells about the ways of joining simple sentences within the compound sentences, and examines certain other questions (such work is usually conducted during the lessons). After the students have done exercises in independently forming one compound sentence out of two or more simple sentences, the teacher confronts them with the reverse problem:

Tea. Let us suppose that we have some kind of sentence before us, and we have to specify what kind it is: simple or compound. What do we have to know to distinguish them?

Stu. Do we have to know the indicative features of simple and compound sentences?

Tea. Here we shall get busy and discover these indicative features.

When the topic "the compound sentence" was studied, the teacher was guided by those knowledges and operations which were formed in the students when they went through the topic "homogeneous parts of the sentence." The students knew that the most important indicative feature of homogeneous parts is the fact that they refer to one and the same part of a sentence. They had assimilated the specific system for the analysis of a sentence. ("At first isolate all the subjects; then all the predicates; then verify whether all the predicates refer to the same subject. Does each subject have homogeneous parts and does each predicate have such parts? Then go on with each of the secondary parts.")

Discovery of the indicative features of compound sentences (as distinct from simple sentences with homogeneous parts) began by examining two sentences:

The cold wind blew as before and did not abate in the least.

The cold wind blew as before, and the freezing weather did not abate in the least.

The students establish the fact that both predicates, *blew* and *abate*, refer to the same subject *wind* in the first sentence, and so they are homogeneous. In the second sentence, they refer to different subjects: the *wind* blew, and the *freezing weather* did not abate.

Tea. Can two predicates be homogeneous if they do not refer to the same subject, i.e., if one of the indicative features of homogeneity is missing?

Stu. No, they cannot.

Tea. If the sentences in which all of the predicates refer to one and the same subject are simple, then what kind of sentences do you think are the ones where the predicates do not refer to one and the same subject?

Stu. They are compound.

Tea. Correct. Sentences which have a subject (or several subjects) and at least one of the predicates does not refer to the subject that the other (others) refer to are compound. This is one of its sufficient indicative features. How can we find out what kind of compound sentence the given one is: compound-coordinate or compound-subordinate? What do we have to pay attention to?

Stu. To the conjunction and to the intonation.

Tea. True. We already talked about the fact that, if there is a coordinating conjunction in the sentence, or if the sentence is joined by coordinating intonation, then such a compound sentence is compound-coordinate. If there is a subordinating conjunction in the sentence, then such a compound sentence is compound-subordinate.[14] Since you and I are only going to be dealing with compound sentences, then in order not to complicate the diagram, we shall not write in the last indicative feature (the presence of a coordinating conjunction or coordinating intonation). But we shall continually keep it in mind.[15]

Then the indicative features are written on the board.

Indicative Features of Compound Sentences Type I

A sentence is compound if
(1) there is a subject in it (or several subjects)
    and
(2) there is a predicate (or several predicates)
    and
(3) at least one of the predicates does not refer to the subject in the sentence.

After this, the teacher shows the students how the structure of the sentences may be depicted diagrammatically. The diagrams of the sentences analyzed are thus:

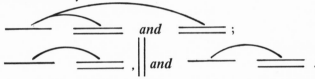

Then the teacher goes on to show the procedural order when the type of the sentences is identified. As soon as the indicative features are established, the method for operating with them is specified.

Tea. What do we have to do when we encounter a sentence, so as to specify whether it is compound or simple? What must be the order of actions for the identification of the sentence's type?

Stu. We must (1) verify whether there is a subject in the sentence.

If yes, then:
(2) isolate all the predicates
(3) specify whether all the predicates refer to the same subject.

If yes, then it is a simple sentence.

If not, then it is compound.

Tea. Let's write down what has been said in the form of a procedural diagram.

He sketches the diagram.

Procedural Diagram for Identifying the Type of a Sentence
(compound or simple with homogeneous parts)

In order to specify the type of a sentence one must:
(1) see whether there is a subject in the sentence

yes   no

(2) isolate all the predicates
(3) see whether all the predicates refer to one subject

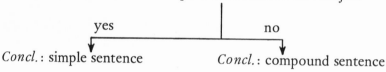

yes   no

Concl.: simple sentence   *Concl.*: compound sentence

Before the teacher goes on to training in the identification of types of sentences, he emphasizes that it is easier to specify what the predicate refers to by going *from the predicate to the subject* and by asking the question from the verb (if the predicate is expressed by a verb). One must ask the question from *each* of the predicates, *who* performed the action (or *what* the given predicate states).

The teacher shows the students how important it is to isolate all the subjects in the sentence from the example of the second sentence: (*The cold wind blew as before, and the freezing weather did not abate in the least*). If one does not set oneself the task of isolating *all* the subjects in the sentence, then one might simply not notice the subject *freezing weather* and think that *did not abate* refers to the word *wind*, all the more so because *did not abate* agrees with the word *wind* and suits it by its meaning. But this would lead to mistakes. A compound sentence would be taken for a simple sentence. This is what often happens when students do not carry out precisely all those operations which are indicated.

Then the students do exercises in identifying the types of sentences and in punctuating them. They use the indicative features for compound sentences which were shown above and execute all the necessary operations. For training, the students are given pairs of sentences, one of which is simple with homogeneous parts and the other compound.

When the experimenter trained students to identify compound sentences type I as distinct from simple sentences with homogeneous parts, he began to work out the elements of the future algorithm. In the process of instruction, the algorithm is put together in parts and formed by elements.

Here are several sentences which were used in training:[16]

1. *The sounds of the hunting horn died away completely and in a little while were heard again in the forest.*

2. *The sounds of the hunting horn completely died away, and in a little while voices were heard.*

3. *Beyond the village was a lake surrounded by willows, and not far away grew a little wood which was overgrown with bushes.*

4. *Beyond the village were a lake surrounded by willows and a little wood which was overgrown with bushes.*[17]

5. *The machine rumbled and drowned out the voice of the technician.*

6. *The machine rumbled, and the voice of the technician was drowned out.*

It can be seen that all the compound sentences included in this exercise have the following structure:*

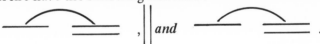

Each of the predicates which does not refer to one of the subjects refers to another one, i.e., each predicate has "its own" subject. As we said above, such a structure is a special case of a more general structure which is reflected in the indicative features shown above:

$$\overbrace{\underline{\quad}\ \underline{\underline{\quad}}}\ ,\ \Big|\Big|and\ \underline{\underline{\quad}}\ .$$

---

*The order of subjects and predicates is not taken into account in the diagrams. This is not important for the specification of the type of the sentence.

What is common in these structures is the fact that one of the predicates does not refer to at least one of the subjects. The difference between these structures consists of the fact that, in sentences having the structure just depicted, one of the predicates does not refer to any subject. It does not have "its own" subject.

Thus, when we studied the compound sentences which "fall under" the sufficient indicative feature type I (see page 475) of compound sentences, we proceeded from more particular cases (known to the students) to more general ones.

We cite an extract from the protocol of the lesson.

Tea. Up until now, you and I have examined compound sentences having a structure like this:

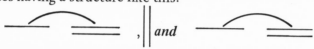

In each of these sentences, each of the predicates did not refer to one of the subjects, but had to refer to another. Each predicate had, as it were, "its own" subject. However, such a structure is not unique for a compound sentence.

Let us take two sentences.

The children had breakfast, and [they] began to dress them for their walk.

The children had breakfast and began to get dressed for their walk.

Carry out all the necessary operations, construct a diagram of each sentence, and specify by the indicative features what kind it is: simple or compound.[18]

The teacher does not show the students a prepared diagram of the new variety of compound sentence, but proposes that they devise such a diagram independently, using the operations which they know.

The students make up the diagrams of these sentences:

It is immediately obvious from the diagram that, in the first sentence, the second predicate does not refer to the subject, and therefore, this sentence is compound. In the second sentence, both predicates refer to the same subject, and therefore it is simple.

Tea. What are the types of simple sentences which make up our compound sentence?

Stu. The first sentence is definite-personal, and the second sentence is indefinite-personal.

As we see, the question of what kind of simple sentences make up the given compound sentence is asked of the students *after* they have specified that the given sentence is compound and have punctuated it. The specification of the types of simple sentences is not the condition for the specification of the type of a sentence (the students identified the type of the sentence without having recourse to the definition of the types of simple sentences). The specification of the types of simple sentences has, one might say, a significance which is "generally educative."

Tea. May a compound sentence having the structure: ⎯ ⎯⎯ , ‖ *and* ⎯⎯ consist of simple sentences of other types?

Stu. Yes it may. A compound sentence of such a structure may consist of a definite-personal and an impersonal sentence.

Tea. Think up an example of such a sentence.

The students propose examples.

Tea. As you see, a sentence can be compound not only when each of the predicates refers to some subject or other, but also when one of the predicates (and there can be several such predicates in the sentence) does not refer to any subject at all. In order to specify that the given sentence is compound, it suffices to establish the fact that one of the predicates does not refer to the subject which is in the sentence. Whether it does or does not refer to some other subject has no significance for specifying the type of sentence.

After this, the teacher trains the students to identify the type of sentence. As before, during training, special attention is given to the correct procedural order.

Here are several of the sentences which were used in training:
1. *Everything came to a standstill and grew quiet.*
2. *Everything came to a standstill, and [it] grew quiet.*
3. *[They] assigned an apartment to them, and they moved into it.*
4. *They got the apartment and moved into it.*
5. *[It] was already growing light, and in the east, the sky was changing its color from one minute to the next.*
6. *The overcoat was a bit too small for me, and [it] was obvious that [it] was not worth buying it.*
7. *The overcoat was made of ordinary cloth and was cheap.*
8. *He felt cold, and he put up his collar.*
9. *[They] got an interesting book for the boy, and he read it with enthusiasm.*
10. *The boy got an interesting book and began to read it with enthusiasm.*

After carrying out this exercise, the students are given exercises of other types whose purpose is not only teaching them to identify sentences on the basis of known indicative features, but also to devise them independently.

The first type of exercise is: to think up a sentence according to the given diagram, for example:

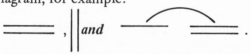

The second type of exercise is of this form. The students are given a definite-personal sentence and a verb in an indefinite form. They have to make up two sentences so that one of them is compound and the other simple with homogeneous parts.

For example, the students are given the sentence and a verb:

I found out about a new book. To get

The following two sentences may be constructed:

I found out about a new book, and soon I succeeded in getting it.

I found out about a new book and soon got it.[19]

We cite an example of work on a sentence when one of the exercises for the identification of the type of a sentence is carried out.

In directing the training exercises for the study of the preceding topics, the teacher asked questions mainly on how and in what sequence one must proceed in order to solve a grammatical problem. This type of question is also fundamental here. The teacher does not give the students any hints, and solves nothing for them. He just consistently draws their attention to the operations and to their sequences. When the students carry out these operations, they work out the elements of the future algorithm. Instruction is conducted in such a way that the algorithm, as we already said, is put together gradually and formulated part by part.

The sentence [*They*] *assigned an apartment to them, and they moved into it* is analyzed.

Tea. What do we have to do first of all in order to specify the type of the sentence?

Stu. We have to isolate all the subjects in the sentence (the subject *they* is isolated).

Tea. What do we have to do after this?

Stu. Isolate all the predicates. (The predicates [*they*] *assigned* and *moved in* are isolated.)

Tea. What do we have to do now?

Stu. Verify whether all the predicates refer to the same subject.

Tea. How is this done?

Stu. We ask the questions from the predicates. Moved in—who? They. And assigned the apartment—who? They? No. This means the predicate *assigned* does not refer to the subject *they*. This is a compound sentence, and we place a comma before *and*.

The students make up a diagram of the sentence:

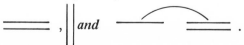

Tea. What are the types of simple sentences which make up this compound sentence?

Stu. The first sentence is indefinite-personal, and the second is definite-personal.

Up to this point, compound sentences having two or more predicates and one subject were analyzed. Now the teacher goes on to another variety of sentences of type I. In sentences belonging to this variety, there are two or more subjects and one predicate. No new knowledge or operations are required of the students in order that they identify this variety of sentences. The sentences are identified on the basis of the same indicative features and the same sequence of operations.

Tea. You and I analyzed sentences which have these structures:

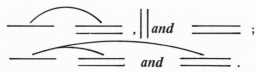

In these sentences, there was one subject and two or more predicates. Now let us examine sentences which have one predicate and two or more subjects, for example:

$$\underline{\quad\quad}\quad\underline{\quad\quad}\quad\underline{\underline{\quad\quad}}\;.$$

What kind of sentences could these be?

Stu. They could be either compound or simple. It all depends on whether the predicate refers to one subject or to both.

Tea. Show diagrammatically what the relations between the subjects and the predicate should be like in order that the sentence be compound and what they should be like in order that the sentence be simple.

The students construct the following diagrams:

(1) ―――― , ‖ ――― ⏜ ════ ;

(2) ―――― , ‖ ―――― , ════ ;

(3) ―― ⌐══⏜ ════ .

Tea. What could be the types of simple sentences which could make up the compound sentences having Structures 1 and 2?

Stu. A sentence having Structure 1 could consist of an elliptical and a definite-personal sentence, and a sentence having Structure 2 of two elliptical and an indefinite-personal or impersonal.

These are written on the board in the following way:

(1) elliptical,   definite-personal
(2) elliptical,   elliptical,   indefinite-personal
                              or
                              impersonal

When we acquainted the students with the compound sentences where there is one subject and two or more predicates, we took the following path: the students were given specific sentences, and they had to determine their structure. When we acquainted students with compound sentences where there was one predicate and two or more subjects, we followed a different path. The number of subjects and predicates in the sentence were shown on the diagram, and the students had to specify what the possible interrelationships were between the elements of the sentence. A considerably higher degree of grammatical and logical abstraction is required to fulfill this assignment, since the students are given no intelligible sentence. Not even one specific word is put before them to see. They have to operate with grammatical abstractions in their "pure form." However, after the practice conducted above, the students, as the experiment showed, can deal with problems of a similar type relatively easily. We believe that problems of a similar type have great developmental significance, and that they should find their place in the practice of instruction.

Tea. Think up examples of each variety of sentences.

The students think up examples. After this, training in the solution of the opposite kind of problem—that of identifying sentences—is conducted.

We show here several of the sentences which were used in the course of training:

1. *The last handshakes, and the trucks start off.*
2. *1920, March, and along the avenues of Tashkent flashes the gold of an early, dry, eastern spring.*
3. *The buses and trucks start off.*
4. *Just one more street, the slope, and we [are] at the top.*

These sentences are analyzed exactly as were the preceding ones.[20]

We cite the analysis of the sentence: *The last handshakes, and the trucks start off* as an example.

Tea. What do we have to do first of all?

Stu. We have to isolate all of the subjects in the sentence. The subjects are isolated: the *handshakes* and *trucks.*

Tea. What do we have to do then?

Stu. We have to isolate all the predicates. There is one predicate here: *start off.*

Tea. Then what?

Stu. We see whether the predicate refers to both subjects. Starts off—who? The trucks. The predicate refers to one subject and not to the other. This means that it is a compound sentence. We place a comma before *and.*

Tea. What are the types of simple sentence which make up this compound sentence?

Stu. The first sentence is elliptical. The second is definite-personal.

Tea. Draw the diagram of this compound sentence. The students draw the diagram.[21]

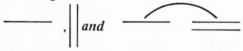

After the students master the system of operations for identifying compound sentences (as distinct from simple ones with homogeneous parts), the teacher gives them compound sentences in which simple sentences with homogeneous parts are included. As a result of this, systems of operations which were formed separately (operations for punctuating a simple sentence with homogeneous parts; operations for punctuating compound sentences) are now united in one single system.

What is the procedure for analyzing compound sentences made up of single sentences with homogeneous parts? It is the same as for the analysis of compound sentences made up of simple sentences without homogeneous parts. The subjects are isolated. The predicates are isolated. Then it is established which predicates refer to which subjects. Once the relations between the subjects and the predicates are established and the groups of each predicate are established, the boundary between the simple sentences becomes clear at once.

The general order of actions for the analysis of compound sentences made up of simple sentences with homogeneous parts is the following:

1. First the boundaries of the simple sentences are established, and punctuation marks are inserted between the sentences. This is done by applying the general diagram for the analysis of sentences (isolate the subjects and predicates and specify which predicates refer to which subjects).

2. After this, each simple sentence is examined separately and punctuated. This is done on the basis of procedural diagrams and rules studied in the section "Homogeneous Parts of the Sentence." Thus, the operations which were formed earlier are included in the new, broader system of operations.

Here are several of the sentences which were used in the process of training:

1. *He sat on the step and washed his long hair and his neck, and the water around him became brown.*

2. *Everyone in the parterre, in the loges, and in the galleries began to scream at the top of their lungs, and the man stopped and began to bow in all directions.*

3. *[Olga] woke up this morning, saw all the light, saw the spring, and joy resounded in my heart; I passionately wanted to go home.*

4. *On the other side of the partition, the machines rumbled and panted, and with each pant, some huge force moved the steamship ponderously and smoothly forward.*

5. *Finally Yekaterina Ivanovna came in dressed in a ballgown, pretty and neat, and Startsev was lost in admiration and was so enchanted that he could not say a single word and just looked at her and laughed.*

We cite the analysis of one of the sentences: *On the other side of the partition, the machines rumbled and panted, and with each pant, some huge force moved the steamship ponderously and smoothly forward* as an example.

Tea. What do we have to do first of all?

Stu. We have to isolate the subjects. There are two of them in this sentence: *machines* and *force*; then we isolate the predicates.

The predicates *rumbled, panted*, and *moved* are isolated.

Tea. What do we have to do then?

Stu. Establish which predicates refer to which subjects: *machines rumbled and panted*, and *force moved*. The sentence is compound.

The diagram is made up:

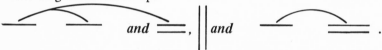

Stu. Then the boundary must be established between the simple sentences, and punctuation marks have to be put in the simple sentences.

The students place a comma before the second *and*.

Tea. What has to be done then?

Stu. The simple sentences must be examined in their turn, and punctuation marks must be put within each one of them. Since the conjunctions *and* which join the homogeneous parts within the simple sentences are not repeated, a comma before the conjunction is not necessary.

After the students are trained to insert punctuation marks in compound sentences which are composed of simple sentences with homogeneous parts, the teacher makes the problem even more complicated and he proposes compound sentences consisting not of two simple sentences, but of three and more. It is expedient to alternate these sentences with simple sentences having homogeneous parts, with compound sentences consisting of simple sentences with homogeneous parts, etc. It is necessary to alternate, in order that the students not know in advance what kind of sentence they are going to encounter and so that they discover them independently, using the operations which they know. The operations which the students have mastered make it possible to analyze any such sentence.

When this exercise is done, it should serve the purpose of generalizing and synthesizing all of the operations studied up until now.

Here are several of the sentences used during training:

1. *The storm howled, the rain pounded, lightning flashed in the darkness, and thunder rumbled continuously, and the winds raged in the thickets.*

2. *The officers regarded the huge corpulent figure of Pierre not without surprise and listened to his stories about Moscow and about the positions of our troops which he had succeeded in getting around.*

3. *Already by the third day, he suffered from fever, and at the same time, he had the chills and aches.*

4. *The day will grow silent, the years will pass, we shall get twice, three times as old, and then a legend will be created about the hero.*

5. *It was beautiful and quiet everywhere, and Syomka was confused.*

6. *The first fifteen years of the nineteenth century in Europe produced an unusual movement of millions of people. People abandon their ordinary pursuits, rush from one end of Europe to the other, pillage, kill each other, triumph and despair, and the whole course of life changes in several years and produces an intensification of the movement which at first increases and then diminishes.*

Up until now, sentences were analyzed where there were either pairs of principal parts or one subject and two or more predicates or one predicate and two or more subjects. Now the teacher goes on to examine the structure of sentences where there is one subject and one predicate, and the predicate does not refer to the subject. Here is the diagram of sentences like this:

$$\text{———} \; , \; || \; and \; \text{═══} \; .$$

We shall write out, one after another, all the structures of sentences examined:

(1)   ——— , || *and* ═══ ;

(2)   ⌒ ═══ , || *and* ═══ ;

(3)   ⌒ ═══ , || *and* ⌒ ═══ .

What is common to all these sentences is the fact that there are subjects and predicates in sentences of any structure even though one of the predicates does not refer to one of the subjects. This is why a compound sentence that has any one of these structures may be identified on the basis of the indicative features given above. In sentences with Structure 1, when the predicate does not refer to the subject in the sentence, it does not refer to any subject at all. Moreover, there is no predicate in the sentence which would refer to some subject. In sentences with Structure 2, there is already such a predicate. In sentences with Structure 3, the second predicate also refers to a specific subject.

It would have been possible not to acquaint the students at all with the varieties of the structures of compound sentences of type I, since the indicative features cited above make possible the identification of a compound sentence having any one of the structures examined. In order to recognize some sentence as compound, it suffices to establish that one of the predicates does not refer to one of the subjects in the sentence. But whether it refers to another subject and whether there is another predicate with the subject to which the given predicate does not refer—all these questions are of no importance for specifying the type of a sentence (simple or compound). We, however, considered it indispensable to acquaint the students with the varieties of sentence structures, since this is important for the development of a general language culture and logical thinking. The student who knows how a sentence is constructed and which variants of the interrelations of its elements exist will not only understand better and have a better feeling for speech, but he will be able to construct it consciously, master different ways of expressing thoughts, and choose the best variants.

We shall now describe the process of acquainting the students with the last variety of compound sentences type I.

Tea. You and I chose sentences with structures like this:

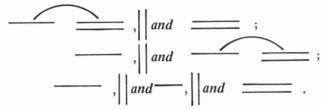

In these sentences, there was either one subject and several predicates or one predicate and several subjects. And now suppose we were to encounter a sentence where there is one subject and one predicate. What kind of a sentence will this be?

The procedure of introducing this variety of compound sentence is the same as the procedure for introducing the preceding variety.

Several students say that it is a simple sentence, but many guess correctly that such a sentence can also be compound.

Tea. In what case could a sentence consisting of one subject and one predicate be compound?

Stu. It could be compound if it consists of two simple sentences, one of which is elliptical, and the other indefinite-personal or impersonal.

Tea. Think up examples of such sentences. (The students think up examples.) What would be the diagram for compound and simple sentences which consist of one subject and one predicate?

The students draw the diagrams:

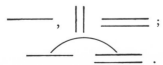

Then training in the identification of sentences having one subject and one predicate that does not refer to the subject is begun.

Here are several of the sentences used in the process of training:

1. *The bell, and [they] began to let [people] into the hall.*
2. *The bell could barely be heard.*
3. *A knock on the door, and [they] came in.*
4. *A fierce, deafening roar, and after several seconds, the rocket could no longer be seen.*
5. *A fierce, deafening roar made everyone tremble.*

We show here a sample of the analysis of one of these sentences. *The bell, and [they] began to let [people] into the hall.*

Tea. How should we act when we analyze the sentence?

Stu. First we have to isolate all the subjects. Here, there is one subject: the *bell*. Then we isolate all the predicates. There is also just one predicate: *began to let in*. Then we see whether the predicate refers to the subject. No. It does not refer to it.

Tea. What are the kinds of simple sentences which make up this compound sentence?

Stu. The first one is an elliptical sentence. The second is indefinite-personal.

Tea. Make up the diagram of this compound sentence.

The students make up the diagram:

$$\underline{\qquad} \; , \Big\| \; and \; \underline{\overline{\qquad}} \; .$$

2. Compound Sentences Type II

After the students are acquainted with the varieties of compound sentences type I, whose characteristic trait is the presence of at least one predicate and one subject, where the predicate does not refer to the subject, the teacher goes on to type II of compound sentences. By type II compound sentences, we mean such compound sentences where there are no predicates when the subject is present. Their structure is:

Tea. We just examined compound sentences having such a structure:

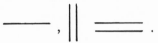

There was one subject and one predicate in these sentences. What would a sentence be, if the predicate were missing in it? Let's "remove" the predicate from our diagram:

―――― .

Stu. That would be a simple sentence.

Tea. What is the type of this simple sentence?

Stu. This is an elliptical sentence.

Tea. And now let us suppose that we were to encounter a sentence where there was not just one subject, but two or more and not one predicate:[22]

―――― ―――― .

What kind of sentence would this be?

Stu. A compound sentence made up of elliptical sentences.

Tea. Think up examples of such sentences. (The students think up such examples.) Can we identify a compound sentence consisting of just subjects on the basis of the indicative features which we have written down?

Stu. No.

Tea. Why?

Stu. Because in those examples it is shown that one of the predicates must not refer to one of the subjects, and here, there are no predicates whatsoever.

Tea. Correct. Just as we cannot identify a definite-personal sentence type II on the basis of the indicative features of definite-personal sentences type I, we cannot identify compound sentences having just subjects on the basis of indicative features of sentences having predicates as well. We have to point out the indicative features proper to such a type of compound sentence.

What will be the indicative features of this type of compound sentence?

Stu. There are two or more subjects and no predicates in such a sentence.

Tea. What conjunction joins the indicative features?

Stu. The conjunction *and*.

Tea. Let's write down these indicative features.

Indicative Features of Compound Sentences Type II

II   (1)   There are at least two subjects
           and
      (2)   there are no predicates

Tea. By what conjunctions should we join the groups of indicative features of compound sentences type I and compound sentences type II?

Stu. By the conjunction *or*.

Tea. Let's put this conjunction into our diagram.

After the conjunction is put into the diagram, it assumes this form:

A sentence is compound if:

I   (1)   there is a subject (or several subjects) in it
           and
      (2)   there is a predicate (or several predicates)
           and
      (3)   at least one of the predicates does not refer to the subject in the sentence.
           OR
II   (1)   there are at least two subjects
           and
      (2)   there are no predicates.

Tea. As we see, the indicative features of compound sentences are "organized" the same way as the indicative features of the "messenger" and like the indicative features of definite-personal sentences. The indicative features of compound sentences type I are, as it were, the indicative features of "Ivan Ivanovich," while the indicative features of compound sentences type II are, as it were, the indicative features of "Pyotr Petrovich."

After the indicative features are written down, the students are given several sentences for training of the type: *The whirlwind, the blizzard, the cold. It is harder and harder to move.* Compound sentences consisting of just elliptical sentences are encountered rather rarely, but nonetheless they are encountered. For example, in Nekrasov: *Trees, and the sun, and the shadows, and the deathly silence of the tomb.*

3. Compound Sentences Type III

After compound sentences which have just subjects (sometimes with dependent secondary parts) are studied, the teacher goes on to the study of compound sentences where there are just predicates (often also with dependent secondary parts).

Tea. You and I examined sentences where there was one subject and one predicate:

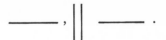

Then we examined sentences where there were just subjects:

$$\underline{\quad\quad}, \Big\|\ \underline{\quad\quad}\ .$$

Now we are going to examine sentences where there are just predicates:

$$\overline{\overline{\quad\quad}}\quad\ \overline{\overline{\quad\quad}}\ .$$

What would be the kind of simple sentence where there are just predicates and no subjects?

Stu. They are definite-personal type II, indefinite-personal, or impersonal.

Tea. Now let us see what kind of sentences those which consist of two predicates could be: simple or compound? Think up sentences where there is no subject, and all of the predicates are expressed only by verbs of the first or second person.

The students propose sentences of the type: [*I*] *walk and sing* and [*You*] *lie and think*.

Tea. What kind are these sentences?

Stu. These are simple sentences with homogeneous parts.

Tea. Why?

Stu. Because both predicates refer to the same subject which can be inserted into the sentence.

(*I*) *walk and sing.*

(*You*) *lie and think.*

Tea. Make up the diagram for these sentences.

The students make up the diagram:

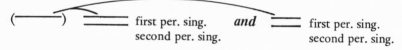

The teacher emphasizes that this diagram is a particular instance of a diagram already known:

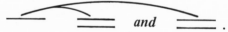

Tea. We examined the case where there are no subjects in the sentence, and *all* the predicates are expressed by verbs of the first or second person. What would be the type of a sentence where not all the predicates are expressed by verbs of the first or second person only? In order to answer this, let us examine two sentences:

I walk through the vast field and take deep breaths.

I walk along the vast field, and [*it*] *is so easy to breathe.*

What kind of sentences are these?

In the previous case, the indicative features of the types of sentences could be inferred by deduction (a sentence having two or several subjects and no predicates is compound consisting of elliptical sentences). In the given case, the indicative features cannot be inferred deductively, and therefore the students are once again given examples whose analysis makes possible the discovery of such indicative features.

This shows in particular how the choice of different methods of instruction—in the given case, methods of introducing new knowledge—are naturally stipulated by the specific conditions and are the functions of these conditions. When one knows these conditions, one can specify precisely which method of instruction must be applied in different cases.[2][3]

Stu. The first is a simple sentence with homogeneous parts. Both predicates in it refer to the same subject which may be inserted:

(I) walk through the vast field and take deep breaths:

(————) ≡≡≡ 1st per. sing. *and* ≡≡≡ 1st per. sing.

The second sentence is compound. The second predicate in it does not refer to the subject *I* which may be inserted:

(I) walk through the vast field, and [*it*] *is so easy to breathe:*

(————) ≡≡≡ , ‖ *and* ≡≡≡ .

Tea. What conclusion may we draw from the examples analyzed? In which cases are sentences without subjects that just have predicates compound and when are they simple? What are the indicative features of compound and simple sentences without subjects?

Stu. If there are no subjects in the sentence, and all the predicates are expressed by verbs of the first or second person, then this is a simple sentence with homogeneous parts. If there are no subjects in the sentence, and not all of the predicates are expressed by verbs of the first or second person, then this is a compound sentence.

Tea. Let's write down these indicative features.

The indicative features are written down.

Indicative Features of Compound Sentences Type III

III   (1)   There are no subjects

               and

     (2)   not all of the predicates are expressed only by verbs of the first or second person.

Tea. Do these indicative features suffice for the identification of compound sentences of this type? Will every sentence that has these indicative features be compound?

The students say yes.

Tea. Actually one must not make this assertion. Everything depends on how one understands the expression "not all the predicates are expressed only by verbs of the first or second person." If this expression is understood in the sense that some predicates are expressed only in the first person or only in the second person, and some are not expressed in this way, then you are correct. Such a sentence will be compound. But it is incorrect to understand this expression this way (and only this way). The expression "not all the predicates are expressed only by verbs of the first or second person (or what amounts to the same thing: it is not true that all the predicates are expressed only by verbs of the first or second person)" signifies either some are so expressed and some are not, or not one of them is so expressed.

When the meaning of negation is clarified for the students, it has a very important significance for fostering the culture of logical thinking in them, since the students (and adults as well) make a great many mistakes because they do not know the laws of the formation of negations and how to use them in different cases. The meaning of negation for the case which we analyzed can be established clearly by using the symbols of mathematical logic.

Suppose that there are just two predicates in the sentence,

and both of them are expressed by verbs of the first person. If we designate the expression "the first predicate is expressed by a verb in the first person" by the letter a_1, and the expression "the second predicate is expressed by a verb in the first person" by a_2, then the expression "all predicates are expressed by verbs in the first person" may be written thus: a_1 & a_2. The negation of this statement has the form $\overline{a_1$ & a_2} which is read: "not all the predicates are expressed by verbs of the first person" or "it is not true that all the predicates are expressed by verbs of the first person." But this statement is equivalent to the statement $\bar{a}_1 \vee \bar{a}_2$ $(\overline{a_1$ & $a_2} \Leftrightarrow \bar{a}_1 \vee a_2)$. The expression $\bar{a}_1 \vee \bar{a}_2$ signifies: the first predicate is not expressed by a verb of the first person, or the second predicate is not expressed by a verb of the first person, or neither the first nor the second predicate is expressed by a verb of the first person (i.e., they both are not expressed by a verb of the first person).

This logical meaning of negation is easy to explain to the students by examples or with the help of logical symbols. Then they easily understand where the misinterpretation of their statement lies—that if not all the predicates are expressed only by verbs of the first or second person, then this means, as it were, that some must be expressed and some not (and therefore, a sentence with such an indicative feature will be compound). Actually, the expression "not all the predicates are expressed only by a verb of the first or second person" assumes that there could be a case where not *one* of the predicates is expressed by a verb of the first or second person. But if not one of the predicates in the sentence is expressed by a verb of the first or second person (when the subject is missing), then such a sentence could be either simple or compound and the combination of indicative features "(1) there is no subject and (2) not all of the predicates are expressed only by verbs of the first or second person" is insufficient in order to draw a conclusion that this is a compound sentence.

Tea. Thus, the indicative feature "not all the predicates are expressed only by verbs of the first or second person" signifies either that some predicates are expressed by such verbs and others are not, or none of the predicates are expressed by such verbs. In the first case, the sentence has to be compound. In the second case, it could be either compound or simple with homogeneous parts (he cites examples of such sentences). An analysis of the sentences shows that sentences where there is no subject and not all of the predicates are expressed only by verbs of the first or second person are always compound with the exception of the case where all the predicates are expressed by verbs of the third person plural. Such sentences, as we shall see later, can be either compound or simple. What indicative feature do we have to add to our diagram in order that the sentence where there is no subject and not all the predicates are expressed only by verbs of the first or second person, would be compound for sure?

Stu. We have to add the indicative feature −"not all the predicates are expressed by verbs of the third person plural."

Tea. Let's write this into the diagram.

The indicative feature is written in.

As we see, the students have already mastered well the method for the construction of the logic diagram of indicative features. They know well what a sufficient indicative feature is and under what conditions some system of indicative features is sufficient for the identification of a phenomenon. They understand that if the indicative features (1) "there is no subject" and (2) "not all the predicates are expressed only by verbs of the first or second person" are in the sentence, it is always compound (with the exception of the case when all the predicates in it are expressed by verbs of the third person plural). For a sentence to be compound it is necessary to introduce an indicative feature that will exclude such a case. This is the indicative feature named by the students "not all the predicates are expressed by verbs of the third person plural."

The diagram of the indicative features for compound sentences type III assumes this form:

Indicative Features of Compound Sentences Type III

III   (1)   There are no subjects
                and

(2) not all the predicates are expressed only by verbs of the first or second person
and
(3) not all the predicates are expressed by verbs in the third person plural.

The general diagram for the indicative features of compound sentences assumes this form:

General Diagram of the Indicative Features of Compound Sentences

A sentence is compound if:

I   (1) it has a subject (or several subjects)
and
(2) it has a predicate (or several predicates)
and
(3) at least one of the predicates does not refer to the subject in the sentence.
OR
II   (1) it has at least two subjects
and
(2) there are no predicates.
OR
III   (1) it has no subjects
and
(2) not all the predicates are expressed only by verbs of the first or second person
and
(3) not all of the predicates are expressed by verbs of the third person plural.

Tea. How must we proceed in order to specify what kind of sentence it would be when there are no subjects?

Here, the teacher reads the following note: "Since compound sentences type II are encountered very rarely, in order to simplify the procedural diagram, we shall leave them out of it. We shall start from the fact that all compound sentences have at least one predicate. It is usually easy to recognize compound sentences without predicates (i.e., consisting of just elliptical sentences).

Stu. First of all, we have to verify whether all of the predicates are expressed only by verbs of the first or second person; if yes, then this is a simple sentence; if not, then we have to verify whether all the predicates are expressed by verbs of the third person plural; if not all of the predicates are expressed by verbs of the third person plural, then this is a compound sentence.

Tea. Let's depict the course of reasoning in the form of a procedural diagram.

A right-hand branch is added to the diagram, and it assumes this form:

Procedural Diagram for Identifying the Type of a Sentence
(compound or simple with homogeneous parts)

In order to specify the type of a given sentence, we must:

1. verify whether there is a subject in the sentence

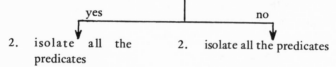

    yes   no

2. isolate all the predicates   2. isolate all the predicates

3. verify whether all the predicates refer to one subject

3. verify whether all the predicates are expressed only by verbs of the first or second person

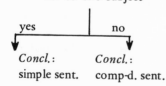

yes   no

    yes   no

Concl.:   *Concl.*:   *Concl.*:   4. verify whether all of the
simple sent.   comp-d. sent.   simple sent.   predicates are expressed
                                             by verbs of the third
                                             person plural

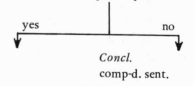

    yes   no

              Concl.
              comp-d. sent.

After this, training in the identification of the types of sentences based on the application of the operations shown in the diagrams begins. The training is conducted in a way which is analogous to that which was described above. During training, the students are presented with sentences where the predicates are expressed not only by verbs of the third person plural, but also by verbs in the past tense of the same number. In so doing, it is shown that, in this case, it suffices just to verify the verb for the indicative feature of its number (is the verb in the plural?). The corresponding additions are brought into the diagram. Then the teacher goes on to study the last type of compound sentence.

4. Compound Sentences Type IV

Tea. Let's go on and study the last type of compound sentences. We saw that if there are no subjects in the sentence and all the predicates are expressed only by verbs of the first or second person, such a sentence is simple. Now, let's examine the case where, in the absence of subjects, all the predicates are expressed by verbs of the third person plural (in the past tense, simply by verbs in the plural). We know that such sentences can be either simple or compound. By what indicative features can we distinguish one type of sentence from another? Let's take two sentences:

[They] took the broken record player to the workshop, and there [they] repaired it.

At the excavation, [they] found a cache of old coins and gave it to the museum.

What kinds of sentences are these?

Several students spot the difference in the types of these sentences at once and point out the indicative features which distinguish them. In the first sentence, the doers of the actions are different people. In the second sentence, they are the same.

Tea. What is there in common between these sentences?

Stu. There are no subjects, and the predicates are expressed by verbs of the past tense in the plural.

Tea. How do they differ?

Stu. In the first sentence, the predicates refer to different subjects which can be conceived, in the second sentence to the same subject.

Tea. Draw the diagrams of these sentences, putting in parentheses the imaginary subjects to which the predicates could refer. The diagrams have this form:

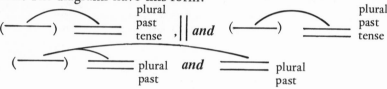

Then the teacher summarizes the discussions which have gone on.

Tea. Let's sum up. If there are two or several predicates in a sentence without subjects, and not all the predicates are expressed only by verbs of the first or second person, and not all are expressed by verbs of the third person (in the past tense, simply by verbs of the plural), then such a sentence is compound.

If all of the predicates are expressed by verbs of the third person plural (in the past tense, simply by verbs in the plural), then one should not draw a conclusion about the sentence, and one must seek to verify an additional indicative feature, i.e., specify who is the doer of the actions.

If the doers of the actions are the same people, then such a sentence is simple. If they are different, then it is compound.

Let's write down the indicative features pointed out.

Indicative Features of Compound Sentences Type IV

IV (1) There are no subjects
         and
    (2) all the predicates are expressed by verbs of the third person plural (in the past tense, simply by verbs in the plural)
         and
    (3) the doers of the actions are different people.

General Diagram for the Indicative Features
of Compound Sentences

A sentence is compound if:

I   (1)   there is a subject (or several subjects)
            and
     (2)   there is a predicate (or several predicates)
            and
     (3)   at least one of the predicates does not refer to the subject in the sentence.
            OR

II (1)   there are at least two subjects
            and
     (2)   there are no predicates.
            OR

III (1)   there are no subjects
            and
     (2)   not all the predicates are expressed only by verbs of the first or second person
            and
     (3)   not all the predicates are expressed by verbs of the third person plural.
            OR

IV (1)   there are no subjects
            and
     (2)   all the predicates are expressed by verbs of the third person plural (in the past tense, simply by verbs in the plural)
            and
     (3)   the doers of the actions are different people.

After this, the teacher has the students formulate the whole sequence of operations necessary for the identification of a compound sentence, i.e., to formulate the algorithm for identifying the type of a sentence (compound or simple with homogeneous parts).

Tea. What must we do and in which sequence in order to

specify whether a given sentence is compound or simple with homogeneous parts?

Stu. We have to begin by verifying whether there are subjects in the sentence. Then we have to verify whether there are predicates.

Tea. Suppose there are subjects and predicates in the sentence. Then what do we have to do?

Stu. Then we have to verify whether all the predicates refer to the same subject.

Tea. Well, what if there are no predicates but just subjects in the sentence?

Stu. Then we draw the conclusion that this sentence is compound made up of elliptical sentences alone.

Tea. What if it is the opposite case—if there are no subjects in the sentence and just predicates?

Stu. Then we have to verify whether all the predicates are expressed only by verbs of the first or second person. If yes, then the sentence is simple. If not, then we have to verify whether all the predicates are expressed by verbs of the third person plural (in the past tense, simply by verbs in the plural). If this is not so, then the sentence is compound.

Tea. Suppose all the predicates in the sentence are expressed by verbs of the third person plural. What should we do then?

Stu. Then we have to ascertain who is doing the acting. If the doers of the actions are the same people, then the sentence will be simple. If they are different, then it will be compound.

Tea. Let's write the last operation into our procedural diagram.

The operations are written out. The diagram becomes like the one depicted on page 470.

As we see, the algorithm is not given to the students all at once, and they do not have to learn it by heart. It proceeds in a natural way from the examination of specific sentences and from the analysis of separate grammatical situations, and it is mastered part by part. Subsequent operations are linked to previous operations which have already been mastered.

After a general procedural diagram for the identification of the sentence types has been formulated, training begins. At first, the students are presented with sentences of types III and IV where there are no subjects (in order to master through practice the indicative features and operations which have been introduced) and then sentences of all the types whose identification requires the application of all the indicative features and operations.

We cite here several of the compound sentences type III and IV and also simple sentences with homogeneous parts which were used in the process of training:

1. *[It] was already dark, and [they] set off multicolored rockets from the bank.*
2. *[It] got dark, and [it] became cold.*
3. *[We] used to gather in the evening and think gloomy thoughts.*
4. *[One] sits down in the morning, and [it] is so easy to think.*
5. *[They] sent off the telegram in the morning and had already received the answer by evening.*
6. *[They] sent off the telegram, and in two hours, [they] had already received it.*
7. *[They] had forgotten about him again, and now [it] makes no difference to him.*
8. *The road was sprinkled with snow, and [it] became difficult to travel along it.*
9. *[They] came up to the village toward evening and spent the night.*
10. *[They] deliver letters coming into the city at the main post office, and there [they] sort them.*
11. *[They] began to build the building in May and finished it in November.*

We shall describe the progress of study involving three sentences as an example.

[It] was already dark, and [they] set off multicolored rockets from the bank.

Tea. With what should we start the analysis of the sentence?

Stu. By verifying whether there are subjects in the sentence. If they are there, we isolate them.

There are no subjects in the sentence. (*Rockets* are not the subject; *they* set them off.)

Tea. What should be our second step?

Stu. We have to isolate all the predicates: [*it*] *was dark* and *set off.*

Tea. What should the third step be (under the condition that there are no subjects in the sentence)?

Stu. We see whether all the predicates are expressed by verbs of the first or second person. No.

Tea. What should we do then?

Stu. We see whether all the predicates are expressed only by verbs of the third person plural (in the past tense, simply by verbs in the plural). No. One predicate is expressed this way. The other is not. This means that it is a compound sentence. Therefore, we separate the simple sentences by a comma.

Tea. What are the kinds of simple sentences which make up this compound sentence?

Stu. The first sentence is impersonal. The second is indefinite-personal.

Tea. Make up a diagram of the compound sentence analyzed. The students make up the diagram:[2 4]

$$ \equiv \quad , \; \| \quad and \quad (\overset{\frown}{\text{———}}) \quad \overset{\frown}{\equiv} \quad . \qquad \begin{matrix}\text{plural}\\\text{past}\end{matrix} $$

[*They*] *sent the telegram in the morning, and in two hours,* [*they*] *had already received it.*

Tea. How should we begin the analysis of the sentence?

Stu. By verifying whether there are subjects in the sentence. If there are, then we must isolate them. There are no subjects in the sentence.

Tea. What is the second action?

Stu. We isolate all the predicates. The predicates here are *sent off* and *received.*

Tea. What are we going to do then (under the condition that there are no subjects in the sentence)?

Stu. We see whether all of the predicates are expressed by verbs of the third person plural (in the past tense, simply by verbs in the plural). They are expressed this way.

Tea. May we draw a conclusion on this basis about what kind of sentence this is (its grammatical type)?

Stu. No, we cannot. We still have to specify who is the doer of the action: the same people or different ones. In this sentence, they are different: some people sent the telegram, and others received it. This means that it is a compound sentence. We separate the simple sentences by a comma.

Tea. What kinds of simple sentence make up this compound sentence?

Stu. It consists of two indefinite-personal sentences.

Tea. Make up the diagram for this sentence.

The students make up the diagram:

[*I am*] *far from home, and my path ahead* [*is*] *difficult.*

Stu. We begin the analysis by verifying whether there are subjects in the sentence, and if there are, we isolate them. Here, there is a subject: *path.*

Tea. What is the second step?

Stu. We isolate all the predicates. Here, there are two: [*I am*] *far* and [*is*] *difficult.*

Tea. If there is even one subject and one predicate in the sentence, what must our third step be?

Stu. We have to verify whether the two predicates refer to that subject. [*Is*] *difficult* refers to it, but [*I am*] *far* does not. This means that it is a compound sentence. Therefore, we separate the simple sentences by a comma.

Tea. What kinds of simple sentences make up the compound sentence?

Stu. The first sentence is impersonal, and the second is definite-personal.

Tea. Make up a diagram of the sentence.

The students make up the diagram:

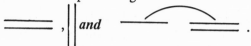

After the students have thoroughly mastered the general procedure for the identification of compound sentences, as distinct from simple sentences with homogeneous parts, the teacher goes on to the study of compound sentences which have a secondary part in common,[25] and then he conducts exercises with the students for the punctuation of such compound sentences where the simple sentences are joined by different coordinating conjunctions.

The reader has undoubtedly paid attention to the fact that, in all the sentences with which the students have dealt so far, the simple sentences within the compound sentence were joined by the conjunction *and* (or intonation). This is not by accident. First of all, the conjunction *and* is one of the most used and "difficult" conjunctions. Second—and this is important—if the students learn to analyze sentences with the conjunction *and* (and punctuate them correctly), they will be able to analyze sentences with any other coordinating conjunctions and also to punctuate them correctly. The method of analysis does not depend on which conjunction joins the simple sentences within the compound sentence. This is why compound sentences where the simple sentences are joined by the conjunction *and* may be considered as models for other compound sentences. The procedure worked out on this model can then easily be applied to compound sentences which have other conjunctions. In the process of working out a procedure, it is expedient to draw the students' attention to unessential details, and conjunctions which join the simple sentences within a compound sentence are just such details (in connection with the method for identifying the type of sentences).

Notes

1. For the most part, here and further on, we shall mean the distinction between compound-coordinate sentences and

simple sentences with homogeneous parts.

2. In the last edition of the textbook [533] the words "fusing them together into one whole in meaning and pronunciation" were added. However, this did not change the validity of the arguments advanced below.

3. The addition in the latest edition of the textbook [533] (namely that these simple sentences fuse in meaning and in pronunciation) does not change the essence of the matter, since, first of all, one may apply in addition the indicative features pointed out only *after* it has been established that the given sentence consists of two or more simple sentences. Second, these additional indicative features also require some indicative features for their identification. "They fuse in meaning and in pronunciation" are indicative features which are far from univocal in meaning. They permit various evaluations and different opinions (this will be particularly evident from the experimental material cited below). In this connection, the question naturally arises: how (by which indicative features) can one determine univocally whether the simple sentences fuse or do not fuse in their meaning and pronunciation? Moreover, is it at all necessary to establish this, if it is already made clear that the given sentence consists of two or more simple sentences? This assertion suffices to draw the conclusion that it is compound.

4. We call the directions for these actions procedures and not algorithms, since these directions are not univocal or specific enough. Different students can arrive (and do arrive) at different outcomes when carrying them out.

5. Here and further on, we shall just examine compound sentences whose simple sentences do not have a secondary part in common and which, therefore, are separated from one another by commas.

6. As we shall see later on, isolating the principal parts is the grammatical operation which enters as an element into an efficient algorithm for the identification of the type of a sentence.

7.   In the terms of logic: from $\overline{p} \Rightarrow \overline{q}$, it does not follow that $p \Rightarrow q$.

8.   The establishment of the fact that the predicate *sent off* does not refer to the subject *polar explorers* (if the polar explorers flew off, then they could not send airplanes with equipment) may be considered an elementary semantic operation. This operation is based on the obvious truth—the original axiom that a person cannot be in two different places at once. Since the semantic operation indicated is elementary for a person, it may be considered an algorithmic operation and be included in the algorithm for the identification of the types of sentences.

9.   Here and below, we proceed from the fact (without mentioning it every time) that there are coordinating conjunctions in the sentence or that its parts are joined by conjunctive intonation. This is an important indicative feature of compound sentences as distinct from compound-complex sentences.

10. Since compound sentences type II (which consist solely of elliptical sentences) are encountered extremely rarely, the indicative features of this type have not been included in the algorithm, in order to simplify it. Also, in order to simplify the algorithm, we did not include in it the indicative feature that simple sentences within a compound sentence are joined by coordinating conjunctions or intonations. We simply presuppose the presence of this feature.

11. It was precisely this algorithm which was cited in [271].

12. The conjunction *and* is used here as a representative of any coordinating conjunction. Compound sentences can be without conjunctions, but since compound sentences with conjunctions are more frequent, the conjunction is included in the diagram.

13. Though, as emphasized above, it is essentially impossible to insert subjects in this kind of sentence (it was a question of indefinite-personal sentences). Subjects change their meaning and make them sentences of a different kind.

14. Compound sentences without conjunctions are not being

examined for the time being.

15. The latter is provoked by the fact that, as we saw, several types of compound sentences exist where each one has this indicative feature. After the indicative features of all the types of compound sentences have been established, it will be expedient to write this feature separately showing that it appears in all types of compound sentences and in all groups of indicative features.

16. Let us recall that the sentences were presented to the students without punctuation marks.

17. The teacher draws the attention of the students to special cases of agreement between subject and predicate when studying the topic "The Subject."

18. The teacher already showed the students that an "operation" is an action and that he will use the word "operation" in those cases where it is not convenient stylistically to use the word "action."

19. Exercises of a similar type are recommended and are often conducted by teachers (see, for example, [707]).

20. The teacher indicates that, in certain instances, simple sentences within a compound sentence may not be separated by commas, but by dashes. Conditions where a dash is used would be examined later.

21. Of course, the teacher does not carry out such a complete, extensive argument with the students when he analyzes each sentence. This argument is carried out as a model from the example of one or two sentences, and then the students proceed independently.

22. The students know that with the principal parts there can be secondary parts which do not influence the type of sentence (compound, simple), and therefore they are not included in the diagram.

23. The importance of the study of the dependencies between the way the instruction proceeds and the conditions of instruction was noted in particular by Danilov (see, e.g., [601]). He advanced the notion of the logic of the educational process, which is precisely what should reflect this dependency.

24. The teacher points out the particular rule and the particular indicative feature: if in a sentence without subjects, the predicates are expressed by different forms (which can be judged from the endings of the verbs) this sentence is always compound. If the predicates are expressed by verbs of the same form, then no conclusion can be drawn. Such a sentence could be either simple or compound, and it has to be verified for the presence of the indicative features shown.

25. This section has not been excluded from the program, but the teacher points out to the students the fact that the punctuation in compound sentences changes depending on the presence in the sentence of a common secondary part.

Chapter Eighteen

Some Aids and Technical Means for Teaching Algorithms

1. "Exercise-books for Independent Study" as a
Means of Forming Algorithmic Processes Operation-by-Operation
One of the essential characteristics of the methods of instruction described previously was the fact that the experimenter formed in the students an integral and single method for the solution of problems. He formed it in them part-by-part as though he were "composing" it from elements. This represented the realization of one of the tasks of experimental instruction—to ensure the *operation-by-operation formation* of the thought processes.

Operations are the original building blocks of intellectual activity, and in order that they be formed consciously and purposefully one must, on the one hand, be able to "manufacture" each of these building blocks, and, on the other hand, "couple" all of them together in a specific way and combine them all into a total, integral system. Such effort presupposes that the teacher not only have a program of knowledge, but also a specific program of operations which the students have to master and a clear idea of which operations must be taught and in which sequence. In order to draw up such a program, it is necessary to make a structural-logical and psychological analysis of both the content of the knowledges mastered by the students and the procedures for their operation, as well as a special psychological analysis of the difficulties and errors of mastery. The grammatical algorithms cited above are nothing more than specific programs of operations. These programs of operations depend on specific

513

systems of indicative features. These are programs of the subject-matters which the teacher has to teach the students. In order that instruction be a systematic and purposeful process, it must include *programming* the students' intellectual activity.[1] But such programming is possible only in the case where the knowledges, habits, and skills to be formed in the students are analyzed from structural-logical and psychological viewpoints and, consequently, when the teacher not only knows what knowledges he must convey to the students and what their structure is, but also which operations guarantee that these knowledges are attained and applied. At present, teachers are not equipped with such programs of knowledges and operations for the majority of subject-matters.

However, in order to successfully teach students the procedures of intellectual work, it is not enough for the teacher to have a program of knowledges and operations. It is necessary to have the effective means present that would guarantee the *reliable and indispensable* formation of the necessary systems of knowledges and operations in the students (in *every* student).

At present, teachers do not have such effective means. Today, when he explains new material and conducts training exercises, a teacher cannot be sure that all the students carry out precisely those operations which are necessary and that they carry them out as they should. Nor can he be sure that these operations are integrated into the system which is required. (The students' mistakes, the difficulties of learning, and the lack of success bear witness to the fact that this is exactly the situation now.) In other words, it is not possible today for a teacher completely to control the formation and progress of the students' intellectual activities. He still cannot always bring about the necessary, correct operations by a precise, calculated action, or correct a defective operation, or stop an unnecessary or incorrect operation. When methods for the operation-by-operation formation of the students' intellectual processes are found, this will mean that the point will be reached where the mastery of knowledges, intellectual skills, and habits will become a precisely calculated, maximally controlled and directed process.

It is impossible, however, to guarantee the successful operation-by-operation formation of intellectual and, in particular, algorithmic processes without the elaboration of methods for *operation-by-operation control* of the course of these processes and of the way the algorithms are "put together" in the students' consciousness.

In order for the teacher to note and correct in time all the deviations from the normal course of the intellectual process which occur during instruction, he has to have constant information on the progress of the student's intellectual processes when the latter solves some problem or other and carries out some activity or other. Without such information—without an *operative feedback* from the student to the teacher—more or less complete control over the processes which take place in the consciousness of the student is not possible.

Currently, when the teacher instructs the students, he does not have the means to obtain such information from all of them. Often he has a very confused idea of what lies behind different, externally expressed actions.

The basic sources of the teacher's knowledge about the progress of the students' assimilation of knowledges, habits, and skills and about their intellectual processes are questions asked in class and various kinds of tests. When the teacher asks questions, if he has mastered the appropriate methods, he can get an idea of how the student thinks and reasons and how certain thought processes take place. But first of all, questioning in class requires considerable expenditure of time. Second, such questioning has an episodical character in relation to each student (only rarely is the teacher able to question each student in detail more than three or four times per term). We do not even discuss the fact that the class is often passive during such detailed questioning. Therefore, under conditions of mass or class-group instruction, it is impossible to obtain *continual* information on the progress of *all* the students' intellectual activity.

In regard to the different types of tests, the teacher has to deal here for the most part with the *results* of the student's intellectual activity (he found the solution or did not, made a mistake or did not). It is impossible for the teacher to understand the *process* of this activity (by what operations did he find the solution?), or to completely control the progress of the intellectual operations, their sequence, etc., from beginning to end. One may also solve a problem by a successful but blind trial, or on the basis of external analogy, or by a mechanical choice of data, or by other means. The correct result of intellectual activity still does not corroborate the fact that the process which produces this result is correct.[2] It requires special methods and special means to recognize the processes by which thinking is carried out.

As we developed the methodology for teaching students how to reason when they solve grammatical problems, we tried to find, on the one hand, the means which would ensure the operation-by-operation formation of their thought processes. On the other hand, these means should make it possible for the teacher to obtain detailed information about each of the stages of formation of these processes and, on this basis, to carry out operation-by-operation control. One of these means was the "Exercise-book for Independent Study" which was specially designed.

Special types of problem-exercises were devised for these exercise-books. Their unique character resides in the fact that, when the student carries out these problems, he has to break down the thought process into separate operations. He is obliged to do all of them and to realize each one of them clearly and precisely. The assignments are so constructed that when the student works on them, he cannot *not do* all of the necessary intellectual operations. He is *obliged* to reason correctly, to think, and to think all the time. At any stage of the process when any operation is being carried out, each of his mistakes can be detected and corrected by the teacher. When this is done—and this is especially important—the mistake can be corrected at that "point" where it was committed.

We briefly characterize one of the typical assignments on the topic "homogeneous parts of a sentence."

On one of the pages of "An Exercise-book for Independent Study," there is a series of sentences without punctuation marks.[3] The final task of the student is to put in the punctuation marks. But in order to specify them, he must perform specific intellectual efforts and carry out a series of operations which he knows. The operations are shown either in oral instruction or in written instruction which accompanies the assignment. This instruction is nothing more than a specific algorithm according to which the student must act.

The basic operations to be executed and worked out when the given assignment is completed are numbered. In order that the student for some reason or other (laziness, not wishing to think, etc.) may not "escape" the completion of the necessary work and may not omit some operation or other or not do it, he has to state the results of each operation in the "exercise-book." Where and in what manner?

Several free lines are placed under each sentence, and at the beginning of each line is a number which designates the number of the operation.

Here is an example:

The flagpole gleamed in the air tilted forward cut through the crowd disappeared in it and in a minute the wide cloth of the soldiers' flag fluttered like a bird over the upturned faces of the people.

(1)
(2)
(3)

Diagram:

When he has carried out the first operation (isolate all the subjects in the sentence), the student has to write down the result of this operation—in the form which is indicated—on the first line. When he performs the second operation (isolate all the predicates), the student has to write down the result of this operation on the second line. On the third line, the student must write down the result of the third operation (as we recall, the third operation consists of establishing which predicate refers to which subject).

Then a diagram of the sentence is drawn up which is graphically depicted on a separate line. After this, the punctuation marks are transferred to the sentence.

This is the form which this sentence has after all the necessary operations are carried out.

The *flagpole gleamed* in the air, *tilted forward, cut through* the crowd, *disappeared* in it, and in a minute the wide *cloth* of the soldiers' flag *fluttered* like a bird over the upturned faces of the people.

1)   *flagpole, cloth*
2)   *gleamed, tilted forward, cut through, disappeared, fluttered*
3)   *flagpole gleamed, tilted forward, cut through, disappeared* and *cloth fluttered*

Diagram :

When the student does this assignment, not only must he find and underline the principal parts (this is usually done), but he must also write them out when he has established the specific relationships among them. As the experiment showed, when the students write out the principal parts on separate lines and show the relationships among them, this is very important for their assimilation of the correct procedure and for a quicker formation of the necessary systems of operations in them.

Originally, in one of the classes on the first three topics, we conducted instruction without the material statement of the results of each operation. The effectiveness of this instruction was lower than expected (especially on the topic "homogeneous parts of a sentence"). Despite the special attention which the teacher paid to the demonstration and disclosure of ways to proceed and operate, many students did not master the operations to a sufficient degree. In a number of cases, they knew how to act, but they did not carry out these actions. This led to mistakes.

At first, it was not clear why the system of operations was

poorly assimilated. The teacher persistently taught the operations, but the students did not learn them as they should have. Finally, an explanation was found.

To think (and to think is, first of all, to carry out specific intellectual operations) is often not easy for children. It is very difficult for many young people to make the necessary intellectual effort and to accomplish all the intellectual operations which are required to solve the problem. The difficulties of intellectual effort engender various ways of "escaping" the assignments—guessing, and proceeding by a method of trial-and-error, etc.

In order to teach the students to think and reason, it is not enough to *acquaint* them with the operations which must be carried out in one or another instance. It also does not suffice to *require* that they do them. It is necessary to create the kinds of conditions and *organize* the instruction so that the student *must* carry out the necessary operations during assimilation of knowledge and during training so that he cannot avoid them and cannot "escape" from the assignment. It is also important, when possible, to make it easy for students to carry out the necessary operations.

Originally, these conditions did not exist. It was difficult for the students to do all of the operations mentally. On the other hand, since the teacher did not control the execution of each operation (he did not have the means for such control), the students did not carry out all the operations despite the teacher's demands. They "escaped" the difficult assignment when this was possible. Thus, despite the fact that the teacher made a great effort to form the necessary systems of operations in the students, these operations were not formed in all the students or were not completely formed. The reasons for this were discovered, and this led to the idea of how to organize the process of instruction and how to create means which would control the students' activity and would oblige them to carry out all the necessary operations, while the teacher could control the execution of each operation. The "Exercise-books for Independent Study" were such a means.

Wherein lies the significance of this method of writing down the necessary parts of the sentence on separate lines?

1. When the student is confronted with the necessity of

filling out all three lines, he must unfailingly carry out all three operations (find the subject, find the predicate, and specify which predicate refers to which subject). With the usual assignment—underline the primary parts of the sentence—the student needs to carry out only two operations. It is not possible to control the execution of the third operation (even if one points out that it must be carried out), since its result cannot be materially stated. Whether the student will mentally carry out this operation and how he will do it is unknown to the teacher. What is not controlled is often not carried out. Meanwhile, the formation of this operation is of decisive importance for the correct differentiation of compound sentences and simple sentences with homogeneous parts.

2. When the results of each operation are stated and intellectual actions are united with physical ones, this makes it easier to carry out intellectual actions (intellectual operations), and at the same time it ensures a more successful formation of the latter.[4]

3. When the results of each operation are stated, this ensures a better realization of both the operations themselves and the sequence in which they must be carried out. This also is of great importance for the successful formation of the necessary abilities and skills.

4. When the principal parts of the sentence are written out, and they are operated upon relatively independently of the other parts of the sentence, the very important operation of covertly ("mentally") isolating the essential elements of the sentence and setting aside the unessential elements is formed in the students. This teaches them to orient themselves just by the essential elements when they solve grammatical problems.[5]

5. When the relationships among the principal parts which are examined apart from the other parts of the sentence are established, this makes it possible to see the *structure* of the sentence well and to work out a *structural approach* to grammatical phenomena. This is very important both for the correct solution of specific grammatical problems and for the development of speech and thought in general.

We shall show from an example how difficult it is in a number of cases to execute all the necessary operations mentally (isolate the essential elements of the sentence and to draw attention away from the unessential elements of the sentence, to establish all the series of homogeneous parts, and to see the structure of the sentence). We shall also show how the solution of all these problems becomes easy when the results of each operation are stated.

Let us take a sentence where students from the control classes usually made mistakes.

Large glass jars with coffee, cinnamon and vanilla, crystal and porcelain tea-caddies and cruets with oil and vinegar were in the kitchen.

The most frequent mistake was placing a comma before *and cruets*. Students often also put a comma before the last *and* (this is acceptable in English, *Trans.*). These mistakes are provoked because students do not know how to determine the structure of this sentence. They "do not see" this structure. However, it suffices to perform the operations indicated above on this sentence—to write out each series of homogeneous parts on a separate line—for the structure of this sentence (and hence the punctuation marks) to become obvious.

(1) *jars, tea-caddies and cruets*
    (a) *jars with coffee, cinnamon and vanilla*
    (b) *crystal and porcelain tea-caddies*
    (c) *cruets with oil and vinegar*
(2) *were*
(3) *jars, tea-caddies and cruets were*

Diagram:[6]

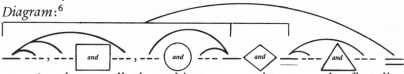

As always, all the subjects are written on the first line (number one). All the predicates are written on the second line (number two). All the subjects with the predicates and indications of their relationships to each other are written on the third line (number three). What is written on the third line shows what kind of a sentence it is at once: simple or compound. After the type of the sentence is specified (the given sentence is simple), one should verify whether the sentence's principal parts that have been isolated have any secondary parts with them. The results of this verification are stated, and the homogeneous parts (if there are any) are written down with those parts of the sentence to which they refer. In our sentence, all the homogeneous parts refer to the subjects.

When the principal parts of the sentence and the series of homogeneous parts are written out on separate lines, it is at once clear how many series of homogeneous parts are in the sentence and whether the conjunction *and* is iterative.[7] When we look at the diagram we see at once that the conjunction before the word *cruets* is not iterative and that a comma is not placed before it. Commas before the other *ands* are not necessary either.

When the students learn how to isolate the principal parts of a sentence and the series of homogeneous parts by overt, physical actions (writing them out on separate lines), this teaches them to pick out these members covertly. It teaches them to carry out all the necessary operations covertly and with awareness.

The following analogy can be established. It can be very difficult for children learning to do sums in their heads to carry out specific arithmetical operations. Then they are given some kinds of material objects, for example, sticks, and they learn arithmetical operations (e.g., addition and subtraction) with the sticks. The actions which must be performed covertly are now done by the children in practice in the form of overt, physical actions. When the children perform arithmetical actions as specific overt, physical actions, they then acquire the ability to do them covertly.

Just as it is difficult for children of five to seven years to perform arithmetical actions covertly, it is difficult for students of the seventh and eighth grades to perform covertly certain grammatical actions, in particular to isolate covertly specific parts of a sentence, establish the corresponding relations between them, etc. Therefore, in order to form these actions as intellectual actions, one must first make it possible for the students to perform them as overt, material actions. When they write out the principal parts of a sentence and the series of homogeneous parts, this is the physical "extraction" of words from a sentence which forms the operation of covertly isolating these words. When they are able covertly to isolate, for example, the subjects and predicates in a sentence (are able to "see" them), this is nothing more than the operation of "extracting" which is just performed covertly.

The method for the formation of intellectual operations by means of physical actions is applied not only when children are taught to do sums in their heads. Many students do not succeed in stereometry, since they cannot mentally operate stereometric figures (e.g., they cannot mentally turn a pyramid around and bisect it in a specific manner, etc.). In order to cultivate in students the ability to intellectually operate stereometric figures, they are often taught to do this from models, i.e., from real, physical objects. The operation of a model makes it possible to perform difficult intellectual operations as physical operations and at the same time it forms these intellectual operations. When the students have learned to break up a pyramid in a specific way, they can then do it covertly.

If, from this standpoint, we compare the method which is usually used—underlining the principal parts—with the method which we applied, namely writing out the principal parts with the subsequent pointing out of the relationships between them, we will see the difference between these methods. To underline the principal parts is not a physical action with a grammatical object (word). It is no more than a mark (compare the designation of some area in a drawing of a pyramid by a letter). When one writes out the words, this is a physical action with a grammatical object, since the words in this case are, as it were, physically extracted from the sentence and are written separately. Different operations are performed with these words outside of the sentence. This apparently explains the greater efficacy of the method of writing out the specific members of the sentence and the series of homogeneous parts as compared to the method of underlining them.

We reviewed the value "Exercise-books for Independent Study" could have for the formation of intellectual activity. We shall point to several more advantages which the use of these exercise-books offers.

When the teacher looks at how the lines are completed, he can rather quickly and easily "diagnose" the state of each student's intellectual activity. He can specify which operations he performed correctly and which incorrectly, and on the basis of this "diagnosis," he can put into effect the necessary and differentiated "therapeutic" influence. There is no need to dwell on the importance of the possibility for a quick diagnosis of the state of intellectual activity, since it is completely clear that no calculated pedagogical influence which is precisely controlled is possible without having a representation of those processes which take place in the student's head when he solves some problem. The analysis of the way students handled assignments in the "Exercise-books for Independent Study" gives the teacher rather complete information on the state of separate links in the thought processes of all the students, and at the same time it creates the possibility for a more complete formation of these processes.

Such a circumstance must be noted. Thanks to "Exercise-books for Independent Study," it is possible for the teacher to carry out some of the work on control, diagnostics, and correction of the discovered defects during the lesson. While the students are working with the "Exercise-books for Independent Study," the teacher walks up and down the aisles and sees how the lines are completed. Owing to the standard form for the completion of the "lines," one can establish very quickly—in several seconds— whether the student performed the operations correctly. Of course, it is sometimes difficult to check the performance of all of the students at a lesson, but it is completely possible to keep the weak students under control. Thus, the use of the "Exercise-books for Independent Study" makes it possible, to some degree, to put into effect the combination of group work with individual work. This is very important for the prevention of mistakes and various deviations from the correct formation of the thought processes.

The following is also essential. Since the formation of intellectual habits proceeds more quickly, easily, and with fewer mistakes with an operation-by-operation mastery of thought processes, it becomes possible to diminish the volume of homework assignments and to achieve a good assimilation of the material and methods of intellectual work during the lesson.

Since the sentences in the "Exercise-books for Independent Study" are printed, the students do not have to waste time in copying many words whose spelling they already know well. (If they encounter words in the sentences which have difficult orthograms, they may set them aside and mark them with dots instead of letters.) In the system of instruction now in operation, in order to practice with a word or specify whether one has to put a comma into the sentence, students often have to write five or six words which have no value from a grammatical standpoint. Much time is taken up with such copying. When "Exercise-books for Independent Study" are applied, one may economize at the expense of copying well known words and considerably increase the number of exercises. The productivity of school work is raised thereby. What has been said should not be interpreted to mean that in general one must try to get the students to write less. It is just a question of freeing them from a useless expenditure of time as they repeatedly copy the same words which they already know.[8]

Finally, the use of "Exercise-books for Independent Study" creates conditions for the realization of self-control and self-verification as well as for mutual verification by the students of each other's work. This not only cultivates a number of important qualities in them, but it also frees the teacher from the excessive load of checking the exercise-books. Such self-verification and mutual verification was used rather widely by us in the process of experimental instruction.

It seems that if similar "Exercise-books for Independent Study" were created for all topics (and perhaps for all subjects) and were printed, published, and introduced into the practice of large schools, this would make it possible not only to save much time for conducting original work in instruction (the time which today is spent unproductively on writing and copying well known words), but would also make it possible to attain, on a large scale, the systematic and purposeful formation of the necessary intellectual operations and their systems and of the correct procedures for intellectual efforts.

2. The Teaching Machine "Tutor I"

It is easy to see that the above description of "Exercise-books for Independent Study" converges on programmed textbooks. They make it possible to carry out the operation-by-operation formation of thought processes and operation-by-operation control. They demand much activity from the students, and they ensure that all the students must develop all the necessary systems of operations. One still may not call these exercise-books programmed textbooks, however, since first of all they do not ensure the immediate reinforcement of the students' responses (in the majority of cases, with the help of these exercise-books, one may just realize delayed operation-by-operation control).[9] Second, they do not permit the individualization of instruction to the fullest extent. Meanwhile, in order to form skills and habits, it is very important that the reinforcement of the students' responses (i.e., telling them whether they performed the responses correctly or incorrectly) is realized immediately after this response is made.[10] This is particularly important when the students perform incorrectly. If these incorrect responses are not rectified in time, an incorrect habit may be formed in the students which then has to be broken. This is very difficult.

The defect pointed out may be avoided if answers for self-verification are inserted into the "Exercise-books for Independent Study" and are programmed. When students are trained in self-verification, this cultivates a number of positive qualities in them. When answers for self-verification are included in the exercise-books, however, this also creates the possibility for a negative phenomenon. It consists of the following: If the students who like to and try to solve problems independently will do this when the answers for self-verification are present in the exercise-books, the intellectually passive students[11] and the students who are too "lazy to think" (and it is especially important for them to think independently) have, in this case, the possibility of "escaping" from the solution of the problem by looking at the correct answer (there is this possibility in all programmed textbooks). Therefore, "Exercise-books for Independent Study" cannot guarantee that all of the students will carry out all of the operations and will solve all of the problems quite independently when the answers for self-verification are present in the exercise-books. Although there are a number of ways to lessen the probability that the students will look at the correct answer and ways to reduce this to a minimum, nevertheless programmed textbooks cannot guarantee that not even one student will ever peek.

The means which make it possible to avoid the defect indicated and to effectively solve a number of other problems of instruction is cybernetic teaching equipment or, as they are sometimes called, "teaching machines."

At the present time, teaching machines are being urgently developed in a number of countries.[12] In our country, one of the first teaching machines of the type "Tutor" ("Tutor I") was created by the author of this book in collaboration with the engineer S.P. Khlebnikov in 1961.[13]

Before we can go on to the description of the work of this machine and its didactic functions,[14] we shall pause briefly at the question of how teaching machines differ from other technical means of instruction. Today, the media of instruction are quite varied (radio, film, television, etc.). The basic reason why teaching machines are put into a special class is the fact that one may speak of instruction only in the proper sense of the word (as distinct from self-instruction) when a two-way process takes place where the reactions of the person taught (student) provoke corresponding actions from the teacher. These, in turn, change the responses of the student, etc. The interactive process in a closed system is realized. Here, the actions of the teacher which have a specific goal, direct and coordinate the actions of the student, according to a specific plan, and to a specific program (realized or unrealized). The various ways to teach the same thing (for example, different methods) are nothing more than the different "programs" by which the "reaction" of the teachers to the actions of the students is realized.

Hence, it is clear that only those mechanisms which fulfill the pedagogue's most important function—the function of a "reverse influence" on the student, the function of correcting, regulating, and controlling his actions and the processes taking place in his consciousness—can be called teaching machines.

When one proceeds from what has been said, it is obvious that the degree of perfection of different types of teaching machines depends directly on how fully and how well they can realize the function of reverse influence on the student and the function of controlling his intellectual actions. It depends on how sensitively they can react to his actions and analyze them in depth. It also depends on how flexibly they can change the program of "response reactions" (adapt) in relation to the process of the students' assimilation of knowledges, skills, and habits. Radio, cinema, and television which are applied in instruction as means of transmitting information do not, however, have the properties of inverse influence on the students, and they cannot react to their actions. It is precisely because they are media of instruction that they are not teaching machines.[15]

The principle of the operation of the "Tutor I" teaching machine may be most simply explained via an example.

Let us suppose that a student has to master the algorithm discussed above for the identification of types of sentences. In order to do this, he has to be trained to carry out the system of operations which enter into the algorithm. The training is conducted with sentences. Each sentence is written on an ordinary cardboard card on which there are several apertures in appropriate places. The card is placed in a perforating mechanism. After this, the machine (owing to the apertures) "knows" which indicative features are in this sentence and, consequently, which operations the student must carry out in order to specify its type.

Let us suppose that the sentence: *[They] brought the broken radio to the workshop, and on the following day [they] repaired it there* is written on the card (the figures designate the numerical codes of the words in the sentence).[16]

The student is confronted with the problem of specifying whether or not a comma should be inserted before *and*. To do this, he must identify the type of sentence: is it compound or simple? If it is compound, then the comma is necessary. If it is simple with homogeneous parts, then the comma is unnecessary.

There are three rows of switches on the machines. Each of the switches of the upper and lower rows (these switches have two positions) may be in one of two positions: "yes" or "no." The switches in the center row can also be in a neutral position (they have three positions). The upper row of switches is intended for the introduction of information on the principal parts of the sentence into the machine (which words are subjects and which are predicates). The second row is for the introduction of information about other indicative features of the sentence which are important for the identification of its type. The third row is for the introduction of the answer to the problem which confronts the student on the type of sentence (i.e., for the introduction of the solution to the problem into the machine).

We shall describe how the analysis of the sentence is accomplished with the help of the machine. First of all, the principal parts of the sentence must be isolated. There are no subjects in the sentence analyzed, and the predicates are the words *brought* and *repaired* (words number 1 and 8). In order to tell the machine that these words are the predicates, the students press switches 1 and 8 (all the switches of the first row are numbered). But let us suppose that the student thought that the word *radio* was a subject and pushed number 3. Right away, a red light lights up (a sign of a mistake), and a special counter above this light counts this mistake.

After the principal parts of the sentence are isolated, it has to be checked for the presence or absence of the indicative features which specify its type. These indicative features are written above the second (middle) row of switches, each indicative feature over one specific switch.

According to the algorithm, one must first of all verify whether there are subjects in the sentence. As we already know, there are no subjects in the sentence. Therefore, we must put the appropriate switch in the "no" position. Then we must verify whether all the predicates are expressed by verbs of the first or second person or not. No, they are not expressed by them. Therefore, we also put the appropriate switch in the "no" position. Then we have to verify whether all of the predicates are expressed by verbs of the third person plural (in the past tense, simply by verbs in the plural). In our example, they are expressed by them. Therefore, the appropriate switch is put in the "yes" position. The last operation remains to be done: check whether the doers of the actions are the same persons. No. Therefore, the last switch is put in the "no" position. (If the student made a mistake at some point, for example, he thought that the doers of the actions were the same persons and therefore put the last switch in the "yes" position, the machine would once again have signaled a mistake, and the counter would have added this mistake to the previous one.)

The analysis of the sentence according to indicative features is ended. Now, the machine must be given the answer (solution). The answers ("compound sentence," "simple sentence") are written under the last row of switches. If the student answered incorrectly that the sentence analyzed was simple, then once again, the red light would go on, and still another mistake would be counted in. If the student answered correctly that the sentence analyzed was compound, a green light would go on, and a second counter would include the correct answer. If an appraisal is desired, the student presses a special switch and the appraisal of his work appears in a little window. The appraisals are set according to a criterion installed in the machine: "five" if there are no mistakes, "four," for one or two mistakes, etc. The block of appraisals is so constructed that depending in what position the special switch is placed, the machine will provide the appraisal either for the solution of one grammatical problem, or two problems, or three, etc., all the way up to eleven problems.

We cited an example of how algorithms for the solution of grammatical problems can be taught with the help of a machine. But the machine may also be used for instruction in other subjects. One need only establish the appropriate algorithms for them. Several such algorithms have been developed. For example, for instruction in the technique of translation from German into

Russian, one may use a number of algorithms devised by Belopolskaya [214].

It is simple to adjust the machine for another program. The whole operation takes less than a minute.

What does instruction with the teaching machine yield? What teaching functions does it perform?

First of all, it permits the control and evaluation not just of the final *result* of the student's intellectual processes (whether or not he solved the problem), but the actual *process* of solution, the actual progress of thinking (by means of control of intermediate results). It analyzes each intellectual operation and informs the student about each step of his work. Thus, an operative feedback loop is introduced into instruction and ensures an immediate reinforcement of the student's responses.

Use of the machine greatly enhances active participation in the process of instruction. When the student works with the machine, he has to think actively all the time and exert all the necessary intellectual effort and perform all the intellectual operations. He cannot, for example, answer the question asked by just relying on intuition or by acting on guesswork. The machine "will not accept" an answer from him until he makes all the necessary judgments and substantiates his answer as he should.

Further, when the machine exercises operation-by-operation control, it corrects the operations at the moment they are accomplished and at the same time prevents the formation and consolidation of defective habits and skills and the formation of incorrect methods of thinking. Since the correct operations are reinforced and incorrect ones are arrested, conditions are created for a quicker cultivation of the necessary skills and habits.

It is very important not only to control the students' actions, but also to cultivate habits of self-control. The machine may also be used to solve this problem. A program of exercises may be set up in such a way that at specific stages of instruction it is not the student's actions when he solves a problem which the machine will control, but it will control the control of his own actions which the student has realized. It will control the self-control.

The individualization of instruction which the machine

makes possible is very essential. At the present time, the machine just permits the individualization of the students' work tempo (self-pacing), but it is possible to modernize it so that it will also take into account certain individual characteristics of assimilation. It is easy to arrange it so that the contents of each subsequent assignment will be "handed out" by the machine depending on the results of how the previous one was executed.

Finally, it is very significant that the teacher is unburdened. He will be freed from what is, to a great extent, mechanical work of checking exercise-books and controlling the students' execution of the necessary operations. He will be able to direct all his attention to the solution of other more complicated and subtle problems which require high qualifications and creative thinking.

At the present time, the machine carries out instructional, diagnostic, and examinational functions. If one were to connect a memory device, it would be able to remember all the incorrect operations for a specific work-period, and after the student had done the assignment, it would "deliver" to the teacher a diagnosis of the state of the student's intellectual activity. This would be like an x-ray of the state of his thought processes. On inspecting it the teacher would at once be able to see which operations the student has not mastered or has poorly mastered and what must be corrected and "adjusted."

The difference between the machine described and many existing machines consists of the fact that it controls (this was already discussed) not only the result of the student's intellectual activity, but the actual process of this activity. It controls not only the final solution of the problem, but the progress of this solution and the correctness of the way in which the thought processes are accomplished. However, much depends on the programming. The same machine may often be used differently, depending on what and how one expects to teach the students. It also depends on the conceptions of teaching and learning that underlie the design and application of the machine. Our approach is based on the systematic formation of intellectual operations and of their structure. We strive to control and to influence the internal mechanisms of intellectual activity as much as possible and to devise methods for the solution of problems.

Let us note that the method of interacting with a machine which we described (the operation-by-operation realization of an algorithmic process and operation-by-operation control) ensures the successful formation of intellectual activity only at a specific stage. This is the stage where all the operations must be accomplished consciously and in detail. When the problem arises at the next stage, where these operations must be "abbreviated" and their execution automatized, it is necessary to renounce the operation-by-operation control (at this stage this may begin to slow down the formation of the necessary qualities of the process). But at this stage, the machine may be switched over to another regime and may control just the final result of the student's activity when he solves the problem. This permits him to solve the problem in an abbreviated and automatized way.

The purpose of the creation and application of teaching machines is not to exclude the teacher from the pedagogical process or to replace him. It is quite the opposite. Application of the machines frees the teacher from much mechanical routine and permits him to concentrate his time and energy on genuinely creative work. It would be extremely unreasonable to try to entrust all of the teacher's functions to the machine. The teacher was and remains the chief link in the educational process. For the process of instruction is not just the transfer of knowledges, abilities, and skills. It is also the process of education, and the educational function of instruction acquires more and more importance each year.

Although the student will work privately with the machine at specific stages, it is not the machine that will make up the program of instruction, but a methodologist and teacher. Therefore, in the final analysis, it is not the machine that teaches, but the teacher and methodologist who teach with the help of the machine. In the hands of the teacher, the machine will be the same medium of instruction as other media; but, in some regards, it is much more complete. With this, the roles and responsibilities of the methodologist and of the teacher grow, since it is very difficult to prepare a good program of instruction, and if any defects occur, it does great damage to all of the students.

In this chapter, we did not intend principally to throw some light on the problem of teaching machines and ways of using them in the process of instruction. We only wished to show the possible ways for the use of machines when intellectual operations are taught and when the necessary methods of thinking are formed in the students. One can hardly doubt that the use of teaching machines for the special instruction in methods of thinking (and not just for those purposes for which they are ordinarily used) has a great future. The use of machines for the instruction in methods of thinking makes it possible to raise the effectiveness of the educational process as a whole.

Notes

1. One should not confuse programming the students' intellectual activity with programmed instruction. The intellectual activity of the students must be programmed even with unprogrammed instruction.
2. Logic, in particular, shows that the correct conclusion (the true conclusion) often may also be obtained from incorrect (false) premises.
3. Each student is given such an exercise-book. It can be prepared typographically and used along with the ordinary exercise-books.
4. As mentioned previously, the value of material expression of intellectual activities and the methods for doing this are shown in the works of Galperin ([573], [577], [578], and others). He developed the theory of the formation of intellectual activities on the basis of physical activities, and he has worked out procedures for the stage-by-stage formation of intellectual activities. We used the achievements of this theory when we constructed the exercises in the "Exercise-books for Independent Study" described above. This theory is a development on a new level of what Vygotsky and his colleagues, Zankov, Leontyev, and Luria,

treated as internal psychic processes based on "exterior mediation" ([566], [569], [629], [692]). Piaget (see, e.g., [749]) and several other psychologists (see, e.g., [554]) develop the idea of the transformation of physical actions into intellectual operations from different positions.

5. It must be noted that, in the class where the instruction was first conducted without writing out the principal parts, considerable difficulties arose for a number of students when they had to isolate the essential elements and to set aside the unessential elements (especially where it was necessary to isolate different series of homogeneous parts). When long sentences with many words were encountered, the students often could not carry out the necessary operations of isolating covertly the series of homogeneous parts. They forgot about the relations of words with the preceding series. This was one of the main reasons for mistakes when sentences with various series of homogeneous parts are punctuated. When the students wrote out the series of homogeneous parts (each series on a separate line) to establish the relationships between the words, this made it considerably easier. It developed the necessary intellectual operations and resulted in a reduction of the number of mistakes.

6. In order to show that the conjunctions link homogeneous parts within the different series of homogeneous parts, and that they must therefore not be considered iterative, they are enclosed in different geometrical figures.

7. Conjunctions which recur within a *given* series of homogeneous parts are considered iterative. When one places punctuation marks within series of homogeneous parts, it depends, according to the rules, essentially on whether the conjunctions which link the homogeneous parts are iterative.

8. In this connection, the "Exercise-books for Independent Study" which were described perform the same role as published exercise-books which at present are published both here and abroad (in particular, the so-called "Tyumensky exercise-books"), although they differ from the latter very

much in terms of the method of instruction embodied in them.

9. During the lesson, it is possible for the teacher to control more or less fully the work of only some of the students. Normally, he takes their exercise-books home and corrects them there. At best, he tells the students the results of his corrections the next day. The control is not as delayed in the case where the students correct each other's work, but this requires that additional time be spent at the lesson. We do not even discuss the fact that if the teacher takes the exercise-books home, his workload of corrections is not diminished.

10. Undoubtedly, it is not expedient in all cases to provide immediate reinforcement; the degree of delay for reinforcement may be different for different cases (this problem requires further research). There is no doubt that in many cases, less delay is better than more delay.

11. This term was introduced by Slavina [792], [794].

12. Voluminous literature is now available on teaching machines (see, for example: [212], [234], [238], [242], [249], [269], [292], [295], [306], [317], [318], [331], [343], [349-354], [356-358], [365], [366], [368], [369], [371-374], [378], [384], [393], [398], [400], [404], [405], [408-412], [414], [416], [418], [419], [423-425], [441], [451], [458-461], [471], [477-480], [486-489], [491], [492], [495], [496], [504-506], [509], and others).

13. Many other forms of teaching machines were created thereafter in different schools and universities. The machine "Tutor I" was manufactured under the direction of engineer M.V. Etkin at the workshop for visual aids of the society "Znaniye."

14. A technical description of the machine is given in our article [279] (together with S.P. Khlebnikov).

15. For more detail on this see, e.g., Shenshev [352].

16. Some words (for example, auxiliary words) were not numbered so as not to overload the perforated card.

Chapter Nineteen

Breaking Down Intellectual Activity into Elementary Intellectual Operations in the Course of Experimental Instruction

We discussed previously the fact that the correct specification of which operations are elementary for the students is very important during the devising of algorithms. We also noted that the notion of the elementarity of an operation is relative and that one and the same operation may be elementary for a person of one age or level of development and not elementary for a person of another age or level of development. Since it is only possible successfully to form a more complex operation when one is guided by operations which have already been established (i.e., operations which are or have become elementary for the student), it is expedient to examine several problems related to the establishment of operations which are elementary for the students and related to means which take them into account when instructional procedures are created.

Until now, we considered each operation for the verification of an indicative feature of a given logic diagram of indicative features as one elementary operation. For example, we considered the operation of isolating the subject elementary when we taught the students the procedural diagram for the identification of types of simple sentences and for the identification of compound sentences as distinct from simple sentences with homogeneous parts.

Actually, if we had turned to the procedural diagram for the identification of a subject (the students made up such a diagram when they studied the topic "the subject"), we would have seen that the identification of a subject is not one simple operation, but a whole system of operations. In such a case, may one consider one system of operations as one simple operation? This question must be answered affirmatively. After the system of operations for the identification of a subject is formed in the students and the corresponding stereotype is synthesized (i.e., one operation begins to provoke another automatically), the execution of each operation ceases to require special attention and effort (and even awareness) and generally ceases to be a *separate* operation for the person. Now, the person operates integral systems of operations and their complexes consciously and at will. They are accomplished as integral entities which are not broken down further. Thus, what was a specific aggregate of operations for the person at one stage of instruction can be an integral single act—one simple operation—at another stage. Now, it suffices to direct the person to this operation (for instance, say: "isolate the subject"), for the whole system of operations which provide the solution to the problem to be carried out. If this were not so, if the person always had to concentrate his attention on each separate elementary operation and deliberately had to provoke and execute each operation, then essentially no activity would be possible. Even when a person solves relatively easy problems, he often has to execute dozens of operations.

The lack of congruence between how one or another operation *is conceived*, how it *is represented*, and how it *actually is* (it may seem that an operation is simple when in reality it is complex and made up of other operations) leads to one important defect encountered in the practice of instruction. It consists of the fact that some operation is often considered simple and elementary, and it is treated as elementary, when it is actually compound (complex) for students of a certain age or level of development. Therefore, it must be formed as a complex operation, i.e., it must be broken down into components, into more elementary operations, and it is these which must be taught. If this is not done, then one cannot succeed in forming the necessary compound (complex) operation. Flaws and defects appear in the intellectual activity which adversely affect the mastery of all subsequent knowledge and the intellectual development of students generally.

In the process of experimental instruction, we paid a great deal of attention to the evaluation of each separate operation from the standpoint of whether it was simple or complex for students of a given age or level of development. Contingent on the complexity, the decision was made of whether or not the given operation should be broken down into the more elementary operations which make it up and to teach them, since all of the components of the complex operation might already be formed in the students.

Research showed that the operations which at first seemed simple and did not seem to have to be broken down into more elementary operations, in practice were not simple for many of the students.[1] It was necessary to break down these operations into the more elementary operations which made them up. When the students had learned them, the necessary complex operations could be formed in them.

We shall cite examples of such complex operations. Among the operations which had to be carried out in order to identify a subject (when the topic "the subject" was studied) was the operation in which the questions *who?* or *what?* were asked of the word, and the word that answers the question *what?* was replaced by pronouns in the nominative case (*he, she, it,* or *they*).

At first, verification of a word for the presence of one of the indicative features of a subject—"does the word answer the question *who?* [nominative case], or correspondingly, *what?* [nom.] seemed to us to be a simple, elementary operation. However, in the experiment, it was discovered that several of the students "did not see" in separate instances all of the words in the sentence which answered the questions *who?* or *what?* [nom.]. The experiment showed that the operation for the verification of this indicative feature consisted of two (successively applied) operations, one of which several students did not apply. In order to isolate the words which answer the questions *who?* or *what?*, one must ask the question of the words (this is one operation) in the process of going from word to word (this is the second operation). It turned out that several students did not master the operation of going from one word to the next, because in many cases, they did not "notice" the words which answered the necessary question. Children (and adults) do not recognize this operation as a special operation. They carry it out automatically in the process of reading. This operation is not formed in some students; it is not carried out automatically, and it has to be especially established.

Further, the operation of the substitution of a word which answers the question *what?* by the pronoun *he* (*she, it, they*) also seemed simple at first. Actually, at the first instant, it is even difficult to imagine how this operation can be broken down into other, more elementary operations and how one can do this at all, since this operation is so elementary. An adult sees at once that in the sentence: *The field was covered with snow*, the word *snow* is replaced by the pronoun *it* [acc.] and not by the word *it* [nom. case].* However, the experiment showed that this was not immediately clear to some students and that the operation of substitution is complex, consisting of several simpler and more elementary operations. When they have not mastered these, it is impossible to make the substitution without mistakes and to specify whether the word is the subject.[2] Thus, during exercises for the application of one of the indicative features of a subject ("the word answering *what?* [nom.] is replaced by the pronouns *he, she, it, they*"),[3] a number of weak students committed the following errors.

1. They tried to substitute all of the pronouns one after another for the word answering the question *what?* In fact, one must use as a substitute only that pronoun which corresponds in number or gender to the noun. For example, if there is the sentence: *The company built the house*, one has to try to replace the word *house* (masculine) by the pronoun *it* [masculine form] and never by *she* [or *it* feminine form] or *it* [neuter form] or *they*. It would seem that this is quite evident and that there is no need, when operations of substitution are taught, to point out:

---

*In Russian, the syntax is the following: *field* is the object in the accusative case of the verb *to cover*, which is in an impersonal neuter form. *Snow* is the agent in the instrumental case. (*Translator.*)

"pay attention to the gender and number of the noun to be replaced." But this was not evident for some students, and they had to be shown this operation which enters as an element into the operation for the substitution of a noun that answers the question *what?* by a pronoun.

2. In one of the sentences, some students with a poorly developed feeling for the language replaced a word in the neuter-accusative case by the neuter-nominative case, and everything "turned out" for them. It is natural to draw the conclusion from this: when students are trained to apply the indicative feature shown, they have to be taught (at least at first) to make a double substitution (*he-him, she-her*, etc.) and choose the pronoun which is most appropriate. This is the second elementary operation which enters into the operation for the verification of the indicative feature which we are examining.

3. When some students replaced the word which answered the question *what?* by a pronoun, they did not read this pronoun with the predicate. In a number of cases, this led to a mistake. For example, in the sentence: [*They*] *expressed gratitude to the association of co-workers*, one student did the following: he found a word which answered the question *what?* (*gratitude* [feminine gender]). Then he replaced this word by the pronoun *it* [fem. gender] (*gratitude-it*) and said "it is replaceable, therefore it is the subject."

Actually, in order to find out which pronoun replaces the noun, one has to read this pronoun along with the predicate: it is impossible to say [*They*] *expressed it* [feminine nominative]. One has to say [*They*] *expressed it* [feminine accusative].

Hence the conclusion: after the noun is replaced by the pronoun, one must read this pronoun with the predicate. This is the third operation which enters into the operation for the verification of the indicative feature that we are examining.

4. There are often adjectives or participles which fulfill the role of attributes along with the nouns that have to be replaced by pronouns, for example: *Our company built a beautiful, large house.* If the nouns are replaced by pronouns and the pronouns are read along with the attributes, nonsense results: *Our company built a beautiful, large it*, or *Our it built a beautiful, large house.*

Therefore, after the noun has been replaced by a pronoun, all of the attributes which refer to the noun must be discarded, and the pronoun must be read with the verb-predicate without attributes. This is the fourth operation which enters into the operation for the verification of the indicative features that we are examining. If it is not carried out, then the substitution is impossible, even though the noun is actually replaced by a pronoun.

5. A substitution seems impossible to certain students for another reason. When they replace the last noun in the sentence: *The birch* [feminine gender] *is a tree* [neuter gender] by a pronoun, and when they read the sentence: *The birch* [fem.] *is it* [neut.], they say: "it does not sound right; it cannot be replaced." In order to avoid similar mistakes, it is necessary to carry out the following operation: when the noun has been replaced by a pronoun, this noun must be put in parentheses after the pronoun and preceded by the words "that is." For example, *The birch is it* (i.e., a *tree*). Or: *Our company built it* (i.e., a *house*). This is the fifth operation which enters into the operation for the verification of the indicative feature which we are examining.

Thus, we see that what seems to be such a simple, elementary operation as the substitution of a noun by a pronoun for some of the students is actually a complex, compound operation. For the majority of the students, this complex operation was simple, and they did not have to be taught the more elementary operations which make up this operation. For some average and weak students, this operation was not simple, and they had to be taught those elementary operations which made up this more complex operation.[4] If this had not been done, one could think that these students would not have been able to apply the given indicative features without mistakes.

One could continue at length the number of examples which show that different operations which seem at first glance to be simple and elementary and even incapable of further reduction are actually complex and composite. The material cited, however, also shows that, before one teaches any operation, it is important to specify whether or not it is elementary, whether it is formed (and how it is formed), and may be included in the algorithm, or whether this operation must first be formed (as we know, for this purpose, it must be broken down into more elementary component operations). The evidence shows that the operations which to an adult seem elementary and incapable of further reduction are actually complex in many cases and have to be specially taught and broken down into components in advance.[5]

We shall pause now at the question of the importance of the role which the "Exercise-books for Independent Study" can play in the formation of these compound (complex) operations.

During experimental instruction, we did not just use the "Exercise-books for Independent Study" to form the systems of operations which make up the logical structure of indicative features and which were shown as simple operations on the procedural diagram of actions. We also used them, when necessary, to form those even more elementary operations into which the former had to be broken down.

We cite one of the exercises on the topic "the subject" whose purpose was to form those elementary operations which make up the operation where the words answering the question *what?* were substituted by the pronouns *he* (*she, it, they*) [nominative case] or correspondingly, *him* (*her, it, them*). The students are given a number of sentences where the words answering the question *what?* are in small capital letters. (For example, [*They*] *gave the boys in our class a MISSION*). The following has to be done:

1. Write this word together with the verb.
2. Replace it first with a pronoun in the nominative case, then with a pronoun in the accusative case. Then write the pronoun together with the verb.
3. Verify which of the pronouns "fits" and which does "not fit." Then cross out the inappropriate one.
4. Go back to the sentence again and write down the pronoun which "fits" as well as the question which the word answers over the emphasized word. Write the conclusion: "subject" or "not the subject" under the word.

Here is the way this sentence looks after all the necessary work has been done with it:

[They] gave the boys in our class a MISSION

(1)   *[They] gave a mission*
(2)   *[they] gave* IT* [nom. case] ; *[they] gave* IT [acc. case]
                                   IT? WHAT?
[They] gave the boys in our class a MISSION.

Despite the seeming difficulty of the instruction, after the students are shown the model of what to do, they immediately remember all of the necessary operations and, as a rule, complete this exercise without mistakes. It should be noted that the execution of this exercise is precisely what ensures the formation of the five operations which were discussed above and which are components of the compound operation for the verification of which pronoun should replace a word answering the question *what?*

"Exercise-books for Independent Study" were also used to form the elementary operations for the topic "homogeneous parts of the sentence." Particular attention was paid to the distinction and inscription of different series of homogeneous parts. This is due to the fact that the inability to carry out the operations for the distinction of homogeneous parts is one of the basic reasons for punctuation mistakes in simple sentences with homogeneous parts.

We cite the model for the execution of one of the tasks whose purpose was to teach the students the operation of isolating series of homogeneous parts.

A long trip around the country visits to cities and villages encounters with workers and farmers enriched me and my friends and will be remembered by us all our lives.

(1)   *trip, visits, encounters*
                    OF WHAT?
       (a)   *visits   cities and villages*

---

*This pronoun does not "fit." We cross it out.

WITH WHOM?

(b) *encounters   workers and farmers*

(2) *enriched and will be remembered*

WHOM?

(a) *enriched   me and friends*

(3) *trip, visits, encounters   enriched and will be remem-bered*

Diagram:

As a result of the analysis, the sentence takes this form:

A long trip around the country, visits to cities and villages, encounters with workers and farmers enriched me and my friends and will be remembered by us all our lives.

We have described the basic tasks and principles of experimental instruction as well as the routes we have taken in carrying them out. Let us now go on to the analysis of the results of experimental instruction.

Notes

1. Let us recall that an operation which we call simple (or elementary) is an operation whose instruction does not require that it be broken down into the more elementary operations which make it up. The evaluation of some operation as simple or complex may be obtained during the experimental study of the intellectual activity of the students of a specific age and class. One of the important indications of whether a certain operation is (or has become) elementary

for a student consists of the fact that he always carries it out uniformly and without mistakes, in response to the appropriate prescription. Here, an essential characteristic of elementary operations is the fact that the operations which are even more elementary—those entering into the composition of the given operation—are no longer realized as separate operations. They have been integrated already into one automatized system which is actualized by one signal or one command.

2.   Here, the condition mentioned above remains in force—the method of substitution of the word which answers the question *what?* by a pronoun is considerably easier and more reliable (and generally "better") than the methods usually applied.

3.   This indicative feature is not sufficient and enters into a system of other indicative features.

4.   The example cited shows the importance of individualized instruction. As mentioned before, programmed instruction creates very favorable conditions for such individualized instruction.

5.   It is not so easy to "track down" the more elementary operations which make them up because adults are not ordinarily aware of these operations.

Part Two: Section Two

Results of Experimental Instruction and Research

Chapter Twenty

Comparative Analysis of the Results of Instruction in Experimental and Control Classes

1. Quantitative Data

We shall present here the quantitative data on the results of instruction in experimental and control classes. The mean number of mistakes per student as the standard criterion task is performed is indicated on the tables.

The criterion test on the topic "The Subject" consisted of 50 sentences and was introduced to the students in the seventh grade immediately after they studied this topic. It was posed to students of the eighth grade after it was reviewed. Each student received two sheets of paper with mimeographed sentences. He either had to underline the subject (if, in his opinion, there was a subject) or to write N.S.—"no subject" (if he thought that there was no subject in the sentence).

Immediately after the topic "Types of Simple Sentences" was completed, a criterion test on this topic was conducted. The same sentences were used in the criterion test on the identification of the subject. In order to exclude the possibility that the type of sentence might be incorrectly specified because of mistakes when the subject and predicate were identified, the principal parts of the sentences shown to the students were underlined.

Controlled practice on the topic "The Compound Sentence" was conducted with students of the eighth grade after they studied this topic. In accordance with the requirement of the academic program, the topic "Homogeneous Parts of the Sentence" was reviewed before this was done. Furthermore, during the study of the topic "The Compound Sentence," as a rule, simple sentences with homogeneous parts were compared with compound sentences. The criterion test was conducted in the following manner. Forty mimeographed sentences which had to be punctuated by the students were distributed to them.

We shall show tables of results with the number of mistakes students from different classes made in the criterion tests on all of the topics indicated (see pages 551 and 552).

Now we shall present tables which show the distribution of students from experimental and control classes according to the number of mistakes they made during the criterion test on the topic "The Compound Sentence."

Distribution of Students from the Experimental Classes
According to the Number of Mistakes They Made

Number of Mistakes on Criterion Test (for 40 sentences)	Number of Students Making Mistakes	Percentage of Students Making Mistakes
0	7	19
1	12	32
2	1	3
3	5	13
4	1	3
5	4	11
6	1	3
7	0	0
8	3	8
9	2	5
from 10 to 15	0	0
16	1	3
Total	37	100

Number of Mistakes Made by Students from Different Classes in the Criterion Test on the Topic "The Compound Sentence"[1]

Class Number	Control Classes							Experimental Class
	No. 1	No. 2	No. 3	No. 4	No. 5	No. 6	No. 7	
Mean Number of Mistakes (per Student) in Class	15.5	15.6	19.1	19.5	20.6	2.5	28.3	3.1
Mean Number of Mistakes (per Student) for Each Group of Classes	20.5							3.1

Number of Mistakes Made by Students from Different Classes in the Criterion Test on the Topic "The Subject"

Class Number	Control Classes																Experimental Classes (7th Grade Classes)				
	7th Grade Classes				8th Grade Classes												No. 1	No. 2	No. 3	No. 4	No. 5
	No. 1	No. 2	No. 3	No. 4	No. 1	No. 2	No. 3	No. 4	No. 5	No. 6	No. 7	No. 8	No. 9	No. 10	No. 11	No. 12					
Mean Number of Mistakes (per Student) in Class	20.6	21.8	22.1	26.9	14	14.8	15.7	16.6	16.8	17.2	17.8	20	21.5	22.2	22.8	23	2.7	2.7	3.1	5.3	5.4
Mean Number of Mistakes (per Student) for Each Group of Classes	22.8				18.5												3.7				

Number of Mistakes Made by Students of Different Classes in the Criterion Test on the Topic "Types of Simple Sentences"

Class Number	Control Classes				Experimental Classes				
	No. 1	No. 2	No. 3	No. 4	No. 1	No. 2	No. 3	No. 4	No. 5
Average Number of Mistakes (per Student) in Class	6.3	8.7	9.2	12.6	1.3	1.5	1.9	2.1	2.5
Mean Number of Mistakes (per Student) for Each Group of Classes	9.2				1.8				

As we see, 19 percent of the students did not make a single mistake; 32 percent made one mistake; 30 percent made from two to five mistakes; and only 19 percent made more than five mistakes. In other words, more than half of the students (51 percent) made no more than one mistake, and about 30 percent from two to five mistakes. It must also be noted that the student who made 16 mistakes, in the opinion of the teachers and the doctor, was mentally retarded and should have been transferred to a special school. The five students who made 8 or 9 mistakes were ill and absent part of the time.

In the control classes, the distribution of mistakes was entirely different. Here are the data on seven control classes whose students performed the same criterion test after they studied the topic "The Compound Sentence."

Distribution of Students in Control Classes
According to the Number of Mistakes They Made

Number of Mistakes in Criterion Test (for 40 Sentences)	By Classes							In All Classes	Percent of Students Making Mistakes Based on the Overall No. of Students in All Classes
	No. 1	No. 2	No. 3	No. 4	No. 5	No. 6	No. 7		
0-5	4	4	1	1	0	0	0	10	4.4
6-10	8	5	1	4	5	1	0	24	10.1
11-15	4	7	3	4	6	1	2	27	11.9
16-20	4	14	8	10	5	7	6	54	23.6
21-25	5	5	8	7	9	8	8	50	21.9
26-30	3	2	1	7	8	5	9	35	15.4
31-35	1	1	2	2	3	4	7	20	8.7
36-40	1	0	0	0	0	1	2	4	1.8
41-47	0	0	0	1	0	1	3	5	2.2
Total	30	38	24	36	36	28	37	229	100

Below appears a summary table which shows the distribution of mistakes by students from control and experimental classes. The percentage of students (in relation to the total number of students in control and experimental classes) who made the given number of mistakes is shown in the table.

Comparative Distributions of Students from Experimental and Control Classes (in %) According to the Number of Mistakes They Made

Number of Mistakes on Criterion Test (for 40 sentences)	Experimental Class	Control Classes
0-5	80	4.4
6-10	17	10.1
11-15	0	11.9
16-20	3	23.6
21-25	0	21.9
26-30	0	15.4
31-35	0	8.7
36-40	0	1.8
41-47	0	2.2
Total	100	100

As we see, while the majority of students from the experimental class (80 percent) made from 0-5 mistakes, the majority of students from the control classes (82.9 percent) made from 6-30 mistakes.

We have presented the quantitative data which show the success of the students from experimental and control classes who solved the grammatical problems. As is evident from the figures cited, the students from the experimental classes made from 5-7 times fewer mistakes than students from the control classes who studied according to the traditional method. This speaks for the great efficacy of the special instruction in general methods of thinking, in particular, algorithms, which was imparted to the students.

2. A Qualitative Analysis

In order to explain the reason for such sharp distinctions between the success of students from the experimental and from the control classes when they solved the problems, we shall analyze the qualitative characteristics of the intellectual activity of both groups. It is known that one of the important results of the application of different methods of instruction is the formation of specific intellectual approaches and specific characteristics of intellectual activity in the students.

The *first characteristic* of intellectual activity found in students from the control classes is the fact that *they cannot give a complete answer to the question of what must be done and in what sequence when solving certain grammatical problems, because they do not know the general method for the solution of these problems.* For example, they do not know what must be done to specify the type of a given simple sentence and whether the given sentence is compound or simple with homogeneous parts, etc. We shall quote the answers of a number of students to the question which is formulated thus: "Let us assume that a person is confronted with a sentence, and he does not know what kind it is: compound, or simple with homogeneous parts. How would you explain to him what he must do and in what order, in order to specify the sentence's type?"

Student Tamara Z.

Instead of answering the question in general form, she tells how she would do it herself.

Stu. I look for the principal parts in the sentence. . . .

Exp. And then?

Stu. If there is a subordinate clause in the sentence, it is compound-subordinate.

Exp. And if there is not?

Stu. If the sentence consists of several independent clauses, then it is compound.

Exp. But how do you know that the sentence consists of several independent clauses or not?

Stu. I look for the principal parts and look at . . . (pause).

Exp. What do you look at?

Stu. I didn't think. . . .

Aside from the fact that one must find the principal parts in the sentence, the student can say nothing. To find the principal parts of a sentence, in itself, does not yield the solution to the problem of the specification of a sentence's type, as we saw. The student does not know the general method for the specification of a sentence's type.

Student Vera M.

Stu. First, one must find the principal parts of the sentence. Then one must specify whether it is compound or simple.

Exp. How does one specify this? What must be done?

The student is silent. She cannot answer the question.

Student Nelly M.

Stu. First of all, one must find the principal parts of the sentence. If there are several principal parts, then it is compound; if not, then it is simple.

Exp. Is it necessary that there must be several pairs of principal parts (several subjects and several predicates) in the sentence?

Stu. No. There could be elliptical sentences or impersonal sentences in it. . . .

Exp. And if there are not several pairs of principal parts, will it necessarily be simple?

Stu. Perhaps not.

Exp. How, then, can one specify whether the sentence is compound or simple with homogeneous parts?

The student cannot answer the question.

Student Lyuda P.

Stu. If the sentence is simple, then it will be personal,

impersonal, indefinite-personal or elliptical. If the sentence is compound, then it has to consist of several simple sentences.

Exp. But how can one find out whether it consists of several simple sentences or not? What does one have to do?

Stu. If the sentence is compound, then there should be conjunctions and enumerative intonation.

Exp. But is it not possible to have conjunctions and enumerative intonation in a simple sentence?

Stu. That is possible.

Exp. Well, then, how does one find out what kind of sentence it is?

Stu. It is difficult to say.

Student Lena L.

Stu. In a compound sentence, the simple sentences are somehow related by the thought.

Exp.. Does this mean that you specify a compound sentence by its thought?

Stu. Yes.

Exp. Well, are not the parts of a simple sentence related by thought?

Stu. They are. . . .

Exp. Well, then how can one distinguish a compound sentence from a simple sentence?

The student cannot answer.

Student Anya I.

Stu. I find the subject and the predicate, and then I see whether there is a complete thought or not.

Exp. What sort of segment of the sentence do you set aside to examine? How do you decide that the given segment must be verified for a complete thought?

Stu. I read until it seems to me that the thought is complete.

An analogous situation takes place in relation to other sentences.

As distinct from students in control classes, the students in experimental classes *formulate a general method for identification precisely and specifically.*

Student Tanya Z.

Stu. First of all, one must find all the subjects and predicates in the sentence. Then one must see which predicate refers to which subject. If all the predicates refer to the same subject, then the sentence is simple; if not, then it is compound. Further, if there are no subjects in the sentence, then one must see whether all of the predicates in it are expressed only by verbs of the first or second person. If they are, then the sentence is simple. If not, then one must see whether all of the predicates are expressed by verbs of the third person plural. If not, then the sentence is simple; if so, then one must see who are the doers of the actions. If the actions are performed by the same persons, then the sentence is simple; if they are performed by different people, then it is compound. That is all.

The other students also gave similar answers. We shall cite still another example—the answer of Student Igor Kh.

Student Igor Kh.

Stu. We find the subject in the sentence, then the predicate. Then we see what refers to which. If all the predicates refer to one and the same subject, then the sentence is simple; if they refer to different subjects, then it is compound. If there are just predicates in the sentence and no subjects, then we see how the predicates are expressed. If they are expressed only by verbs of the first or second person, then the sentence is simple. If they are expressed neither by verbs of the first nor the second person, nor by verbs of the third person plural, then the sentence is compound. If they are all expressed by verbs of the third person plural, then it all depends on who is the doer of the actions. If the same person performs the actions, the sentence is simple. If different persons perform the actions, then it is compound.

Thus, we see that both students indicate a complete and general system of operations which makes it possible to identify a compound sentence as distinct from a simple sentence with homogeneous parts. This system of operations is a general method which permits the solution of the majority of problems from a given class. The operations which the students indicate correspond entirely to the procedural diagram which was shown above. The students' answers differ only in their verbal formulations.

The second characteristic of the intellectual activity typical for students in control classes is *the great heterogeneity of the different students' methods for the solution of one and the same problem.* Students in the same class with the same teacher often follow different routes in solving one and the same problem. Many of these routes are inefficient, incomplete, or simply incorrect and lead to mistakes. This is the direct result of the fact that the students were not taught a general method for the solution of grammatical problems of a specific type and that efficient systems of intellectual operations were not formed in them.[2]

For example, when the subject is identified, some students use the double question *who? what?*, and others do not use it. Some students ascertain the case of the word which answers the question *what?*. Other students do not and immediately draw the incorrect conclusion that the given word is the subject. There are students who, when they have asked the question about the given word, read the word together with the predicate, but there are those who do not (which leads them to err). Thus, one student (Anya F.) in the sentence [*They*] *brought skis for the boy*, asked the questions *who? what?* of the word *skis*, and since she did not relate this noun to the predicate-verb, concluded: "This is the subject." If she had performed the indicated operation of correlation, she would have seen that [*they*] *brought* requires the questions *whom? what?* [acc.] and not *who? what?* [nom.] and that *skis* are not the subject. But the student did not apply this operation.

Some of the students perform operations on the analysis of formal-grammatical indicative features. Some others analyze semantic indicative features for the most part, etc.

This sort of heterogeneity was observed when students from the control classes solved different grammatical problems. Examples from the topic "The Compound Sentence" were cited above.

If we examine the methods students from the experimental classes use to solve grammatical problems from this standpoint, their characteristic is their *unanimity about the procedures to be applied.* When different students solve a problem of the same type, they apply the same procedures—the correct and efficient ones. The corresponding procedures were described above. Therefore we shall not repeat them here.

The third characteristic of the intellectual activity of students from the control classes is the fact that *when the student analyzes one kind of sentence, he follows one route. When he analyzes another—often completely analogous—he follows another route* although in reality the procedural methods should be the same general ones. Thus, not only do students of the same class fail to master the general method but also, each individual student fails to do so. This explains what seems at first glance to be a completely incomprehensible phenomenon, i.e., when the student solves one problem correctly but another one like it incorrectly. Here are several examples from the topic "The Subject." The sentences shown below were either right next to each other or not far from each other in the assignment.

1. *Mornings,* [*they*] *brought newspapers.* (The student writes "N.S.," i.e., "no subject," which is correct.) [*It*] *covered the field with snow.* (Here the same student underlines the word *field* as the subject, which is incorrect.)

2. *A triangle is the polygon with the least number of sides.* (The student correctly underlines the word *triangle* as the subject.) *The polygon with the least number of sides is the triangle.* (The very same student underlined the same word *triangle* as the subject and therefore committed an error.)

The absence of one single general method is particularly obvious. This is seen when the data obtained from an individual experiment are analyzed. Let us examine the operations which one of the best students who went through the individual experiment, Lyuda V., applied for the identification of the subject.

When she analyzed the sentence: *Another person would have gladly done it,* this student applied the double question *who?*

what? But when she analyzed a completely analogous sentence: *Not many take that difficult route*, she did not ask the double question, but only *who?* When she analyzed the sentence: *A courageous person can endure everything*, she neither asked the double question *who? what?*, nor the question *who?*, but for some reason just *what?* This led her to error. She underlined the word *everything* as the subject. When this same student analyzed the sentence: *The triangle is the polygon with the least number of sides*, she did not apply the operation of asking the questions, but tried to find the answer by intuition.

Stu. (after some hesitation) Here the subject could be either *triangle* or *polygon*, but it is more likely to be *triangle*.

When the experimenter, who is unsatisfied with the answer obtained by intuition, asks the student to give a basis for her opinion, the student has recourse to the operation which recalls one of the indicative features she thinks is in the textbook (though this indicative feature is not there).

Stu. It says in the textbook that the subject comes first and then the composite predicate. We have the same case here.

What is most interesting here is the fact that when she analyzes the sentence which follows this one and is completely analogous, *The polygon with the least number of sides is the triangle*, the student examined does not refer to this indicative feature, does not apply it, and says with complete assurance that the subject can be either *polygon* or *triangle*.

Then comes the sentence: *To follow the thoughts of a great man is the most interesting of sciences.* When the student tested analyzes it, she does not use any of the operations used earlier. Instead, she tries a completely new one, which happens to be an incorrect method. She reverses the first and second parts of the sentence (*The most interesting science* [*is*] *to follow the thoughts of a great man*) and incorrectly underlines the word *science* as the subject. The student answers yes to the question of whether she is sure that *science* is the subject.

We shall not describe the other methods used by the girl questioned. The examples cited suffice to show that even the best student lacks a general procedure for the identification of a subject and that she has not mastered the kind of general method of operations which would permit her to correctly find the subject in any sentence.

Even more striking is the failure of students to master a general procedure (and not know of it) when they solve the problem of the identification of a compound sentence as distinct from a simple sentence with homogeneous parts.

We shall cite parts as an example, the way student Galya K. analyzed three sentences.

The train left, and its lights soon disappeared.

Stu. We look for the subject and the predicate. (The student correctly points out the subject and the predicate in the sentence.) This is a compound sentence. A comma is necessary.

Exp. How do you know?

Stu. Well, obviously, here we have a subject and a predicate and another subject and predicate. One can see right away that it consists of two simple sentences.

The night was foggy, and the light of the moon pierced through the fog.

Stu. (She answers right away, as soon as she reads the sentence.) It is simple. No comma is necessary.

Exp. Why do you think it is simple?

Stu. The night was foggy, and the moonlight pierced through the fog. Everything is joined together. It is one thought.

Exp. But why did you decide that there was one thought here?

Stu. The night is foggy (the student emphasizes the word *foggy* by intonation), and the moonlight pierced through this fog. One cannot separate one from the other.

Exp. And did you look for the subjects and predicates in this sentence?

Stu. No.

Exp. Why?

Stu. One can see at once that it is a simple sentence.

Exp. Well, try and find them.

The student finds the subjects and predicates and underlines them.

Stu. This is a compound sentence. There is a subject and a predicate here and another subject and predicate. Why did I not notice that at once?

The birch was the first to bud, and it was the first to turn yellow.

Stu. (After a long pause) I would say that it is a simple sentence.

Exp. Why?

Stu. One and the same birch tree is spoken about.

Exp. But did you find the subjects and the predicates?

Stu. I found them.

Exp. Well, here we also have a subject and a predicate and then another subject and predicate.

Stu. I do not even know. . . . On the one hand, it seems to be compound, while on the other, simple. After all, it is one and the same birch tree! No, I would still say that it is a simple sentence.

As we see, when this student analyzes sentences which are of completely analogous structure, she acts differently. When she analyzed the first sentence, she applied the operation of isolating the principal parts and their relationships. When she had established the structure of the principal parts ("a subject and a predicate and another subject and predicate"), the student said with assurance that the sentence was compound. When she analyzed the second sentence, she did not perform these operations and followed a completely different route. She analyzed the thought of the sentence, and only the thought, and tried to establish the number of thoughts contained in the sentence. When the student analyzed the third sentence—although she isolated the subjects and predicates—she paid more attention to whether one or more things were spoken about in the sentence. When it happened that, on the one hand, there were complete pairs of principal parts in agreement and, on the other hand, that one and the same subject was spoken about, the student did not know what to do and by what indicative features she should have been guided. Of the two indicative features which seemed to contradict each other in the student's opinion, she chose the incorrect one and made a mistake.

We could cite many analogous examples. As distinct from students of the control classes, students from the experimental classes *solved grammatical problems of one type by a single method*. They applied the necessary system of operations to sentences which differed according to their specific content.

Here is how, after experimental instruction, Tanya Z., mentioned above, acted and reasoned when she solved a problem on the identification of the subject.[3]

Another [person] would have gladly done it.

Stu. We verify the first indicative feature: we see whether there are words which answer the question *who?* There are: *another [person]*. Then we see whether this word enters into the composition of a compound predicative nominative. No. This means that *another [person]* is the subject: both indicative features are present.

Few go along this difficult route.

Stu. We verify the first indicative feature. The word *few* answers the question *who?* This word does not enter into the composition of a compound predicative nominative. Therefore it is the subject.

A triangle is the polygon with the least number of sides.

Stu. We verify the first indicative feature. No word in the sentence answers the question *who?* We verify the second indicative feature. Two words in the sentence answer the question *what?—triangle* and *polygon.* We substitute them with pronouns:

triangle—it, polygon—it; then we see which of them enters into the composition of the compound predicate nominative. The predicate is *is a polygon*. Therefore, *triangle* is the subject.

The polygon with the least number of sides is the triangle.

Here the student reasons in exactly the same way and correctly isolates the word *polygon* as the subject.

To follow the thoughts of a great man is the most interesting science.

Stu. We verify the first indicative feature. No word answers the question *who?* Therefore, we see which word answers the question *what? What?—*science. We substitute*—is it*. Does *science* enter into the composition of the compound predicate nominative? It does. This means that it is not the subject. Then we have to see whether the subject is expressed by an indefinite form of the verb. *To do what?—to follow. To follow is a science. To follow* is the subject.

As we see, the student acts according to a single method, independently of the specific content of the sentence. She applies a general system of operations to different grammatical objects. The other students analyze the sentences in a similar way.

We shall not present examples of how the students applied a single method for the identification of types of sentences, since such examples were cited above.

The fourth characteristic of the activities of a number of students from the control classes is *their uncertainty in their actions and in their solutions*. This brought about the following circumstance.

A general method for the solution of problems, as we saw, is that system of operations which is applied to any problem of a given class and which makes it possible to solve them one way. Since the indicative features and methods used by the students from control classes are in many cases not general, of course, the students often cannot correctly solve all the problems of a given class.

Actually, the method of asking the double question *who?* *what?* makes it possible to identify the subject in sentences of the type: *The parents bought skates for their son*. But it can no longer ensure the identification of the subject in the sentence: *The triangle is the polygon with the least number of sides*, since both the word *triangle* and the word *polygon* answer the question in the nominative case. The application of this method to the sentence: *To follow the thoughts of a great man is the most interesting science* invariably leads to a mistake. Here, the word *science*, which is not the subject, answers the questions *who? what?*

The same may be said of the procedure in which compound sentences are identified by picking out the pairs of principal parts in agreement. This method, which ensures the solution of problems with sentences of the type: *The transparent forest grows dark alone, and the fir tree shows green through the frost, and the river gleams beneath the ice*, no longer gives the solution to problems of the type: *The polar explorers flew off to the wintering station, and [they] sent airplanes and supplies after them*, since there are no complete pairs of primary members in agreement in these sentences.

The method of breaking down the sentence into parts which are independent and complete in relation to thought is an analogous matter. If one can easily specify the type of a sentence like: *The sun rose, and the detachment moved on* by this method, it can lead to a mistake (and in fact usually does) when it becomes necessary to specify the type of a sentence like: *The bookcase was large, and in it were many books*. The second simple sentence in the composition of this compound sentence cannot exist without the first one, since it is not understood where the many books are contained. Nevertheless, this simple sentence, which enters into the composition of the compound sentence, is "grammatically independent and can stand alone" in the grammatical context.

If the student has not mastered complete systems of indicative features and general methods for the solution of grammatical problems, when he encounters sentences where the known indicative features and his own particular methods seem inapplicable, he often has to proceed by trial and error. He strays onto the path of guesswork and tries to solve the problem by intuition.

For example, when student Zina D. first read the sentence: *The train left, and its lights soon disappeared*, she put a comma before *and*. Then, after much hesitation, she crossed it out.

Exp. Why did you cross out the comma?

Without saying anything, the girl puts it back in.

Exp. What makes you doubt?

Stu. I really do not know how it should be here. . . . It seems as though a comma is necessary if one acts according to the rules, but I feel as though a comma is not necessary here.

Exp. Why?

Stu. The train left and its lights soon disappeared. There is one thought here and one sentence. It is quite impossible to put a comma here.

Exp. Does this mean that simply because the lights are related to the train, it must follow that this is one simple sentence and a comma is unnecessary?

Stu. I do not know, but I feel that a comma does not have to be put in here.

Student Katya T. thinks for a long time about the sentence: [*It*] *grew dark, and* [*it*] *began to grow cold.*

Exp. What is bothering you here?

Stu. I do not know what to do about a comma here . . . There are no subjects here, just predicates. . . . At first I put in a period. It seemed there were two separate sentences: [*It*] *grew dark.* [*It*] *began to grow cold.* But on the other hand, it all seems to go together. It grew dark, and right away it began to grow cold. It is not possible to put a comma here. It seems to be unnatural.

Exp. How should one reason in order to answer the question of whether or not a comma is necessary?

The student is silent. She does not know what to say.

Exp. Here you have encountered a sentence. What must you do to find out whether a comma is necessary before *and* or not?

Stu. At first I put in a period. But this is not always the way. . . . I really do not know what to do here.

The student is at a loss. She does not know what must be done with the sentence in order to solve the problem before her. She does not know the method of reasoning and the operations which must be carried out in order to arrive at the specific answer. This provokes uncertainty, hesitation, and attempts to solve the problem by depending solely on flair ("it seems unnatural").

Student Tanya P. has to punctuate the sentence: *He walked in the woods for a long time and finally caught sight of a lake gleaming in the distance.* First she put a comma before *and*; then she crossed it out.

Exp. Why did you cross out the comma?

Stu. According to the rule, these are homogeneous parts, but if one reads the sentence, there is a pause before *and*, and one must put in a comma. At first, I put in a comma in accordance with the pause, but then I decided that it was better to go by the rules. . . . Nevertheless, I very much feel the need to put one here.

The student acted according to the rules and crossed out the comma. Nonetheless, she expressed a certainty that a comma was necessary here.

The next characteristic—*the fifth*—typical of a number of students in the control classes is a *distrust of rules and definitions* which are in the textbook and explained by the teacher. The result of this distrust is *a lack of conviction about the correctness of those definitions and rules which the student knows.* A lack of correspondence and even opposition arises between what the student knows and his convictions, as well as between what the student knows and his actions.

If student Tanya P., after much hesitation, nevertheless acted according to the rule and crossed out the comma, a number of students have more confidence in their own flair and experience, and they act in opposition to the rules and definitions which they know.

Student Irina K. underlined the word *suddenly* as the subject in the sentence: *Suddenly "whoa" was heard close by.*

Exp. Why do you think *suddenly* is the subject?

Stu. *Suddenly* is talked about.

Exp. But such an indicative feature—"the subject is that which is talked about"—is not in the textbook.

Stu. Well, what of it? It could still be the subject since it is talked about.

Student Vladimir P. underlines the pronoun *me* in the sentence: *It is absolutely necessary for me to talk with you.*

Stu. For whom? *Me.* For whom is it necessary to talk? *Me.* This is the subject.

Exp. But do you not see that *me* does not answer the questions *who? what?*

Stu. Nonetheless it is the subject.

The statement of Katya G., a very capable and intelligent, outstanding student, is very interesting.

Stu. I do not always obey the rules. It often seems to me that a comma is necessary. I imagine a person who is writing. Where there is a pause—that is where a comma is absolutely necessary. And I put one in.

Such a disparity between those indicative features which the students are taught and which they know, and those indicative features by which *they in fact orient themselves* in their actions is quite typical for many students. Sometimes the students are unaware of this discrepancy, but often they are aware of it.

Thus, student Irina K., whom we mentioned before, names the indicative features for the subject which are cited in the textbook ("the independent part of the sentence which answers the questions *who?* or *what?* is called the subject"). But, in fact, she is often guided by completely different indicative features. Thus, in the example cited, she was guided by the indicative feature "the subject is that which is talked about in the sentence." When she analyzed the sentence [*I*] *would begrudge nothing to a friend*, this student underlined the word *nothing* because she was guided by the indicative feature of intonation.

Exp. Why do you think *nothing* is the subject?

Stu. Because [*I*] *would begrudge* NOTHING *to a friend.* (She emphasizes the word *nothing* by intonation.)

Exp. Yes, but one could also read it: *To a* FRIEND [*I*] *would begrudge nothing.* What then—would the subject be *friend*?

The student is silent. She does not know what to answer.

It is characteristic that the student is quite aware of the disparity between the indicative features which she names and those by which she is guided. She answered the question of what must be done to find the subject in the sentence:

Stu. First one must find the word which is talked about in the sentence.

Exp. Then?

Stu. This word has to be in the nominative case.

Exp. How does one specify the nominative case?

Stu. One must ask the questions *who? what?*

Exp. Do you act in this way yourself?

Stu. No. Right away I orient myself. For example, in the sentence: *Nothing frightened him,* I saw who was being talked about[4] —about him.

Exp. And what kind of question does this word answer?

Stu. The question *whom?*

Exp. Do you consider, therefore, that *him* is the subject?

Stu. Yes.

Exp. But *him* does not answer the questions *who? what?*

Stu. Nevertheless, since the sentence talks about *him, him* is the subject.

As we see, the indicative features which the student was taught and which she names when she defines the meaning of the subject "do not work" in fact. The knowledge obtained in the learning process does not influence the student's course of action and does not regulate the process of identification. There is no interaction between the sphere of obtained knowledge and the sphere of operations. The definitions given in the textbook and explained by the teacher are one thing (the student knows them, she learned them by heart in order to answer the teacher at an oral quiz), but. . .the methods of reasoning and the factual knowledge by which she is guided and on which her convictions and actions are based are completely different.

The reason for such phenomena, characteristic of many

students, is the fact that the knowledge which they receive in the course of instruction is often not connected with the operations which must be carried out in order to apply this knowledge. When the teacher gives the indicative features of grammatical phenomena as a rule, he does not say how one must operate with these indicative features. This is not discussed in the textbook. It does not form those operations by which the process of reasoning and the solution of the problems is achieved. This leads to the spontaneous—and often incorrect—formation of operations and methods of reasoning in the students.

All this is a manifestation of that defect which we mentioned earlier: during instruction, the necessary attention is not paid to the formation of general methods of reasoning in the students. The problem of the formation of such methods—as a special, particular problem—is not usually put before the teacher.

In a number of cases, the students' lack of knowledge of a general and complete system of indicative features for grammatical phenomena and of the corresponding operations leads to the fact that, since they have no effective means at hand for solving the problem, *they are obliged to "invent" some kinds of indicative features and methods of their own.* Many of these features and methods are completely untenable. We come now to *the sixth characteristic* typical of students from the control classes.

We saw that the indicative features of a subject which the students know do not suffice in order to identify it in sentences like this, for example: *A triangle is the polygon with the least number of sides* or *To follow the thoughts of a great man is the most interesting science.* But the subject must be found in the sentences—such is the assignment. There can be only one way out for the student—to solve the problem on the basis of the indicative features and operations which might come to his mind at the moment he solves the problem and which might seem to him to "fit" in the given case. But such indicative features and operations which are not verified with many examples by special methods are untenable in the majority of cases.

For example, let us examine the indicative features which the students thought up as they mistakenly underlined the word *doctor* as subject in the sentence: *She [is] a doctor.*

Student Kolya K.

Stu. The accent falls on the word *doctor. She [is] a* DOCTOR.

(He emphasizes the word *doctor* by intonation.)

Thus the indicative feature which the student "invents" is the following: "The subject is the word on which the intonation falls."

Student Nina B.

Stu. Doctor is more important.

The indicative feature of the subject thought up by this student is: "The subject is that word which is the most important one."

Student Sergei M.

Stu. I am not sure that *doctor* is the subject; but it is most likely *doctor.*

Exp. Why?

Stu. Here, something specific is indicated—a person's profession—whereas there it is simply *she.*

The indicative feature thought up by Sergei M. is: "The subject is the word which indicates something specific, such as a person's profession."

There is no need to prove the unsoundness of these indicative features. The picture we observe in relation to operations is analogous. When the student does not know the procedure and does not know what to do with the sentence in order to solve the problem before him, he begins to apply operations which are completely arbitrary, groundless, and invented. We cited one of the examples above in a different context. When the student Lyuda V. did not know what must be done to specify the subject in the sentence: *To follow the thoughts of a great man is the most interesting science,* she applied this operation: she changed the positions of the first and second parts of the sentence and on that basis came to the conclusion that *science* was the subject. The unsoundness of such a method for the specification of the subject is obvious.

We shall cite a few more examples.

Student Lenya Ya. mistakenly placed a comma before *and* in the sentence: *They went across the bridge into the village of Borodino, from there [they] turned left and past a large number of troops and cannons and came out by a high kurgan.*

Exp. Why did you put a comma before *and*?

Stu. This is a compound sentence. It consists of two simple sentences: *They went across the bridge into the village of Borodino, from there [they] turned left. And past a large number of troops and cannons and came out by a high kurgan.*

Exp. Is it possible that *passed a large number of troops and cannons and came out by a high kurgan* is an independent sentence?

Stu. But the subject THEY *came out by a high kurgan* can be put in.

The operation which the student applies for the specification of the sentence's type is incorrect. The application of such an operation cannot give the correct solution of the problem.

Student Marina K. mistakenly placed a comma before *and cruets* in the sentence: *Large glass jars with coffee, cinnamon, and vanilla, crystal and porcelain tea-caddies and cruets with oil and vinegar were in the kitchen.* [This is acceptable in English, *Trans.*]

Exp. Why do you think that a comma is necessary here? How did you reason when you analyzed the sentence?

Stu. What is talked about here? About jars with coffee, cinnamon and vanilla, crystal and porcelain tea-caddies, and about cruets with oil and vinegar. There is an enumeration here. Three objects are indicated, and we separate them by commas.

The student did not apply certain important operations necessary for the correct punctuation in the sentence with homogeneous parts. Instead of that, she applied the operation for the enumeration of objects which of itself does not lead to the solution of the problem.[5]

Many more similar examples could be cited.

The mistakes of students from the control classes are provoked by the fact that, in a number of cases, they apply incorrect, faulty operations. Mistakes also often arise in the cases where the students do apply the correct operations. This happens when the students do not know *all* the operations which always lead to the solution of the problem (*the system of operations is incomplete*) and therefore do not carry them out, or else when *they do not carry out the operations in the sequence which is necessary.*

There is a *seventh characteristic* typical of the intellectual activity of students from the control classes. Examples of mistakes prompted by the fact that students do not apply all the operations necessary for the solution of a problem were cited above. We shall pause and look at mistakes provoked by the fact that the students do not carry out operations in the necessary order.

Student Galya S. underlined the word *house* in the sentence: *From the high embankment, Ptakha caught sight of a familiar house and only then believed that he was safe.*

Exp. Why do you think that *house* is the subject?

Stu. Ptakha caught sight of *what?* The house. It answers the question *what?*

Exp. And are there words in the sentence which answer the question *who?*

Stu. (She reads the sentence through again.) There is: *Ptakha. Ptakha caught sight and believed.* Here the subject is *Ptakha.*

If the student had begun by asking the question *who?*, she would not have committed an error in the given sentence. But she began with the question *what?*, which should be asked after it turns out that there are no words in the sentence which answer the question *who?* This produced a mistake.

Similar examples are not rare. But the incorrect sequence of operations for the solution of a problem where different types of sentences are identified (compound or simple with homogeneous parts) produces an especially large number of mistakes.

Punctuation before coordinating conjunctions (e.g., before

the conjunction *and*) depends on what these conjunctions join. In the sentence: *The transparent forest grows dark alone, and the fir tree shows green through the frost*, a comma is necessary before *and*, since the conjunction unites simple sentences within the compound sentence. In the sentence: *The river gleams dimly and rushes over the stones on the bank* a comma is not necessary before the conjunction, since it joins homogeneous parts within a simple sentence and is not repeated. (If the conjunction joins homogeneous parts within a simple sentence, the punctuation before the conjunction depends on whether the conjunction is repeated or not.)

Whether or not one places a comma before the conjunction depends first of all on what it joins together. Therefore, it is quite understandable that, before one carries out the operation for the verification of the conjunction for repetition, one must carry out the operation (or combination of operations) to establish the type of the sentence.[6] The omission of the first operation (here, we shall consider the combination of operations to establish the type of a sentence as one operation) and the execution of the second operation as the first can lead—and usually does lead—to mistakes. Since attention is often not paid to the formation of a specific sequence of operations, i.e., the appropriate algorithmic process (when necessary) in the students, mistakes produced for reasons of this kind are very frequent.

Student Lyuda P. did not place a comma before the conjunction *and* in the sentence: [*They*] *took the broken radio to the workshop, and there* [*they*] *repaired it.*

Exp. Why did you not place a comma before *and?*

Stu. Because the conjunction *and* is used just once.

Stu. (She thought it over.) Now I see that it is compound.

Exp. And up until then did you think about whether it was compound or simple?

Stu. No. I saw that the conjunction was not repeated, and I did not put in a comma.

As we see, the student began the analysis of the sentence directly with the second operation (with the verification of the conjunction for repetition), without carrying out the first operation. She did not specify the type of the sentence, although she could have done it and did it after the experimenter's question.

Student Vera K. did not place a comma before *and* in the sentence: *Night, fog, and nothing can be seen.*

Exp. Why did you not place a comma before *and?*

Stu. It is not necessary here.

Exp. Why?

Stu. The conjunction *and* is not repeated.

Exp. And what kind of sentence is this?

Stu. Oh! It is a compound sentence. I did not think it over.

Student Tamara B. incorrectly placed a comma before the second *and* (*and the billows*) in the simple sentence: *You will harken to the roll of thunder, and the voice of the storm and the billows, and the cry of the village cocks.*

Exp. Why did you place a comma before the second *and?*

Stu. The conjunction *and* is repeated.

We could cite many other examples of mistakes resulting from the fact that complete systems of the necessary operations have not been formed in the students or the correct sequence of these operations has not been established.

The defects shown above in the intellectual processes of students from the control classes as they solved grammatical problems were absent in students from the experimental classes. And this is understandable, since the basic reason which engendered these mistakes was eliminated.

Since the students from the experimental classes mastered a complete system of general indicative features and operations which ensured the solution of any problem from a given class, they did not have to resort to the "discovery" of "their own" indicative features and to operations which did not correspond to the nature of the grammatical phenomena and were therefore incorrect. Since the students from the experimental classes were taught the application of indicative features, no contradiction arose between their knowledge and their actions and the conviction of the correctness of the knowledge which they had. The knowledge was linked right away to the necessary operations, and at the same time it regulated the operational sphere of reasoning. Since special attention was paid to the formation of the correct sequence of operations in the process of experimental instruction, the reasons for mistakes connected with a lack of knowledge or a lack of completion of some operations or the destruction of their order were eliminated. The students' actions in experimental classes were confident. The students saw precisely what must be done and in what sequence in order to solve the problem which confronted them. At the same time, solutions were excluded which were based on guesswork and trial and error.

A characteristic trait which distinguished students from the experimental classes was the fact that they not only used the correct methods, but they could substantiate strictly and prove the correctness of their actions and solutions. The students from control classes as distinct from students of the experimental classes—and this is their *eighth characteristic*—most often either *could not prove their different statements* or *proved them incorrectly*. The basic mistake (or defect) of their proofs consisted of the fact that, when they substantiated their statements, they referred to insufficient indicative features. Sometimes, these insufficient indicative features were part of a system of sufficient indicative features. Sometimes, they were simply thought up by the students and actually could not serve as a means of identification of the corresponding grammatical phenomena. Here, two types of cases were encountered: (1) the solution of the problem was correct, but the reasoning and proof were incorrect (in this case, the correct solution often turned out to be accidental); and (2) both the solution and the way it was substantiated were incorrect.

A typical example of the first case is the incorrect proof by way of reference to insufficient indicative features. Thus, in the sentence: *The newspapers brought important news*, student Igor S. correctly underlined *newspapers* as the subject.

Exp. Why do you think *newspapers* is the subject?

Stu. It answers the question *what*?

The proof is not true, since the indicative feature cited by the student is insufficient. It does not have to be the subject which answers the question *what?* The correct solution in the given instance turned out to be purely accidental. This came to light as soon as the student went on to the next sentence: [*They*] *brought the newspapers in the morning*. Here, he also underlined the word *newspapers* as the subject, because it answered the question *what?*

This example shows convincingly that the correct solution of a problem is by no means an index of whether the student knows the indicative features of the corresponding grammatical phenomenon and the correctness of that procedure which helps him to solve the problem. The correct answer, as was already noted, can often be accidental.

Another example is the incorrect proof through reference to *incorrect* indicative features.

A student from the control class, Natasha N., correctly underlined as subject the word combination *The director together with the assistant director* in the sentence: *The director together with the assistant director go to the factory shop.*

Exp. Why not just *director?*

Stu. Because of the meaning. They went together.

The indicative feature of meaning is not the determining one in this case, and therefore, the reasoning is incorrect. The correct solution of the problem turned out to be accidental. This becomes obvious as soon as the student starts to analyze the sentence: *A woman with a child goes into the house.* In this sentence, the student also underlined the word combination *a woman with a child* as the subject, which in this case, is incorrect.

Exp. Why do you think that the subject is *a woman with a child*, and not just *woman?*

Stu. Because she does not go in alone, but with a child.

It is quite clear that it does not follow from the fact that the woman went into the house with a child and not alone, that in the given sentence the subject is a word combination.[7]

We shall cite the example of another instance where both the solution and the means of its substantiation are incorrect. Galya S. did not place a comma before *and*, although it was necessary in the sentence: *We came up to the house, we knocked on the door, and [they] invited us to come in.*

Exp. How many sentences have we here?

Stu. Two.

Exp. Which?

Stu. We came up to the house is one thing. *We knocked on the door and [they] invited us to come in* is another.

Exp. Why do you think that *we knocked on the door and [they] invited us to come in* is a single simple sentence?

Stu. Because we knocked and *right away* they asked us to come in. Therefore it is one sentence here.

The indicative feature which the student cites as the basis for the solution is incorrect. It does not follow from the fact that one event follows immediately after another that the sentence which describes these events is simple.

Many similar erroneous explanations and incorrect reasonings given by students could be cited.

Unlike the students from the control classes, the overwhelming majority of students from the experimental classes *proceeded in their actions from correct foundations, and they could correctly substantiate their solutions.* This was the result of the fact that the students of the experimental classes were trained to recognize the sufficiency of indicative features and to make a decision only after they had established the presence of all the conditions sufficient for a conclusion.

We shall cite the answers of several students from the experimental classes which concern their substantiation (proof) of their solutions.

Student Viktor K. correctly underlined the pronoun *she* as subject in the sentence: *She [is] a doctor.*

Exp. Why do you think that the subject here is *she* and not *doctor?*

Stu. Because this word answers the question *who?* and does not enter into the composition of a predicate nominative.

The student cites two indicative features as substantiation. They are sufficient when combined. It follows from this that *she* is the subject.

The same student wrote that there was no subject in the sentence: [*It*] *covered the field with snow.*

Exp. Why could not the subject be *field*? It does answer the question *what?*

Stu. This does not suffice. It is still necessary that the word *field* can be replaced by the pronoun *he,* (*she, it, they*) and that it does not enter into the composition of a predicate nominative. But here, the word *field* is replaced by *it* [acc.] [*It*] *covered it with snow.* Therefore, *field* is not the subject.

As we see, the student can not only substantiate the fact that the word belongs to a specific category (explain why *she* is the subject), but also prove why the word does not belong to a specific category (why *field* is not the subject).

Student Tamara S. correctly punctuated the sentence: *They went across the bridge into the village of Borodino, from there [they] turned left and past a large number of troops and cannons and came out by a high kurgan.* She said this was a simple sentence.

Exp. Why did you decide that this was a simple sentence?

The student at once isolated the parts of the sentence essential for the solution of this problem and pointed out the sufficient indicative features that showed this was a simple sentence.

This same student gave her reasons why she placed a comma before *and* in the sentence: *The bookcase was large, and there were many books contained in it.*

Stu. The *bookcase* was *large,* and *were contained* does not refer to *bookcase.*

Exp. What do you mean, it does not refer to it? After all, they were contained in the bookcase.

Stu. That is as it refers to things. But the predicate *were contained* does not refer to the subject the *bookcase.* It is impossible to say the *bookcase were contained.*

Thus, the student clearly realizes the difference between the semantic and formal-grammatical relations and correctly refers to the sufficient, formal-grammatical indicative features when she substantiates her conclusion.

It is much more difficult for the students to prove the correctness of their solutions when they analyze sentences like:

[*It*] *grew dark, and* [*it*] *became cold*, since it is necessary here to use negative indicative features.[8] However, in general, the students also dealt successfully with these proofs.

Here is how student Nadya M. explains why the sentence [*It*] *grew dark, and* [*it*] *became cold* is compound.

Stu. There are no subjects, and the predicates are expressed neither by verbs of the first and second person, nor by verbs of the third person plural. Therefore the sentence is compound.

The proof is absolutely correct. The student pointed out the combination of indicative features which sufficed to consider the sentence compound.

Student Vladimir P. explains in the following way why he considers the sentence: [*They*] *took the broken radio to the workshop, and there* [*they*] *repaired it* compound.

Stu. There are no subjects here. *Radio* is not the subject: [*they*] *took it*. The predicates are expressed by verbs in the plural, but the doers of the actions are different people. The owners took the radio, and the technicians repaired it. Therefore, the sentence is compound.

Here, as we see, all of the indicative features whose combination suffices to identify the given sentence are also shown.

The overwhelming majority of students from the experimental classes substantiated their solutions in an analogous way.

There was one defect, however, in the reasoning and proof of different students from the experimental classes. They considered the presence of some indicative features in the grammatical phenomena obvious, and therefore they did not mention them in the process of proof, although they took these indicative features into account as they solved the problem, and they were guided by them.

Here is one of the examples. Student Sergei P. correctly punctuated the sentence: [*It*] *was getting dark, and* [*it*] *began to grow cold*. He said that it was compound.

Exp. Why do you consider this sentence compound?

Stu. Here, the predicates are not expressed by verbs of the first or second person, nor by verbs of the third person plural.

Exp. But is this a sufficient indicative feature?

Stu. No. There are also no subjects here.

Exp. Well, why did you not mention this indicative feature right away?

Stu. It is just obvious that there are none.

Thus, the student is guided by two indicative features and by their sufficient combination, but he mentions only one indicative feature as substantiation.

In similar situations, the students had to be shown once more that, when they give explanations and proofs, they must not assume that some things are obvious, and they must point out all the indicative features which are the grounds for one or another conclusion.[9] *A propos*, let us note that it is not easy to teach the students this, since the habit of assuming that something is obvious is very strong in them (this has been indicated many times in psychological and methodological literature). The origin of this habit is related to the fact that often the teacher does not pay attention (except, perhaps a mathematics teacher) to whether the students prove their statements in a strictly logical way and observe a very important requirement in the process of proof. This requirement is to refer to that combination of indicative features which is sufficient for one or another conclusion as a basis for their statements.

We have analyzed the characteristics of the intellectual processes in students from control and experimental classes. We have pointed out the defects in the reasoning procedures of students from the control classes that led to a large number of mistakes when they solved grammatical problems. The defects pointed out explain in full the previously presented quantitative results from the controlled assignments which they carried out.

Let us go on now to an analysis of the reasons for mistakes which students from experimental classes committed. Although their mistakes were considerably fewer, an analysis of them is of great interest.

Notes

1. While the quantitative data presented in these tables follow a uniform pattern which attests to a *very substantial* superi-

ority of experimental class-groups over their controls, the question of statistical significance might be raised. Prof. Landa's comment on this issue is that Russian policy for publications such as this book—at the time of original publication (1966)—was not to overwhelm the intended readership with a great deal of statistical detail. That the statistical significance of the data matched their obvious material significance is demonstrated by the following example. The mean number of mistakes made on the criterion test for "The Compound Sentence" by the experimental class-group is 3.1 with a standard deviation of 0.8. The comparable values for Control Class No. 1, which achieved the fewest mean number of mistakes, is 15.5 with a standard deviation of 2.2. A student's t-test computed for the difference indicates a probability ($p < .01$) of less than one chance in a hundred that these means do not, in fact, differ. (*Editor.*)

2.   The differences in the approach to the solution of the very same problems are natural and normal if these problems do not have efficient algorithms of solution, or if these algorithms are unknown. In relation to the problems which we are examining, we cannot call such a situation normal. If there are efficient and economical methods for their solution, why should each student try to solve the problem "in his own way," often following the path of accidental trials and errors? (What has been said does not mean that, as they solve the same type of problems, the students absolutely must apply the very same method and arrive at the solution by following one path. If there exist different effective methods for the solution of problems of a given type, different students could apply those methods which are more convenient for them or "more acceptable," but they must know and master these methods.)

3.   In eighth grade of School No. 312, during the time set aside for review, the topics, "The Subject," "Types of Simple Sentences," and "Homogeneous Parts of the Sentence," were

studied once again with the students.

4. In this sentence, the student mistakenly underlined the pronoun *him* as the subject, whereas the subject is *nothing*.

5. This and the two preceding examples may be interpreted in the same way as the examples cited just before, i.e., as examples of the "invention" by the students of "their own" indicative features. But, in the latter examples, it is especially evident that the "invention" of their own indicative features leads to the "invention" of "their own" corresponding operations.

6. This example shows how important it is, for the correct solution of a problem, to carry out operations in a specific sequence and, consequently, how important it is in corresponding instances to teach a specific order of operations (algorithmic processes) and not just the operations themselves.

7. The indicative feature by which one must be guided in similar cases is the number of a verb. In the first sentence, the subject cannot be *director*, since the predicate is expressed by a verb in the plural ([*they*] *go*). Conversely, in the second instance, the subject cannot be *the woman with the child*, since the verb is in the singular ([*she*] *goes*). If the author had wanted to make the subject a word combination, he would have said [*they*] *go*.

8. A negative indicative feature is one which consists of the absence of an indicative feature. We used negative indicative features several times in the structural diagrams of indicative features.

9. A complete and strict substantiation of their solutions was required of students during their lessons.

Chapter Twenty-One

Characteristics of the Way Students in Experimental Classes Assimilated Algorithms

1. The Causes of Student Errors in Experimental Classes

We conducted the most systematic observations and experiments with the students of the eighth grade in School 312 of Moscow. All of the students in this grade were included in the individual experiment. Therefore, we present the analysis of a problem of interest relating to the experimental data obtained from that class. Results presented derive from criterion tests on the topics "Homogeneous Parts of a Sentence" and "Compound Sentences."

The mistakes of students from the experimental class were produced by the fact that, when they analyzed some sentences, they did not carry out all of the operations which they were taught and which they knew. In several instances, the algorithm which they had mastered did not work: why did this happen?

One may assume that the reason resides in the fact that those incorrect stereotypes of intellectual processes already formed in the students during previous instruction, often already in the beginning classes, were not sufficiently inhibited. Evidently, as a result of the individual characteristics of the type of higher nervous activity, these stereotypes were so entrenched in a number of students that it was not possible to completely "break" them in the process of experimental instruction, and under certain conditions they became manifest.

The student Irina M., for example, provides a model. These stereotypes were very persistent in her. There were five mistakes in the controlled work of Irina M., in addition to five self-corrections. The individual experiment showed that, despite the fact that she knew the required procedure well, her first reaction to the sentence—at that point in time, she did not particularly control herself—was the habitual reaction which had been formed earlier as a result of her previous experience.[1] In the majority of cases, the self-control formed in the process of experimental instruction "was turned on" in the student before she could make a mistake; but sometimes it happened later (and then she would correct the mistake she had made). Sometimes self-control did not take place, and then the mistake remained unnoticed by her. But it sufficed to draw her attention to the correct way of proceeding, and she would notice her mistake and quickly correct it.

Here are several mistakes from the records which show well the dynamics of the interaction and the struggle between the correct method and the incorrect stereotype.

At first this student incorrectly placed a comma before the last *and*; then she herself crossed it out in the sentence: *On the way to Vladivostok we went through Sverdlovsk and Novosibirsk, Chita and Khabarovsk and many other towns* [a comma before the last *and* is acceptable in English, *Trans.*].

Exp. Why did you at first place a comma before the last *and* and then cross it out?

Stu. At first I put in the comma according to intonation; then I verified it, saw that I had made a mistake, and crossed it out.

The student's first reaction was her habitual one. It was only after the habitual, incorrect stereotype "began to operate" that she began to check herself, i.e., to act according to the correct method. This led to the fact that she noticed and corrected the mistake.

In the sentence: *Many engineers and scientists came here and stayed for the rest of their lives*, the student at first mistakenly put a comma before the second *and* and then crossed it out.

Stu. I put in the comma according to intonation and then checked myself and corrected my mistake.

In the sentence: *The heavy fetters will fall off, the dungeons crash down, and freedom will greet us at the entrance, and our brothers will give us back our swords* the student mistakenly did not place a comma before the first *and.*

Exp. Is a comma necessary here or not?

Stu. (Attentively reads the sentence, thinks.) It is necessary. A second simple sentence begins.

Exp. Why did you not put it in right away?

Stu. From the intonation. But then I checked myself.

In the sentence: *They went over the bridge into the village of Borodino, from there [they] turned left and past a large number of troops and cannons and came out by a high kurgan* the student incorrectly placed a comma before the first *and.* She answered the question as to why she acted in this way:

Stu. I analyzed the sentence thus: *They went over the bridge into the village of Borodino* is the first section, a complete sentence. *From there [they] turned left* is another sentence. *And past a large number of troops and cannons [they] came out by a high kurgan* is still another sentence.

Exp. But what is the correct way to analyze the sentence? What should be the order of actions?

The student names the necessary order of actions and, at the request of the experimenter, applies all of the necessary operations. She correctly analyzes the sentence, discovers her mistake, and corrects it herself.

Thus, the reason for the student's mistakes is not a defect in the method which she was taught, but lies in the fact that, as a result of her specific individual characteristics, the stereotypes of intellectual activity hindered the application and "turning on" of this method which was new for her.

We observe an analogous situation in a number of other students.

Student Misha G. (who committed nine errors) mistakenly did not place a comma before *and* in the sentence: *The lake became quite shallow, and [it] was not at all difficult to get across it.*

Exp. Why do you consider that no comma is necessary here?

Stu. We are talking about the lake here. . . . But generally, if we go by the rules, a comma is necessary. *To get across* does not refer to *lake.*

Exp. Why did you not think of that at once?

Stu. I acted from habit.

Thus, at first, the student "acted from habit." He oriented himself by the thought and did not apply the necessary operations. However, it sufficed to point out his mistake to him, and he immediately applied the necessary procedure and correctly solved the problem. Misha G. was a rather inert, phlegmatic boy. Undoubtedly the stereotypes were formed in him with particular tenacity, and this explains the circumstance that he had the greatest number of mistakes, if Olga G. was not considered.

In some students, the habitual, incorrect stereotype is sometimes achieved so quickly that they do not even have the time to be aware of it in their actions. This trait was also characteristic of Misha G.

In the sentence: *Large glass jars with coffee, cinnamon, and vanilla, crystal and porcelain tea-caddies, and cruets with oil and vinegar were in the kitchen* he incorrectly placed a comma before *and cruets* [acceptable in English, *Trans.*]. As soon as the experimenter pointed to this sentence, the student said:

Stu. A comma is not necessary here: *jars, tea-caddies,* and *cruets* are homogeneous subjects.

Exp. Why then did you put a comma in at first?

Stu. Mechanically. I did not even think.

An analysis of the mistakes of Misha G. and some other students makes us suppose that the success in the assimilation of methods of reasoning depends not only on the flexibility of their nervous processes but on such an individual characteristic of the students as the level of their intellectual activeness. This individual trait has been researched very little,[2] but each teacher knows that some students like intellectual work and solve intellectual problems easily and with pleasure, while for other students it is difficult to exert their minds and to carry through some chain of intellectual operations. Such students try "to escape" intellectual work and to avoid straining their minds. The level of intellectual activity of the former and the latter is different.[3] In some students, intellectual passivity coincides with the tenacity of stereotypes which have been formed in them (for example, in student Misha G.).[4]

In Misha G., the tendency not to carry out intellectual operations in those instances where it seemed to him that one could not carry them out became apparent in the following case.

He mistakenly put a comma before the second *and* in the sentence: *Many engineers and scientists came here and stayed for the rest of their lives.*

Exp. Why do you think a comma is necessary here?

Stu. It is not necessary here. I put it in according to intonation.

Then he executed all the necessary operations and proved that the comma was unnecessary.

Exp. Why did you not establish that at once?

Stu. I did not go through the necessary steps.

Exp. But did you do them in other instances?

Stu. Yes.

Exp. Why did you do them in some instances and not in others?

Stu. When the sentence is long and difficult, then I do them in order to analyze it, but here it seemed to be clear.

Thus, it sufficed that the sentence seemed easy to the student and that he could solve it, and he at once dispensed with the intellectual effort.

Intellectual passivity and the unwillingness to make intellectual efforts (and this is evidently linked with the fact that it is difficult for the student to make these efforts) become apparent in still another characteristic mistake. Misha G. did not put a comma before *and* in the sentence: *I did not want to sleep, the punch was finished, and yet, there was nothing to do.* After the experimenter's question, he independently noticed the mistake and corrected it. Since he knew how to reason, he performed all necessary operations and saw at once that this was a compound sentence made up of simple ones.

Exp. Why did you not put in the comma right away?

Stu. I thought that *was finished* and *to do* were homogeneous parts. Both ended in the same letter.

The student did not make the necessary effort to distinguish between the verb endings. As soon as it seemed to him that *dopit* [drunk up, i.e., finished, a passive past participle, *Trans.*] and *delat'* [to do, infinitive, *Trans.*] were homogeneous parts according to their form, he immediately drew the conclusion that this was a simple sentence. Once he had come to this conclusion, he did not even attempt to perform the operation of relating the predicates to the subjects. He did not verify whether *to do* referred to the word *punch*.

We see an analogous picture with the student Rina V. She also tries to avoid intellectual effort and does not perform the necessary operations when she thinks this is possible.

For example, she mistakenly placed a comma before *and cruets* in the sentence: *Large glass jars with coffee, cinnamon, and vanilla, crystal and porcelain tea-caddies, and cruets with oil and vinegar were in the kitchen.* She acted in a manner which was habitual and easy for her.

Stu. Crystal and porcelain tea-caddies is one thing, and cruets with oil and vinegar is another.

She did not perform any of the operations for the differentiation of the series of homogeneous parts, although she knew them all and carried them out in easier cases. The experimenter had to interfere in order that this student force herself to make the necessary intellectual effort and correctly analyze the sentence.

The student did not place a comma before *and children rejoice* in the sentence: *All the paths in the world are open to us, and the earth salutes us, the flowers grow, and children rejoice, and the rich fields are heavy with grain.*

Exp. Why do you think that a comma is unnecessary here?

Stu. It is necessary here. I did not think about what *grow* and *rejoice* refer to. So, at first glance, it seemed that they were homogeneous parts. Now I see that they refer to different subjects. Therefore, they are not homogeneous.

As soon as it seemed to the student that *grow* and *rejoice* were homogeneous parts, she did not take the trouble to perform the operation and check what these predicates referred to.

Several students (in experiments on other topics) frankly said that they did not carry out operations because they "did not feel like it."

The overwhelming majority of mistakes (around 90 percent) were due to the two reasons indicated—the tenacity of incorrect pre-formed stereotypes of intellectual processes and intellectual passivity which led to the fact that the students did not carry out the operations which they knew well.

The remaining 10 percent of mistakes were made for the following reasons: (1) lack of mastery of certain operations which should have been formed in previous instruction, (2) an imprecise knowledge of the rules.

Individual characteristics influenced the mastery of the required procedures not only by those students who made mistakes, but also by those who did not make mistakes. Observations during lessons showed that different students mastered the general methods of reasoning at different speeds. There were students in the class (for example, Tanya Z.) for whom it sufficed to analyze two or three sentences to recall and master all the necessary indicative features and operations solidly and well. But there were also students for whom the mastery of the very same things required a much greater number of exercises. When some students had composed a procedural diagram, they practically did not refer to it at all afterward. Others used the diagrams for greater or lesser periods of time.

Important differences were observed also in the speed of automatization of intellectual habits. Students Tanya A., Vitya K., Nadya L., and a number of other students passed on very quickly from extensive reasoning to a direct differentiation of the necessary elements of the sentence and of their relationships. When asked the question of whether or not she carried out extensive reasoning when she analyzed a sentence, Nadya L., for example, said: "No. I read the sentence and see at once where the subject and predicate are and whether they refer to each other." To the same question Vitya K. answered: "At first, I analyzed everything in detail, but now I see at once what kind of a sentence it is."

But there were students who recalled up to the very last moment what had to be done and in what sequence in order to correctly specify the type of sentence and the punctuation. If, for one reason or another, they did not completely carry out the correct procedural order, then, as we saw, this was explained by the fact that they "fell back" on the habitual procedure and made mistakes.

Whereas about a third of the students completed the test in 20-25 minutes, there were some who needed 40-45 minutes. Two students handed in their papers after the bell. This also testifies to the different degree of automatization of the habits formed in the students.

Despite the important individual differences in the mastery and application by the students of the methods of reasoning taught them, all the students mastered the necessary systems of indicative features and applied them to the solution of the majority of problems.[5]

We must note that the experimental method of instruction ensured not only a higher level of knowledge and habits in students from the experimental class, but it also brought about an improvement *in their intellectual development*. An index thereto was the generally successful transfer of the assimilated methods of reasoning from one subject-matter to another (because all of the sentences presented to the students in the course of controlled tasks were new to them). Another index of progress in intellectual development was the fact that during instruction students learned to compare independently grammatical phenomena of different content, to discover the common and different indicative features, to specify the logical structure of the indicative features, and to formulate an efficient and general system of operations (algorithms). It is precisely all this which testifies to the development of qualities of reasoning, since the students applied the intellectual habits and abilities pointed out, when they studied different topics and when they operated upon different subject-matter.

2. The Students' Attitudes Toward
the Methods of Experimental Instruction

In order to evaluate correctly one or another method of instruction, not only is the quality of the students' mastery of knowledge, skills, and habits and the level of their mastery of reasoning very important, but what is also important is the attitude evoked in the student by the new way of instruction. How does it influence motivation to learn and what interest in the subject does it provoke, etc.? The most important task of a school is the formation of a proper motivation for learning, an interest in knowledge, and the desire to study. Hence, the quality itself of the mastery of knowledge, skills, and habits depends on how successfully this task is accomplished.

As we conducted experimental instruction, we were interested, of course, not only in how the algorithmic instruction influenced the development of the students' intellectual abilities but we were also interested in what attitude it established in the students, and how it influenced the motivation to study and the interest in it. It was also important to find out to what extent students could easily and quickly assimilate the material with the given method, whether the algorithmic instruction complicated the study process, and whether it introduced any additional difficulties.

Observations during lessons showed that the algorithmic methods of instruction produced great interest and much initiative in the students. The teachers noted that they began "not to recognize" many students who were considered passive, lazy, and not interested in the subject. Also, the students told us many times in conversations that "the Russian language, it turns out, is an interesting subject." In order to obtain more precise data on the attitude of the students toward the algorithmic method of instruction, we asked the students of the two seventh grades (Moscow Schools No. 328 and 613), after they studied the topic "The Subject," to express their views in writing. We said that, in order to perfect the methods of instruction, it was necessary to know to what extent each method was acceptable and what were its merits and defects. We noted that the students' opinions on this question were very important.

We asked the students the following questions:

(1) Did you like the new way of studying grammar?

(2) What were its advantages and defects in comparison with the usual method?

(3) Would you like to study the remaining topics in the same way?

We emphasized that it was especially important to know the defects of the new way and that the students should not hesitate to write about them. They were told that it was a completely voluntary affair to write down "replies" (they wrote them at home) and that they did not have to sign their names to them. In a few days we received 61 replies (some of the students wrote their replies collectively). A positive attitude toward the method used was expressed in all the replies. In several replies, which we shall review more closely later on, they spoke of a desire to simplify the method for the identification of the subject.

We cite several replies.

"Recently we studied the topic 'The Subject.' We did not study it the usual way, by learning rules according to the textbook, but we studied according to indicative features. It seems to me that it is easier and more comprehensible to learn this way. I would like very much to study all the rest of the topics in this way."*

"I really liked the new topic 'Indicative Features of the Subject.' When indicative features were applied to the words, it became easy to specify the subject in the sentence. It would also be easy to specify other parts of speech if it were made as simple as finding the subject."

"The method by which we found the subject is very good. Everything is mastered quickly. This method is very clear and easy."

"The fact that we learned the indicative features of the subject is very good because, thanks to them, it was very easy to recognize the subject in the sentence. The teacher explained the material clearly. I liked such a system of explanation. I would very much like to have all the other topics explained in the same way."

"Recently we studied a new topic 'The Subject.' But we did not study it the usual way. We did not learn the rules by heart, but did exercises on sheets of paper. I liked this method of study

---

*The replies are cited verbatim. They retain the style of their authors.

better. I understood everything well. It is more interesting and easier to study this way."

"I really liked the new method for the specification of the subject. It makes the search for the subject considerably easier. In the test exercises which we wrote, this method speeded up its completion considerably. I would be very interested to find out whether the rules could be made even shorter."

"When they told us that we were going to specify the subject in a new way, I was very glad. I very much like to specify the subject in this way. This method is very easy and easy to remember, but you must take it seriously. If you make a mistake, everything goes to pot. But at the same time, it is nonetheless easier than the preceding methods, and it is somehow fun to specify the subject."

"Several weeks ago, they began to teach the Russian language in our class according to a new system. It is easier to learn according to this system. Thanks to simple and intelligible examples, it is easy to remember the rules. Instead of the usual examples, they gave us examples from life, where a person's indicative features were compared with the indicative features of a subject. With the help of these examples, I understood the explanations better. I like this method of instruction."

"At one of the lessons, the teacher of Russian language came to our class and acquainted us with a new method for the specification of the subject in the sentence. With this method, which consists of special tables, it is easy to find the subject according to indicative features. I liked this way because you always have the indicative features of the subject right before your eyes, and therefore, they are easily mastered. The teacher explained this method to us very well and also prepared us for the test exercises. It is advisable that this method be applied when grammar is taught."

"It is very easy to work according to indicative features, easier than by the textbook. I like to work according to the new method. We are given less homework, and now we do much more work in class than we did earlier."

"I like the new method of instruction. Now it is impossible

to make a mistake in finding the subject. When the new teacher came to class, I thought that the lesson would not be interesting, but it went off very well. We were given examples. For instance, we had to recognize a person by the indicative features of clothing. After two lessons in class, test exercises on new material were given. Our class did the test exercises better than the other classes.* We didn't have one 'F.' After the introduction of the new material, the lessons were much more interesting."

"What I like about the new system of instruction is that it is clear and understandable for everyone. It has now become easier to recognize the subject according to indicative features."

"I like the new system of instruction. It is easier and more understandable. I wish they taught like that all the time."

The remaining replies are analogous to those cited. As we see, the students note the following characteristics of assimilation of material with the new methods of instruction: they are intelligible, easy, interesting, simple, quickly mastered and remembered, require less homework and more active class work ("we are given less homework, and now we do much more work in the class than we did earlier"). The assimilation is of a higher quality ("our class did the test exercises better than the other classes"). There is no need to learn the rules by heart ("we did not learn the rules by heart but did exercises on sheets of paper"). Interest is awakened in the efficient "arrangement" of the rules ("I would be interested to know whether the rules could be made even shorter"). The necessity for a serious attitude toward the method is noted ("you must take it seriously. If you make a mistake, everything goes to pot"). A positive emotion appears ("somehow it is fun to specify the subject"). Almost all of the students indicated a desire to study other topics by the same method.

Out of all the students who handed in "replies," two students noted that it made no difference to them what method was used to identify the subject. One student expressed a desire to find a means to specify the subject "by an even quicker way." One indicated that one type of subject "gave him a bit of trouble," and

*He means the control classes of this school.

one student noted the "detailedness and complexity of the work."
We cite the statements of these students.

"The method we used to learn the subject was not bad. But I already understood everything well. So, for me it was all the same what method we learned by."

"Recently we learned a new method to find the subject. I liked this method. In my opinion, it is easy and easy to remember. But before this, it was not hard for me to find the subject. I would also like, if possible, to learn how to specify the other parts of a sentence and parts of speech just as easily and especially how to distinguish short adjectives from adverbs. Then we could make fewer mistakes in dictation."

"I like the new method of instruction. It has become much easier to find the subject. But I think that the process of finding the subject is long. Would it not be possible, with the help of these same indicative features, to find the subject in the sentence much more quickly?"

"It seems to me that the topic 'The Subject' is not a difficult topic. I had trouble understanding it when the subject is a verb of indefinite form. In the lower grades, we were still not acquainted with such subjects. As for the rest, it was comparatively easy. I only confuse the subject in the case of sentences where the subject is some kind of word combination."

"This work is very interesting, but very detailed and complicated."

Thus, even those students for whom the method of instruction was a matter of indifference (these are, as a rule, the bright students) and those who encountered some difficulties noted that they reacted favorably to the new method. One of the students who wanted the process of finding the subject to be simpler and who encountered some difficulties when he assimilated the method for the identification of the subject, nevertheless noted that "it became much easier to find the subject with the help of indicative features."

When we sum up a review of the students' replies, we can, therefore, say that the algorithmic instruction not only raises the quality of knowledge, skills, and habits of the students and makes the process of assimilation easier, but it forms in them an interest in the subject and generates positive motives for learning.

Notes

1. One must take into consideration the fact that the topics "The Subject," "Homogeneous Parts of the Sentence," and "The Compound Sentence" were already studied in beginning grades and that specific methods for the identification of these grammatical phenomena and specific methods for approaching the sentence are formed in the students, though these are often incorrect.

2. For example, Slavina drew attention to it [792], [794]. As we already mentioned, she distinguished among the students a group of children whom she designated as intellectually passive.

3. Everyone is familiar with differences in children's physical activeness. Some children are very energetic, lively, and restless. They run during recess, they "fidget" during lessons, and they like gym and sports. Others do not have these needs. Moreover, for them, the necessity to show physical activity is related to overcoming certain difficulties. The differentiation of children by the degree of their activeness exists, obviously, not only in the motor sphere, but also in the intellectual sphere. Differentiation of the latter type evidently influences the mastery of knowledge, skills, and habits. It conditions the individual characteristics of the children's intellectual efforts.

4. A more precise answer to the question of how these two individual characteristics are interconnected (and whether they are connected at all) requires further research which ought to be conducted according to its own particular methodology.

5. The student Olga G. is an exception. She is a girl (according to the teachers and a doctor) with specific deficiencies in mental development, for whom the task of correctly analyzing 40 sentences in 45 minutes was beyond her capabilities. Moreover, in an individual experiment after the criterion test, when she was asked to analyze those sentences in which she made mistakes and was given unlimited time, Olga G.

independently discovered her mistakes and corrected them in some of the sentences. It is also important to note that many students in the control classes made considerably more than 16 mistakes (these were classes where the average number of mistakes per student exceeded 20). This is in no way an index of their mental development. The number of mistakes, irrespective of the method of instruction, still tells absolutely nothing.

Conclusion

An attempt was made in this book to shed some light on the problem of teaching students algorithms (algorithmic prescriptions and processes), and an experiment was described in which students were taught several grammar topics with the help of algorithms.

The question is whether or not such instruction entails a complication of the process of instruction. Does not the method proposed presuppose that, besides grammar, the students must also become proficient in logic? Does not the present author treat logic as a superstructure of grammar? These questions must be answered in the negative. It must be emphasized, first of all, that if we were to succeed in giving the students some notions of logic, not only would there be nothing detrimental in this, but on the contrary, it would be beneficial to the students. Logic is an important science whose significance for school is greatly underestimated at the present time. The purpose of our research, however, was not to form ways of teaching elements of logic while grammar was studied (this problem still awaits solution). Our purpose was to place the instruction in grammar itself on a firm, logical basis and to make the grammatical thinking of the students really logical.

The instruction of logical operations can never be a superstructure on any subject, since logical operations do not exist separately from grammatical, mathematical, and other operations. They exist *in them* and are realized *through them*. The concept of a "logical operation" is abstract. Its great importance consists of the fact that it makes it possible to determine what is general in a person's thought, independently of the content with which he operates. The logical is that which is general in grammatical, mathematical, and any other reasoning, as soon as this reasoning correctly reflects reality.

Logical operations do not exist in pure form. They always appear in the form of some kind of material (as opposed to formal, *Ed.*) operations. This is why, when we taught students methods of logical thought, we did not teach logic *along with* grammar. We did not form logical thought in them *in addition* to grammatical thought. This cannot be done. We taught them logical operations in the form of grammatical operations, *the logical structure of grammatical knowledge*, and logic for grammatical material. In brief, we taught them the logic of grammar. But since the logical structures of grammatical knowledge have common elements with the logical structures of any other knowledge, when we taught general methods of grammatical thought, we taught the students some general methods of thinking.

Should students be taught intellectual operations? Does it not suffice that they study grammar rules?

Grammar rules, like laws, theorems, and definitions in other sciences, serve as a means of solving problems only in the case where they are correctly applied. The application of rules, particularly the recognition of situations where the rule is applicable, is achieved by means of special operations. Just as it is impossible to solve a manufacturing problem (for example, to make something) without carrying out specific component manipulations (operations), it is also impossible to solve an intellectual problem (a grammatical, mathematical problem or one pertaining to physics, etc.) without carrying out specific intellectual operations. The execution of a specific aggregate of intellectual operations to solve a problem is an objective necessity. But if this is so, then it is incorrect to think that some problems may be solved without executing the operations. Also untenable is the opinion sometimes expressed as: "Why, for example, carry out some operations in order to identify the subject in a sentence or specify the type of a sentence, if it is evident in itself—evident without any operations?"

The opinions that something is "obvious at once," that

something "is immediately grasped" without any operations, "comes to mind of itself," etc., are illusions engendered by the fact that there is no awareness of many operations because of their automatization. Generally speaking, not only is there no awareness of automatized intellectual operations, but of many other psychic and physiological processes as well. Thus, for example, under ordinary circumstances, a person is not aware of the contractions of his heart muscles. But it does not follow from the fact that there is no awareness of the muscle contractions, that the heart works without contractions. There is an essential difference, however, between the work of the heart and the "work" of thinking. The "operations" which the heart performs are innate (although they are perfected in the process of its vital activity). The operations which a person performs in the process of thinking are not innate, and he must be specially taught to do them.

Those who are unaware of their operations, and believe that they solve some problems "without any operations," draw their conclusion from this premise: if one may solve problems without operations, then why "split hairs" by teaching some sort of operations? Why waste time and energy on this? Is it necessary to complicate simple things?

The opinion that one may solve problems "without operations" is untenable. An objective analysis of the structure of knowledge, in particular, grammatical knowledge, shows that intellectual processes which seem at first glance to be simple and not to require any special operations, are actually very complex and break down into a considerable number of operations. If this were not so, if it did not just seem simple to identify, for example, the subject, then all the students would identify this part of the sentence without mistakes. This is not so in reality. We do not even refer to other, more complex problems. Generally, if a simple knowledge of grammar rules sufficed for grammatically correct writing, then there would be no failures in school (the majority of the students know the rules, and it is not that difficult for them to learn the rules). Meanwhile, the Russian language (along with mathematics) is one of the most difficult subjects and one in which failures are very frequent.

It is difficult to find a person today who would doubt that knowledge must be specially taught. Regarding the teaching of intellectual operations, however, people often consider it unnecessary to teach operations and believe that the instruction in the operations only complicates the learning process. But operations are just as necessary a component of thinking as knowledge (i.e., content of knowledge, *Ed.*). Without operations, the process of thinking is as impossible as it is without knowledge. As for the complexity in the mastery of operations, the question arises: If it is complicated to master operations under the conditions of organized instruction, how much more complicated must it be to master these operations independently and spontaneously when operations are not specifically taught?

We saw that it is not as complicated as it might seem at first glance to master specific operations under conditions of specially tailored instruction. But even if it really were a complicated matter to master some system of operations, this is in no way an argument against the special instruction in operations. Quite the contrary, the more complex the system of operations is, the more important it is to teach it and the more difficult it is to master this system of operations in the absence of such instruction. The complexity of the operations is not lessened simply by the fact that we refuse to teach them. Not to teach them because they are complex is to ignore the complexity and place all the difficulties on the students' shoulders.

In pedagogy, there exist two ways to form habits of logical thinking. The first way is the unconscious mastery of logical methods by the students. This takes place during the process in which specific material from the textbooks is learned and during the solution of problems by practice. The second way is that of conscious mastery of those methods when the teacher specially draws the students' attention to those logical means with whose help the solutions of problems are achieved.

The formation of operations with the first approach to instruction is realized, as Leontyev [693] expresses it, through an "adjustment" of actions to the conditions of the object (or by way of simple imitation). When the students do this as they reason, they usually fail to realize how their thought process takes place. They do not know which intellectual operations they perform.* They often cannot answer the questions of what must be done in order to solve a specific problem and how they must think and reason when they do this. Intellectual operations which are formed in this way are poorly realized, and, most important, they are poorly controlled and regulated. They are actualized only when directed at objects with which they were connected in experience (or ones similar to them) and cannot be freely and consciously controlled or transferred from one subject-matter to another.

However, should this means of instruction be excluded from educational practice? No, it should not, because comparatively simple systems of operations may take shape even without special instruction. Moreover, the formation of certain initial, elementary operations is achieved only in this way. But this method becomes ineffective as well as unsatisfactory when there is the problem of forming more complex systems of operations. When the student does not know how to act in order to solve some problem, in this case, he usually stops at the method of blind trials, guesses, and random actions, or actions by analogy (and the latter is the only means of finding the solution under these conditions!). Since some of these actions lead to the solution of the problem, they are reinforced in the student's consciousness. This is how specific intellectual skills and habits are formed.

Such a means, however, for the formation of intellectual

---

*Intellectual abilities formed in this way during instruction may be compared, for example, with the ability to button a button. One is not usually taught how to button a button, and the actions which make up this ability are formed spontaneously in the process of adapting to the object at which they are directed. A characteristic feature of the ability to button a button is the fact that a person usually does not know which operations make up this ability. He cannot say what must be done in order to button a button. The ability to button a button is not broken down into operations in his consciousness, and, as a rule, he does not realize them. A similar situation takes place with intellectual abilities which are spontaneously formed in the process of adapting to the object at which they are directed.

operations is long, inefficient, and often painful. Its important defect lies in the fact that the systems of operations formed spontaneously and unconsciously in the process of adjusting to some object are often imperfect and not general enough, or just simply incorrect. (The experimental results cited earlier showed this quite clearly.) The application of these operations, although it helps solve the problem in some usually more or less limited circumstances, cannot ensure the solution of problems in many other circumstances (e.g., it does not ensure the correct identification of any type of sentence presented to a student).

Progressive-minded pedagogues of the past always attributed great importance to the special instruction of students in methods of logical thinking. The majority of contemporary pedagogues also admit the necessity of such instruction. Then why is the first way still the most prevalent one in practice up to the present day? This is related to the fact that the special conscious, and purposeful instruction of intellectual operations presupposes a knowledge of the mechanisms (dynamics, *Ed.*) of thinking, a reduction of intellectual activity into specific components, and a precise determination and calculation of the structure of these elements, of their internal connections, dependencies, and interactions. Such knowledge cannot be obtained simply through individual empirical experience. It is the result of the development of a science, first of psychology and logic, and at the present time, of cybernetics.

The level of development in psychology and logic, even several decades ago, was such that the creation of a scientifically based general theory for the instruction in systems of intellectual operations, in particular, algorithms of intellectual activity, was hardly possible. At the present time, thanks to new ideas which have appeared in psychology, logic, and cybernetics, not only is it possible to raise the question about the creation of such a theory, but it is indispensable. Present-day ideas in the field of psychology, logic, and cybernetics not only make it possible to create a scientific theory for the instruction in intellectual operations and their systems, but also to raise the question about the types of these systems.

The instruction in algorithms, like the instruction in methods of non-algorithmic character, may be achieved in different ways. One way is to present the algorithms in a ready-made form so that the students have only to learn them, and then to reinforce these algorithms during exercises. With another method, algorithms are not handed to the students in prepared form, but the students discover them themselves. The instruction is arranged in such a way that the students independently find the necessary and, at the same time, efficient systems of operations. It is important, when this is done, that the assimilation of these systems of operations is achieved not by rote-learning, but as a result of properly designed exercises. It is clear that, in specific cases, the first form of instruction may turn out to be expedient. But, in our opinion, the second method should be the important, basic, and leading one in instruction. It is this method which we used in experimental instruction. We did not impose algorithms on the students. We did not provide them in advance or present them in ready-made form. The students did not have to specially learn them. The systems of operations forming the basis of the algorithms were gradually assembled by the students in the process of independent, active, practical, linguistic actions. The verbal formulation of the algorithms or of their separate elements and the representation of their diagrams was only the result, the sum of the formation of separate operations.

Algorithms do not represent some sort of supra-programmed material. They are not some sort of supplement to the operations which the students perform. These operations are embodied in the algorithms. The algorithms organize them in a specific manner. The precise verbal formulation of operations makes it possible to comprehend them better and therefore to execute them freely and to direct the course of one's own intellectual activity. Just as a strict and good organization of any effort cannot make it more difficult but only facilitates it, so the strict organization of intellectual operations cannot create additional difficulties. An additional load and difficulties are not created for the students by the fact that a specific order and a system are introduced into their intellectual processes, but rather because such an order and system are lacking. The results of the experiments described previously, in fact, showed this.

Algorithms are not an addition to grammatical rules. They are not something external in relation to them. They are a specific, clearly realized, and precisely formulated means to the systematization of rules and to the organization of intellectual activity by their application. They are a means which have their origin in the logical structure of the rules. The dependency described earlier of the logical structure of the algorithms on the logical structure of the indicative features of grammatical (and other) phenomena showed the internal link between the organization of grammatical notions and rules, on the one hand, and the intellectual operations carried out by the student, on the other hand.

Not only is it unnecessary to learn algorithms by heart in the process of instruction, but, in the final analysis, it is unnecessary to remember them. Just as the knowledge of grammar rules is necessary, not for the sake of the rules themselves, but in order to write correctly, so the students' knowledge of algorithms is necessary, not for the sake of the algorithms themselves, but for the mastery of correct procedures. Usually, in the process of writing, a person who has learned to write correctly does not recall those rules by which he writes, though the knowledge of these rules is precisely what made him literate at one time. He can even forget them because he now "has them in his possession." It is the same with algorithms. A person who has mastered correct procedures through algorithms usually does not even remember them as he acts; moreover, he can forget them altogether. But the real significance of the instruction in algorithms consists of the fact that he at one time formed these correct procedures, and when he formed them, they became, as it were, embodied in him and materialized in him.

As it was said above, when one teaches students methods of thinking, this does not amount to the same thing as the instruction in algorithms. The concept "a method of thinking" is broader than the concept of "algorithm." When a person is able to establish the indicative features which are common to objects and phenomena, to specify their logical structures, to find the correct and efficient order of operations, to construct independently a reasoning process, to substantiate and prove his statements in a strict and logical way, etc., then he has mastered specific methods of thought. But they are not all algorithms.

As we saw in experimental instruction, a great deal of attention was paid not only to teaching the students to *apply* algorithms, but also to teaching them how to *devise* these algorithms. Mastery of an algorithm is the mastery of a specific, general procedure of thought, but the method for the devising of algorithms is not itself an algorithm. The application of an algorithm is an algorithmic process. The design of an algorithm is a process which, in many cases, is not algorithmic. It is not always possible to indicate the algorithm for the design of an algorithm. Herein lies one of the manifestations of that methodological principle that it is impossible, in principle, to algorithmize all thinking. It is possible to algorithmize only some facets of it and some systems of operations. What has been said does not in the least detract from the great importance of algorithms; rather, it emphasizes the fact that they do not suffice for the solution of all thought problems which arise in the practice of human thought.

Out of all the results of experimental instruction, the formation in the 'students of methods of thinking of non-algorithmic character has a particularly great significance. This is connected with the fact that the number of objects and phenomena with which the student must deal in his practical and theoretical activities are infinitely varied. No instruction is able to teach everything and foresee all those specific problems which confront a person during his life and which he should be able to solve. Therefore, to prepare students for life means, first of all, to equip them with the kind of general methods of activity (and most important, intellectual activity, since it regulates the other forms of human activity) which will make it possible for the students to solve the different problems which life will place before them.

As we said above, problems from different subject-matters have features in common among them. They can be attributed, by specific indicative features, to the same type, and can therefore be solved through the application of identical methods (e.g., problems of life as well as scientific, grammatical, mathematical problems, etc.). It is difficult to overestimate the importance of this fact. The establishment of the general indicative features in problems of the most varied content and the discovery of general methods for their solution thus makes it possible to devise a process of instruction such that when, for example, a student masters a method for the solution of a specific type of mathematical problems, he is simultaneously prepared for a quicker and easier mastery of methods for the solution of specific types of grammatical, geographical, and other problems. If the method for the solution of some kind of grammatical problems is the same in principle as the method for the solution of analogous kinds of mathematical problems, then when one learns this general method for one subject, one can simply transfer it to another subject and not spend the time to learn this method again. (Examples of the transferral of methods of thinking from some subjects to others were cited above.) At the present time, it is difficult to foresee what economies of the students' and teachers' time and effort instruction could bring, if it were possible to discover sufficiently general methods for the solution of problems from "different subjects" and, as these methods are formed, to make use of their transferral from one subject field to another.

The elaboration of a whole sphere of problems relative to this is an important task of didactics as well as of pedagogical psychology and pedagogical logic, which unfortunately does not exist as an independent branch of pedagogy. Almost no one in the Soviet Union studies the logical foundations of instruction. The more sciences differentiate—and this is inevitably reflected in the subject-matter of instruction and, through this subject-matter, in methods—the greater the importance assumed by didactics as a general theory of instruction. This is so, because it is precisely didactics which are called upon to establish and formulate the common features contained in the subject-matter of knowledge, types of problems, and methods studied "within" separate academic subjects. Only after what is general has been established can one teach it and at the same time consciously and purposefully develop the general intellectual abilities of the student by transferring methods from one subject to another. It is in this way that they will be assured a good preparation for life.

The instruction in general methods of thinking does not just facilitate and hasten the process of assimilation of knowledge and of operations. It also forms such an extremely important quality of thinking as the ability to look for and find what is common to the most diversified and mutually distant things, phenomena, and processes. It leads to the fact that, when students solve problems, they are in a position to apply one, general method. They are able to independently transfer these methods from one subject-matter to another. The history of scientific discoveries and inventions shows the tremendous importance of this quality of thinking. A large number of scientific discoveries and inventions were associated precisely with the fact that the scientist or inventor was able to examine, from a general standpoint, objects and phenomena which were disparate and seemed to have nothing in common with each other. This was possible because he applied to his analysis of one subject field methods formed in another subject field and up until that specific moment used only in the latter.

The tremendous success of cybernetics, only recently formed as an independent science, bears witness to the importance of transferring methods from one subject-matter to another. These successes of cybernetics turned out to be possible because, in cybernetics, from a single and common standpoint, one is able to examine certain processes (namely control processes) which take place in nature, in society, and in thinking. Cybernetics established that methods formed within separate, specific sciences, and up until a certain time, applied within specific subject fields, could be transferred to subject fields which were adjacent and even remote from one another. As a result, sciences which earlier were only faintly connected with one another began to interact closely and began to use each other's methods in their own fields; they began to develop these methods through joint effort. (This is obvious from the example of the interaction between linguistics and mathematics, genetics and information theory, logic and electrical engineering, etc.)

The characteristics of the present-day direction in the development and analysis of science has a major significance for didactics as a general theory of instruction. The demands which science and life today make on the quality of thinking by the young people who are finishing school are different from what they were two decades ago. Perhaps the greatest demand consists of the fact that the person who finishes school must master the kind of methods of thinking which will make it possible for him to examine the most diverse processes and phenomena from a general standpoint. Not only must he understand what is common, what today links different (and until recently remote from one another) branches of science and technology, but he must be in a position to discover and apprehend this independently and to use the results of his learning in practical activity.

The study of general methods of thinking and the means of their instruction is also very important for the development of didactics itself. We tried to show above how the presence of general didactic tasks in the process of instruction—which are dictated, in particular, by the generality of methods of thinking which must be taught to the students—engenders a general system of methods of instruction itself. We saw that it is possible to teach knowledge from the most varied subject fields by a single, general method as soon as the conditions of instruction and the didactic tasks which must be solved are one and the same. One could show that this refers not only to the instruction of different topics within one subject (in our experiment—the grammar of the Russian language), but also to the instruction in various subjects. If general methods for the solution of problems by students are possible, then why not assume the possibility of general methods of instruction for these general methods? We tried to show that such general methods are feasible and that they must be further elaborated. This must be one of the most important tasks of didactics.

When we speak of general methods of instruction, we mean that concept of a method which was presented at the beginning of this book: the concept of a method as an aggregate of specific operations, as their specific system. This concept of a method does not correspond (or rather does not completely correspond) to what is usually understood by methods of instruction (a method of discussion, a method of laboratory work, etc.). Reflected in it is a somewhat different—and therefore, very prospective—aspect of the study of methods.

If the specific study of methods of thinking (and generally, any methods of activity) presuppose their structural-operational analysis, i.e., the reduction of methods into components (operations) and the establishment of the internal links between these components and their efficient orderings, then such an approach must be applied to the study of methods of instruction. The structural-operational approach to the study of methods of instruction must lead to the construction of efficient algorithms of instruction and to the elaboration of means for the specification of optimal systems of instructional operations. The discovery and design of efficient algorithms of instruction are a new and very actual problem of the present day. They arise within the structural-operational approach to methods of instruction.

The elaboration of efficient algorithms of instruction is very closely related to the problem of programmed instruction. The experiment described above, which was to teach students algorithms for the solution of grammatical problems, represents an attempt to devise an instructional process as a process where specific qualities of thinking are formed in the students on the basis of a specific program of actions which the teacher and the students must carry out during instruction. It is possible that the given attempt is still incomplete, but there are all the grounds for assuming that the approach to instruction as well as to the process of the formation of students' specific qualities of thinking in relation to the given program opens up broad perspectives for pedagogy and, in particular, will be important for programmed instruction. Such an approach opens up the possibility of applying the powerful apparatus of contemporary mathematical logic, information theory, cybernetics, and operations research to the analysis of the instructional process (and, evidently, in the future, to the process of education).

At the present time, the choice of ways and methods of instruction which suit different conditions is usually achieved intuitively by the teacher and on the basis of experience. Today, neither methodology nor didactics make available to the teacher methods for the specification of one or another optimal combination of ways of instruction for each totality of conditions. It is natural that such a choice of ways and methods of instruction on the basis of experience and intuition can only coincide with the optimal in a small percentage of cases. The installation in pedagogy of methods for the design of efficient algorithms and of contemporary methods for the calculation of optimal variants for actions in different ratios makes it possible to create in pedagogy itself the kinds of methods which permit the calculation of the best methodology of instruction under specific conditions.

Even such a problem as this can be solved if one follows the route outlined above. We spoke repeatedly of the fact that the effective instruction in methods of thinking for students presupposes the distinction of general types of problems from among the most varied subject fields (and, therefore, in academic subjects) and the transferral of methods for solution from one subject field to another. But, in order to achieve this, the teacher must not only know his subject well, but other subjects also. He must be able to analyze the types of problems encountered there and the methods for their solution. It is obvious, however, that the teacher cannot know different subjects equally well. What is the way out of this situation? From our viewpoint, it lies in the installation of methods of formalization within pedagogy.

We showed above what aspect formalization could assume in grammar and how it is possible to describe symbolically grammar rules and the structures of the indicative features of linguistic phenomena. Since the structures of operations for the application of indicative features depends on the logical structure of the indicative features, the structures of the operations shown and the methods for the solution of grammatical problems may also be described symbolically.

A symbolic description makes it possible to abstract separate indicative features and operations from a specific subject-matter, and because of this, it is possible to establish their most general structures in their "pure form."

Once the structures of specific indicative features and operations are depicted with the help of symbols, they assume a general significance. If, for example, the structure of the indicative features of a circle is symbolically described in a specific "language of symbols," then the form in which it is presented is not only a reflection of the structure of the indicative features of the given specific geometrical figure, but of any indicative features of any objects and phenomena which have the given structure of indicative features. When the teacher knows the symbolic description of a given structure, he should be able to discover the same structure of indicative features in another object as soon as that structure is formalized.

We have shown that students from the seventh grades easily mastered the structure of indicative features which were diagrammatically represented by circles, and they operated successfully with these abstract, logical diagrams of the structures of indicative features. There is nothing, however, to prevent one from representing indicative features not by circles, but by letters, and in this way arriving at the maximally convenient form for the description of the structures of indicative features (and, if necessary, of operations). This in no way impedes understanding, since, from the standpoint of the facility of mastery, there is no difference between the fact that an indicative feature is represented by a circle or a letter.

The fact that students of the lower grades—as research carried out at the present time shows—can already master beginning algebra without significant difficulties is a guarantee that not only the teachers, but also the students easily familiarize themselves with symbolisms. A specific of algebra (in comparison with arithmetic) is that, in the case of the former, a figure is represented by specific signs—letters—and the operations on numbers are replaced by operations on letters. It is quite obvious that, if the symbolical representation of numbers and operations with letters is not difficult for students, then the symbolic operation of indicative features of things and phenomena as well as the mastery of elements of symbolic (mathematical) logic will also not be difficult.

The mastery of the logical structures of indicative features and operations expressed as symbols (letters, logical operators, etc.) can play an important role for the development of abstract thinking in the students and for the development of their intellectual activity. As soon as identical structures of the indicative features of different phenomena (from different subject fields) are symbolically represented, the students see what they have in common at once and are in a position independently to transfer methods of thinking from one subject field to another.

The symbolic representation of the structures of indicative features of different phenomena leads to a quicker formation of the students' notion of the way human knowledge is constructed. This has great significance for the formation in them of efficient ways of cognition and for the development of their scientific *Weltanschauung*. One must also mention the fact that the mastery of this kind of symbolism will prepare the students to grasp contemporary scientific language.

In our work, just the first steps toward the elaboration of the problems outlined above were made. The research in this direction, which was described earlier herein, showed that the solution of the problems indicated would have great significance both for the improvement of instruction in practice and for pedagogical theory itself.

A structural-operational analysis of general methods of thinking as well as of the actual methods of instruction is the prerequisite for the creation of the kind of theory of instruction which will make it possible to introduce elements of precise calculation and prediction into the design of the teaching process. It will equip the teacher with the means for making a valid determination, design, and selection of the most efficient and effective methods of instruction. Herein lies one of the basic tasks of didactics and pedagogical psychology.

Bibliography*

I. Problems of Logic and the
Theory of Knowledge
1. Alekseyev, M.N. *Dialekticheskaya logika* (Dialectical
    Logic). Moscow, 1960. (All references to Moscow hence-
    forth will be abbreviated to M.)
2. Andreyev, I.D. *O metodakh nauchnovo poznania* (On
    Methods of Scientific Cognition). M., 1964.
3. Biryukov, B.V. "Idealizatsia" (Idealization). *Filosofskaya
    enstiklopedia* (Philosophical Encyclopedia), Vol. 2. M.,
    1962.
4. Vecker, L.M. and Lomov, B.F. "O chuvstvennom obraze
    kak izobrazhenii" (On the Sensual Image as Representa-
    tion). *Voprosy filosofii* (Problems of Philosophy), IV,
    1961.
5. Voishvillo, Ye. K. *Predmet i znacheniye logiki* (The
    Subject and Meaning of Logic). M., 1960.
6. Vostrikov, A.V. *Teoria poznania i dialekticheskovo mate-
    rializma* (The Theory of Knowledge and Dialectical Mate-
    rialism). M., 1965.
7. Gorsky, D.P. *Voprosy abstraktsii i obrazovaniye ponyaty*
    (Problems of Abstraction and the Representation of
    Concepts). M., 1961.
8. Gruzenberg, S.O. *Geny i tvorchestvo. Osnovy teorii i*

*Works noted by an asterisk (*) are those which the author consulted during
the writing of this book, but which are not cited or referred to in the text.

psikhologii tvorchestva (Genius and Creation. Foundations of the Theory and Psychology of Creation). Leningrad, 1924. (Henceforth, Leningrad will be abbreviated to L.)

9. *Dialektika i logika. Zakony myshlenia* (Dialectics and Logic. The Laws of Reasoning). M., 1962.

10. Carnap, P. *Znacheniye i neobkhodimost* (Meaning and Necessity). M., 1959 (In Russian translation of work published originally elsewhere. Henceforth to be indicated by "Trans.").

11. Kedrov, B.M. *Yedinstvo dialektiki, logiki i teorii poznania* (The Unity of Dialectics, Logic, and the Theory of Knowledge). M., 1963.

12. Kopnin, P.V. *Dialektika kak logika* (Dialectics as Logic). Kiev, 1961.

13. Kursanov, G.A. *Dialektichesky materializm o ponyatii* (Dialectical Materialism on Understanding). M., 1963.

14. Lapshin, I.I. *Filosofia izobretenia i izobreteniye v filosofii* (The Philosophy of Invention and Invention in Philosophy). Prague, 1922.

15. *Logika* (Logic). Eds. D.P Gorsky and P.V. Tavanets. M., 1956.

16. Maltsev, V.I. *Ocherk po dialekticheskoi logiki* (An Essay on Dialectical Logic). M., 1964.

17. Medvedev, N.V. *Teoria otrazhenia i yeyo yestestvenno— nauchnoye obosnovaniye* (The Theory of Reflection and Its Foundation in Natural Science). M., 1963.

18. Osipov, I.N. and Kopnin, P.V. *Osnovnye voprosy teorii diagnoza* (Basic Problems of the Theory of Diagnostics). M., 1951.

19. Pavlov, T. *Teoria otrazhenia* (The Theory of Reflection). M., 1949. (Trans.)

20. *Problemy logiki nauchnovo poznania* (Problems of the Logic of Scientific Cognition). Ed. P.V. Tavanets. M., 1964.

21. *Problemy myshlenia v sovremennoi nauke* (Problems of Reasoning in Modern Science). Eds. P.V. Kopnin and M.B. Vilnitsky. M., 1964.

22. *Problemy formalizatsii semantiki yazyka. Tezisy nauchnoi konferentsii* (Problems of Formalizing Language Semantics. Theses of the Scientific Confederation). M., 1964.

23. Poincaré, A. *Mathematicheskoye tvorchestvo. Psikhologichesky etyud* (Mathematical Creation. A Psychological Study). Yuryev, 1909. (Trans.)

24. Poincaré, A. *Nauka i gipoteza* (Science and Hypothesis). M., 1904. (Trans.)

25. Reznikov, L.O. *Ponyatiye i slovo* (Concept and the Word). L., 1958.

26. Reznikov, L.O. *Gnoseologicheskiye voprosy semiotiki* (Gnoseological Problems of Semiotics). L., 1964.

27. Rozental, M.M. *Printsipy dialekticheskoi logiki* (Principles of Dialectical Logic). M., 1960.

28. Serebryannikov, O.F. "Yevristicheskiye vozmozhnosti metodov formalnoi logiki" (Heuristic Possibilities of the Methods of Formal Logic). *Nekotorye voprosy metodologii nauchnovo issledovania* (Certain Problems of the Methodology of Scientific Research), No. 1, 1965.

29. Spirkin, A.G. *Proiskhozhdeniye soznania* (The Origin of Consciousness). M., 1960.

30. Starchenko, A.A. *Logika v sudebnom issledovanni* (Logic in Juridical Research). M., 1958.

31. Tyukhtin, V.S. *O prirode obraza (psikhicheskoye otrazheniye v svete idei kibernetiki)* (On the Nature of Image [Psychological Reflection in the Light of Ideas from Cybernetics]). M., 1963.

32. Cherkesov, V.I. *Materialisticheskaya dialektika kak logika i teoria poznania* (Materialistic Dialectics as the Logic and Theory of Cognition). M., 1962.

33. Schaff, A. *Vvedeniye v semantiku* (Introduction to Semantics). M., 1963. (Trans.)

34. Khaskhachikh, F.I. *O poznavayemosti mira* (On the Knowability of the World). M., 1950.

35. Carnap, R. *Introduction to Semantics and Formalization of Logic*. Cambridge, Mass., 1959.

36. Kemeny, J.G. "New Approach to Semantics. Part One."

The Journal of Symbolic Logic, XXI (1), 1956.
37. Stegmüller, W. *Das Wahrheitsproblem und die Idee der Semantik*. Wien, 1957.

II. Mathematics, Mathematical Logic, and the Theory of Algorithms. Cybernetics and Certain of Its Applications

38. Hadamard, J. *Elementarnaya geometria* (Elementary Geometry), Part 1. M., 1957. (Trans.)
39. Aizerman, M.A. and others. *Logika, avtomaty, algoritmy* (Logic, Computers, and Algorithms). M., 1963.
40. Amosov, N.M. *Kibernetika i meditsina* (Cybernetics and Medicine). M., 1963.
41. Amosov, N.M. and Shkabara, Ye. A. "Opyt postanovki diagnoza pri pomoshchi diagnosticheskikh mashin" (An Experiment in Establishing a Diagnosis with the Help of Diagnostic Machines). *Eksperimentalnaya khirurgia i anesteziologia* (Experimental Surgery and Anesthesiology), IV, 1961.
42. Anokhin, P.K. "Fiziologia i kibernetika" (Physiology and Cybernetics). *Voprosy filosofii* (Problems of Philosophy), IV, 1957.
43. Artobolevsky, I.I. and others. "Terminologia osnovnykh ponyaty avtomatiki" (Terminology of the Fundamental Concepts of Automation). *Trudy Mezhdunarodnoi federatsii po avtomaticheskomu upravleniyu* (Transactions of the International Federation on Automatic Control), Vol. 3. M., 1961.
44. Artobolevsky, I.I., Vishnevsky, A.A., and Bykhovsky, M.L. "Avtomaticheskaya sistema otyskania klinicheskovo pretsedenta" (An Automatic System for Finding Clinical Precedents). *Eksperimentalnaya khirurgia i anesteziologia* (Experimental Surgery and Anesthesiology), III, 1962.
45.* Akhmanova, O.S. and others. *O tochnykh metodakh issledovania yazyka* (On Precise Methods of Language Research). M., 1961.
46. Babsky, Ye. B. and Parin, V.V. *Fiziologia, meditsina i tekhnichesky progress* (Physiology, Medicine, and Techni-

cal Progress). M., 1965.

47. Berg, A.I. "O nekotorykh problemakh kibernetiki" (On Certain Problems of Cybernetics). *Voprosy filosofii* (Problems of Philosophy), V, 1960.

48. Berg, A.I. "Kibernetika i nekotorye tekhnicheskiye problemy upravlenia narodnym khozyaistvom" (Cybernetics and Certain Technical Problems of Control in the National Economy). *Voprosy filosofii* (Problems of Philosophy), II, 1961.

49. Berg, A.I. "Soyuz matematiki i elektroniki" (The Union of Mathematics and Electronics). *Radio*, VI, 1962.

50. Berg, A.I. "Nauka velichaishikh vozmozhnostei" (The Science of the Greatest Probabilities). *Priroda* (Nature), VII, 1962.

51. Berg, A.I. "Kibernetika i obshchestvennye nauki" (Cybernetics and the Social Sciences). *Metodologicheskiye problemy nauki* (Methodological Problems of Science). M., 1964.

52. Berkeley, E. *Simvolicheskaya logika i razumniye mashiny* (Symbolic Logic and Thinking Machines). M., 1961. (Trans.)

53. Bir, St. *Kibernetika i upravleniye proizvodstvom* (Cybernetics and Control of Industry). M., 1963.

54. Biryukov, B.V. "Avtomatizatsia i obshchestvo" (Automation and Society). *Vestnik istorii mirovoi kultury* (The World Culture Herald), IV, 1959.

55. Biryukov, B.V. and Spirkin, A.G. "Filosofskiye problemy kibernetiki" (Philosophical Problems of Cybernetics). *Voprosy filosofii* (Problems of Philosophy), IX, 1964.

56. Biryukov, B.V. and Konoplyankin, A.A. "Matematika i logika" (Mathematics and Logic). *Dialektichesky materializm i voprosy yestestvoznania* (Dialectical Materialism and Problems of Natural Science). M., 1964.

57. Boiko, Ye. I. "Mozhet li mashina myslit?" (Can a Machine Think?). *Voprosy Psikhologii* (Problems of Psychology), I, 1965.

58. Bongard, M.M. "Modelirovaniye protsessa uznavania na

tsifrovoi vychislitelnoi mashine" (Simulating the Recognition Process on a Computer). *Biofizika* (Biophysics), VI (2), 1961.

59. Bongard, M.M. "Modelirovaniye protsessa uznavania" (Simulating the Recognition Process). *Nauka i zhizn* (Science and Life), VI, 1965.

60. Braynes, S.N. "Neirokibernetika" (Neurocybernetics). *Kibernetika na sluzhbu kommunizmu* (Cybernetics at the Service of Communism) Ed. A.I. Berg, Vol. I. M., 1961.

61. Braynes, S.N., Napalkov, A.V., and Svechinsky, V.B. *Neirokibernetika* (Neurocybernetics). M., 1962.

62. Brodsky, I.N. *Elementarnoye vvedeniye v simvolicheskuyu logiku* (Elementary Introduction to Symbolic Logic). L., 1964.

63. Bykhovsky, M.L. "Veroyatnostnaya logika postroyenia samoobuchayushchevosya diagnosticheskovo protsessa na matematicheskikh mashinakh" (Probability Logic of the Construction of Self-teaching Process on Mathematical Machines). *Eksperimentalnaya khirurgia i anesteziologia* (Experimental Surgery and Anesthesiology), I, 1962.

64. Bykhovsky, M.L. "Metod fazovovo intervala v probleme diagnostiki" (The Method of the Phase Interval in the Problem of Diagnostics). *Eksperimentalnaya khirurgia i anesteziologia* (Experimental Surgery and Anesthesiology), II, 1962.

65. Bykhovsky, M.L. "Matematicheskiye metody v medetsine" (Mathematical Methods in Medicine). *Matematika v shkole* (Mathematics in School), V, 1963.

66. Bykhovsky, M.L., Vishnevsky, A.A., and Kharnas, S. Sh. "Voprosy postroyenia diagnosticheskovo protsessa pri pomoshchi matematicheskikh mashin" (Problems of Constructing a Diagnostic Process with the Help of Mathematical Machines). *Eksperimentalnaya khirurgia i anesteziologia* (Experimental Surgery and Anesthesiology), IV, 1961.

67.* Vecker, L.M. "K postanovke problemy modelirovania sensornykh funktsy" (On Stating the Problem of Simulation of Sensory Functions). *Leningrad University Herald*, XVII, 1963 (Series on Economics, Philosophy, and Law), No. 3.

68. Vecker, L.M. *Vospriatiye i osnovy yevo modelirovania* (Perception and the Foundations for Its Simulation). L., 1964.

69. Venikov, V.A. "Nekotorye metodologicheskiye voprosy modelirovania" (Certain Problems of Simulation). *Voprosy filosofii* (Problems of Philosophy), XI, 1964.

70. Ventsel, Ye. S. *Elementy teorii igr* (Elements of the Theory of Games). M., 1959.

71. Ventsel, Ye. S. *Vvedeniye v issledovaniye operatsy* (An Introduction to Operations Research). M., 1964.

72. Ventsel, Ye. S. *Teoria veroyatnostei* (The Theory of Probabilities). M., 1964.

73. Williams, J.D. *Sovershenny strateg, ili Bukvar po teorii strategicheskikh igr* (The Compleat Strategist, or a Handbook for the Theory of Strategic Games). M., 1960. (Trans.)

74. Wiener, N. *Kibernetika ili upravleniye i svyaz v zhivotnom i mashine* (Cybernetics or Control and Communication in the Animal and the Machine). M., 1958. (Trans.)

75. Wiener, N. *Kibernetika i obshchestvo* (Cybernetics and Society). M., 1958. (Trans.)

76. Wiener, N. "Nauka i obshchestvo" (Science and Society). *Voprosy filosofii* (Problems of Philosophy), VII, 1961. (Trans.)

77. Vishnevsky, A.A. and others. "Kibernetika v khirurgii" (Cybernetics in Surgery). *Eksperimentalnaya khirurgia* (Experimental Surgery), I, 1959.

78. Gaase-Rapoport, M.G. *Avtomaty i zhivye organizmy* (Computers and Living Organisms). M., 1961.

79. Gastev, Yu. "Massovaya problema" (The Universal/Mass Problem). *Filosofskaya entsiklopedia* (Philosophical Encyclopedia), Vol. 3. M., 1964.

80. Gelfand, I.M. and Tsetlin, M.L. "Printsip nelokalnovo poiska v sistemakh avtomaticheskoi optimizatsii" (The

Principle of Non-local Search in Systems of Automatic Optimization). *Doklady ANSSSR* (Lectures of the USSR Academy of Sciences), Vol. 137 (2), 1961.

81. Gelfand, I.M. and Tsetlin, M.L. "O nekotorykh sposobakh upravlenia slozhnymi sistemami" (On Certain Ways to Control Complex Systems). *Uspekhi matematicheskikh nauk* (Successes in the Mathematical Sciences), XVII (1), 1962.

82. Getmanova, A.D. *Vyrazheniye deduktivnykh umozaklyucheny v traditsionnoi i simvolicheskoi logike* (The Expression of Deductive Conclusions in Traditional and Symbolic Logic). Ed. B.V. Biryukov. Murmansk, 1962.

83. Hilbert, D. and Akkerman, V. *The Foundations of Theoretical Logic*. M., 1947 (Trans.).

84. Glushkov, V.M. *Teoria algorithmov* (The Theory of Algorithms). Kiev, 1961.

85. Glushkov, V.M. *Sintez tsifrovykh avtomatov* (A Synthesis of Numerical Computers). M., 1962.

86. Glushkov, V.M. "Modelirovaniye myslitelnykh protsessov" (Simulation of the Thought Processes). *Priroda* (Nature), II, 1963.

87. Glushkov, V.M. "Kibernetika i myshleniye" (Cybernetics and Reasoning). *Voprosy filosofii* (Problems of Philosophy), I, 1963.

88. Glushkov, V.M., Grishchenko, N.M., and Stogny, A.A. "Algoritm raspoznavania osmyslennykh predlozheny" (An Algorithm of Identification of Intelligible Sentences). *Printsipy postroyenia samoobuchayushchikhsya sistem* (Principles for the Construction of Self-teaching Systems). Kiev, 1962.

89. Gnedenko. B.V. *Kurs teorii veroyatnostei* (A Course in the Theory of Probabilities). M., 1954.

90.* Gnedenko, B.V. "O roli matematicheskikh metodov v biologicheskikh issledovaniakh" (On the Role of Mathematical Methods in Biological Research). *Voprosy filosofii* (Problems of Philosophy), I, 1959.

91. Gnedenko, B.V. "Nekotorye voprosy kibernetiki i statis-

tiki" (Several Problems of Cybernetics and Statistics). *Kibernetika na sluzhbu kommunizmu* (Cybernetics at the Service of Communism), Vol. 1. M., 1961.

92. Gnedenko, B.V. and Khinchin, A. Ya. *Elementarynoye vvedeniye v teoriyu veroyatnostei* (Elementary Introduction to the Theory of Probabilities). M., 1961.

93. Goldman, S. *Teoria informatsii* (The Theory of Information). M., 1957. (Trans.)

94. Goncharenko, M. *Kibernetika v voyennom dele* (Cybernetics in Warfare). M., 1960.

95.* Grandstein, I.S. *Pryamaya i obratnaya teoremy* (The Direct and Reverse Theorem). M., 1959.

96. Grenyovsky, G. *Kibernetika bez matematiki* (Cybernetics without Mathematics). M., 1964.

97.* Gulin, F.F. *Primeneniye electronnykh vychislitelnykh mashin dlya upravlenia proizvodstvennymi protsessami* (The Application of Electronic Computers for the Control of Industrial Processes). M., 1961.

98. Gutenmacher, L.I. *Electronnye informatsionno-logicheskiye mashiny* (Electronic Informational-logical Machines). M., 1960.

99. Zhinkin, N.I. "Zvukovaya kommunikativnaya sistema obezyan" (The Sound Communication System of Monkeys). *Izvestia APNRSFSR* (News of the Academy of Pedagogical Sciences of the Russian SSR—henceforth abbreviated to APSRSSR), No. 113, 1960.

100. George, F. *Mozg kak vychislitelnaya mashina* (The Brain as Computer). M., 1963. (Trans.)

101. Zholkovsky, A.K., Leontyeva, N.N., and Martemyanov, Yu. S. "O printsipialnom ispolzovanii smysla pri mashinnom perevode" (On the Principal Use of Meaning with Mechanical Translations). *Trudy Instituta tochnoi mekhaniki i vychislitelnoi tekhniki ANSSSR* (Transactions of the Institute of Precise Mechanics and Computer Technology of the Academy of Sciences of the USSR), No. 2. M., 1961.

102. Ivakhnenko, A.G. *Samoobuchayushchiyesya sistemy s*

polozhitelnymi obratnymi svyazyami (Self-teaching Systems with Positive Reverse Feedback). Kiev, 1963.

103. Kaluzhnin, L.A. "Ob algoritmizatsii matematicheskikh zadach" (On the Algorithmization of Mathematical Problems). *Problemy kibernetiki* (Problems of Cybernetics), No. 2. M., 1959.

104. Kaluzhnin, L.A. *Chto takoye matematicheskaya logika* (What is Mathematical Logic). M., 1964.

105. Klaus, G. *Vvedeniye v formalnuyu logiku* (Introduction to Formal Logic). M., 1960. (Trans.)

106. Klaus, G. *Kibernetika i filosofia* (Cybernetics and Philosophy). M., 1963. (Trans.)

107. Klini, S.K. *Vvedeniye v metamatematiku* (Introduction to Metamathematics). M., 1957. (Trans.)

108. Kolbanovsky, V.N. "O nekotorykh spornykh voprosakh kibernetiki" (On Certain Disputed Problems of Cybernetics). *Filosofskiye problemy kibernetiki* (Philosophical Problems of Cybernetics). M., 1961.

109. Kolmogorov, A.N. "Avtomaty i zhizn" (Computers and Life). *Nauka i zhizn* (Science and Life), X and XI, 1961.

110. Kolmogorov, A.N. "Algoritm" (The Algorithm). *Bolshaya sovetskaya enstiklopedia* (The Large Soviet Encyclopedia), Vol. 2. M., 1950.

111.* Kolmogorov, A.N. and Uspensky, V.A. "K opredeleniyu algoritma" (Towards a Definition of an Algorithm). *Uspekhi matematicheskikh nauk* (Successes in the Mathematical Sciences), Vol. 13, No. 4, 1958.

112. Kolman, A. "Chto takoye kibernetika" (What Is Cybernetics?) *Filosofskiye voprosy sovremennovo yestestvoznania* (Philosophical Problems of Modern Natural Science). M., 1961. (Trans.)

113. Kopnin, P.V. "Ponyatiye myshlenia i kibernetika" (Cybernetics and a Concept of Reasoning). *Voprosy filosofii* (Problems of Philosophy), II, 1961.

114. Cossa, P. *Kibernetika* (Cybernetics). M., 1958. (Trans.)

115. Kraismer, L.P. *Tekhnicheskaya kibernetika* (Technical Cybernetics). M., 1958.

116. Kraismer, L.P. *Bionika* (Bionics). M., 1962.
117.* Kulagina, O.S. "Ob operatornom opisanii algoritmov perevoda i avtomatizatsii protsessa ikh programmirovania" (On the Operative Description of Translation Algorithms and the Automation of Their Programming Process). *Problemy kibernetiki* (Problems of Cybernetics). No. 2. M., 1959.
118. Laidly, R.S. and Lasted, L.B. "Obyektivnye osnovania diagnoza" (Objective Foundations of Diagnosis). *Kiberneticbesky sbornik* (A Collection on Cybernetics), II. M., 1961. (Trans.)
119. Li Yao-tsu and Vandervelde, U.I. "Teoria nelineinykh samonastraivayushchikhsya sistem" (The Theory of Nonlinear Self-adjusting Systems). *Trudy Mezhdunarodnoi federatsii po avtomaticheskomu upravleniyu* (Transactions of the International Federation on Automatic Control), Vol. 2. M., 1961.
120. Lyapunov, A.A. "O nekotorykh obshchikh voprosakh kibernetiki" (On Certain Universal Problems of Cybernetics). *Problemy kibernetiki* (Problems of Cybernetics), No. 1. M., 1958.
121. Lyapunov, A.A. and Kitov, A.I. "Kibernetika v tekhnike i ekonomike" (Cybernetics in Technology and Economics). *Voprosy filosofii* (Problems of Philosophy), IX, 1961.
122. Lyapunov, A.A. and Shestopal, G.A. "Ob algoritmicheskom opisanii protsessov upravlenia" (On the Algorithmic Description of Control Processes). *Matematicheskoye prosveshcheniye* (Mathematical Education), II, 1957.
123. Lyapunov, A.A. and Yablonsky, S.V. "Teoreticheskiye problemy kibernetiki" (Theoretical Problems of Cybernetics). *Problemy kibernetiki* (Problems of Cybernetics), No. 9. M., 1963.
124. Luce, P.D. and Raifa, Kh. *Igry i reshenia* (Games and Decisions). M., 1961. (Trans.)
125.* Malinovsky, B.N. *Tsifrovye upravlyayushchiye mashiny i avtomatizatsia proizvodstva* (Numerical Machines for Control and the Automation of Industry). M., 1963.
126. Markov, A.A. "Nevozmozhnost nekotorykh algoritmov v

teorii assotsiativnykh sistem" (The Impossibility of Certain Algorithms in the Theory of Associative Systems). *Doklady ANSSSR* (Lectures of the Academy of Sciences of the USSR), ILV (7), 1947.

127. Markov, A.A. "Teoria algoritmov" (The Theory of Algorithms). *Trudy Matematicheskovo instituta im. V.A. Steklova* (Transactions of the V.A. Steklov Mathematical Institute), Vol. 42. M., 1954.

128.* Melchuk, I.A. "O standartnykh operatorakh dlya algoritma avtomaticheskovo analiza russkovo nauchnovo teksta" (On Standard Operators for the Algorithm of Automatic Analysis of a Russian Scientific Text). *Mashinny perevod* (Machine Translation). *Trudy instituta tochnoi mekhaniki i vychislitelnoi tekhniki* (Transactions of the Institute of Precise Mechanics and Computer Technology), No. 2, 1961.

129.* Melchuk, I.A. "O standartnoi forme i kolichestvennykh kharakteristikakh nekotorykh lingvisticheskikh opisany" (On the Standard Form and Quantitative Characteristics of Certain Linguistic Descriptions). *Voprosy yazykoznania* (Problems of Linguistics), I, 1963.

130. Mikulich, L.I. "Nekotorye voprosy mashinnoi evristiki (obzor)" (Certain Problems of Machine Heuristics [A Survey]). *Zarubezhnaya radioelektronika* (Foreign Radio-electronics), X, XI, 1964.

131. Minsky, M. "Ha puti k iskusstvennomu intellektu" (Steps Toward Artificial Intelligence). *Trudy instituta radioinzhenerov* (Transactions of the Institute of Radio Engineers), LXIX, No. 1, 1961.

132. Moiseyev, V.D. *Voprosy kibernetiki v biologii i meditsine* (Problems of Cybernetics in Biology and Medicine), M., 1960.

133. Morse, F. and Kimball, J. *Metody issledovania operatsii* (Methods of Operations Research). M., 1956. (Trans.)

134. Napalkov, A.V. "Izucheniye printsipov pererabotki informatsii golovnym mozgom" (A Study of the Principles of Information Processing in the Brain). *Kibernetiku—na*

sluzhbu kommunizmu (Cybernetics at the Service of Communism). Ed. A.I. Berg, Vol. I. M., 1961.

135. Napalkov, A.V. and Orfeyev, Yu. V. "Aktualnye voprosy razvitia evristicheskovo programmirovania" (Current Problems on the Development of Heuristic Programming). *Voprosy filosofii* (Problems of Philosophy), VI, 1965.

136. Neiman, J. "Vychislitelnaya mashina v mozg" (Computer into Brain). *Kiberneticbesky sbornik* (A Collection on Cybernetics), I. M., 1960. (Trans.)

137. Novik, I.B. *Kibernetika. Filosofskiye i sotsiologicheskiye problemy* (Cybernetics. Philosophical and Sociological Problems). M., 1963.

138. Novikov, P.S. "Ob algoritmicheskoi nerazreshimosti problemy tozhdestva slov v teorii grupp" (On the Algorithmic Unsolvability of the Problem of Identical Words in the Theory of Groups). *Trudy Matematicheskovo instituta im. V.A. Steklova* (Transactions of the V.A. Steklov Mathematical Institute), Vol. 44. M., 1955.

139. Novikov, P.S. *Elementy matematicheskoi logiki* (Elements of Mathematical Logic). M., 1959.

140. Newell, A. and Simon, H. "Modelirovaniye chelovecheskovo myshlenia na vychislitelnoi mashine" (Computer Simulation of Human Thinking and Problem Solving). *Kibernëtika i zhivoi organizm* (Cybernetics and the Living Organism). Kiev, 1964. (Trans.)

141. Newell, A., Shaw, D., and Simon, H. Empirical Exploration with the Logic Theory Machine: A Case Study in Heuristics. In E. Feigenbaum and J. Feldman (Eds.) *Computers and Thought*. N.Y.: McGraw-Hill, 1963.

142. Orlov, V.B. "O modelirovanii shakhmatnoi igry na elektronnykh mashinakh" (Simulation of Chess Games on Electronic Computers). *O nekotorykh voprosakh sovremennoi matematiki i kibernetiki* (Certain Problems of Modern Mathematics and Cybernetics). M., 1965.

143. Parin, V.V. "Kibernetika y fiziologii i meditsine" (Cybernetics in Physiology and Medicine). *Voprosy filosofii* (Problems of Philosophy), X, 1961.

144. Parin, V.V. "Matematika zhizni" (The Mathematics of Life). *Priroda* (Nature), VII, 1962.

145. Poletayev, I.A. *Signal* (The Signal). M., 1958.

146. Pushkin, V.N. "Nekotorye voprosy psikhologii upravlenia proizvodstvennym protsessom na zheleznodorozhnom transporte" (Certain Problems of Control Psychology for the Industrial Process of Rail Transport). *Voprosy psikhologii* (Problems of Psychology), III, 1959.

147. Pushkin, V.N. "K ponimaniyu evristicheskoi deyatelnosti v kibernetike i psikhologii" (Towards an Understanding of Heuristic Activity in Cybernetics and Psychology). *Voprosy psikhologii* (Problems of Psychology), I, 1965.

148. Pushkin, V.N. *Operativnoye myshleniye v bolshikh sistemakh* (Operative Reasoning in Large Systems). M., 1965.

149.* Revzin, I.I. and Rozentsveig, V. Yu. *Osnovy obshchevo i mashinovo perevoda* (Foundations of Universal and Machine Translations). M., 1964.

150. Reitman, V.R. "Razrabotka programm dlya reshenia intellektualnykh problem" (The Processing of Programs for the Solution of Intellectual Problems). *Zarubezhnaya radioelektronika* (Foreign Radio-electronics), I, 1962.

151. Rovensky, Z., Uyemov, A., and Uyemova, Ye. *Mashina i mysl* (The Machine and Thought). M., 1960.

152. Rozanov, Yu. A. "Teoria veroyatnosti i yeyo prilozhenia" (The Theory of Probabilities and Its Applications). *O nekotorykh voprosakh sovremennoi matematiki i kibernetiki* (Certain Problems of Modern Mathematics and Cybernetics). M., 1965.

153. Siforov, V.I. "Obshchiye tendentsii razvitia sovremennovo yestestvoznania" (General Tendencies in the Development of Modern Natural Science). *Voprosy filosofii* (Problems of Philosophy), IV, 1963.

154. Smirnov, V.A. "Algoritmy i logicheskiye skhemy algoritmov" (Algorithms and the Logical Diagrams of Algorithms). *Problemy logiki* (Problems of Logic). M., 1963.

155. Sobolev, S.L., Kitov, A.I., and Lyapunov, A.A. "Osnovnye cherty kibernetiki" (The Fundamental Characteristics of

Cybernetics). *Voprosy filosofii* (Problems of Philosophy), IV, 1955.

156. Sobolev, S.L. and Lyapunov, A.A. "Kibernetika i yestest-voznaniye" (Cybernetics and Natural Science). *Voprosy filosofii* (Problems of Philosophy), V, 1958.

157. Sokolovsky, Yu. I. *Kibernetika nastoyashchevo i budushchevo* (Cybernetics of the Present and the Future), Kharkov, 1959.

158. Solodovnikov, V.V. *Nekotorye cherty kibernetiki* (Certain Characteristics of Cybernetics). M., 1956.

159. Stolyar, A.A. *Elementarnoye vvedeniye v matematicheskuyu logiku* (Elementary Introduction to Mathematical Logic). M., 1965.

160. Tarski, A. *Vvedeniye v logiku i metodologiyu deduktivnykh nauk* (Introduction to Logic and the Methodology of the Deductive Sciences). M., 1958. (Trans.)

161. Trapeznikov, V.A. "Avtomatika i chelovechestvo. Doklad na plenarnom zasedanii kongressa" (The Computer and Mankind. Lecture at the Plenary Session of the Congress). *Trudy I Mezhdunarodnovo kongressa Mezhdunarodnoi federatsii po avtomaticheskomu upravleniyu* (Transactions of the First International Congress of the International Federation on Automatic Control), Vol. 1. M., 1960.

162. Teplov, L.P. *Ocherki po kibernetike* (Essays on Cybernetics). M., 1963.

163. Trakhtenbrot, B.A. *Algoritmy i mashinnoye resheniye zadach* (Algorithms and the Mechanical Solution of Problems). M., 1957.

164. Turing, A. *Mozhet li mashina myslit?* (Can a Machine Think?). M., 1960. (Trans.)

165. Uspensky, V.A. *Lektsii o vychislimykh funktsiakh* (Lectures on Calculable Functions). M., 1960.

166. Uspensky, V.A. "Algoritm" (The Algorithm). *Filosofskaya entsiklopedia* (The Philosophy Encyclopedia), Vol. 1. M., 1960.

167. Feldbaum, A.A. "O primenenii vychislitelnykh ustroistv v avtomaticheskikh sistemakh" (On the Application of

Computer Equipment in Automatic Systems). *Avtomatika i telemekhanika* (Automation and Telemechanics), XI, 1956.

168. Feldbaum, A.A. "Avtomatichesky optimizator" (The Automatic Optimizer). *Avtomatika i telemekhanika* (Automation and Telemechanics), VIII, 1958.

169. Feldbaum, A.A. "Vystupleniye na I Mezhdunarodnom kongressa Mezhdunarodnoi federatsii po avtomaticheskomu upravleniyu" (Presentation at the First International Congress of the International Federation on Automatic Control). *Trudy I Mezhdunarodnovo kongressa Mezhdunarodnoi federatsii po avtomaticheskomu upravleniyu* (Transactions of the First International Congress of the International Federation on Automatic Control), Vol. 3. M., 1961.

170. Feldbaum, A.A. *Osnovy teorii optimalnykh avtomaticheskikh sistem* (Foundations for the Theory of Optimal Automatic Systems). M., 1963.

171. *Filosofskiye voprosy kibernetiki* (The Philosophical Problems of Cybernetics). M., 1961.

172. Kharkevich, A.A. "O tsennosti informatsii" (On the Value of Information). *Problemy kibernetiki* (Problems of Cybernetics), No. 4. M., 1961.

173. Chavchanidze, V.V., Toronzhadze, A.F., and Bukreyev, I.N. "Kibernetika kak nauka" (Cybernetics as a Science). *Trudy Instituta kibernetiki ANGruz SSR* (Transactions of the Institute of Cybernetics of the Academy of Sciences of the Georgian SSR). Tbilisi, 1963.

174.* Chernov, G. and Moses, L. *Elementarnaya teoria staticheskikh resheny* (The Elementary Theory of Statistical Solutions). M., 1962. (Trans.)

175. Church, A. *Vvedeniye v matematicheskuyu logiku* (Introduction to Mathematical Logic), Vol. 1. M., 1960. (Trans.)

176. Chichinadze, V.K. "O nekotorykh voprosakh postroyenia samonastraivayushchikhsya i samoobuchayushchikhsya sistem avtomaticheskovo upravlenia osnovannykh na printsipakh sluchainovo poiska" (About Several Problems on

the Construction of Self-adjusting and Self-teaching Systems of Automatic Control Based on the Principles of Arbitrary Search). *Trudy I Mezhdunarodnovo kongressa Mezhdunarodnoi federatsii po avtomaticheskomu upravleniyu* (Transactions of the First International Congress of the International Federation on Automatic Control), Vol. 2. M., 1961.

177.   Shalyutin, S.M. "Algoritmy i vozmozhnosti kibernetiki" (Algorithms and the Possibilities of Cybernetics). *Voprosy filosofii* (Problems of Philosophy), VI, 1962.

178.   Shtoff, V.A. *Rol modelei v poznanii* (The Role of Models in Cognition). L., 1963.

179.   Ashby, W.R. *Vvedeniye v kibernetiku* (Introduction to Cybernetics). M., 1959. (Trans.)

180.   Ashby, W.R. *Konstruktsia mozga* (Design for a Brain). M., 1964. (Trans.)

181.   Yaglom, A.M. and Yaglom, I.M. *Veroyatnost i informatsia* (Probability and Information). M., 1960.

182.   Yaglom, I.M. "Kibernetika i teoria informatsii" (Cybernetics and the Theory of Information). *O nekotorykh voprosakh sovremennoi matematiki i kibernetiki* (Certain Problems of Modern Mathematics and Cybernetics). M., 1965.

183.   Yablonsky, S.V. "Osnovnye ponyatia kibernetiki" (Basic Concepts of Cybernetics). *Problemy kibernetiki* (Problems of Cybernetics), No. 2. M., 1959.

184.   Yanovskaya, S.A. "Ischisleniye" (Calculus). *Filosofskaya enstiklopedia* (The Philosophy Encyclopedia), Vol. II. M., 1962.

185.* Yanovskaya, S.A. "O filosofskikh voprosakh matematicheskoi logiki" (On the Philosophical Problems of Mathematical Logic). *Problemy logiki* (Problems of Logic). M., 1963.

186.   Church, A. An unsolvable problem of elementary number theory. *American Journal of Mathematics*, Vol. 58, 1936.

187.   Gelernter, H.L. Realization of a geometry theorem proving machine. *Proc. Intern. Conf. on Information Processing*.

UNESCO, Paris, 1960.

188.   Gelernter, H.L. and Rochester, N. Intelligent behavior in problem-solving machines. *IBM Journal*, October, 1958.

189.   Gödel, K. Über formal unentscheidbare Sätze der Principia Mathematica und verwandter Systeme, I. *Monatshefte fur Mathematik und Physik*, 1931, Bd. 38.

190.   Kletsky, E.J. An application of the information theory approach to failure diagnosis. *IRE Trans.* 1960, Dec., PRQC—9, No. 3, 1961, No. 9.

191.   Minsky, M.L. Some methods of artificial intelligence and heuristic programming. In: *Mechanization of Thought Processes*, Vol. I. London, 1959.

192.   Newell, A. and Shaw, J.C. Elements of a theory of human problem solving. *Psychological Review*, Vol. 65, No. 3, 1958.

193.   Newell, A., Shaw, J.C., and Simon, H.A. Report on a general problem-solving program. *Proc. Intern. Conf. on Information Processing.* UNESCO, Paris, 1960.

194.   Paycha, F. Medical diagnosis and cybernetics. In: *Mechanization of Thought Processes*, Vol. II. London, 1959.

195.   Post, E.L. Recursive unsolvability of a problem of Thue. *The Journal of Symbolic Logic*, Vol. 12, 1947.

196.   Stachowiak, H. Denken und Erkennen im Problemfeld der Kybernetik. Wien, Springer-Verlag, 1964.

197.   Steinbuch, K. Lernende Automaten. *Elektronische Rechenanlagen*, Bd. I, H. 4 und 5, 1959.

198.   Steinbuch, K. Automat und Mensch. Berlin (Göttingen), Heidelberg, 1961.

199.*   Thiele, H. *Klassische* und "moderne" Algorithmen-Begriffe. *Mathematische und physikalisch-technische Probleme der Kybernetik*, Berlin, 1963.

200.   Turing, A.M. On computable numbers with an application to the Entscheiduns-problem. *Proceedings of the London Mathematical Society*, Ser. 2, Vol. 42, 1936.

201.   Zemanek, H. Automaten und Denkprozesse. Braunschweig, 1962.

III. The Application of Cybernetics and Logic in Pedagogy.

202.* Aleksandrov, G.N. "Razumno sochetat novoye s traditsionnym" (An Intelligent Combination of the New with the Traditional). *Vestnik vysshei shkoly* (University Herald), IV, 1964.

203.* Aleksandrov, G.N. "O nekotorykh voprosakh programmirovannovo obuchenia" (On Certain Problems of Programmed Instruction). *Programmirovannoye obucheniye* (Programmed Instruction). Kuibyshev, 1965.

204.* Aleksandrov, N.V. "Problemy programmirovannovo obuchenia" (Problems of Programmed Instruction). *Sovetskaya pedagogika* (Soviet Pedagogy), VI, 1965.

205.* Alekseyev, M.N. *Logika i pedagogika* (Logic and Pedagogy). M., 1965.

206. Alekseyev, N.G. "Pravomeren li 'algoritmichesky' podkhod k analizu protsessov obuchenia" (Is the "Algorithmic" Approach to the Analysis of the Instruction Process Right?). *Sovetskaya pedagogika* (Soviet Pedagogy), III, 1963.

207.* Alekseyev, O.G. *Ispolzovaniye obuchayushchikh mashin pri obuchenii slepykh uchashchikhsya* (The Use of Teaching Machines for the Instruction of Blind Students). Sverdlovsk, 1964.

208. Artyomov, V.A. "Kibernetika, teoria kommunikatsii i shkola" (Cybernetics, the Theory of Communication, and School). *Sovetskaya pedagogika* (Soviet Pedagogy), I, 1962.

209.* Artyomov, V.A. "Ob obuchayushchikh mashinakh i programmirovannom obuchenii inostrannym yazykam" (On Teaching Machines and Programmed Instruction of Foreign Languages). *Inostrannye yazyki v shkole* (Foreign Languages in School), VI, 1962.

210.* Arkhangelsky, S.I. "Novaya tekhnika v uchebnom protsesse" (New Technology in the Educational Process). *Vestnik vysshei shkoly* (The University Herald), VIII, 1962.

211. Atutov, P.R. "O primenenii kibernetiki v pedagogike" (On the Application of Cybernetics in Pedagogy). *Sovetskaya pedagogika* (Soviet Pedagogy), IX, 1962.

212. Ball, G.A. "Kriterii tselesoobraznosti tekhnicheskovo uslozhnenia obuchayushchikh mashin" (Criteria for the Expediency of Making Teaching Machines Technically Complex). *Radyanska shkola* (Soviet School), XII, 1964. (Trans. into Ukrainian.)

213.* Dovgyallo, A.M. and Mashbits, Ye. I. "Teoretichesky analiz obuchayushchikh programm" (A Theoretical Analysis of Teaching Programs). *Novye issledovania v pedagogicheskikh naukakh* (New Experiments in the Pedagogical Sciences), No. IV. M., 1965.

214. Belopolskaya, A.R. "Opyt primenenia obuchayushchikh algoritmov" (An Experiment in Applying Algorithms of Instruction). *Vestnik vysshei shkoly* (The University Herald), VI, 1963.

215. Belopolskaya, A.R. and Krylova, V.A. "Snachala—algoritmy, potom mashina (Opyt programmirovannovo obuchenia inostrannym yazykam v Leningradskom gos. un-te im. A.A. Zhdanova)" (At First, Algorithms, then A Machine [An Experiment in the Programmed Instruction of Foreign Languages at the A.A. Zhdanov State University of Leningrad]). *Vestnik vysshei shkoly* (The University Herald), VIII, 1964.

216. Belopolskaya, A.R. and Landa, L.N. "Ob ispolzovanii kino v obuchenii tekhnike perevoda" (On the Use of Cinema in the Instruction of Translation Techniques). *Primeneniye tekhnicheskikh sredstv i programmirovannovo obuchenia v srednei i vysshei shkole* (The Application of Technological Means and of Programmed Instruction in Secondary Schools and Universities), Vol. II. M., 1963.

217.* Berg, A.I. *Kibernetika—nauka ob optimalnom upravlenii* (Cybernetics: The Science of Optimal Control). M.—L., 1964.

218.* Berg, A.I. "Kibernetika i obshchestvennye nauki" (Cybernetics and the Social Sciences). *Materialy sessii obshchevo* *sobrania ANSSR, 19-20 oktyabra 1962 g.* (Materials of the Session of the General Assembly of the Academy of Sciences of the USSR). M., 1963.

219.* Berg, A.I. "Vooruzhat kadry peredovoi naukoi" (O programmirovannom obuchenii) (Training Specialists by Progressive Science [On Programmed Instruction]). *Avtomatizatsia sevodnya i zavtra* (Automation Today and Tomorrow). M., 1963.

220.* Bespalko, V.P. "Chto takoye programmirovannoye obucheniye" (What Is Programmed Instruction). *Narodnoye obrazovaniye* (Popular Education), V, 1963.

221. Bespalko, V.P. *Problematika i materialy k issledovatelskoi rabote po programmirovannomu obucheniyu* (Problems and Materials for Research Work on Programmed Instruction). M., 1965.

222.* Birilko, Yu. I. and Saburova, G.G. "Realizatsia nekotorykh psikhologicheskikh printsipov v obuchayushchikh mashinakh v SShA" (The Realization of Certain Psychological Principles in Teaching Machines in the U.S.A.). *Voprosy psikhologiki* (Problems of Psychology), IV, 1962.

223. Vlasenkov, A.I. *Materialy k issledovaniyu po teme "ispolzovaniye algoritmov v obuchenii orfografii"* (Materials for Research on "Use of Algorithms for the Instruction of Orthography"). M., 1965.

224.* Vlasenkov, A.I. *Formirovaniye orfograficheskikh navykov s ispolzovaniyem algoritmov obuchenia* (The Formation of Orthographic Skills Through the Use of Algorithms of Instruction). Dissertation for the Degree of Candidate of Pedagogical Sciences. M., 1965.

225.* Volodin, N.V. "V zashchitu vyborochnoi sistemy vvoda otvetov" (In Defense of the Selective System for the Input of Answers). *Vestnik vysshei shkoly* (The University Herald), V, 1964.

226.* Volodin, N.V. *Programmirovannoye obucheniye inostrannym yazykam (metodicheskiye ukazania)* (Programmed Instruction of Foreign Languages [Methodological Directions]). M., 1965.

227.* *Voprosy teorii i praktiki optimalnovo upravlyayemovo (programmirovannovo) obuchenia* (Problems of the Theory and Practice of Optimal Controlled [Programmed] Instruction). Eds. M.G. Yaroshevsky and L.M. Freidman, Dushanbe, 1963.

228. Galperin, P. Ya. "Programmirovannoye obucheniye i zadachi korennovo usovershenstvovania metodov obuchenia" (Programmed Instruction and Problems of a Profound Mastery of Methods of Instruction). *Programmirovannoye obucheniye. Metodicheskiye ukazania* (Programmed Instruction. Methodological Instructions). M., 1964.

229.* Galperin, P. Ya. "O psikhologicheskikh osnovakh programmirovannovo obuchenia" (On the Psychological Foundations of Programmed Instruction). *Novye issledovania v pedagogicheskikh naukakh* (New Experiments in the Pedagogical Sciences), No. IV. M., 1965.

230.* Galperin, P. Ya. and Reshetova, Z.A. "Programmirovannoye obucheniye proizvodstvennym navykam" (Programmed Instruction of Industrial Skills). *Izvestia APNRSFSR* (News of the APSRSSR), No. 133, 1964.

231. Gastev, Yu. A. "O metodicheskikh voprosakh ratsionalizatsii obuchenia" (On the Methodological Problems of Making Instruction Efficient). *Kibernetika, myshleniye, zhizn* (Cybernetics, Thought, Life). M., 1964.

232.* Gluskin, V.M. and Zilberg, L.I. "Primeneniye kiberneticheskikh ustroistv v protsesse obuchenia" (The Application of Cybernetic Equipment in the Instruction Process). *Inostrannye yazyki v shkole* (Foreign Languages in School), VI, 1964.

233. Glushkov, V.M. "Kibernetika i pedagogika" (Cybernetics and Pedagogy). *Nauka i zhizn* (Science and Life), II, 1964.

234. Glushkov, V.M. "O nekotorykh perspektivakh razvitia i primenenia obuchayushchikh mashin" (On Certain Perspectives of the Development and Application of Teaching Machines). *Izvestia vysshikh uchebnykh zavedeny* (News of Institutions of Higher Learning), VI (4), 1963.

235.* Glushkov, V.M., Dovgyallo, M.A., Syomin, V.P., and
        Yushchenko, Ye. L. *K voprosu o programmirovannom
        obuchenii programmirovaniyu na E.V.M.* (More About the
        Problem of Programmed Instruction of Electronic Com-
        puter Programming). Kiev, 1963.
236.* Gnedenko, B.V. "O programmirovannom obuchenii" (On
        Programmed Instruction). *Morskoi sbornik* (Marine Collec-
        tion), IX, 1963.
237.* Gnedenko, B.V. "Simvol progressivnykh idei i metodov v
        pedagogike" (The Symbol of Progressive Ideas and
        Methods in Pedagogy). *Vestnik vysshei shkoly* (The Uni-
        versity Herald), V, 1965.
238.   Golodnyak, A.T. and Morozov, M.B. *Programmirovannoye
        obucheniye i obuchayushchiye mashiny* (Programmed In-
        struction and Teaching Machines). Kiev, 1964.
239.* Gokhlerner, M.M. and Eigner, G.V. "Opredeleniye gram-
        maticheskikh modelei optimalnovo ritma pri program-
        mirovannom obuchenii inostrannomu yazyku" (The Speci-
        fication of Grammatical Models of Optimal Rhythm
        During the Programmed Instruction of a Foreign Lan-
        guage). *Tezisy dokladov zonalnoi otchotno-nauchnoi kon-
        ferentsii po pedagogike i psikhologii* (The Theses of
        Lectures at the Zonal, Precise Scientific Congress on
        Pedagogy and Psychology). Kharkov, 1964. (In Ukrainian.)
240.   Granik, G.G. "Priyomy raboty po razlicheniyu chastits 'ne'
        i 'ni'" (Methods for Work on the Distinction of the
        Particles "Not" and "No"). *Russky yazyk v shkole* (The
        Russian Language in School), I, 1964.
241.   Granik, G.G. *Formirovaniye u shkolnikov priyomov
        umstvennoi raboty v protsesse vyrabotki orfograficheskovo
        navyka* (The Formation in Students of Methods of Mental
        Work During the Elaboration of Spelling Skills). Reference
        to a Dissertation for the Degree of Candidate of Pedagogi-
        cal Sciences (in Psychology). M., 1965.
242.   Greben, I.I. and Dovgyallo, A.M. *Avtomaticheskiye ustro-
        istva dlya obuchenia (obuchayushchiye mashiny)* (Auto-

matic Equipment for Instruction [Teaching Machines]).
Kiev, 1965.

243.* Dovgyallo, A.M. *Klassifikatsia i printsipy postroyenia obuchayushchikh mashin* (The Classification and Principles of Construction for Teaching Machines). Kiev, 1963.

244.* Doroshkevich, A.M. "Pervye rezultaty raboty s programmirovannym uchebnikom" (The First Results of Work with a Programmed Textbook). *Vestnik vysshei shkoly* (The University Herald), VIII, 1963.

245.* Doroshkevich, A.M. *Sredstva programmirovannovo obuchenia (Metodicheskiye posobiye dlya prepodavatelei tekhnikumov)* (Means of Programmed Instruction [Methodological Appliances for Teachers in Technical Schools]). M., 1965.

246.* Zaichik, M. Yu. "Programmirovannaya laboratornaya rabota s samokontrolem deistvia uchashchevosya" (Programmed Laboratory Work with Auto-control of a Student's Action) *Izvestia APNRSFSR* (News of the APSRSSR), No. 128, 1963.

247.* Zaliznyak, A.A. "Opyt obuchenia anglo-russkomu perevodu s pomoshchyu algoritma" (An Experiment in Teaching English-Russian Translations with the Help of an Algorithm). *Pytannya prykladnoi lyngvystyky. Tezi dopovidei mizhvuzivskoi naukovoi konferentsy* (Problems of Practical Linguistics. Topics of the Lectures at the Inter-University Scientific Congress). Chernivtsi, 1960.

248.* Zinovyev, S.I. "O nekotorykh pedagogicheskikh problemakh programmirovannovo obuchenia" (On Certain Pedagogical Problems of Programmed Instruction). *Vestnik vysshei shkoly* (The University Herald), XII, 1963.

249. Ivanov, A.A. *Primeneniye obuchayushchikh mashin* (The Application of Teaching Machines). Kiev, 1964.

250.* Ilyina, T.A. "O pedagogicheskikh osnovakh programmirovannovo obuchenia" (On the Pedagogical Foundations of Programmed Instruction). *Sovetskaya pedagogika* (Soviet Pedagogy), VIII, 1963.

251.* Ilyina, T.A. "O teorii i praktike programmirovannovo

obuchenia" (On the Theory and Practice of Programmed Instruction). *Sovetskaya pedagogika* (Soviet Pedagogy), I, 1964.

252.* Ilyina, T.A. and Ogorodnikov, I.T. "Organizatsia eksperimentalnoi proverki metodiki programmirovannovo obuchenia" (The Organization of Experimental Verification of Methods of Programmed Instruction). *Sovetskaya pedagogika* (Soviet Pedagogy), II, 1965.

253.* Isayev, L.N. "Pedagogicheskaya effektivnost programmirovannovo obuchenia (Iz opyta eksperimentalnoi raboty)" (The Pedagogical Effectiveness of Programmed Instruction [From the Results of Experimental Work]). *Sovetskaya pedagogika* (Soviet Pedagogy), XI, 1963.

254.* Isayev, L.N. "Opyt programmirovannovo obuchenia russkomu yazyku" (An Experiment in the Programmed Instruction of the Russian Language). *Novye issledovania v pedagogicheskikh naukakh* (New Experiments in the Pedagogical Sciences), No. IV. M., 1965.

255.* *Ispolzovaniye tekhnicheskikh sredstv v uchebnom protsesse. Materialy pedagogicheskikh issledovany* (The Use of Technical Means in the Educational Process. Materials from Pedagogical Research). Ed. N.M. Shakhmayev. M., 1963.

256. Itelson, L.B. "Ob ispolzovanii matematicheskikh i kiberneticheskikh metodov v pedagogicheskikh issledovaniakh" (On the Use of Mathematical and Cybernetic Methods in Pedagogical Research). *Sovetskaya pedagogika* (Soviet Pedagogy), IV, 1962.

257. Itelson, L.B. "O nekotorykh problemakh teorii programmirovannovo obuchenia" (On Several Problems of the Theory of Programmed Instruction). *Sovetskaya pedagogika* (Soviet Pedagogy), IX, 1963.

258. Itelson, L.B. *Matematicheskiye i kiberneticheskiye metody v pedagogike* (Mathematical and Cybernetic Methods in Pedagogy). M., 1964.

259. Itelson, L.B. *Matematicheskiye metody v pedagogike i pedagogicheskoi psikhologii* (Mathematical Methods in

Pedagogy and Pedagogical Psychology). Dissertation for the Degree of Doctor of Pedagogical Sciences (in Psychology). M., 1965.

260.* Itelson, L.B. and Kreimer, A. Ya. "O sravnitelnoi effektivnosti razlichnykh struktur izlozhenia uchebnovo materiala" (On the Comparative Effectiveness of Different Structures for the Presentation of Educational Material). *Sovetskaya pedagogika* (Soviet Pedagogy), IV, 1965.

261.* Karimov, Kh. K. and Yaroshevsky, M.G. "Nekotorye psikhologicheskiye voprosy programmirovania uchebnovo materiala dlya obuchenia vtoromu yazyku" (Certain Psychological Problems of Programming Educational Material for the Instruction of a Second Language). *Voprosy teorii i praktiki optimalno upravlyayemovo (programmirovannovo) obuchenia* (Problems of the Theory and Practice of Optimally Controlled [Programmed] Instruction). Eds. M.G. Yaroshevsky and L.M. Freidman. Dushanbe, 1963.

262. Kerimov, D.A. "O kibernetiko-programmirovannom obuchenii v pravovedenii" (On Cybernetic-Programmed Instruction in Jurisprudence). *Vestnik Leningradskovo universiteta* (The Leningrad University Herald), XXIII (4), 1962.

263.* Keilman, E.I. "Opyt izlozhenia temy 'Logarifmy' iz kursa algebry srednei shkoly dlya programmirovannovo obuchenia" (An Experiment in Stating the Topic "Logarithms" from a Course in Secondary School Algebra for Programmed Instruction). *Voprosy teorii i praktiki optimalno upravlyayemovo (programmirovannovo) obuchenia* (Problems of the Theory and Practice of Optimally Controlled [Programmed] Instruction). Eds. M.G. Yaroshevsky and L.M. Freidman. Dushanbe, 1963.

264.* Klimov, Ye. *Chetyre zadachi programmirovannovo obuchenia* (Four Tasks of Programmed Instruction). Kazan, 1965.

265.* Kostyuk, G.S. "O psikhologicheskikh osnovakh programmirovania obuchenia" (On the Psychological Bases of Programming Instruction). *Radyanska shkola* (The Soviet

School), V, 1964. (In Ukrainian.)

266.* Kostyuk, G.S., Menchinskaya, N.A., and Smirnov, A.A. "Aktualnye zadachi shkoly i problemy psikhologii obuchenia" (Present-day Tasks of School and the Problems of the Psychology of Instruction). *Voprosy psikhologii* (Problems of Psychology), V, 1963.

267.* Kulik, V.T. "Gruppovye zanyatia s primeneniyem programmirovannykh tekstov" (Group Studies with the Application of Programmed Texts). *Vestnik vysshei shkoly* (The University Herald), VII, 1963.

268.* Kushelyov, Yu. N. "Mashina prikhodit na pomoshch prepodavatelyu" (Machines Come to the Aid of the Teacher). *Vestnik vysshei shkoly* (The University Herald), I, 1963.

269. Kushelyov, Yu. N., Landa, L.N., Uskov, V.G., and Shenshev, L.V. "Obuchayushchaya mashina s issledovatelskim i funktsiami" (A Teaching Machine with Research Functions). *Primeneniye tekhnicheskikh sredstv i programmirovannovo obuchenia v srednei, spetsialnoi, i vysshei shkole* (The Application of Technical Means and Programmed Instruction in Secondary and Technical Schools and Universities). Ed. V.M. Taranov. M., 1965.

270.* Ladanov, I.D. Programmirovannoye obucheniye i bikheviorizm" (Programmed Instruction and Behaviorism). *Sovetskaya pedagogika* (Soviet Pedagogy), VII, 1964.

271. Landa, L.N. "Obucheniye uchashchikhsya obshchim metodam myshlenia i problema algoritmov" (Teaching Students General Methods of Reasoning and the Problem of Algorithms). *Voprosy psikhologii* (Problems of Psychology), I, 1961.

272. Landa, L.N. "Opyt primenenia matematicheskoi logiki i teorii informatsii k nekotorym problemam obuchenia" (An Experiment in the Application of Mathematical Logic and the Theory of Information to Certain Problems of Instruction). *Voprosy psikhologii* (Problems of Psychology), II, 1962.

273. Landa, L.N. "Kibernetika i pedagogika" (Cybernetics and

Pedagogy). *Nauka i zhizn* (Science and Life), III, 1962.

274. Landa, L.N. "O kiberneticheskom podkhode k teorii obuchenia" (On the Cybernetic Approach to the Theory of Instruction). *Voprosy filosofii* (Problems of Philosophy), IX, 1962.

275.* Landa, L.N. "K voprosu o matematicheskikh metodakh postroyenia i otsenki algoritmov raspoznavania" (On the Problem of Mathematical Methods for the Construction and Evaluation of Algorithms of Identification). Presentations 1 and 2. *Izvestia APNRSFSR* (News of the APSRSFSR), No. 129, 1963.

276. Landa, L.N. "Algoritmichesky podkhod k analizu protsessov obuchenia pravomeren" (The Algorithmic Approach to an Analysis of Instruction Procedures Is Justifiable). *Voprosy psikhologii* (Problems of Psychology), IV, 1963.

277. Landa, L.N. "Logicheskoye modelirovaniye myslitelnykh protsessov kak metod ikh issledovania" (Logical Simulation of Thought Processes as a Method to Examine Them). *Tezisy dokladov na II syezde Obshchestva psikhologov* (Topics of Lectures at the Second Session of the Society of Psychologists), No. 5. M., 1963.

278. Landa, L.N. *Algoritmy i programmirovannoye obucheniye. Nekotorye voprosy teorii i metodiki programmirovania* (Algorithms and Programmed Instruction. Several Problems of the Theory and Methodology of Programming). M., 1965.

279. Landa, L.N. and Khlebnikov, S.P. "Uchebnoye ustroistvo 'Repetitor I'" (The Educational Apparatus "Tutor I"). *Ispolzovaniye tekhnicheskikh sredstv v uchebnom protsesse* (The Use of Technical Means in the Educational Process). *Izvestia APNRSFSR* (News of the APSRSSR), No. 128, 1963.

280.* Landa, L.N., Orlova, A.M., and Granik, G.G. "Nekotorye printsipy programmirovannovo obuchenia russkomy yazyku" (Several Principles of the Programmed Instruction of the Russian Language). *Russky yazyk v shkole* (The Russian Language in School), III, IV, V, 1965.

281.* Lebedev, P.D. "Eta problema nuzhdayetsya v glubokom izuchenii" (This Problem Needs Profound Study). *Vestnik vysshei shkoly* (The University Herald), III, 1963.

282.* Leontyev, A.N. and Galperin, P. Ya. "Teoria usvoyenia znany i programmirovannoye obucheniye" (The Theory of Knowledge Assimilation and Programmed Instruction). *Sovetskaya pedagogika* (Soviet Pedagogy), X, 1964.

283.* Leontyev, L.N. and Galperin, P. Ya. "Psikhologicheskiye problemy programmirovannovo obuchenia" (Psychological Problems of Programmed Instruction). *Izvestia APNRSFSR* (News of the APSRSSR), No. 138, 1965.

284.* Malirzh, F. (Editor in Chief), Tsikha, V., Jelinek, S., and Purm, R. *Metodika prepodavania russkovo yazyka v 6-9 klassakh chekhoslovatskoi osnovnoi devyatiletnei shkole* (Methodology for the Instruction of Russian in Grades 6-9 in Czechoslovak Elementary 9 Year Schools). Prague, 1965.

285.* Maslova, G.G. *O programmirovannom obuchenii matematike* (The Programmed Instruction of Mathematics). M., 1964.

286.* Makarova, G.I. "V chom sushchnost novovo metoda?" (Of What Does the New Method Consist?). *Vestnik vysshei shkoly* (The University Herald), XII, 1964.

287.* Makarova, G.I. "O vozmozhnostyakh perekhoda na programmirovannoye obucheniye na kafedrakh russkovo yazyka, rabotayushchikh so studentami-inostrantsami" (On the Possibilities of Switching to Programmed Instruction in Departments of Russian for Foreign Students). *Programmirovannoye obucheniye v tekhnicheskom vuze* (Programmed Instruction in Technical Universities). Kiev, 1965.

288.* Mashbits, Ye. I. "Issledovaniye sravnitelnoi effektivnosti konstruktivnykh i vyborochnykh otvetov" (Research on the Comparative Effectiveness of Constructive and Selective Answers). *Radyanska shkola* (The Soviet School), V, 1965. (In Ukrainian.)

289.* Mashbits, Ye. I. and Bondarovskaya, V.M. "Osnovnye

napravlenia programmirovannovo obuchenia za rubezhom"
(Basic Directions in Programmed Instruction Abroad).
Radyanska shkola (The Soviet School), IX, 1963. (In
Ukrainian.)

290.* Matyushkin, A.M. "Sopostavleniye lineinoi i razvetvlennoi
sistem programmirovannovo obuchenia" (The Comparison
of Linear and Ramified Systems of Programmed Instruction). *Izvestia APNRSFSR* (News of the APSRSSR), No.
133, 1964.

291.* Matyushkin, A.M. "Ob odnom effekte obuchenia pri
nekotorykh tipakh razvetvleny v programmirovannykh
uchebnykh materialakh" (On the Effect of Instruction
with Several Types of Ramifications in Programmed
Educational Materials). *Izvestia APNRSFSR* (News of the
APSRSSR), No. 138, 1965.

292. Netushil, A.V., Kushelyov, Yu. N., Uskov, V.G., Bud-
yonny, I.P., and Sviridov, A.P. "Avtomaticheskiye ustro-
istva dlya kontrolya tekushchei uspevayemosti studentov"
(Automatic Equipment for the Control of Students'
Continued Success). *Izvestia vysshikh uchebnykh zavedeny* (News of Institutes of Higher Learning). *Radiotekhnika* (Radiotechnology), Vol. VI (4), 1963.

293. *Obuchayushchiye mashiny Sverdlovskovo pedagogichesk-
ovo instituta i ikh primeneniye* (Teaching Machines of the
Sverdlovsk Pedagogical Institute and Their Application).
Ed. D.I. Penner. Sverdlovsk, 1965.

294.* Ovchinnikov, A.A. and Pushinsky, V.S. "Primeneniye
metodov logicheskikh diagramm v planirovanii i organizatsii uchebnovo protsessa" (The Application of Methods of
Logical Diagrams in the Planning and Organization of the
Educational Process). *Izvestia ANNSSSR* (News of the
Academy of Sciences of the USSR), *Tekhnicheskaya
kibernetika* (Technical Cybernetics), III, 1964.

295. Ozhogin, V. Ya. and Denisov, A. Ye. *Obuchayushchiye
mashiny. Metodicheskiye osnovy postroyenia logicheskikh
skhem. Metod. posobiye dlya obshchetekhn. fak-tov.*
(Teaching Machines. Methodological Foundations for the

Construction of Logical Diagrams. Methodological Means for Technological Faculties). Kiev, 1964.

296.* Orlova, A.M. "Opyt sostavlenia programmirovannykh materialov po kursu russkovo yazyka" (An Experiment in Compiling Programmed Materials for a Course in the Russian Language). *Novye issledovania v pedagogicheskikh naukakh* (New Research in the Pedagogical Sciences), No. IV. M., 1965.

297.* Penner, D.I., Komsky, D.M., and Kolpakov, M.F. "Programmirovannoye obucheniye v uchebnykh zavedeniakh Sverdlovskoi oblasti" (Programmed Instruction in Educational Institutions of the Sverdlovsk Region). *Sovetskaya pedagogika* (Soviet Pedagogy), I, 1965.

298.* Platonov, K.K. "Psikhologicheskiye voprosy teorii trenazherov" (The Psychological Problems of the Theory of Trainers). *Voprosy psikhologii* (Problems of Psychology), IV, 1961.

299.* "Primeneniye tekhnicheskikh sredstv i programmirovannovo obuchenia v srednei i vysshei shkole" (The Application of Technical Means and Programmed Instruction in Secondary School and the University). *Materialy 1-y Vsesoyuznoi, 2-y i 3-y Vserossyskikh konferentsy* (Materials for the First All Union and the Second and Third All Russian Conferences), Vols. 1-2. M., 1963.

300.* *Primeneniye tekhnicheskikh sredstv i programmirovannovo obuchenia v srednoi, spetsialnoi i vysshei shkole* (The Application of Technical Means and Programmed Instruction in Secondary and Technical Schools and Universities). Ed. V.M. Taranov. M., 1965.

301.* *Programmirovannoye obucheniye i kiberneticheskiye obuchayushchiye mashiny* (Programmed Instruction and Cybernetic Teaching Machines). Ed. A.I. Shestakov. M., 1963.

302.* "Programmirovannoye obucheniye v shkole" (Programmed Instruction in School). *Uchonye zapiski Moskovskovo gosudarstvennovo pedagogicheskovo instituta im. V.I. Lenina* (Learned Transactions of the Moscow State V.I.

Lenin Pedagogical Institute), No. 228, 1964.

303.* *Programmirovannoye obucheniye v tekhnicheskom vyze* (Programmed Instruction in the Technological Institute). Ed. Yu. F. Chubuk. Kiev, 1965.

304.* *Programmirovannoye obucheniye* (Programmed Instruction). Ed. G.N. Aleksandrov. Kuibyshev, 1965.

305.* Protasova, G.N. and Shenshev, L.V. "O razlichnom ponimanii programmirovannovo obuchenia inostrannym yazykam" (On a Different Concept of Programmed Teaching of Foreign Languages). *Izvestia APNRSFSR* (News of the APSRSSR), No. 138, 1965.

306. Prokofyev, A.V. *Programmirovannoye obucheniye. Programmirovanniye uchebniki. Mashiny dlya obuchenia* (Programmed Instruction. Programmed Textbooks. Machines for Instruction). M., 1965.

307.* Rakityanskaya, Z.I. "Obucheniye leksike na osnove programmirovannovo posobia" (Vocabulary Instruction on the Basis of Programmed Appliances). *Inostrannye yazyki v shkole* (Foreign Languages in School), IV, 1964.

308.* Regelson, L.M. "Matrichny kontrol usvoyenia kursa" (Matrix Control of Course Mastery). *Vestnik vysshei shkoly* (The University Herald). I, 1964.

309. Redko, V.N. and Yushchenko, Ye. L. *K voprosu klassifikatsii i minimizatsii logicheskikh graf-skhem obuchenia* (More on the Problem of Classification and Minimization of Logical Graph-Diagrams of Instruction). Kiev, 1965.

310.* Reshetova, Z.A. "Upravleniye protsessom formirovania proizvodstvennykh umeny i yevo programmirovaniye" (Control of the Process of Formation of Industrial Skills and Its Programming). *Programmirovannoye obucheniye. Metodicheskiye ukazania* (Programmed Instruction. Methodological Directions). M., 1964.

311. Reshetova, Z.A. "Programmirovannoye obucheniye proizvodstvennym navykam" (Programmed Instruction of Industrial Skills). *Izvestia APNRSFSR* (News of the APSRSSR), No. 138, 1965.

312. Reshetova, Z.A. and Kaloshina, I.P. "Programmirovannoye

obucheniye proizvodstvennym navykam" (Programmed Instruction of Industrial Skills). *Novye issledovania v pedagogicheskikh naukakh* (New Research in the Pedagogical Sciences), No. IV. M., 1965.

313.* Rogova, G.V. "Organizatsia priyoma informatsii" (The Organization of Information Reception). *Inostrannye yazyki v shkole* (Foreign Languages in School), IV, 1964.

314. Rozenberg, M.I. "Ispolzovat dostizhenia kibernetiki v nauchno-pedagogicheskikh issledovaniakh i shkolnoi praktike" (Using the Achievements of Cybernetics in Scientific-Pedagogical Research and in School Practice). *Sovetskaya pedagogika* (Soviet Pedagogy), IV, 1962.

315.* Rozenberg, N.M. "Obucheniye algoritmam umstvennykh i prakticheskikh deistvy" (Teaching Algorithms for Mental and Practical Activities). *Sovetskaya pedagogika* (Soviet Pedagogy), VIII, 1965.

316.* Rostunov, T.I. "Sushchnost programmirovannovo metoda obuchenia" (The Essentials of the Programmed Method of Instruction). *Programmirovannoye obucheniye i kiberneticheskiye obuchayushchiye mashiny* (Programmed Instruction and Cybernetic Teaching Machines). M., 1963.

317. Rostunov, T.I. "Obuchayushchy kompleks" (The Teaching Complex). *Avtomatizatsia proizvodstva i promyshlennaya elektronika* (Automization of Production and Industrial Electronics), Vol. 2. M., 1963.

318. Rostunov, T.I. and Sokolinsky, I. Ya. "Klassifikatsia, trebovania i printsipy postroyenia prosteishikh obuchayushchikh mashin" (The Classification, Requirements and Principles of the Construction of the Most Simple Teaching Machines). *Izvestia vysshikh uchebnykh zavedeny* (News of Institutions of Higher Education). *Radiotekhnika* (Radio Technology), Vol. VI. (4), 1963.

319.* Ryakhovsky, G.D. "Metod sostavlenia programmirovannykh materialov" (Method for Compiling Programmed Materials). *Radyanska shkola* (The Soviet School), II, 1964. (In Ukrainian.)

320.* Samarin, Yu. A. and Esaulov, A.F. "Psikhologichesky

aspekt programmirovannovo obuchenia" (The Psychological Aspect of Programmed Instruction). *Materialy po programmirovannomu obucheniyu* (Materials on Programmed Instruction). L., 1964.

321.* Sankovsky, Ye. A. "Sokrashchaya vremya i kontrol na konsultatsii" (Reducing Time and Control at a Consultation). *Vestnik vysshei shkoly* (The University Herald), I, 1963.

322.* Smirnov, A.A. "Metodika programmirovannovo obuchenia" (The Methodology of Programmed Instruction). *Sredneye spetsialnoye obrazovaniye* (Special Secondary Education), IX, 1964.

323.* Solovyova, Ye. Ye. "Opyt primenenia mashinnoi tekhniki v pedagogicheskom issledovanii" (An Experiment in Applying Machine Technology in Pedagogical Research). *Sovetskaya pedagogika* (Soviet Pedagogy), VIII, 1962.

324. Solovyova, Ye. Ye. "O kiberneticheskom podkhode k issledovaniyu problem ispolzovania tekhnicheskikh sredstv obuchenia" (On the Cybernetic Approach to Research on Problems of Using Technical Means of Instruction). *Izvestia APNRSFSR* (News of the APSRSSR), No. 128, 1963.

325.* Sidelkovsky, A.P. "Algoritmichesky podkhod k analizu protsessov obuchenia pravomeren" (The Algorithmic Approach to the Analysis of Teaching Processes Is Lawful). *Voprosy psikhologii* (Problems of Psychology), V, 1964.

326.* Stolyar, A.A. *Logicheskiye problemy prepodavania matematiki* (The Logical Problems of Teaching Mathematics). Minsk, 1965.

327.* Talyzina, N.F. "Aktualnye problemy programmirovannovo obuchenia" (Current Problems in Programmed Instruction). *Radyanska shkola* (The Soviet School), IX, 1963. (In Ukrainian.)

328. Talyzina, N.F. "Programmirovaniye distsiplin matematicheskovo tsikla" (Programming Disciplines from the Mathematical Cycle). *Programmirovannoye obucheniye. Metodicheskiye ukazania* (Programmed Instruction.

Methodological Instructions). M., 1964.

329.* Taranov, V.M. "Mozhet li mashina uchit?" (Can a Machine Teach?). *Mozhet li mashina uchit?* Ed. V.M. Taranov. Gorky, 1963.

330.* Tersky, L.N. "Opyt programmirovannovo obuchenia" (An Experiment in Programmed Instruction). *Sovetskaya pedagogika* (Soviet Pedagogy), X, 1964.

331. Tikhonov, I.I. "Kak klassifirovat tekhnicheskiye sredstva pri programmirovannom obuchenii" (How to Classify Technical Means When Programmed Instruction is Used). *Vestnik vysshei shkoly* (The University Herald), X, 1964.

332.* Tikhonov, I.I. "Opyt organizatsii eksperimentalnykh programmirovannykh zanyaty" (An Experiment in the Organization of Experimental Programmed Studies). *Sovetskaya pedagogika* (Soviet Pedagogy), VI, 1965.

333.* Trubetskoi, M.N. "Opyt ispolzovania programmirovannykh kart v obuchenii" (An Experiment in the Use of Programmed Cards in Teaching). *Sovetskaya pedagogika* (Soviet Pedagogy), IX, 1965.

334.* Trubetskoi, M.N. and Kolmogortsev, G.G. *Nekotorye prosteishiye sredstva programmirovannovo obuchenia. Posobiye dlya uchitelei* (Several Very Simple Means of Programmed Instruction. Appliances for Teachers). Krasnoyarsk, 1964.

335.* Tumanova, Ye. I. "Diagnostika pitania rasteny po ikh vneshnemu vidu" (Diagnosis of the Feeding of Plants According to Their External Appearance). *Shkola i proizvodstvo* (School and Production), I, 1965.

336. Farber, V. G. "O logicheskikh sredstvakh shkolnoi grammatiki" (On the Logical Means of School Grammar). *Logiko-grammaticheskiye ocherki* (Logical-grammatical Essays). M., 1961.

337. Farber, V.G. "K razrabotke uproshchonnoi punktuatsii" (Toward an Elaboration of a Simplified Punctuation). *Russky yazyk v shkole* (The Russian Language in School), IV, 1965.

338. Farber, V.G. "Ob odnom podkhode k sovershenstvovaniyu

shkolnoi grammatike" (On One Approach to the Improvement of School Grammar). *Russky yazyk v natsionalnoi shkole* (The Russian Language in the National School), VI, 1963.

339.* Fyodorov, P.A. "Raspoznavaniye osnovnykh mineralnykh udobreny" (The Identification of Basic Mineral Fertilizers). *Shkola i proizvodstvo* (School and Production), I, 1965.

340.* Freidson, I.R. and Gaziyev, A.A. "Primeneniye programmirovannovo metoda obuchenia prigotovke spetsialistov flota" (The Application of a Programmed Method of Instruction During the Preparation of Specialists for the Fleet). *Morskoi sbornik* (The Marine Collection), XII, 1963.

341. Freidman, L.M. "Uchebnye algoritmy raspoznavania" (Educational Algorithms of Identification). *Izvestia APN-RSFSR* (News of the APSRSSR), No. 129, 1963.

342. Freidman, L.M. "Logiko-matematicheskaya model raspoznavania v uchebnoi deyatelnosti" (A Logical-mathematical Model for Identification in Educational Activity). *Voprosy teorii i praktiki optimalno upravlyayemovo (programmirovannovo) obuchenia* (Problems of the Theory and Practice of Optimally Controlled [Programmed] Instruction). Eds. M.G. Yaroshevsky and L.M. Freidman. Dushanbe, 1963.

343. Kharkovsky, Z.S. "O roli obuchayushchikh mashin v uchebnom protsesse pri programmirovannom obuchenii" (On the Role of Teaching Machines in the Educational Process During Programmed Instruction). *Sovetskaya pedagogika* (Soviet Pedagogy), III, 1965.

344. Chentsov, A.A. "Sposoby otyskania ratsionalnykh algoritmov dlya vypolnenia prakticheskikh rabot" (Methods for Finding Efficient Algorithms for the Execution of Practical Work). *Sovetskaya pedagogika* (Soviet Pedagogy), III, 1965.

345.* Chilikin, M.G. "Osnovnye zadachi programmirovannovo obuchenia" (The Basic Tasks of Programmed Instruction).

Programmirovannoye obucheniye i kiberneticheskiye obuchayushchiye mashiny (Programmed Instruction and Cybernetic Teaching Machines). M., 1963.

346.* Chubuk, Yu. F. "Izuchat pedagogicheskuyu i psikhologicheskuyu storonu novovo metoda" (Studying the Pedagogical and Psychological Sides of the New Method). *Vestnik vysshei shkoly* (The University Herald), III, 1964.

347.* Shapovalenko, S.G. "O programmirovannom obuchenii khimii" (On Programmed Instruction of Chemistry). *Khimia v shkole* (Chemistry in School), V, 1963.

348.* Shapovalenko, S.G. *Teoreticheskiye problemy programmirovannovo obuchenia* (The Theoretical Problems of Programmed Instruction). M., 1965.

349. Shakhmayev, N.M. "Nekotorye problemy svyazennye s ispolzovaniyem novykh tekhnicheskikh sredstv v obuchenii" (Certain Problems Connected with the Use of New Technical Means in Instruction). *Izvestia APNRSFSR* (News of the APSRSSR), No. 128, 1963.

350. Shakhmayev, N.M. *Ispolzovaniye tekhnicheskikh sredstv v prepodavanii fiziki* (The Use of Technical Means in the Teaching of Physics). M., 1964.

351. Shenshev, L.V. "Ob ispolzovanii magnitografa v kachestve obuchayushchei mashiny" (On the Use of a Taperecorder as a Teaching Machine). *Izvestia APNRSFSR* (News of the APSRSSR), No. 133, 1964.

352. Shenshev, L.V. *K voprosu o roli i meste obuchayushchikh mashin v sisteme programmirovannovo obuchenia (Materialy k Vserossyskoi konferentsii po programmirovannomu obucheniyu i primeneniyu tekhnicheskikh sredstv* (On the Problem of the Role and Place of Teaching Machines in the System of Programmed Instruction [Materials for the All-Russian Congress on Programmed Instruction and the Application of Technical Means]). M., 1965.

353. Shenshev, L.V. "O realizatsii printsipov programmirovannovo obuchenia v prepodavanii inostrannykh yazykov i o nekotorykh obuchayushchikh ustroistvakh" (On the Realization of Principles of Programmed Instruction in the

Teaching of Foreign Languages and on Certain Teaching Equipment). *Primeneniye tekhnicheskikh sredstv i programmirovannovo obuchenia v srednei shkole* (The Application of Technical Means and Programmed Instruction in Secondary School). Novosibirsk, 1965.

354. Shestakov, A.I. "Opyt primenenia obuchayushchikh mashin v SShA" (An Experiment in the Application of Teaching Machines in the USA). *Sovetskaya pedagogika* (Soviet Pedagogy), XII, 1962.

355. Erndiyev, P.M. "Kiberneticheskiye ponyatiye i problemy didaktiki" (Cybernetic Concepts and Problems of Didactics). *Sovetskaya pedagogika* (Soviet Pedagogy), XI, 1963.

356. Abel, H. Lehrmaschinen und programmierter Unterricht. *Berufspädagogische Zeitschrift*, 1964, Heft 6.

357. Annet, J., Kay, H., and Sime M. Teaching Machines. *Discovery*, May 1961.

358. Auswick, K. Teaching Machines and Programming. Oxford, Pergamon Press, 1964.

359.* Bakovljev, M. Sustina programirane nastave i pitanje potrebe i mogucnosti njenog proucavanja kod nas. *Pedagogija*, 1964. No. 3. Summary in Russian.

360.* Becher, K.E. Programme und Programmieren. *Berufspädagogische Zeitschrift*, 1964, Heft 6.

361.* Becker, J.L. A Programmed Guide to Writing Autoinstructional Programs. N.Y., 1963.

362.* Bera, M.A. Programmatique. *L'Education Nationale*, 1964, No. 37.

363.* Besset, J., Metais, C. L'automatisme dans l'enseignement. *Automatisme*, No. 4, April 1964.

364.* Biancheri, A. Qu'est-ce que l'enseignement programmé? *La Pedagogie Cybernetique*, Vol. 2, No. 2, July 1964.

365. Bitzer, D. and Braunfeld P. Computer Teaching Machine Project: PLATO on Illiac. *Computers and Automation*, v. XI, No. 2, 1962.

366. Blyth, J.W. La machine à enseigner et l'être humain. In: *Ou en est l'enseignement audio-visuel. Coll. Etudes et Documents d'Education*, No. 50, 1963, Paris, UNESCO.

367.* Bock, H. und Walsch W. Über die Erarbeitung von Unterrichts-algorithmen. *Berufsbildung*, 1963, No. 12.

368. Braunfeld, P.G. Problems and Prospects of Teaching With a Computer. *Journal of Educational Psychology*, 1964, No. 4.

369. Briggs, L.J. Teaching Machines for Training of Military Personnel in Maintenance of Electronic Equipment. In: E. Galanter, *Automatic Teaching: The State of the Art*, N.Y., 1959.

370.* Bunescu, V., Beldescu, G., Ionescu-Miciora, E., and Popescu, S. Instruirea programata si posibilitatile ei de aplicare in scoala noastra. *Revista de pedagogie*, 1964, No. 5. Summary in Russian.

371. Bushnell, D.D. Computers in the Classroom. *Data Processing*, Vol. 4, No. 4, 1962.

372. Bushnell, D.D. Computers in Education. *Computers and automation*, Vol. 12, No. 3, 1963.

373. Canac, H. L'educateur devant la machine. *L'Education Nationale*, 1965, No. 15-16.

374. Carpenter, F. and Hutchcroft, R. For More Effective Learning, Use Teaching Machines in Industrial Education. *School Shop*, Vol. 21, No. 5, 1962.

375. Claus, G. Zur Anwendung der Informationstheorie auf lernpsychologische Probleme. *Pädagogik*, 1965, No. 1.

376. Claus, G. Zur Handlungsanalyse durch Algorithmen und ihre Anwendung im Unterricht. *Pädagogik*, 1965, No. 4.

377. Claus, G. Denkpsychologie und Kybernetik. *Berufsbildung*, 1963, No. 5.

378. Cogniot, G. L'enseignant et la machine. *Europe*, Revue Mensuelle, No. 433-434, May-June 1965.

379.* Corell, W. Verhaltenspsychologische Grundlagen des programmierten Lernens. *Programmiertes Lernen und programmierter Unterricht*, 1964, No. 2.

380.* Corell, W. Pädagogische Verhaltenspsychologie. Ernst Reinhart Verlag, München/Basel, 1965.

381. Couffignal, L. La mécanisation de la pédagogie, l'enseignement programmé. *La Pedagogie Cybernetique*, Vol. 2, No.

2, July 1964.

382. Couffignal, L. La pédagogie cybernétique. *L'Education Nationale*, 1965, No. 15-16.

383. Couffignal, L. La pédagogie cybernétique. *Europe*, Revue Mensuelle, No. 433-434, May-June 1965.

384. Coulson, J.E. (Ed.) *Programmed Learning and Computer-based Instruction*. N.Y., Wiley, 1962.

385.* Cram, D. *Explaining Teaching Machines and Programming*. San Francisco, Fearon, 1961.

386.* Cros, L. Programmation et éducation. *L'Education Nationale*, 1965, No. 15-16.

387.* Crowder, N.A. On the Differences Between Linear and Intrinsic Programming. *Phi Delta Kappan*, Bloomington, Indiana, 1963, No. 6.

388.* Crowder, N.A. Automatic Tutoring By Intrinsic Programming. In: A.A. Lumsdaine and R. Glaser, *Teaching Machines and Programmed Learning: A Source Book*. Washington, NEA, 1960.

389. Cube, F. Zur Theorie des mechanischen Lernens. Hamburg, Schnelle, 1960.

390. Cube, F. Kybernetische Grundlagen des Lernens und Lehrens. Stuttgart, Ernst Klett, 1964.

391.* DeCecco, J.P. (Ed.) *Educational Technology: Readings in Programmed Instruction*. N.Y., Holt, Rinehart and Winston, 1964.

392.* Decote, G. Vers l'enseignement programmé. Paris, Gauthier-Villars, 1963.

393. Delchet, R. and Lefevre, L. Valeur pédagogique des machines à enseigner (seminaire de recherche du troisieme cycle). *L'Education Nationale*, No. 23, June 20, 1963.

394.* Desamais, R. Techniques d'auto-instruction. *Cooperation Pedagogique*, No. 2-3, April-September 1963.

395.* Descombes, A. Instruction programmée. *Hommes et Techniques*, No. 230, January 1964.

396.* Deterline, W.A. *An Introduction to Programmed Instruction*. N.Y., Prentice-Hall, 1962.

397. Dietz, A. Unterrichtsforschung und Kybernetik. *Pädagogik*, 1963, No. 1.

398. Dieuzeide, H. Les machines à apprendre. *L'Education Nationale*, No. 24, September 19, 1963.

399.* Ecke, P. Programmierter Unterricht in der Unterstufe. *Pädagogik*, 1965, No. 6.

400. Evans, L.H. and Arnstein, G. Automation and the Challenge to Education. Washington, National Educational Association, 1962.

401.* Evans, J.L. and Glaser, R. The Development and the Use of a *Standard* Program for Investigating Programmed Verbal Learning. *American Psychologist*, 1960, No. 15.

402.* Fekete, J. A programozott oktatas nehany kerdese. *Pedagogiai Szemie*, 1965, No. 2. Summary in Russian.

403.* Filep, R.T. *Prospectives in Programming*. N.Y., Macmillan, 1963.

404. Fine, B. *Teaching Machines*. N.Y., Sterling, 1962.

405. Finn, J.D. and Perrin, D.G. Teaching Machines and Programmed Learning, 1962: A Survey of the Industry. Washington, National Education Association, 1962.

406. Fleszner, J. O mozliwosciach i perspektywach myslenia cybernetycznego w pedagogice. *Kwartalnik pedagogiczny*, 1964, No. 1. Summary in Russian.

407.* Franck, R. La semi-programmation. *L'Education Nationale*, 1965, No. 15-16.

408. Frank, H. Kybernetische Grundlagen der Pädagogik. Eine Einführung in die Informationspsychologie. Baden-Baden, Agis, 1962.

409. Frank, H. (Hrsg.) Lehrmaschinen in kybernetischer und pädagogischer Sicht. Referate der ersten deutshen Lehrmaschinentagung. Verlage Klett und Oldenburg, Stuttgart und München, 1963.

410. Frank, H. (Hrsg.) Lehrmaschinen in kybernetischer und pädagogischer Sicht. Referate des zweiten Nürtinger Symposions über Lehrmaschinen. Ernst Klett Verlag Stuttgart, R. Oldenburg Verlag München, 1964.

411. Freinet, C. Machines enseignantes et programmation.

L'Educational Nationale, No. 34, November 28, 1963.

412. Freinet, C. Le travail à l'ecole Freinet selon la nouvelle technique des bandes enseignantes. *L'Educateur*, 1964, No. 4.

413.* Freinet, C. Des méthodes actives à la programmation. *L'Education Nationale*, 1965, No. 15-16.

414. Fry, E.B. *Teaching Machines and Programmed Instruction. An Introduction.* N.Y./London: McGraw-Hill, 1963.

415.* Frey, P. Das Problem der Fragen und Antworten im programmierten Text. *Berufsbildung*, 1965, No. 3.

416. Gagne, R.M. Training Devices and Simulators: Some Research. *American Psychologist*, Vol. 9, March 1954.

417.* Gal, R. Perspectives nouvelles sur l'enseignement programmé. *L'Education Nationale*, No. 28, October 15, 1964.

418. Galanter, E.H. (Ed.) *Automatic Teaching: The State of the Art.* N.Y., Wiley, 1959.

419. Galanter, E.H. Mechanization of Teaching. *Bulletin of the National Association of Secondary School Principals*, No. 44, April 1960.

420.* Gauthier, A. and Pauliat, P. Structuralisme et programmation. *Cahiers pedagogique*, 1964, No. 47.

421. Gentilhomme, I. Optimisation des algorithmes d'enseignement. *La Pedagogie Cybernetique*, 1964, No. 4.

422. Gentilhomme, I. Enseignement cybernétique du russe scientifique. *Europe*, Revue Mensuelle, No. 433-434, May-June 1965.

423. Goldsmith, M. (Ed.) *Mechanization in the Classroom. An Introduction to Teaching Machines and Programmed Learning.* London, Souvenir Press, 1963.

424. Goodman, R. *Programmed Learning and Teaching Machines: An Introduction.* London, English Univ. Press, 1962.

425. Goodman, R. Computer Controlled Teaching Machines. *Technology*, No. 6, September 1962.

426. Graff, K. Kybernetische Pädagogik. *Die Deutsche Berufs- und Fachschule*, 1962, Heft 9.

427.* Green, E.J. *The Learning Process and Programmed Instruction*. N.Y., Holt, Rinehart and Winston, 1962.

428.* Gourevitch, M. Essai d'enseignement programmé des mathématiques. *La Pedagogie Cybernétique*, Vol. 2, No. 2, May 1964.

429. Guillaumaud, J. Puissance et valeur de la pédagogie cybernétique. *Europe*, Revue Mensuelle, No. 433-434, May-June 1965.

430.* Hignette, M. L'enseignement programmé et l'éducation des adultes. *L'Education Nationale*, 1965, No. 15-16.

431.* Hinze, K. Erste Erfahrungen bei der Entwicklung moderner Lehrhilfsmittel. *Pädagogik*, 1963, No. 8.

432.* Hinze, K. Pädagogisch-psychologische Probleme des verzweigten programmierten Unterrichts. *Pädagogik*, 1964, No. 3.

433.* Holland, J.G. Evaluating Teaching Machines and Programs. *Teachers College Record*, Vol. 63, No. 1, October 1961.

434.* Holland, J.G. and Skinner, B.F. *The Analysis of Behavior: A Program for Self-Instruction*. N.Y., McGraw-Hill, 1961.

435.* Hoang Van Chi. La programmation et les mathématiques. *La Pedagogie Cybernétique*, Vol. 2, No. 2, May 1964.

436.* Hughes, J.L. *Programmed Learning: A Critical Evaluation*. Chicago, Educational Methods, 1963.

437. Iffland, E. Die Anwendung mathematischer Methoden in der Methodik nichtmathematischer Fächer. *Pädagogik*,, 1964, No. 5.

438.* Iffland, E. Neue Wege zur Steigerung der Effektivität des Unterrichts. *Pädagogik*, 1964, No. 3.

439. Kelbert, H. Über die Anwendung der Algorithmen von Ljapunow in der Berufspädagogik. In: *Mathematische und physikalisch-technische Probleme der Kybernetik*, Berlin, 1963.

440. Kelbert, H. Kybernetik und Berufspädagogik. In: *Kybernetik in Wissenshaft, Technik und Wirtschaft der DDR*. Berlin, 1963.

441. Kelbert, H. Programmierter Unterricht und Anwendung moderner Lehrhilfsmittel. *Pädagogik*, 1963, No. 4.

442.* Kelbert, H. Aufgaben und Probleme des programmierten Unterrichts. *Pädagogik*, 1964, No. 3.

443.* Kirchberger, A. L'enseignement programmé dans le monde. *L'Education Nationale*, 1965, No. 15-16.

444.* Kiss, A. Programozott tanitas a gyakorlatban. *Pedagogiai szemle*, 1965, No. 7-8. Summary in Russian.

445.* Klaus, D. The Art of Auto-instructional Programming. *Audio-Visual Communications Review*, Vol. 9, 1961.

446. Klix, F. Bemerkungen uber einige mathematisch-kybernetische Probleme in der psychologischen Forschung. In: *Mathematische und physikalisch-technische Probleme der Kybernetik* Berlin, 1963.

447. Klix, F. Über Zweck und Zielsetzung der kybernetischen Behandlung psychologischer Probleme und einige Aufgaben dieser psychologischen Forschungen in der Deutschen Demokratischen Republik. In: *Kybernetik in Wissenschaft, Technik und Wirtschaft der DDR*, Berlin, 1963.

448. Knezu, V. Prvni zkusenosti a programovani uciva. *Pedagogika*, 1963, No. 4. Summary in Russian.

449.* Knezu, V. K problematice sestavovani programy. *Pedagogika*, 1964, No. 5. Summary in Russian.

450. Köler, R. and Reiners, K.H. Kybernetik als Hilfe didaktischer Forschung. *Wissenschaftliche Zeitschrift TU Dresden*, 1963, No. 4.

451. Komoski, P.K. Teaching Machines. *Instructor*, No. 70, March 1961.

452.* Kulic, V. Experimentalni analyza procesu programovaneho uceni a nekterych jeho principu. *Pedagogika*, 1963, No. 6. Summary in Russian.

453.* Labin, E. Défense de l'instruction programmée. *L'Education Nationale*, No. 8, February 20, 1964.

454. Lange, W. Kybernetik und Verbesserung der Bildungs und Erziehungsarbeit. *Pädagogik*, 1963, No. 2.

455.* *Le courrier de la recherche pedagogique*. L'enseignement programmé. Numéro spécial, January 1965. Publication de l'Institut pédagogique nationale.

456.* Leith, G.O., Peel, E.A., and Curr, W. A Handbook of

Programmed Learning. Birmingham, 1964.

457.* Leplat, J. Formation professionnelle et aménagement du travail. *Bulletin de Psychologie*, March 20, 1962.

458. Leplat, J. L'enseignement automatisé. Caractéristiques générales et possibilités d' application à la formation professionnelle. *Bulletin du Centre D'Etudes et Recherches Psychotechniques*, No. 1, January-March 1963.

459. Lewis, B.N. and Pask, G. The Theory and Practice of Adaptive Teaching Systems. In: R. Glaser (Ed.) *Teaching Machines and Programmed Learning, Data and Directions.* National Education Association, 1964.

460. Lumsdaine, A.A. Teaching Machines and Programmed Instruction. In: *New Methods and Techniques in Education. Educational Studies and Documents*, No. 48, UNESCO, 1963.

461. Lumsdaine, A.A. and Glaser, R. *Teaching Machines and Programmed Learning: A Source Book.* Washington, Dept. of Audio-Visual Instruction, National Education Association, 1960.

462.* Lysaught, J.P. and Williams, C.M. *A Guide to Programmed Instruction.* New York/London, Wiley, 1963.

463.* Mager, R.F. *Preparing Objectives for Programmed Instruction.* San Francisco, Fearon, 1960.

464.* Mcloughlin, H.G. A Program for Programmed Instruction. *Technical Education*, June 1963.

465.* Margulies, S. and Eigen, L.D. (Eds.) *Applied Programmed Instruction.* N.Y., Wiley, 1962.

466.* Markle, S.M. *Good Frames and Bad: A Grammar of Frame Writing.* N.Y., Wiley, 1964.

467.* Markle, S.M., Eigen, L.D., and Komoski, P.K. *A Programmed Primer on Programming.* N.Y., Center for Programmed Instruction, 1961.

468.* Mazur, M. Nauczanie programowane. *Kwartalnik pedagogiczny*, 1964, No. 1. Summary in Russian.

469.* Melet, R. Pédagogie et programmation. *L'Education Nationale*, 1965, No. 15-16.

470. Metais, C. Données actuelles d'une pédagogie cyber-

nétique. *La Pedagogie Cybernetique*, No. 1, March 1963.

471.   Meyer, G. Automatiesierungstechnik im Unterricht. *Poly-technische Bildung und Erziehung*, 1963, No. 8-9.

472.*   Metev, V. Ponyatia i printsipy na kibernetikata i v'zmozh-nosti za teknnikovo prilozheniye v pedagogikata (Notions and Principles of Cybernetics and the Possibility of Its Technical Application in Pedagogy). *Narodna prosveta* (Popular Education), VII, 1964.

473.*   Milan, M. K otazkam programovania uciva. *Jednotna skola*, 1964, No. 1. Summary in Russian.

474.*   Milton, O. and West, L. P.I.: What It Is and How It Works. N.Y., Harcourt, Brace and World; London, Hart-Davis, 1963.

475.*   Novicicov, E. and Negoescu, V. Instruirea programata si masinile de instruire. *Revista de pedagogie*, 1964, No. 3. Summary in Russian.

476.*   Okon, W. Nauczanie *podajace* a nauczanie programowane. *Kwartalnik pedagogiczny*, 1963, No. 4. Summary in Russian.

477.   Oleron, P. Léleve, le programme et la machine. *L'Education Nationale*, 1965, No. 15-16.

478.   Pask, G. Teaching Machines. *New Scientist*, Vol. 10, No. 234, 1961.

479.   Pask, G. Interaction Between a Group of Subjects and Adaptive Automation to Produce a Self-Organizing System for Decision-Making. In: H. von Foerster and G.W. Zopf (Eds.) *Principles of Self-Organization*. London, Pergamon Press, 1962.

480.   Pask, G. Self-Organizing Systems Involved in Human Learning and Performance. In: *3rd Bionics Symposium*, Dayton, Ohio, 1963.

481.   Pernin, D. Des machines à enseigner a l'instruction programmée. *Hommes et Techniques*, No. 230, January 1964.

482.*   Pernin, D. Qu'est-ce que l'instruction programmée? *Hommes et Techniques*, No. 230, January 1964.

483.*   Pernin, D. Domaines d'application et coût de l'instruction programmée. *Hommes et Techniques*, No. 230, January 1964.

484.   Pohl, L. Die methodische Behandlung des Participe passe. *Wissenschaftliche Zeitschrift der Friedrich-Schiller-Universität Jena*, 1964, Jahrgang 13, Gesellschafts-und sprachwissenschaftliche Reihe, Heft 2.

485.* Popescu-Neveanu, P., Constantinescu-Stoleru, P., Cretu, T., and Zlate, M. Programarea invatamintului si problemele cercetarilor de psihologie-pedagogie. *Revista de pedagogie*, 1964, No. 1. Summary in Russian.

486.   Porter, D. A Critical Review of a Portion of the Literature on Teaching Devices. *Harvard Educational Review*, Vol. 27, No. 2, 1957.

487.   Porter, D. Teaching Machines. In: Lumsdaine, A.A. and Glaser, R. *Teaching Machines and Programmed Learning: A Source Book*. Washington, NEA, 1960.

488.   Pressey, S.L. Development and Appraisal of Devices Providing Immediate Automatic Scoring of Objective Tests and Concomitant Self-Instruction. *Journal of Psychology*, Vol. 29, 1950.

489.   Pressey, S.L. Teaching Machines (and Learning Theory) Crisis. *Journal of Applied Psychology*, Vol. 47, No. 1, 1963.

490.* Radu, I, Krau, E., and Cozonac, S. Instruirea programata si problemele aplicarii ei in scoala. *Revista de pedagogie*, 1964, No. 6. Summary in Russian.

491.   Richmond, W.K. Teachers and Machines. An Introduction to the Theory and Practice of Programmed Learning. London and Glasgow, 1965.

492.   Roe, A. Automated Teaching Methods Using Linear Programs. *Journal of Applied Psychology*, Vol. 46, No. 3, June 1962.

493.* Sander, M. Der programmierte Unterricht in der allgemeinbildenden Schule. Bad Neuenahr, Mars-Lehrmittelverlag, 1964.

494.   Sauvan, J. Modeles de fonctions de la mentalité. *La Pedagogie Cybernetique*, 1964, No. 4.

495. Schirm, R.W. *Lehrmaschinen* und programmierte Unterweisung. Darmstadt, 1963.
496. Schramm, W. The Newer Educational Media in the United States. *New Methods and Techniques in Education*. Educational Studies and Documents. No. 48, UNESCO, 1963.
497.* Schramm, W. The Research on Programmed Instruction. Washington, U.S. Dept. of Health, Education and Welfare, 1964.
498.* Schröter, G. Automation und Lehrmaschinen. *Berufspädagogische Zeitschrift*, 1964, Heft 6.
499.* Schuffenhauer, H. Über die Entwicklung kybernetischer Unterrichtsmittel und das Problem der Programmierung von Unterrichtsabschnitten. *Pädagogik*, 1963, No. 11.
500.* Shakmaev, N., Jinkine, N., and Petrouchine, S. Recherches sur l'utilisation des moyens techniques dans l'enseignement en U.R.S.S. In: *Nouvelles Methods et Moyens d'Education*. Paris, UNESCO, 1963.
501.* Silberman, H.F. The Automation of Teaching. *Behavioral Science*, Vol. 5, 1960.
502.* Silberman, H.F. What Are the Limits of Programmed Instruction? *Phi Delta Kappan*, 1963, Vol. 44, No. 6.
503.* Skinner, B.F. The Science of Learning and the Art of Teaching. *Harvard Educational Review*, Vol. 24, No. 2, 1954.
504. Skinner, B.F. Teaching Machines. *Science*, Vol. 128, No. 3330, 1958.
505. Skinner, B.F. Why Do We Need Teaching Machines? *Harvard Educational Review*, Vol. 31, No. 4, 1961.
506. Smallwood, R.D. A Decision Structure for Teaching Machines. Cambridge, 1962.
507.* Smith, W.J. and Moore, J.W. *Programmed Learning: Theory and Research. An Enduring Problem in Psychology. Selected Readings*. Princeton, 1962.
508. Stancin, S. Pedagogia si cibernetica. *Revista de pedagogie*, 1963, No. 10. Summary in Russian.
509. Stolurow, L.M. *Teaching by Machine*. Washington, 1961.
510.* Stolurow, L. Let's Be Informed on Programmed Instruc-

tion. *Phi Delta Kappan*, Vol. 44, No. 6, 1963.

511.* Stolurow, L.M. Model the Master Teacher or Master the Teaching Model. University of Illinois. Training Research Laboratory Technical Report, No. 3, July 1964.

512.* Thelen, H.A. Programmed Materials Today: Critique and Proposals. *The Elementary School Journal*, Vol. 63, No. 4, 1963.

513.* Thomas, C.A., Davies, I.K., Openshaw, D., and Bird, J.B. *Programmed Learning in Perspective*. London, City Publicity Services, 1963.

514.* Tollingerova, D. O vyznamu matematicke teorie her a programovani pro psychologickou analyzu uceni. *Pedagogika*, 1962, No. 12. Summary in Russian.

515.* Tollingerova, D. Programovane uceni jako svetovy problem. *Pedagogika*, 1964, No. 4. Summary in Russian.

516. *Über die Rolle der Kybernetik in Lehr- und Lernprozes*. Material für Fachschullehrer. Hrsg. v. Inst. für Fachschulwesen der DDR. Karl-Marx-Stadt, 1964.

517. Tsvetkov, D. *Principi i metodi na programiraneto na uchebnia material* (Principles and Methods for Programming Textbook Materials). *Narodna prosveta* (Popular Education), VII, 1964.

IV. The Physiology of Higher Nervous Activity. General and Pedagogical Psychology. Didactics and the Methodology of Instruction. Linguistic Foundations for Teaching a Native Language

518. Abakumov, S.I. *Metodika punktuatsii* (The Methodology of Punctuation). M., 1954.

519. Aidarova, L.I. "Formirovaniye lingvisticheskovo otnoshenia k slovu u mladshikh shkolnikov" (The Formation of a Linguistic Attitude Toward Words in Young Schoolchildren). *Voprosy psikhologii* (Problems of Psychology), V, 1964.

520. Algazina, N.N. *Izucheniye pravil pravopisania okonchania prilagatelnykh i prichasty v vosmiletnei shkole* (Study of the Rules for Spelling the Endings of Adjectives and

Adverbs in Grammar School). M., 1960.
521.* Algazina, N.N. *Preduprezhdeniye oshibok v postroyenii slovosochetany i prilozheny* (Prevention of Mistakes when Word Combinations and Sentences are Constructed). M., 1962.
522. Ananyev, B.G. *Psikhologia chuvstvennovo poznania* (The Psychology of Sensual Knowledge). M., 1960.
523. Ananyev, B.G. "Trud kak vazhneisheye usloviye razvitia chuvstvitelnosti" (Work as the Most Important Condition for the Development of Perceptivity). *Voprosy psikhologii* (Problems of Psychology), I, 1955.
524. Anokhin, P.K. "Osobennosti afferentnovo apparata uslovnovo refleksa i ikh znacheniye dlya psikhologii" (The Characteristics of the Afferent Apparatus of the Conditioned Reflex and Their Significance for Psychology). *Voprosy psikhologii* (Problems of Psychology), VI, 1955.
525. Anokhin, P.K. "Operezhayushcheye otrazheniye deistvitelnosti" (Anticipatory Reflection of Reality). *Voprosy psikhologii* (Problems of Psychology), VII, 1962.
526.* Anokhin, P.K. "Refleks tseli kak obyekt fiziologicheskovo analiza" (The Reflex of Purpose as Object of Physiological Analysis). *Zhurnal vysshei nervnoi deyatelnosti* (Journal of Higher Nervous Activity), XII (1), 1962.
527.* Antsyferova, L.I. "Rol analiza v poznanii prichinno-sredstvennykh otnosheny" (The Role of Analysis in the Cognition of Cause-Effect Relationships). *Protsess myshlenia i zakonomernosti analiza* (The Thought Process and the Laws of Analysis, Synthesis, and Generalization). Ed. S.L. Rubenstein. M., 1960.
528. Arana, L. "Vospriyatiye kak veroyatny protsess" (Perception as a Probability Process). *Voprosy psikhologii* (Problems of Psychology), V, 1961.
529. Babushkin, I.I. "O razvitii myshlenia uchashchikh na urokakh" (On the Development of Students' Reasoning During Lessons). *O povyshenii soznatelnosti uchashchikhsya v obuchenii* (On Raising the Consciousness of Students During Instruction). Ed. M.A. Danilov. M., 1957.

530.* Barkhudarov, S.G. "O sostoyanii metodiki prepodavania russkovo yazyka" (On the State of Methodology for Russian Language Instruction). *Russky yazyk v shkole* (The Russian Language in School), IV, 1959.

531.* Barkhudarov, S.G. and Kryuchkov, S. Ye. "Ob uchebnike sintaksisa dlya VI-VIII Klassov" (On a Textbook of Syntax for Classes VI-VIII). *Russky yazyk v shkole* (The Russian Language in School), IV, 1962.

532. Barkhudarov, S.G. and Kryuchkov, S. Ye. *Uchebnik russkovo yazyka* (A Textbook for the Russian Language). Parts I and II. M., 1961.

533. Barkhudarov, S.G. and Kryuchkov, S. Ye. *Uchebnik russkovo yazyka* (A Textbook for the Russian Language). Parts I and II. M., 1965.

534. Bernstein, N.A. "Nekotorye nazrevayushchiye problemy regulyatsii dvigatelnykh aktov" (Certain Imminent Problems on the Regulation of Motor Acts). *Voprosy psikhologii* (Problems of Psychology), VI, 1957.

535. Bernstein, N.A. "Ocherednye problemy fiziologii aktivnosti" (Recurrent Problems in the Physiology of Activity). *Problemy kibernetiki* (Problems of Cybernetics), No. 6. M., 1961.

536. Bernstein, N.A. "Puti i zadachi fiziologii aktivnosti" (Paths and Problems of the Physiology of Activeness). *Voprosy filosofii* (Problems of Philosophy), VI, 1961.

537. Bernstein, N.A. "Novye linii razvitia v fiziologii i ikh sootnosheniye s kibernetikoi" (New Lines of Development in Physiology and Their Relation to Cybernetics). *Voprosy filosofii* (Problems of Philosophy), VIII, 1962.

538. Bernstein, N.A. "Novye linii razvitia v sovremennoi fiziologii" (New Lines of Development in Modern Physiology). *Materialy konferentsii po metodam fiziologicheskikh issledovany cheloveka* (Materials for the Congress on Methods of Physiological Research of Mankind). Eds. A.A. Letayev and V.S. Farfel. M., 1962.

539.* Blinov, G.I. *Sopostavlenia pri obuchenii punktuatsii* (Comparisons During the Instruction of Punctuation). M., 1959.

540.* Blinov, G.I. *Izucheniye svyazi slov na urokakh russkovo yazyka. Posobiye dlya uchitelya* (The Study of Word Relations at Russian Language Lessons. Appliances for the Teacher). M., 1963.

541.* Bogoyavlensky, D.N. *Psikhologia usvoyenia orfografii* (The Psychology of the Mastery of Spelling). M., 1957.

542. Bogoyavlensky, D.N. "Formirovaniye priyomov umstvennoi raboty uchashchikhsya kak put razvitia myshlenia i aktivizatsii uchenia" (Formation of Methods of Mental Work for Students as a Way to Develop Reasoning and to Enliven Study). *Voprosy psikhologii* (Problems of Psychology), IV, 1962.

543. Bogoyavlensky, D.N. and Menchinskaya, N.A. *Psikhologia usvoyenia znany v shkole* (The Psychology of Mastering Knowledges in School). M., 1959.

544. Bogoyavlensky, D.N. and Fedorenko, L.P. "Voprosy aktivizatsii obuchenia grammatike i orfografii v vechernei (smennoi) shkole" (Problems of Enlivening the Instruction of Grammar and Spelling in Night [Shift] School). *Psikhologia aktivizatsii obuchenia v vechernei srednei shkole* (The Psychology of Enlivening Instruction in Secondary Night School). Ed. D.N. Bogoyavlensky. M., 1963.

545. Bodrov, V.A., Genkin, A.A. and Zarakovsky, G.A. "Nekotorye zakonomernosti reaktsii cheloveka na testovye zadachi, modeliruyushchiye prinyatiye odnovo iz dvukh vozmozhnykh resheny" (Certain Laws of Human Reactions to Test Problems, Which Simulate the Reception of One of Two Possible Solutions). *Doklady APNRSFSR* (Lectures of the APSRSSR), V, 1961; II, IV, 1962.

546. Bozhovich, L.I. "Psikhologichesky analiz upotreblenia pravila na bezudarnye glasnye kornya" (A Psychological Analysis of the Requirements for a Law on Unaccented Vowel Roots). *Sovetskaya pedagogika* (Soviet Pedagogy), V-VI, 1937.

547. Bozhovich, L.I. "Psikhologichesky analiz formalizma v usvoyenii shkolnykh znany" (A Psychological Analysis of

Formalism in the Mastery of School Knowledges). *Sovet-skaya pedagogika* (Soviet Pedagogy), XI, 1945.

548.* Bozhovich, L.I. "Znacheniye osoznania yazykovykh obobshcheny" (The Significance of the Realization of Language Generalizations). *Izvestia APNRSFSR* (News of the APSRSSR), No. 3, 1946.

549. Bozhovich, L.I., Leontyev, A.N., Morozova, N.G., and Elkonin, D.B. *Ocherki psikhologii detei (mladshy shkolny vozrast)* (Essays on Child Psychology [Primary School Age]). M., 1950.

550. Boiko, Ye. I. *Vremya reaktsii cheloveka. Istoria, teoria, sovremennoye sostoyaniye i prakticheskoye znacheniye khronometricheskikh issledovany* (The Time of Human Reaction. The History, Theory, Modern Status, and Practical Significance of Chronometric Research). M., 1964.

551.* Bruner, J. *Protsess obuchenia* (The Teaching Process). M., 1962. (Trans.)

552.* Brushlinsky, A.V. *Issledovaniye napravlyonnosti myslitel-novo protsessa* (Research on the Direction of the Thought Process). Dissertation for the Degree of Candidate of Pedagogical Sciences (in Psychology). M., 1964.

553.* Brushlinsky, A.V. "Rol analiza i abstraktsii v poznanii kolichestvennykh otnosheny" (The Role of Analysis and Abstraction in the Cognition of Quantitative Relations). *Protsess myshlenia i zakonomernost analiza, sinteza i obobshchenia* (The Thought Process and the Law of Analysis, Synthesis, and Generalization). Ed. S.L. Ruben-stein. M., 1960.

554. Vallon, A. *Ot deistvia k mysli* (From Act to Thought). M., 1956. (Trans.)

555. Veispal, O.V. "Elementy formalnoi logiki na urokakh russkovo yazyka i literaturnovo chtenia v VII klasse" (Elements of Formal Logic at Lessons of the Russian Language and Readings of Literature for the VII Class). *Trudy pervoi nauchno-pedagogicheskoi konferentsii uchitelei g. Leningrada* (Transactions of the First Scien-

tific-Pedagogical Congress of Teachers in Leningrad). L., 1940.

556.   Vecker, L.M. "O sravnitelnoi kharakteristike predmetnykh deistvy i operatsy upravlenia" (On the Comparative Characteristics of Object Activities and Operations of Control). *Materialy IV nauchnoi konferentsii po fiziologii truda, posvyashchonnoi pamyati A.A. Ukhtomskovo* (Materials of the Fourth Scientific Congress on the Physiology of Work Dedicated to the Memory of A.A. Ukhtomsky). L., 1963.

557.* Vecker, L.M. "K sravnitelnomu analizu psikhicheskoi regulyatsii regulirovania v avtomatakh" (On the Comparative Analysis of Psychic Regulation and Regulation in Computers). *Voprosy filosofii* (Problems of Philosophy), II, 1963.

558.* Vetrov, A.A. "Produktivnoye myshleniye i assotsiatsia" (Productive Thought and Association). *Voprosy psikhologii* (Problems of Psychology), VI, 1959.

559.   Vishnepolskaya, A.G. "Vlianiye rasprostranyonnosti opredelyonnykh orfogramm v yazyke na usvoyenii uchashchimisya pravopisania bezudarnykh glasnykh" (The Influence of the Prevalence of Specific Orthograms in the Language on Students' Mastery of Correct Spelling of Unaccented Vowels). *Voprosy psikhologii obuchenia i vospitania v shkole* (Problems of the Psychology of Instruction and Education in School). Ed. Z.I. Kalmykova. M., 1956.

560.   Vishnepolskaya, A.G. "Vlianiye chtenia na pravopisaniye uchashchikhsya" (The Influence of Reading on Students' Correct Spelling). *Voprosy psikhologii* (Problems of Psychology), III, 1959.

561.* *Voprosy psikhologii usvoyenia grammatiki i orfografii* (Problems of the Psychology of Mastering Grammar and Spelling). Ed. D.N. Bogoyavlensky. M., 1959.

562.   Vorobyov, G.V. *Voprosy prepodavania geometrii v VI-VII klassakh v svyazi s rabotoi uchashchikhsya v shkolnykh masterskikh* (Problems of Teaching Geometry in the VI-VII Classes in Connection with Students' Work in

School Workshops). M., 1962.

563. Vorobyov, G.V. "Priyom utochnenia uchashchimisya usvaivayemykh ponyaty" (The Method of Having Students Make the Concepts They Have Learned More Precise). *Obucheniye shkolnikov priyomam samostoyatelnoi raboty* (Teaching Students Methods of Independent Work). Eds. M.A. Danilov and B.P. Yesipov. M., 1963.

564.* Voronin, M.T. *Upotrebleniye zapyatoi v predlozheniakh s soyuzom "i" (Osnovnye sluchai)* (The Use of the Comma in Sentences with the Conjunction "And" [Basic Instances]). M., 1960.

565.* Voronin, L.G. *Analiz i sintez slozhnykh razdrazhitelei u vysshikh zhivotnykh* (The Analysis and Synthesis of Complex Stimuli in Higher Animals). L., 1952.

566. Vygotsky, L.S. "Razvitiye vysshikh form vnimania v detskom vozraste" (The Development of Higher Forms of Attention in Childhood). *Izbrannye psikhologicheskiye issledovania* (Collected Psychological Research). M., 1956.

567.* Vygotsky, L.S. *Izbrannye psikhologicheskiye issledovania* (Collected Psychological Research). M., 1956.

568.* Vygotsky, L.S. *Razvitiye vysshikh psikhicheskikh funktsii* (The Development of Higher Psychic Functions). M., 1960.

569. Vygotsky, L.S. and Luria, A.R. *Etyudy po istorii povedenia* (Essays on the History of Behavior). M.-L., 1930.

570. Galkina-Fedoruk, Ye. M. *Bezlichnye predlozhenia v sovremennom russkom yazyke* (Impersonal Sentences in Modern Russian). M., 1958.

571. Galperin, P. Ya. "Opyt izuchenia formirovania umstvennykh deistvy" (An Experiment in the Study of the Formation of Mental Actions). *Doklady na soveshchanii po voprosam psikhologii* (Lectures at the Conference on Problems of Psychology). M., 1954.

572. Galperin, P. Ya. "O formirovanii umstvennykh deistvy i ponyaty" (On the Formation of Mental Actions and Concepts). *Vestnik Moskovskovo universiteta* (The Moscow University Herald)—Series on Economics, Philosophy,

and Law, IV, 1957.

573. Galperin, P. Ya. "O formirovanii chuvstvennykh obrazov i ponyaty" (On the Formation of Sensual Images and Concepts). *Materialy soveshchania po psikhologii (iyul, 1955)* (Materials of the Conference on Psychology [July, 1955]). M., 1957.

574. Galperin, P. Ya. "Umstvennoye deistviye kak osnova formirovania mysli i obraza" (Mental Action as the Basis for the Formation of a Thought and an Image). *Voprosy psikhologii* (Problems of Psychology), VI, 1957.

575. Galperin, P. Ya. "Tipy oriyentirovki i tipy formirovania deistvy i ponyaty" (Types of Orientation and Types of Formation of Actions and Concepts). *Doklady APN-RSFSR* (Lectures of the APSRSSR), II, 1959.

576. Galperin, P. Ya. "Razvitiye issledovany po formirovaniyu umstvennykh deistvy" (The Development of Research on the Formation of Mental Actions and Concepts). *Psikho-logicheskaya nauka v SSSR* (Psychological Science in the USSR), Vol. I. M., 1959.

577. Galperin, P. Ya. *Osnovnye rezultaty issledovany po prob-leme "Formirovaniye umstvennykh deistvy i ponyaty"* (The Basic Results of Research on the Problem "The Formation of Mental Actions and Concepts"). Lecture for the Degree of Doctor of Pedagogical Sciences (in Psychology). M., 1965.

578. Galperin, P. Ya. and Georgiyev, L.S. "K voprosu o formirovanii nachalnykh matematicheskikh ponyaty" (On the Problem of the Formation of Beginning Mathematical Concepts). *Doklady APNRSFSR* (Lectures of the APSRSSR), I, III, IV, V, 1960; I, 1961.

579. Galperin, P. Ya. and Dubrovina, A.N. "Tip oriyentirovki v zadanii i formirovaniye grammaticheskikh ponyaty" (A Type of Orientation in Assignment and Formation of Grammatical Concepts). *Doklady APNRSFSR* (Lectures of the APSRSSR), III, 1957.

580. Galperin, P. Ya., Zaporozhets, A.V., and Elkonin, D.B. "Problema formirovania znany i umeny u shkolnikov i

novye metody obuchenia v shkole" (The Problem of Forming Knowledges and Skills in Students and New Methods of Instruction in School). *Voprosy psikhologii* (Problems of Psychology), V, 1963.

581. Galperin, P. Ya. and Talyzina, N.F. "Formirovaniye nachalnykh geometricheskikh ponyaty na osnove organi-zovannovo deistvia uchashchikhsya" (The Formation of Elementary Geometry Concepts on the Basis of Students' Organized Actions). *Voprosy psikhologii* (Problems of Psychology), I, 1957.

582. Ganelin, Sh. I. *Didakticbesky printsip soznatelnosti* (The Didactic Principle of Consciousness). M., 1962.

583.* Gmurman, V. Ye. "O grammaticheskikh voprosakh" (On Grammar Problems). *Russky yazyk v shkole* (The Russian Language in School), V-VI, 1946.

584. Gmurman, V. Ye. "K voprosu o psikhologii usvoyenia orfograficheskikh pravil" (On the Problem of the Psychology of Mastering Orthographic Rules). *Sovetskaya pedagogika* (Soviet Pedagogy), IV-V, 1946.

585.* Gmurman, V. Ye. "Usvoyeniye uchashchimisya nauchnoi terminologii" (Students' Mastery of Scientific Terminology). *Sovetskaya pedagogika* (Soviet Pedagogy), X, 1950.

586. Govorkova, A.F. "Opyt izuchenia nekotorykh intellektual-nykh umeny" (An Experiment in the Study of Certain Intellectual Skills). *Voprosy psikhologii* (Problems of Psychology), II, 1962.

587. Golant, Ye. Ya. *Metody obuchenia v sovetskoi shkole* (Methods of Instruction in the Soviet School). M., 1957.

588. Golomshtok, I. Ye. "Psikhologichesky analiz arifmetich-eskikh deistvy uchashchikhsya pervovo klassa" (A Psychological Analysis of the Arithmetical Actions of Students of Class I). *Uchonye zapiski Moskovskovo oblastnovo peda-gogicheskovo instituta im. N.K. Krupskoi* (Learned Trans-actions of the N.K. Krupskaya Moscow Regional Pedagogi-cal Institute, Vol. 91, 1960.

589. Gorskaya, G.I. *Priyomy aktivizatsii poznavatelnoi deyatel-nosti uchashchikhsya na urokakh russkovo yazyka*

(Methods for Enlivening Cognitive Activities of Students During Russian Language Lessons). Lipetsk, 1961.

590. Gorskaya, G.I. *Novoye na urokakh russkovo yazyka. Iz opyta raboty uchitelya g. Lipetska* (New Things for Russian Language Lessons. From the Experience of the Work of a Lipetsk Teacher). M., 1963.

591. Granik, G.G. "Primeneniye razbora dlya zakreplenia navykov v shkole rabochei molodyozhi" (The Application of Selection to Consolidate Habits in a School for Young Workers). *Russky yazyk v shkole* (The Russian Language in School), V, 1962.

592. Granik, G.G. "Priyomy umstvennoi raboty pri usvoyenii orfografii" (Methods of Mental Work During the Mastery of Orthography). *Psikhologia aktivizatsii obuchenia v vechernei srednei shkole* (The Psychology of Enlivening Instruction at Secondary Night School). Ed. D.N. Bogoyavlensky. M., 1963.

593. Granik, G.G. "Rabota nad punktuatsiyei pri pisme pod diktovku (Opyt organizatsii umstvennoi deyatelnosti uchashchikhsya)" (Work on Punctuation During Dictations [An Experiment in Organizing Students' Mental Activity]). *Russky yazyk v shkole* (The Russian Language in School), I, 1965.

594. Grashchenkov, N.I. and others. "Osnovnye voprosy struktury reflektornovo deistvia i ikh metodologicheskaya otsenka" (Basic Problems on the Structure of Reflex Action and Their Metholodogical Evaluation). *Voprosy filosofii* (Problems of Philosophy), VIII, 1962.

595. Gruzdev, P.N. "Vospitaniye myshlenia v protsesse obuchenia" (Cultivating Thought During the Instruction Process). *Voprosy vospitania myshlenia v protsesse obuchenia* (Problems of Cultivating Thought During the Instruction Process). M., 1949.

596. Gurova, L.L. "Osoznavayemost myslitelnykh operatsii pri reshenii prostranstvennykh zadach" (The Realizability of Mental Operations During the Solution of Spatial Problems). *Myshlenia i rech* (Thought and Speech). Eds. N.I.

Zhinkin and F.N. Shemyakin. M., 1963.

597. Gurova, L.L. "Sposoby formirovania soznatelno reguliruyemykh myslitelnykh operatsii pri reshenii prostranstvennykh zadach" (Ways to Form Consciously Regulated Thought Operations During the Solution of Spatial Problems). *Ibid.*

598. Gusarskaya, G. Yu. *Moi opyt vospitania v starshikh klassakh interesa i razvitia sklonnosti uchashchikhsya k nauchnym znaniam* (My Experience in Cultivating in Students of the Upper Classes an Interest in Scientific Knowledges and Developing a Skill for Them). Kazan, 1961.

599. Davydov, V.V. "Obrazovaniye nachalnovo ponyatia o kolichestve u detei" (The Cultivation in Children of an Elementary Notion of Quantity). *Voprosy psikhologii* (Problems of Psychology), II, 1957.

600. Davydov, V.V. "O psikhologicheskom analize soderzhania deistvy" (On the Psychological Analysis of the Content of Actions). *Tezisy dokladov na II syezde Obshchestva psikhologov* (Topics of the Lectures at the Second Session of the Psychologists' Association), No. 5. M., 1963.

601. Danilov, M.A. *Protsess obuchenia v sovetskoi shkole* (The Teaching Process in Soviet Schools). M., 1960.

602. Danilov, M.A. "Umstvennoye vospitaniye shkolnikov" (The Mental Education of Students). *Sovetskaya pedagogika* (Soviet Pedagogy), XII, 1964.

603. Danilov, M.A. and Yesipov, B.P. *Didaktika* (Didactics). M., 1957.

604. Doblayev, L.P. "Myslitelnye protsessy pri sostavlenii uravneny" (The Mental Processes During the Compilation of Equations) *Izvestia APNRSFSR* (News of the APSRSSR), No. 80, 1957.

605. Dobromyslov, V.A. "O razvitii logicheskovo myshlenia uchashchikhsya V-VII klassov na zanyatiakh po russkomu yazyku" (On Developing Logical Thought in Students of the V-VII Classes During Russian Language Lessons). M., 1956.

606. Dobromyslov, V.A. *Izucheniye grammaticheskikh opredeleny i pravil v V-VII klassakh* (The Study of Grammar Definitions and Rules in Classes V-VII). M., 1951, pp. 15-37.

607. Dobronravov, B.K. "Razvitiye myshlenia na urokakh matematiki" (The Development of Thought at Mathematics Lessons). *Trudy pervoi nauchno-pedagogicheskoi konferentsii uchitelei g. Leningrada* (Transactions of the First Scientific-Pedagogical Congress of Teachers of the City of Leningrad). L., 1940.

608. Dobrynin, N.F. "Problema znachimosti v psikhologii" (The Problem of Significance in Psychology). *Materialy soveshchania po psikhologii* (Materials for the Conference on Psychology). M., 1957.

609. Dobrynin, N.F. "O znachimosti poluchayemykh uchashchimisya znany" (On the Significance of Knowledges Obtained by Students). *Voprosy psikhologii* (Problems of Psychology), I, 1960.

610. Dodonov, B.I. "Protsess kategorialnovo uznavania grammaticheskovo materiala" (The Process of Categorical Recognition of Grammar Material). *Voprosy psikhologii* (Problems of Psychology), II, 1959.

611. Dolin, A.O. "Novye fakty k fiziologicheskomu ponimaniyu assotsiatsii u cheloveka" (New Facts for a Psychological Understanding of Association in Man). *Arkhiv biologicheskikh nauk* (The Biological Sciences Archive), No. 1-2, 1936.

612. Dubovis-Aranovskaya, D.M. "O nekotorykh usloviakh ponimania struktury teksta uchashchimisya" (On Certain Conditions for Students' Comprehension of the Structure of a Text). *Voprosy psikhologii* (Problems of Psychology), I, 1962.

613. Dudnikov, A.V. *Metodika punktuatsii v svyazi s izucheniyem sintaksisa slozhnovo predlozhenia* (The Methodology of Punctuation in Relation to the Study of the Syntax of a Complex Sentence). Dissertation for the Degree of Candidate of Pedagogical Sciences. M., 1952.

614.* Dudnikov, A.V. *Punktuatsia slozhnovo predlozhenia* (The Punctuation of a Complex Sentence). M., 1958.

615. Yerastov, N.P. *Psikhologicheskiye osnovy formirovania navyka vyrazhenia myslei svoimi slovami u uchashchikhsya* (The Psychological Foundations for the Formation in Students of the Habit of Expressing Thoughts in Their Own Words). Dissertation for the Degree of Candidate of Pedagogical Sciences (in Psychology). M., 1955.

616. Yeritsyan, M.S. "Materialy k psikhologii deduktivnykh umozaklyucheny" (Materials on the Psychology of Deductive Conclusions). *Izvestia APNRSFSR* (News of the APSRSSR), No. 120, 1962.

617. Yesipov, B.P. "Aktivizatsia myshlenia uchashchikhsya v protsesse obuchenia" (Enlivening Students' Thought During Instruction). *Izvestia APNRSFSR* (News of the APSRSSR), No. 20, 1949.

618. Yesipov, B.P. *Samostoyatelnaya rabota uchashchikhsya na urokakh* (Students' Independent Work at Lessons). M., 1961.

619. Zhekulin, S.A. "Razvitiye intellektualnykh operatsy pri reshenii zadach" (The Development of Intellectual Operations During the Solution of Problems). *Voprosy psikhologii* (Problems of Psychology), II, 1965.

620. Zhilina, Ye. M. "Skhema kak sredstvo vospitania myshlenia na urokakh russkovo yazyka" (The Diagram as a Means of Cultivating Thought During Russian Language Lessons). *Voprosy vospitania myshlenia v protsesse obuchenia* (Problems of Cultivating Thought During Instruction). M., 1949.

621.* Zhinkin, N.I. "Razvitiye pismennoi rechi uchashchikhsya III-VII klassov" (The Development of Written Speech in Students of the III-VII Classes). *Izvestia APNRSFSR* (News of the APSRSSR), No. 78, 1956.

622.* Zhinkin, N.I. *Mekhanizm rechi* (Speech Mechanism). M., 1958.

623. Zhuikov, S.F. "K psikhologii formirovania orfograficheskikh navykov" (On the Psychology of Forming Ortho-

graphic Habits). *Izvestia APNRSFSR* (News of the APSRSSR), No. 80, 1957.

624. Zhuikov, S.F. *Psikhologia usvoyenia grammatiki v nachalnykh klassakh* (The Psychology of Mastering Grammar in the Elementary Classes). M., 1964.

625. Zhuikov, S.F. Formirovaniye orfograficheskikh deistvy (u mladshikh shkolnikov). (The Formation of Orthographic Actions [in Primary School Children]). M., 1965.

626. Zabolotnev, M.I. "Formirovaniye priyomov samostoyatelnovo myshlenia uchashchikhsya" (The Formation in Students of Methods of Independent Thought). *Voprosy psikhologii* (Problems of Psychology), II, 1964.

627. Zabuga, I.S. "Zakrepleniye navykov pisma po russkomu yazyku" (The Consolidation of Writing Habits for Russian). *Russky yazyk v shkole* (The Russian Language in School), VI, 1956.

628.* Zavalishina, D.N. and Pushkin, V.N. "O mekhanizmakh operativnovo myshlenia" (On the Mechanics of Operative Thought). *Voprosy psikhologii* (Problems of Psychology), III, 1964.

629. Zankov, L.V. "Razvitiye pamyati umstvenno otstalovo rebyonka" (Developing Memory in a Mentally Retarded Child). *Umstvenno otstaly rebyonok* (The Mentally Retarded Child). Eds. L.S. Vygotsky and I.I. Danyushevsky, Vol. I (1). M., 1935.

630. Zankov, L.V. *Naglyadnost i aktivizatsia uchashchikhsya v obuchenii* (Visual Methods and Enlivening Students in Instruction). M., 1960.

631.* Zankov, L.V. *O predmete i metode didakticheskikh issledovany* (On the Subject and Methods of Didactic Research). M., 1962.

632. Zaporozhets, A.V. *Razvitiye proizvolnykh dvizheny* (The Development of Voluntary Movements). M., 1960.

633. Zakharov, A.N. "Sopostavleniye teoreticheski vozmozhnovo i realnovo khoda reshenia zadach" (A Comparison Between a Theoretically Possible Course of Solution for Problems and a Real Course). *Voprosy psikhologii* (Prob-

lems of Psychology), VI, 1959.

634.   Zakharov, A.N. "Ob ispolzovanii informatsii v zadachakh, reshayemykh s pomoshchyu prob" (On the Use of Information in Problems Which Are Solvable With the Help of Trials). *Myshleniye i rech* (Thought and Speech). Eds. N.I. Zhinkin and F.N. Shemyakin. M., 1963.

635.   Zevald, L.O. "Materialy k voprosu o sistemnosti" (Materials on the Problems of Systemics). *Trudy fiziologicheskikh laboratorii im Akad. I.P. Pavlova* (Transactions of the I.P. Pavlov Physiological Laboratory), Vol. X, 1941.

636.   Zeiliger-Rubenstein, Ye. O. "K voprosu o vospitanii u uchashchikhsya priyomov myshlenia" (On the Problem of Cultivating Methods of Thought in Students). *Voprosy vospitania myshlenia v protsesse obuchenia* (Problems of Cultivating Thought During Instruction). Eds. P.N. Gruzdev and Sh. I. Ganelin. M.-L., 1949. (This entry is more complete than 620.)

637.   Zelinsky, K. "Kamo gryadeshi" (Where Are We Going). *Literaturnaya gazeta* (The Literary Gazette), March 10, 1960.

638.   Zinchenko, V.P. and others. "Stanovleniye i razvitiye pertseptivnykh deistvy" (The Formation and Development of Perceptive Actions). *Voprosy psikhologii* (Problems of Psychology), III, 1962.

639.   Zinchenko, V.P. "Teoreticheskiye problemy psikhologii vospriatia" (The Theoretical Problems of the Psychology of Perception). *Inzhenernaya psikhologia* (Engineering Psychology). Eds. A.N. Leontyev, V.P. Zinchenko, and D. Yu. Panov. M., 1964.

640.   Zinchenko, V.P. *Neproizvolnoye zapominaniye* (Involuntary Recollection). M., 1961.

641.   Zykova, V.I. *Ocherki psikhologii usvoyenia nachalnykh geometricheskikh znany* (Essays on the Psychology of Elementary Geometry). M., 1955.

642.   Zykova, V.I. "Psikhologichesky analiz primenenia geometricheskikh znany k resheniyu zadach s konkretnym soderzhaniyem" (A Psychological Analysis of the Applica-

tion of Geometrical Knowledges to the Solution of Problems with Specific Content). *Psikhologia primenenia znany k resheniyu uchebnykh zadach* (The Psychology of Applying Knowledges to the Solution of School Problems). Ed. N.A. Menchinskaya. M., 1958.

643. Zykova, V.I. *Formirovaniye prakticheskikh umeny na urokakh geometrii* (The Formation of Practical Abilities at Geometry Lessons). M., 1963.

644.* Zyubin, L.M. "Issledovaniye umstvennoi aktivnosti uchashchikhsya V klassa pri vosproizvedenii i prakticheskom primenenii znany" (Research on the Mental Activity of Students of the V Class During the Repetition and Application of Practical Knowledges). *Sovetskaya pedagogika* (Soviet Pedagogy), V, 1957.

645. Ivanov-Smolensky, A.G. "O vzaimodeistvii pervoi i vtoroi signalnykh sistem v norme i v patologii" (The Interaction of the Primary and Secondary Signal Systems Under Normal and Pathological Conditions). *Opyt obyektivnovo izuchenia raboty i vzaimodeistvia signalnykh sistem golovnovo mozga* (An Experiment in the Objective Study of the Workings and Interactions of the Brain's Signal Systems). M., 1963.

646. Indik, N.K. *Myslitelnye protsessy pri formirovanii novovo deistvia* (The Mental Processes at the Formation of a New Action). Dissertation for the Degree of Candidate of Pedagogical Sciences (in Psychology). M., 1951.

647. Kabanova-Meller, Ye. N. "Psikhologichesky analiz primenenia geograficheskikh ponyaty i zakonomernostei" (A Psychological Analysis of the Application of Geographic Concepts and Laws). *Izvestia APNRSFSR* (News of the APSRSSR), No. 28, 1950.

648. Kabanova-Meller, Ye. N. "Formirovaniye priyomov umstvennoi deyatelnosti u shkolnikov" (The Formation of Methods of Mental Activity in Students). *Sovetskaya pedagogika* (Soviet Pedagogy), VI, 1959.

649. Kabanova-Meller, Ye. N. *Psikhologia formirovania znany i*

navykov u shkolnikov (The Psychology of Forming Knowledges and Habits in Students). M., 1962.

650. Kalmykova, Z.I. "Protsessy analiza i sinteza pri reshenii arifmeticheskikh zadach" (Processes of Analysis and Synthesis During the Solution of Arithmetical Problems). *Izvestia APNRSFSR* (News of the APSRSSR), No. 71, 1955.

651. Kalmykova, Z.I. "Urovni primenenia znany pri reshenii fizicheskikh zadach" (Standards for the Application of Knowledges During the Solution of Physics Problems). *Psikhologia primenenia znany k resheniyu uchebnykh zadach* (The Psychology of Applying Knowledges to the Solution of School Problems). Ed. N.A. Menchinskaya. M., 1958.

652. Kalmykova, Z.I. "Effektivnost primenenia znany po fizike v zavisimosti ot razlichnykh uslovy ikh usvoyenia" (The Effectiveness of the Application of Knowledges in Physics Depending on the Different Conditions of Their Mastery). *Primeneniye znany v uchebnoi praktike shkolnikov* (The Application of Knowledges in School Practice of Students). Ed. N.A. Menchinskaya. M., 1961.

653. Kaminsky, S.D. and Maiorov, F.P. "O dinamicheskom stereotipe u obezyan" (The Dynamic Stereotype in Monkeys). *Fiziologichesky zhurnal SSSR* (The Physiological Journal of the USSR), XXVII (6), 1939.

654. Karpinsky, G.K. and others. *Fizika. Uchebnik dlya VI klassa srednei shkoly* (Physics. A Textbook for the VI Class of Secondary School). M., 1957.

655. Kildyushevsky, B.F. *Razvitiye aktivnovo myshlenia uchashchikhsya na urokakh fiziki* (The Development of Student's Active Reasoning During Physics Lessons). Kuibyshev, 1952.

656. Kitayev, N.N. *Rabota s uchashchimisya otstayushchimi po russkomu yazyku* (Work with Students Who Are Backward in Russian). M., 1962.

657. Kirillova, G.D. "Obucheniye uchashchikhsya umeniyu sravnivat" (Teaching Students the Ability to Compare).

Sovetskaya pedagogika (Soviet Pedagogy), VII, 1957.

658.* Kirillova, G.D. "O vospitanii u shkolnikov samostoyatel-nosti" (Cultivating Independence in Students). *Sovetskaya pedagogika* (Soviet Pedagogy), I, 1961.

659. Kirsanov, A.A. "Individualizatsia protsessa obuchenia kak sredstvo razvitia poznavatelnoi aktivnosti i samostoyatel-nosti uchashchikhsya" (Individualization of the Instruction Process as a Means of Developing Cognitive Activity and Independence in Students). *Sovetskaya pedagogika* (Soviet pedagogy), V, 1963.

660. Kiselyov, A.P. *Arifmetika. Uchebnik dlya 5-vo i 6-vo klassov srednei shkoly* (Arithmetic. A Textbook for Classes 5 and 6 of Secondary School). M., 1955.

661. Knysh, L.P. "Priyomy sochetania frontalnoi i individualnoi raboty na uroke" (Methods to Combine Collective and Individual Work at Lessons). *Russky yazyk v shkole* (The Russian Language in School), IV, 1958.

662.* Kobyzev, A.I. *Individualnye zadania po russkomu yazyku v 5-7 klassakh* (Individual Assignments in the Russian Language for Classes 5-7). Ed. M.A. Danilov, M., 1957.

663. Kossov, B.B. "O nekotorykh metodakh sposobstvuyu-shchikh vydeleniyu sushchestvennykh priznakov vosprini-mayemykh obyektov" (On Certain Methods Which Facilitate the Differentiation of the Essential Indicative Features of Perceivable Objects). *Voprosy psikhologii* (Problems of Psychology), I, 1960.

664. Kossov, B.B. "Ob usloviakh opredelyayushchikh strukturu vospriatia" (On the Conditions Which Specify the Structure of Perception). *Izvestia APNRSFSR* (News of the APSRSSR), No. 120, 1962.

665. Kossov, B.B. "Sravnitelnaya deyatelnost formy i prostran-stvennovo polozhenia vosprinimayemykh obyektov" (The Comparative Efficacy of the Forms and Spatial Positions of Perceivable Objects). *Zritelnye vospriatia* (Visual Perceptions). Ed. P.A. Shevaryov. M., 1964.

666.* Kostyuk, G.S. "Voprosy psikhologii ponimania: (Problems of the Psychology of Understanding). *Sovetskaya pedagog-*

ika (Soviet Pedagogy), IX, 1948.

667. Kostyuk, G.S. "O razvitii myshlenia u detei" (On the Development of Reasoning in Children). *Doklady na XIV Mezhdunarodnom kongresse po psikhologii* (Lectures at the XIV International Congress on Psychology). M., 1954.

668. Kostyuk, G.S. "Voprosy psikhologii myshlenia" (Problems of the Psychology of Reasoning). *Psikhologicheskaya nauka v SSSR* (Psychological Science in the USSR), Vol. I. M., 1959.

669. Kostyuk, G.S., Prokoliyenko, L.N., and Sinitsa, I. Ye. "O putyakh rukovodstva umstvennym razvitiyem uchashhikhsya v protsesse obuchenia" (On Ways to Control the Mental Development of Students During the Process of Instruction). *Tezisy dokladov na II syezde Obshchestva psikhologov* (Topics of the Lectures at the II Session of the Society of Psychologists), No. 5. M., 1963.

670.* Koul, M., Korzh, N., and Keller, L. "Obucheniye veroyatnostyam pri dlitelnoi trenirovke" (The Instruction of Probabilities During Prolonged Training). *Voprosy psikhologii* (Problems of Psychology). II, 1965. (Trans.)

671. Krinchik, Ye. P. "Izucheniye protsessa pererabotki informatsii chelovekom v situatsii vybora" (A Study of How Humans Process Information in a Situation of Choice). *Doklady APNRSFSR* (Lectures of the APSRSSR), IV, 1961, II, III, 1962.

672.* Krutetsky, V.A. "Opyt analiza sposobnostei k usvoyeniyu matematiki u shkolnikov" (An Experiment in Analyzing Students' Abilities to Assimilate Mathematics). *Voprosy psikhologii* (Problems of Psychology), I, 1959.

673.* Krutetsky, V.A. "O nekororykh osobennostyakh myshlenia shkolnikov, malosposobnykh k matematike" (On Certain Characteristics of the Reasoning of Students Who Are Not Gifted for Mathematics). *Voprosy psikhologii* (Problems of Psychology), V, 1961.

674.* Kryuchkov, S. Ye. "O sostoyanii metodiki prepodavania russkovo yazyka" (On the Condition of Methodology for the Instruction of the Russian Language). *Russky yazyk v* *shkole* (The Russian Language in School), IV, 1960.

675.* Kryuchkov, S. Ye. "O sisteme i tipe uchebnikov po russkomu yazyku" (On the System and Type of School Textbooks on the Russian Language). *Russky yazyk v shkole* (The Russian Language in School), II, 1961.

676. Kudryavtsev, T.V. "Vzaimootnosheniye teoreticheskikh znany i prakticheskikh deistvy (pri vypolnenii shkolnikami elektromontazhnykh rabot)" (The Interrelation of Theoretical Knowledge and Practical Actions [During the Students' Execution of Electrical Work]). *Primeneniye znany v uchebnoi praktike shkolnikov* (Application of Knowledges in Students' School Practice). Ed. N.A. Menchinskaya. M., 1961.

677. Kudryavtsev, T.V. "Opyt psikhologicheskoi kharakteristiki primenenia znany po mashinovedeniyu k resheniyu tekhnicheskikh zadach" (An Attempt at a Psychological Characterization of the Application of Knowledges in the Operation of Machines for the Solution of Technical Problems). Ed. N.A. Menchinskaya. M., 1961.

678. Kudryavtsev, T.V. and Yakimanskaya, I.S. *Razvitiye tekhnicheskovo myshlenia uchashchikhsya* (The Development of Students' Technical Reasoning). M., 1964.

679. Kudryavtseva, Ye. M. "Razvitiye logicheskovo myshlenia pri obuchenii biologii" (The Development of Logical Reasoning During the Instruction of Biology). *Biologia v shkole* (Biology in School), VI, 1957.

680. Kudryavtseva, Ye. M. "Usvoyeniye i primeneniye znany o zhizni rasteny" (The Mastery and Application of Knowledges on the Life of Plants). *Primeneniye znany v uchebnoi praktike shkolnikov* (The Application of Knowledges in Students' School Practice). Ed. N.A. Menchinskaya. M., 1961.

681.* Kulak, I.A. *Formirovaniye slozhnykh sistem vremennykh svyazei u cheloveka* (The Formation in Man of Complex Systems of Temporary Links). Minsk, 1962.

682.* Kulibaba, I.I. "K izucheniyu slozhnosochininyonnovo predlozhenia v VII klasse" (Toward the Study of the

Complex Sentence in Class VII). *Russky yazyk v shkole* (The Russian Language in School), VI, 1961.

683. Kurashov, I.V. "Poznavatelnaya samostoyatelnost uchashchikhsya v protsesse izuchenia novykh znany kak rezultat organizatsii ikh deyatelnosti uchitelem" (Students' Cognitive Independence During the Study of New Knowledges as a Result of Organization of Their Activities by the Teacher). *Ob usloviakh razvitia poznavatelnoi samostoyatelnosti i aktivnosti uchashchikhsya na urokakh* (On the Conditions for the Development of Students' Independence and Liveliness at Lessons). Ed. M.A. Danilov. Kazan, 1963.

684. Landa, L.N. *K psikhologii formirovania metodov rassuzhdenia (na materiale reshenia geometricheskikh zadach na dokazatelstvo uchashchimisya VII-VIII klassov)* (On the Psychology of the Formation of Methods for Reasoning [on Material of Students' Solution of Geometrical Problems by Proof]). Autoreference to Dissertation for the Degree of Candidate of Pedagogical Sciences (in Psychology). M., 1955.

685. Landa, L.N. "O roli poiskovykh prob v protsesse myshlenia" (On the Role of Searching Trials in the Thought Process). *Tezisy dokladov na soveshchanii po voprosam psikhologii poznania (20-22 maya, 1957).* (Topics of the Conference on Problems of the Psychology of Cognition). M., 1957.

686. Landa, L.N. "O formirovanii u uchashchikhsya obshchevo metoda myslitelnoi deyatelnosti pri reshenii zadach" (On the Formation in Students of a General Method for Mental Activity During the Solution of Problems). *Voprosy psikhologii* (Problems of Psychology), III, 1959.

687. Landa, L.N. "O nekotorykh nedostatkakh umstvennoi deyatelnosti uchashchikhsya samostoyatelnoye resheniye zadach" (On Certain Defects in Students' Mental Activity Which Hinder the Independent Solution of Problems). *Izvestia APNRSFSR* (News of the APSRSSR), No. 115, 1961.

688. Landa, L.N., and Belopolskaya, A.R. "Formirovaniye u uchashchikhsya obshchikh skhem umstvennykh deistvy kak usloviye effektivnovo obuchenia metodam umstvennoi raboty" (The Formation in Students of General Diagrams of Mental Actions as a Condition for the Effective Instruction of Methods of Mental Work). *Tezisy dokladov na I syezde Obshchestva psikhologov* (Topics of the Lectures at the I Session of the Society of Psychologists). M., 1959.

689.* Levitov, N.D. *Detskaya i pedagogicheskaya psikhologia* (Child and Pedagogical Psychology). M., 1964.

690.* Leites, N.S. *Ob umstvennoi odaryonnosti* (On Intellectual Giftedness). M., 1960.

691.* Lemberg, R.G. *Didakticheskiye ocherki* (Didactic Essays). Alma-Ata, 1964.

692. Leontyev, A.N. *Razvitiye pamyati. Eksperimentalnoye issledovaniye vysshikh psikhicheskikh funktsii* (Memory Development. Experimental Research on Higher Psychic Functions). M.-L., 1931.

693. Leontyev, A.N. "O soznatelnosti uchenia" (On the Consciousness of Learning). *Izvestia APNRSFSR* (News of the APSRSSR), No. 7, 1947.

694. Leontyev, A.N. "O materialisticheskom, reflektornom, i subyektivno-idealisticheskom ponimanii psikhiki" (On the Materialistic, Reflexive, and Subjective-idealistic Understanding of the Psyche). *Sovetskaya pedagogika* (Soviet Pedagogy), VII, 1952.

695. Leontyev, A.N. "Priroda i formirovaniye psikhicheskikh svoistv i protsessov cheloveka" (The Nature and Formation of Man's Psychic Characteristics and Processes). *Voprosy psikhologii* (Problems of Psychology), I, 1955.

696. Leontyev, A.N. "Obucheniye kak problema psikhologii" (Instruction as a Psychological Problem). *Voprosy psikhologii* (Problems of Psychology), I, 1957.

697. Leontyev, A.N. "O formirovanii sposobnostei. Doklad na I syezde Obshchestva psikhologov v Moskve (29 iyunya-4 iyulya 1959)" (The Formation of Abilities. A Lecture at

the I Session of the Society of Psychologists [June 29-July 4, 1959] in Moscow). *Voprosy psikhologii* (Problems of Psychology), I, 1960.

698. Leontyev, A.N. "Biologicheskoye i sotsialnoye v psikhike cheloveka. Doklad na XVI Mezhdunarodnom psikhologicheskom kongresse avgust 1960 g." (The Biological and Social in Man's Psyche. A Lecture at the XVI Session of the International Psychology Congress, August 1960). *Voprosy psikhologii* (Problems of Psychology), I, 1960.

699. Leontyev, A.N. *Problemy razvitia psikhiki* (Problems of Psychological Development). M., 1965.

700. Leontyev, A.N. and Krinchik, Ye. P. "Primeneniye teorii informatsii v psikhologicheskikh issledovaniakh" (The Application of the Information Theory in Psychological Research). *Voprosy psikhologii* (Problems of Psychology), V, 1961.

701. Leontyev, A.N. and Krinchik, Ye. P. "O nekotorykh osobennostyakh protsessa pererabotki informatsii chelovekom" (On Certain Characteristics of Human Information Processing). *Voprosy psikhologii* (Problems of Psychology), VI, 1962.

702. Leontyev, A.N. and Krinchik, Ye. P. "Pererabotka informatsii chelovekom v situatsii vybora" (Human Processing of Information in a Situation of Choice). *Inzhenernaya psikhologia* (Engineering Psychology). Eds. A.N. Leontyev, V.P. Zinchenko, and D. Yu. Panov. M., 1964.

703.* Leontyev, K.L., Lerner, A. Ya., and Oshanin, D.A. "O nekotorykh zadachakh issledovania sistemy 'chelovek i avtomat' " (On Certain Problems of the Research on the System "Man and Computer"). *Voprosy psikhologii* (Problems of Psychology), I, 1961.

704. *Lipetsky opyt ratsionalnoi organizatsii uroka* (The Lipetsk Experiment in Efficiently Organizing a Lesson). Eds. M.A. Danilov, V.P. Strekozin, and I.A. Ponomarev. M., 1963.

705. Lipkina, A.I. "Abstragirovaniye uchashchimisya svoistv obyektov nezhivoi prirody" (Students' Abstraction of the Characteristics of Inanimate Objects). *Psikhologia primene-*

nia znany k resheniyu uchashchimisya zadach (Students' Application of Knowledges for the Solution of Problems). Ed. N.A. Menchinskaya. M., 1958.

706. Lipkina, A.I. *Razvitiye myshlenia na urokakh obyasnitel-novo chtenia* (The Development of Reasoning at Lessons of Explanatory Reading). M., 1961.

707. Lomizov, A.F. *Metodika punktuatsii v svyazi s izucheni-yem sintaksisa* (Punctuation Methods in Relation to Syntax Study). M., 1964.

708.* Lomov, B.F. *Chelovek i tekhnika. Ocherki inzhenernoi psikhologii* (Man and Technology. Essays on Engineering Psychology). L., 1963.

709.* Lordkinanidze, D.O. *Printsipy, organizatsia i metody obuchenia* (Principles, Organization, and Methods of Instruction). M., 1957.

710.* Luria, A.R. "Razvitiye konstruktivnoi deyatelnosti do-shkolnika" (Developing a Pre-schooler's Constructive Activity). *Voprosy psikhologii rebyonka doshkolnovo vozrasta* (Problems in the Psychology of the Pre-school Child). Eds. A.N. Leontyev and A.V. Zaporozhets. M.-L., 1948.

711. Luria, A.R. and Tsvetkova, L.S. "Programmirovaniye konstruktivnoi deyatelnosti pri lokalnykh porazheniakh mozga" (Programming Constructive Activity in Cases of Localized Diseases of the Brain). *Voprosy psikhologii* (Problems of Psychology), II, 1965.

712.* Lyublinskaya, A.A. "Prichinnoye myshleniye rebyonka v deistvii" (Causal Reasoning of a Child in Action). *Izvestia APNRSFSR* (News of the APSRSSR), No. 17, 1948.

713.* Lyublinskaya, A.A. *Ocherki psikhicheskovo razvitia re-byonka* (Essays on the Psychological Development of a Child). M., 1959.

714. Lyapin, N.N. *Mysli o rabote uchitelya* (Thoughts on a Teacher's Work). M., 1965.

715.* Matyushkin, A.M. "Issledovaniye psikhologicheskikh zakonomernosti protsessa analiza" (Research on the Psychological Laws of the Analytical Process). *Voprosy psikhologii* (Problems of Psychology), III, 1960.

716. Mashbits, Ye. I. "Formirovaniye obobshchonnykh operatsii kak put podgotovki uchashchikhsya k samostoyatelnomu resheniyu geometricheskikh zadach" (The Formation of Universal Operations as a Way to Prepare Students for the Independent Solution of Geometrical Problems). *Izvestia APNRSFSR* (News of the APSRSSR), No. 129, 1963.

717. Menchinskaya, N.A. "Psikhologia usvoyenia ponyaty" (The Psychology of the Mastery of Concepts). *Izvestia APNRSFSR* (News of the APSRSSR), No. 28, 1950.

718. Menchinskaya, N.A. "K probleme psikhologii usvoyenia znany" (More on the Problem of the Psychology of Mastering Knowledges). *Izvestia APNRSFSR* (News of the APSRSSR), No. 61, 1954.

719. Menchinskaya, N.A. "Nekotorye voprosy primenenia znany na praktike" (Several Questions on the Application of Knowledges in Practice). *Voprosy psikhologii* (Problems of Psychology), I, 1955.

720. Menchinskaya, N.A. *Psikhologia obuchenia arifmetike* (The Psychology of Teaching Arithmetic). M., 1955.

721. Menchinskaya, N.A. Introduction to the Collection *Primeneniye znany k resheniyu uchebnykh zadach* (The Application of Knowledges for the Solution of School Problems). Ed. N.A. Menchinskaya. M., 1958.

722. Menchinskaya, N.A. Introduction to the Collection *Primeneniye znany v uchebnoi praktike shkolnikov* (The Application of Knowledges in Students' School Practice). Ed. N.A. Menchinskaya. M., 1961.

723. Mikulinskaya, M. Ya. "K voprosu ob oriyentirovochnykh priznakakh glavnykh chlenov predlozhenia" (On the Problem of the Orienting Indicative Features of a Sentence's Primary Members). *Doklady APNRSFSR* (Lectures of the APSRSSR), II, 1960.

724. Miller, G.A., Galanter, E., and Pribram, K.H. *Plany i struktura povedenia* (Plans and the Structure of Behavior). M., 1964. (Trans.)

725.* Mileryan, Ye. A. "Psikhologicheskiye osobennosti reshenia nekotorykh konstruktivnykh zadach v starshikh klassakh srednei shkoly" (The Psychological Characteristics of the Solution of Certain Constructive Problems in the Upper Classes of Secondary School). *Voprosy psikhologii* (Problems of Psychology), II, 1964.

726. Mingazov, E.G. "Voprosy usilenia samostoyatelnosti i aktivnosti uchashchikhsya na urokakh matematiki" (Problems in the Reinforcement of Students' Independence and Liveliness During Mathematics Lessons). *Ob usloviakh razvitia poznavatelnoi samostoyatelnosti i aktivnosti uchashchikhsya na urokakh* (On Conditions for the Development of Students' Cognitive Independence and Liveliness During Lessons). Ed. M.A. Danilov. Kazan, 1963.

727. Moskalenko, K.A. *Kommentirovannye uprazhnenia* (Annotated Exercises). Lipetsk, 1961.

728.* Nazarova, L.K. *Aktivizatsia obuchenia pravopisaniyu na osnove uchota individualnykh osobennostei uchashchikhsya v III i IV klassa* (Enlivening Spelling Instruction with Regard to the Individual Characteristics of Students in Classes III and IV). M., 1962.

729.* Nazarova, L.K. *Obucheniye gramote na osnove uchota individualnykh osobennostei uchashchikhsya* (Instruction of Reading and Writing on the Basis of Students' Individual Characteristics). M., 1965.

730. Nepomnyashchaya, N.I. "Rol obuchenia v kompensatsii nekotorykh neirodinamicheskikh defektov u umstvenno otstalykh detei" (The Role of Instruction in Compensating for Certain Neurodynamic Defects in Mentally Retarded Children). *Voprosy psikhologii* (Problems of Psychology), II, 1957.

731. Nepomnyashchaya, N.I. "K voprosu o psikhologicheskikh mekhanizmakh formirovania umstvennovo deistvia" (On the Problem of the Psychological Mechanisms of the Formation of a Mental Action). *Doklady APNRSFSR* (Lectures of the APSRSSR), II, 1957.

732. Nikitin, N.N. *Geometria. Uchebnik dlya 6-8 klassov* (Geometry. A Textbook for Classes 6-8). M., 1964.

733. Okon, V. *Protsess obuchenia* (The Instruction Process). M., 1962.

734. "O povyshenii uspevayemosti shkolnikov i prinstip znachimosti v psikhologii" (On Increasing Students' Progress and the Principle of Signification in Psychology). *Uchonnye zapiski Moskovskovo gorodskovo pedagogicheskovo instituta im. V.P. Potyomkina* (Learned Writings of the Moscow V.P. Potyomkin Pedagogical Institute), Vol. 69 (4), Part I. M., 1958.

735. Orlova, A.M. "Psikhologia ovladenia ponyatiyem 'podlezhashcheye' " (The Psychology of Mastering the Concept "The Subject"). *Izvestia APNRSFSR* (News of the APSRSSR), No. 28, 1950.

736. Orlova, A.M. "Psikhologicheskiye uslovia differentsirovania uchashchimisya glavnykh tipov prostovo predlozhenia" (Psychological Conditions for the Differentiation by Students of the Main Types of Simple Sentences). *Izvestia APNRSFSR* (News of the APSRSSR), No. 78, 1956.

737. Orlova, N.M. *Usvoyeniye sintaksicheskikh ponyaty uchashchimisya (ocherki).* (Students' Mastery of Syntactical Concepts [Essays]). M., 1961.

738.* Oshanin, D.A. and Panov, D. Yu. "Chelovek v avtomaticheskikh sistemakh upravlenia" (Man in Automatic Control Systems). *Voprosy filosofii* (Problems of Philosophy), V, 1961.

739. Pantina, N.S. "Formirovaniye dvigatelnovo navyka pisma v zavisimosti ot tipa oriyentirovki v zadanii" (Formation of Motor Skill of Writing in Relation to the Type of Assignment Orientation). *Voprosy psikhologii* (Problems of Psychology), IV, 1957.

740. Pavlov, I.P. "Dinamicheskaya stereotipia vysshevo otdela golovnovo mozga." *Polnoye sobraniye sochineny* (The Dynamic Stereotype of the Upper Section of the Brain. Complete Works), Vol. 3, Book 2. M., 1951.

741. Pavlov, I.P. "Fiziologia vysshei nervnoi deyatelnosti" (The Physiology of Higher Nervous Activity), Vol. 3, Book 2. M., 1951.

742. Pavlov, I.P. "Fiziologia i patologia vysshei nernvoi deyatelnosti" (The Physiology and Pathology of Higher Nervous Activity), Vol. 3, Book 2. M., 1951.

743. Pavlov, I.P. "Kharakteristika korkovoi massy bolshikh polushary s tochki zrenia izmeneny vozbudimosti eyo otdelnykh punktov" (The Characteristics of the Cortex of the Cerebrum from the Standpoint of Changing the Stimulability of Its Separate Points), Vol. 3, Book 2. M., 1951.

744. Pavlov, I.P. "Lektsii o rabote bolshikh polushary golovnovo mozga" (Lectures on the Workings of the Cerebrum), Vol. 4, Book 2. M., 1951.

745. *Pavlovskiye sredy* (Pavlov's Environments), Vol. 2. M.-L., 1949.

746. *Ibid.*, Vol. 3.

747.* Petrova, Ye. N. *Materialy k izucheniyu slozhnovo predlozhenia v VII klasse* (Materials for the Study of Complex Sentences). L., 1938.

748.* Petrova, V.I. "O sootnoshenia orfograficheskovo pravila i navyka" (On the Relationship of Orthography Rules and Habits). *Voprosy psikhologii* (Problems of Psychology), II, 1957.

749. Piaget, J. and Inhelder, B. *Genezis elementarnykh logicheskikh struktur. Klassifikatsii i seriatsii* (The Genesis of Elementary Logical Structures. Classification and Serialization). M., 1963. (Trans.)

750. Polya, G. *Kak reshat zadachu* (How to Solve It). M., 1961. (Trans.)

751. Pokrovsky, V.K. "Razvitiye u uchashchikhsya umenia proizvodit prosteishiye logicheskiye operatsii" (Development in Students of the Ability to Perform the Simplest Logical Operations). *Voprosy psikhologii obuchenia i vospitania v shkole* (Problems of the Psychology of Teaching and Education in School). Ed. Z.I. Kalmykova. M., 1956.

752. Polyakova, A.V. "O vyrabotke u uchashchikhsya navykov samostoyatelnovo primenenia orfograficheskikh pravil"

(On the Elaboration in Students of Habits of Independently Applying Spelling Rules). (Based on Materials for the Instruction of Similar Rules in the V-VI Classes). *Izvestia APNRSFSR* (News of the APSRSSR), No. 115, 1961.

753.   Polyakova, A.V. "Formirovaniye u shkolnikov ratsionalnykh sposobov myshlenia v protsesse usvoyenia znany (na materiale russkovo yazyka)" (The Formation in Students of Efficient Means of Reasoning During the Assimilation of Knowledges [from Materials on the Russian Language]). *Obucheniye shkolnikov priyomam samostoyatelnoi raboty* (Teaching Students Methods of Independent Work). Eds. M.A. Danilov and B.P. Yesipov. M., 1963.

754.* Ponomarev, Ya. A. *Psikhologia tvorcheskovo myshlenia* (The Psychology of Creative Thought). M., 1960.

755.   Potapova, A.K. "Razvitiye logicheskovo myshlenia uchashchikhsya na urokakh grammatiki v III klasse" (The Development of Students' Logical Reasoning at Grammar Lessons of Class III). *O povyshenii soznatelnosti uchashchikhsya v obuchenii* (On Raising Students' Consciousness During the Assimilation of Knowledges). Ed. M.A. Danilov. M., 1957.

756.   Pototsky, A.V. *O pedagogicheskikh osnovakh obuchenia matematiki* (On the Pedagogical Foundations of Mathematics Instruction). M., 1963.

757.* Prangishvili, A.S. "Psikhologicheskaya nauka v Gruzii (k 40-letiyu ustanovlenia Sovetskoi vlasti)" (Psychological Science in Georgia [for the 40th Anniversary of the Establishing of Soviet Rule]). *Voprosy psikhologii* (Problems of Psychology), IV, 1961.

758.* Preobrazhenskaya, L.F. "Ob aktivizatsii metodov obyasnenia novovo materiala po grammatike" (On Enlivening Methods for the Explanation of New Grammar Material). *Iz opyta raboty uchitelei russkovo yazyka* (From the Working Experience of Russian Language Teachers). M., 1953.

759.* Preobrazhenskaya, Ye. P. and Adrianova, M. Ye. *Preduprezhdeniye orfograficheskikh i punktuatsionnykh oshibok v*

VI-VII klassakh (iz opyta raboty) (Preventing Spelling and Punctuation Mistakes in Classes VI-VII [from Work Experience]). M., 1959.

760.* *Problemy sposobnostei* (Problems of Abilities). Ed. V.N. Myasitsev. M., 1962.

761. Provotorova, V.N. "Put povyshenia effektivnosti urokov matematiki" (Ways to Raise the Effectiveness of Mathematics Lessons). *Lipetsky opyt ratsionalnoi organizatisii uroka* (The Lipetsk Experiment in Efficient Organization of Lessons). Eds. M.A. Danilov and others. M., 1963.

762. *Psikhologia detei doshkolnovo vozrasta* (The Psychology of Preschool Children) (Glava V, "Razvitiye myshlenia") (Chapter V, "The Development of Reasoning"). Eds. A.V. Zaporozhets and D.B. Elkonin. M., 1964.

763.* *Psikhologia reshenia uchashchimisya proizvodstvenno-tekhnicheskikh zadach* (The Psychology of Students' Solutions for Industrial-Technical Problems). Ed. N.A. Menchinskaya. M., 1965.

764. *Psikhologia usvoyenia grammatiki i orfografii* (The Psychology of the Mastery of Grammar and Spelling). Ed. D.N. Bogoyavlensky. M., 1961.

765.* *Razvitiye uchashchikhsya v protsesse obuchenia* (The Development of Students During Instruction). Ed. L.V. Zankov. M., 1963.

766.* Razumovskaya, M.K. *Uroki russkovo yazyka v sedmom klasse. Metodicheskiye razrabotki* (Russian Language Lessons in Class Seven. Methodological Elaborations). M., 1957.

767. Repkin, V.V. "Formirovaniye orfograficheskovo navyka kak umstvennovo deistvia" (The Formation of Spelling Skill as a Mental Action). *Voprosy psikhologii* (Problems of Psychology), II, 1960.

768. Reshetnikov, V.I. *Formirovaniye priyomov abstragirovania i umstvennoye razvitiye uchashchikhsya* (The Formation of Methods of Abstraction and Students' Mental Development). Dissertation for the Degree of Candidate of Pedagogical Sciences (in Psychology). M., 1964.

769. Reshetova, Z.A. "Tipy oriyentirovki v zadanii i tipy proizvodstvennovo obuchenia" (Types of Orientation in Assignments and Types of Industrial Instruction). *Doklady APNRSFSR* (Lectures of the APSRSSR), V, 1959.

770.* Rozhdestvensky, N.S. *Obucheniye orfografii v nachalnoi shkole* (Spelling Instruction in Elementary School). M., 1960.

771.* Roshka, Al. "Uslovia sposobstvuyushchiye abstragirovaniyu i obobshcheniyu" (Conditions Which Favor Abstraction and Generalization). *Voprosy psikhologii* (Problems of Psychology), VI, 1958.

772. Rubenstein, S.L. "Problema deyatelnosti i soznania v sisteme sovetskoi psikhologii" (The Problem of Activity and Consciousness in the System of Soviet Psychology). *Uchonye zapiski MGU* (Learned Transactions of Moscow State University), No. 90, M., 1945.

773. Rubenstein, S.L. *Osnovy obshchei psikhologii* (Foundations of General Psychology). M., 1946.

774.* Rubenstein, S.L. *Bytiye i soznaniye. O meste psikhologicheskovo vo vseobshchei vzaimosvyazi yavleny materialnovo mira* (Being and Consciousness. On the Role of the Psychological in the Universal Interrelations of the Phenomena of the Material World). M., 1957.

775. Rubenstein, S.L. *O myshlenii i putyakh yevo issledovania* (On Reasoning and Ways to Investigate It). M., 1958.

776. Rusov, Yu. V. *Psikhologichesky analiz protsessov reshenia geometricheskikh zadach na postroyeniye* (A Psychological Analysis of the Processes for the Solution of Geometry Problems of Construction). Dissertation for the Degree of Candidate of Pedagogical Sciences (in Psychology). Gorky, 1955.

777. Saburova, G.G. "Psikhologichesky analiz primenenia orfograficheskikh pravil raznovo tipa (A Psychological Analysis of the Application of Different Kinds of Orthography Rules). *Doklady APNRSFSR* (Lectures of the APSRSSR), IV, 1958.

778. Saburova, G.G. "Psikhologicheskiye osobennosti etapov

primenenia nekotorykh orfograficheskikh pravil uchashchimisya nachalnykh klassov" (Psychological Characteristics
of the Stages of Beginning Students' Application of
Certain Spelling Rules). *Psikhologia usvoyenia grammatiki
i orfografii* (The Psychology of the Mastery of Grammar
and Orthography). Ed. D.N. Bogoyavlensky. M., 1961.

779. Samarin, Yu. A. "Stil umstvennoi raboty starshikh shkolnikov" (Older Students' Style of Mental Work). *Izvestia
APNRSFSR* (News of the APSRSSR), No. 17, 1948.

780. Samarin, Yu. A. *Ocherki psikhologii uma* (Essays on the
Psychology of the Mind). M., 1962.

781.* Sverchkova, R.T. "K kharakteristike postanovki tekhnicheskovo diagnoza" (On the Characteristic of Formulating
a Technical Diagnosis). *Voprosy psikhologii* (Problems of
Psychology), II, 1963.

782. *Svyaz obuchenia s zhiznyu, s trudom uchashchikhsya* (The
Relationship of Instruction to Life and to Students'
Work). Ed. P.R. Atutor. M., 1963.

783. *Svyaz obuchenia v vosmiletnei shkole s zhiznyu* (The
Relationship of Instruction in Eight Year School to Life).
Eds. E.I. Monoszon and M.N. Skatkin. M., 1962.

784. Semyonovich, A.F. and Vorobyov, G.V. *Pervye uroki
geometrii* (iz opyta raboty uchitelya) (First Geometry
Lessons [from a Teacher's Work Experience]). M., 1958.

785. Skatkin, M.N. *Formalizm v znaniakh uchashchikhsya i puti
yevo preodolenia* (Formalism in Students' Knowledge and
Ways to Overcome It). M., 1947.

786. Skatkin, M.N. *O didakticheskikh osnovakh svyazi obuchenia s trudom uchashchikhsya* (On the Didactic Foundations for the Relationship of Instruction with Students'
Work). M., 1960.

787. Skatkin, M.N. *Aktivizatsia poznavatelnoi deyatelnosti
uchashchikhsya v obuchenii* (Enlivening Students' Cognitive Activity in Instruction). M., 1965.

788. Skipin, G.V. "O sistemnosti v rabote bolshikh polushary"
(On Systematics in the Workings of the Cerebrum). *Trudy
fiziologicheskikh laboratory im. ak. I.P. Pavlova* (Trans-

actions of Academician I.P. Pavlov Physiological Laboratories), Vol. 3, 1938.

789.* Skobelev, G.N. "Predvaritelnye samostoyatelnye raboty uchashchikhsya po matematike v starshikh klassakh" (Preliminary Independent Work by Students of the Upper Classes). *Samostoyatelnaya rabota uchashchikhsya na urokakh* (Students' Independent Work at Lessons). Ed. B.P. Yesipov. M., 1960.

790. Skripchenko. A.V. "Formirovaniye obobshchonnykh sposobov reshenia arifmeticheskikh zadach u mladshikh shkolnikov" (The Formation of Abilities to Generalize in Younger Students as They Solve Arithmetic Problems). *Voprosy psikhologii* (Problems of Psychology), IV, 1963.

791.* Slavina, L.S. "Psikhologichesky analiz obuchenia punktuatsii" (A Psychological Analysis of Punctuation). *Sovetskaya pedagogika* (Soviet Pedagogy), I, 1939.

792. Slavina, L.S. "Psikhologicheskiye uslovia povyshenia uspevayemosti u odnoi iz grupp otstayushchikh shkolnikov I klassa" (The Psychological Conditions for the Improvement of Results in One Group of Backward Students of Class I). *Izvestia APNRSFSR* (News of the APSRSSR), No. 36, 1951).

793.* Slavina, L.S. "O nekotorykh osobennostyakh umstvennoi raboty neuspevayushchikh shkolnikov" (On Certain Characteristics of the Mental Work of Backward Students). *Doklady na soveshchanii po voprosam psikhologii* (Lectures at a Conference on Problems of Psychology). M., 1954.

794. Slavina, L.S. *Individualny podkhod k neuspevayushchim i nedistsiplinirovannym uchenikam* (The Individual Approach to Backward and Undisciplined Students). M., 1958.

795.* Slavskaya, K.A. "Protsess myshlenia i aktualizatsia znany" (The Process of Reasoning and the Actualization of Knowledges). *Voprosy psikhologii* (Problems of Psychology), III, 1959.

796.* Slavskaya, K.A. "Protsess myshlenia i ispolzovania znany"

(The Process of Reasoning and Using Knowledges). *Protsess myshlenia i zakonomernosti analiza, sinteza i obobshchenia* (The Reasoning Process and the Laws of Analysis, Synthesis, and Generalization). Ed. S.L. Rubenstein. M., 1960.

797.* Smirnov, A.A. "Voprosy usvoyenia ponyaty shkolnikami" (Problems of Students' Mastery of Concepts). *Sovetskaya pedagogika* (Soviet Pedagogy), VIII-IX, 1946.

798. Smirnov, A.A. "Vlianiye napravlyonnosti i kharaktera deyatelnosti na zapominaniye" (The Influence of the Direction and the Character of Activity on Memorization). *Trudy instituta psikhologii ANGruzSSR* (Transactions of the Institute of Psychology of the Georgian SSR Academy of Sciences). Tbilisi, 1945.

799. Smirnov, A.A. *Psikhologia zapominania* (The Psychology of Memorization). M., 1948.

800.* Sokolov, A.N. "Graficheskoye sopostavleniye logicheski predpolagayemovo i fakticheskovo khoda reshenia zadach" (Comparison by Graph of the Logically Assumed and Factual Progress of Problem Solving). *Voprosy psikhologii* (Problems of Psychology), VI, 1961.

801. Sokolov, Ye. N. *Vospriatiye i uslovny refleks* (Perception and the Conditioned Reflex). M., 1958.

802. Sokolov, Ye. N. "Veroyatnostnaya model vospriatia" (A Probability Model for Perception). *Voprosy psikhologii* (Problems of Psychology), II, 1960.

803.* Solovyov, I.M. "K psikhologii uznavania" (On the Psychology of Recognition). *Sovetskaya pedagogika* (Soviet Pedagogy), II, 1943.

804.* *Sposobnosti i interesy* (Abilities and Interests). Eds. N.D. Levitov and V.A. Krutestkovo. M., 1962.

805.* Stepanov, A.V. *K voprosu o matematicheskom razvitii shkolnika* (On the Problem of a Student's Mathematical Development). Dissertation for the Degree of Candidate of Pedagogical Sciences (in Psychology). M., 1952.

806. Stolyarov, I. "Razvitiye logicheskovo myshlenia uchashchikhsya v protsesse zanyaty po grammatike" (The Develop-

ment of Students' Logical Reasoning During the Study of Grammar). *Russky yazyk v shkole* (The Russian Language in School), V, 1952.

807.   Strezikozin, V.P. *Organizatsia protsessa obuchenia v shkole* Organization of the Instruction Process in School). M., 1964.

808.   Sultankhudzhanova, L.M. "Individualny podkhod v rabote s uchashchimisya I-II klassov" (The Individual Approach to Work with Students of Classes I-II). *Voprosy psikhologii obuchenia i vospitania* (Problems of the Psychology of Instruction and Education). Ed. Z.I. Kalmykova. M., 1956.

809.   Sukhobskaya, G.S. "Ob avtomatizatsii umstvennykh deistvy" (On the Automation of Mental Activities). *Voprosy psikhologii* (Problems of Psychology), III, 1958.

810.   Talyzina, N.F. "Osobennosti umozaklyucheny pri reshenii geometricheskikh zadach" (Characteristics of Deductions During the Solution of Geometry Problems). *Izvestia APNRSFSR* (News of the APSRSSR), No. 80, 1957.

811.   Talyzina, N.F. "Usvoyeniye sushchestvennykh priznakov ponyaty pri organizatsii deistvia ispituyemykh" (Mastery of Essential Indicative Features of Concepts During Organization of Actions of Students Who Are Being Tested). *Doklady APNRSFSR* (Lectures of the APSRSSR), II, 1957.

812.   Talyzina, N.F. "Puti formirovania nachalnykh nauchnykh ponyaty" (Ways to Form Elementary Scientific Concepts). *Doklady APNRSFSR* (Lectures of the APSRSSR), IV, 1960.

813.   Talyzina, N.F. and Nikolayeva, V.V. "Zavisimost formirovania geometricheskikh ponyaty ot iskhodnoi formy deistvia" (Formation of Geometrical Concepts Depends on an Initial Form of Action). *Doklady APNRSFSR* (Lectures of the APSRSSR), VI, 1961.

814.   Talyzina, N.F. and Butkin, G.A. "Opyt obuchenia geometricheskomu dokazatelstvu" (An Experiment in Teaching Geometry Proofs). *Izvestia APNRSFSR* (News of the APSRSSR), No. 133, 1964.

815.* Tekuchev, A.V. *Metodika prepodavania russkovo yazyka v srednei shkole* (The Methodology of Russian Language Instruction in Secondary School). M., 1958.

816. Tekuchev, A.V. *Grammatichesky razbor v shkole* (Grammar Analysis in School). M., 1963.

817. Terekhova, O.P. "K voprosu o formirovanii nachalnykh fizicheskikh ponyaty" (On the Problem of Forming Elementary Physics Concepts). *Voprosy psikhologii* (Problems of Psychology), II, 1962.

818. Tikhomirov, O.K. "Resheniye myslitelnykh zadach kak veroyatnostny protsess" (The Solution of Mental Problems as a Probability Process). *Voprosy psikhologii* (Problems of Psychology), V, 1961.

819. Tikhomirov, O.K. "Issledovaniye optimalnykh sposobov proverki gipotez v norme i patologii" (Research on Optimal Means of Verifying Hypotheses under Normal and Pathological Conditions). *Doklady APNRSFSR* (Lectures of the APSRSSR), IV, V, VI, 1961; I, 1962.

820. Tikhomirov, O.K. "Raspoznavaniye sistem signalov" (Identification of Signal Systems). *Doklady APNRSFSR* (Lectures of the APSRSSR), IV, 1962.

821. Tikhomirov, O.K., Belik, Ya. Ya., Poznanskaya, E.D., and Turchenkova, N. Kh. "Opyt primenenia teorii informatsii k analizu protsessa reshenia myslitelnykh zadach chelovekom" (An Experiment in Applying the Information Theory to an Analysis of the Process of How Man Solves Mental Problems). *Voprosy psikhologii* (Problems of Psychology), IV, 1964.

822. Thorndike, E.L. *Protsess uchenia u cheloveka* (Human Learning). M., 1935. (Trans.)

823. Ustritsky, I.V. "Bolnoi vopros. O sisteme i metodike obuchenia punktuatsii" (A Pressing Problem. On the System and Methodology of Punctuation Instruction). *Russky yazyk v shkole* (The Russian Language in School), IV, 1936.

824.* Ushakov, M.V. *Metodika pravopisania. Posobiye dlya*

uchitelei (Methodology for Reading and Writing. Appliances for Teachers). M., 1959.

825.* Fatkin, L.V. "Obshchiye ponyatia teorii informatsii i ikh primeneniye v psikhologii i psikhofiziologii" (General Concepts of the Information Theory and Their Application in Psychology and Psychophysiology). *Inzhenernaya psikhologia* (Engineering Psychology). Eds. A.N. Leontyev, V.P. Zinchenko, and D. Yu. Panov. M., 1964.

826. Fedorenko, L.P. *Uprazhnenia dlya ukreplenia znany po grammatike* (Exercises to Consolidate Grammar Knowledges). M., 1955.

827. Fedorenko, L.P. *Printsipy i metody obuchenia russkomu yazyku* (Principles and Methods for Russian Language Instruction). M., 1964.

828. Feigenberg, I.M. "Veroyatnostnoye prognozirovaniye v deyatelnosti mozga" (Probability Prognosis in the Activity of the Brain). *Voprosy psikhologii* (Problems of Psychology), II, 1963.

829. Feigenberg, I.M. and Levi, V.L. "Veroyatnostnaya prognozirovaniye i eksperimentalnoye issledovaniye yevo pri patologicheskikh sostoyaniakh" (Probability Prognosis and Experimental Research in It Under Pathological Conditions). *Voprosy psikhologii* (Problems of Psychology), I, 1965.

830. Firsov, G.P. *Nablyudenia nad zvukovoi i intonatsionnoi storonoi rechi na urokakh russkovo yazyka* (Observations on the Spoken and Intonated Word at Russian Language Lessons). M., 1959.

831. Firsov, G.P. *Znacheniye raboty nad intonatsiyei dlya usvoyenia sintaksisa i punktuatsii v shkole* (The Importance of Work on Intonation for the Mastery of Syntax and Punctuation in School). M., 1962.

832. Fleshner, E.A. "Psikhologia usvoyenia i primenenia shkolnikami nekotorykh fizicheskikh ponyaty" The Psychology of Students' Mastery and Application of Certain Physics Concepts). *Psikhologia primenenia znany k resheniyu shkolnykh zadach* (The Psychology of Applying Knowl-

edges to the Solution of School Problems). Ed. N.A. Menchinskaya. M., 1958.

833.* Fleshner, E.A. "Vozrastnye osobennosti abstragirovania v protsesse primenenia znany" (Age Characteristics of Abstraction in the Process of Applying Knowledges). *Voprosy psikhologii* (Problems of Psychology), II, 1964.

834. Freidkin, I.S. "K metodike prepodavania vtorovo zakona Nyutona v srednei shkole" (On the Methodology of Teaching Newton's Second Law in Secondary School). *Doklady APNRSFSR* (Lectures of the APSRSSR), III, 1962.

835. Freidkin, I.S. "K metodike prepodavania temy 'Teplota i rabota' " (On the Methodology of Teaching the Topic "Heat and Work"). *Izvestia APNRSFSR* (News of the APSRSSR), No. 138, 1965.

836.* Khlebnikova, A.V. *Organizatsia i metodika prepodavania russkovo yazyka v V-VIII klassakh* (The Organization and Methodology of Teaching the Russian Language in Classes V-VIII). M., 1960.

837.* Chebysheva, V.V. "O nekotorykh osobennostyakh myslitelnykh zadach v trude rabochikh" (On Certain Characteristics of Mental Problems in Workers' Jobs). *Voprosy psikhologii* (Problems of Psychology), II, 1963.

838. Shakshinskaya, Ye. N. "Opyt provedenia samostoyatelnoi raboty uchaschikhsya po uchebniku pri izuchenii novovo materiala" (An Experiment in Verifying Students' Independent Work According to a Textbook During the Study of New Material). *Ob usloviakh razvitia poznavatelnoi samostoyatelnosti i aktivnosti uchashchikhsya na urokakh* (On Conditions for the Development of Students' Cognitive Independence and Activeness at Lessons). Ed. M.A. Danilov. Kazan, 1963.

839. Shardakov, M.N. *Myshleniye shkolnika* (A Student's Reasoning). M., 1963.

840. Shevaryov, P.A. "Nekotorye zamechania k probleme assotsiatsii" (Certain Observations on the Problem of

Association). *Izvestia APNRSFSR* (News of the APSRSSR), No. 80, 1957.

841.   Shevaryov, P.A. *Obobshchonnye assotsiatsii v uchebnoi rabote shkolnika* (General Associations in a Student's School Work). M., 1959.

842.   Shevaryov, P.A. "K voprosu o strukture vospriaty" (On the Problem of the Structure of Perceptions). *Izvestia APNRSFSR* (News of the APSRSSR), No. 120, 1962.

843.   Shevaryov, P.A. "Nekotorye voprosy psikhologii zritel-nykh vospriaty" (Certain Problems of the Psychology of Visual Perception). *Zritelnye vospriatia* (Visual Perceptions). Ed. P.A. Shevaryov. M., 1964.

844.*   Shemyakin, F.N. "Nekotorye problemy sovremennoi psikhologii myshlenia i rechi" (Certain Problems of the Modern Psychology of Thought and Speech). *Myshleniye i rech* (Thought and Speech). Eds. N.I. Zhinkin and F.N. Shemyakin. M., 1963.

845.   Shenshev, L.V. "Obshchiye momenty myshlenia v pro-tsessakh usvoyenia matematiki i inostrannovo yazyka" (Moments in Common in Reasoning During the Mastery of Mathematics and a Foreign Language). *Voprosy psikhologii* (Problems of Psychology), IV, 1960.

846.   Shekhtyor, M.S. "Izucheniye mekhanizmov simultannovo uznavania" (A Study of the Mechanics of Simultaneous Recognition). *Doklady APNRSFSR* (Lectures of the APSRSSR), II, V, 1961; I, 1963.

847.   Shekhtyor, M.S. "Nekotorye teoreticheskiye voprosy psikhologii uznavania" (Certain Theoretical Problems of the Psychology of Recognition). *Voprosy psikhologii* (Problems of Psychology), IV, 1963.

848.*   Shif, Zh. I. *Ocherki psikhologii usvoyenia russkovo yazyka glukho-nemymi shkolnikami* (Essays on the Psychology of Deaf-mute Students' Mastery of the Russian Language). M., 1954.

849.*   Shchedrovitsky, G.P. and Alekseyev, N.G. "O vozmozh-nykh putyakh issledovania myshlenia kak deyatelnosti" (On Possible Ways of Investigating Thought as an Action).

Doklady APNRSFSR (Lectures of the APSRSSR), III, 1957.

850.* Shchedrovitsky, G.P. "Issledovaniye myshlenia detei na materiale resheny arifmeticheskikh zadach" (A Study of Children's Reasoning Based on Material from the Solutions of Arithmetic Problems). *Razvitiye poznavatelnykh i volevykh protsessov u doshkolnikov* (The Development of Cognitive and Volitional Processes in Pre-schoolers). Eds. A.V. Zaporozhets and Ya. Z. Neverovich. M., 1965.

851. *Eksperimentalnye issledovania po psikhologii ustanovki* (Experimental Research on the Psychology of Purpose). Tbilisi, Vol. I, 1958; Vol. II, 1963.

852. Elinzon, V.B. "Formirovaniye priyomov umstvennoi raboty v kurse geografii" (Formation of Methods of Mental Work in a Geography Course). *Nash opyt uchebno-vospitatelnoi raboty v shkole* (Our Experience in Academic-Educational Work in School). Eds. T.I. Danyushevskaya and V.I. Samokhvalovaya. M., 1962.

853.* Elkin, D.G. "Vospriatiye vremeni i operazhayushcheye otrazheniye (The Perception of Time and Anticipatory Reflection). *Voprosy psikhologii* (Problems of Psychology), III, 1964.

854. Elkonin, D.B. "Formirovaniye umstvennovo deistvia zvukovovo analiza slov u detei doshkolnovo vozrasta" (The Formation of the Mental Action of Oral Word Analysis in Preschool Children). *Doklady APNRSFSR* (Lectures of the APSRSSR), I, 1957.

855. Elkonin, D.B. *Detskaya psikhologia* (Child Psychology). M., 1960.

856.* Elkonin, D.B. "Eksperimentalny analiz nachalnovo etapa obuchenia chteniyu" (An Experimental Analysis of the Beginning Stage of Reading Instruction). *Voprosy psikhologii uchebnoi deyatelnosti mladshikh shkolnikov* (Problems of the Psychology of Beginning Students' School Activity). M., 1962.

857. Erdniyev, P.N. *Razvitiye navykov samokontrolya pri obuchenii matematike* (The Development of Habits of Self

Control During the Instruction of Mathematics). M., 1957.

858. Erdniyev, P.N. *Sravneniye i obobshcheniye pri obuchenii matematike* (Comparison and Generalization During Mathematics Instruction). M., 1960.

859. Yakobson, P.M. "Osobennosti myshlenia uchashchikhsya pri vypolnenii tekhnicheskikh zadany" (Characteristics of Students' Reasoning During the Completion of Technical Problems). *Psikhologia primenenia znany k reshenii uchebnykh zadach* (The Psychology of Applying Knowledges for the Solution of School Problems). Ed. N.A. Menchinskaya. M., 1958.

860. Yakovlev, F.I. "Resheniye poznavatelnykh prakticheskikh zadach na uroke (na opyte prepodavania fiziki). (The Solution of Cognitive Practical Problems at Lessons [Based on Experience of Teaching Physics]). *Sovetskaya pedagogika* (Soviet Pedagogy), IX, 1959.

861. Yakimanskaya, I.S. "Vospriatiye i ponimaniye uchashchimisya chertezha i uslovia zadachi v protsesse yeyo reshenia" (Students' Perception and Understanding of the Sketches and Conditions of a Problem During Its Solution). *Primeneniye znany v uchebnoi praktike shkolnikov* (The Application of Knowledges in Students' School Practice). Ed. N.A. Menchinskaya. M., 1961.

862. Yaroshchuk, V.L. "Rol osoznania tipovykh priznakov pri reshenii arifmeticheskikh zadach opredelyonnovo tipa" (The Role of Recognition of Model Indicative Features When Arithmetic Problems of a Specific Type Are Solved). *Voprosy psikhologii* (Problems of Psychology), I, 1959.

863. Hick, W.E. On the Rate of Gain of Information. *Quart. j. exper. psychol.* Vol. IV, No. 1, 1952.

864. Hyman, R. Stimulus Information as a Determinant of Reaction Time. *J. exper. psychol.* Vol. 45, No. 3, 1953.

865. Linhart, J. Psychologicke problemy teorie uceni. Praha, 1965. Summary in Russian.

866.* Quastler, H. (Ed.) *Information Theory in Psychology.* Glencoe, 1955.

867. Woodworth, R.S. and Schlosberg, H. Experimental Psychology. N.Y./London, 1955.

Author Index

Abakumov, S.I., 347
Alekseyev, M.N., 165
Ananyev, B.G., 320
Anokhin, P.K., 319, 323, 334
Artobolevsky, I.I., 74, 90, 93, 101
Ashby, W.R., 30
Ashkenuze, V.G., 143

Barkhudarov, S.G., 118
Bazhenov, N., 104
Belopolskaya, A.R., 184, 531
Bernstein, N.A., 74, 90, 93, 101, 330, 334
Biryukov, B.V., 47, 104, 207
Bongard, M.M., 103
Botkin, S.P., 27
Braynes, S.N., 52, 54
Bulgakov, N., 74

Chentsov, A.A., 316
Church, A., 213

Danilov, M.A., 511
Dobrynin, N.F., 333
Dodonov, B.I., 130, 144

Dudnikov, A.V., 346

Farber, V.G., 213
Feigenberg, I.M., 330
Feldbaum, A.A., 74, 89, 99
Fillippovich, N., 74
Firsov, G.P., 449
Freidman, L.M., 110, 308

Galanter, E., 102
Galperin, P. Ya., 43, 143, 386, 534
Gastev, Yu., 29
Gavrilov, N., 74
Gentilhomme, I., 316
Glushkov, V., 29, 45, 105, 110
Gnedenko, B.V., 314
Goedel, K., 29, 48
Gorsky, D.P., 153, 212
Granik, G.G., 116, 183, 184
Grashchenkov, N.I., 330

Itelson, L.B., 330
Ivakhnenko, A.G., 96

Kaloshina, I.P., 383
Kaluzhnin, L.A., 47, 105
Khramoi, N., 74
Kimball, J., 147
Klaus, G., 104
Kletsky, E.J., 314
Klini, S.G., 47
Kolmogorov, A.N., 14
Konoplyankin, A.A., 47
Kopnin, P.V., 27
Krinchik, Ye. P., 329, 334, 335
Kryuchkov, S. Ye., 118

Landa, L.N., 31, 52, 72, 103, 308
Latash, N., 330
Leibnitz, G.W., 16
Leontyev, A.N., 184, 320, 329, 334, 335, 536, 603
Lerner, N., 74
Luria, A.R., 534
Lyapunov, A.A., 52, 55, 70, 301, 308

Markov, A.A., 16, 29, 58
Menchinskaya, N.A., 386
Meyerov, N., 74
Miller, G.A., 102
Morse, F., 147

Napalkov, A.V., 52, 54
Newell, A., xv, 70
Nikitin, N.N., 61
Novikov, P.V., 16

Orlov, V.B., 117

Orlova, A.M., 390, 440, 444
Osipov, I.N., 27

Pavlov, I.P., 319, 321, 333
Piaget, J., 535
Pohl, L., 316
Poltayev, I.A., 308
Polyakova, A.V., 116, 119, 121
Post, E.L., 16
Pribram, K.H., 102
Pushkin, V.N., 52

Reshetova, Z.A., 383
Rubenstein, S.L., 320

Schaff, A., 49
Shaw, J.C., xv, 70
Shenshev, L.V., 333, 536
Shestopal, G.A., 52, 55
Shevaryov, P.A., 51, 145, 167
Shorygin, N., 74
Simon, H.A., xv, 70
Slavina, L.S., 536, 598
Smirnov, A.A., 320
Spirkin, A.G., 104
Starchenko, A.A., 25
Sukhov, N., 74
Svechinsky, V.B., 52, 54

Talyzina, N.F., 383
Tavanets, P.V., 153, 212
Trakhtenbrot, B.A., 46, 59
Trapeznikov, V.A., 98

Uspensky, V.A., 29, 47, 50, 145

Ustritsky, I.V., 347
Uznadze, N., 329

Vecker, L.M., 104
Ventsel, Ye. S., 147
Vishnepolskaya, A.G., 131
Vlasenkov, A.I., 116, 184
Vorobyov, G.V., 61
Vygotsky, L.S., 534

Wiener, N., viii

Yablonsky, S.V., 70
Yaglom, A.M., 327
Yaglom, I.M., 327
Yanovskaya, S.A., 145

Zankov, N., 534
Zaporozhets, A.V., 143
Zelinsky, K., 436
Zhinkin, N.I., 52

Subject Index

Ability, 170

Action (*see also* Operation), 10, 20, 57, 359

Algorithm, 10, 11, 33, 47, 50, 71, 78, 145, 148, 170, 257, 342, 399, 529, 600, 606f
  application of, 51
  classification of, 73, 83, 105
  controlling, 74
  design of (devising of), 88, 154, 304, 608
  diagnostic, 31
  difficulty and ease, 137, 306
  discovery of, 149, 152, 606
  efficiency of, 150, 168, 244f, 286, 291, 295f, 612
  evaluation of, 139, 303f
  functional, 74
  generality property, 17, 33
  of identification, 106, 155, 174, 188, 225, 245
  knowledge of, 161
  mastery of, 161, 607
  normal, 58
  requirements for, 34
  resultivity property, 18, 33
  rigidity of, 63, 98, 104
  search algorithm, 86, 90, 101, 103, 133
  solution algorithm, 100, 126, 141, 151, 185
  specificity property of, 17, 33
  teaching of, 152, 156, 169, 183, 342f
  of transformation, 107, 175
  universal, 16

Algorithmic description, 52, 59

Algorithmic process, 50, 51, 54, 127, 133, 171
  execution of, 54
  forming of, 159, 515

Algorithmization, 68, 122

Ambiguity, 35

Assimilation, 65, 92, 198, 515, 606, 610

Associations, 51

Automatic control, 71, 73, 91

Bi-conditional (⇔), 177

Bit, 23, 281

Brain centers, 320f

Cognition, 31, 196

Cognitive activity, 285

Computer-administered instruction, ix

Concept forming, 153

Conditioned reflex, 319, 322

Constancy, 37

Control, 53, 65, 76, 84, 86, 163, 515, 519

Control classes, 345, 549f, 568, 576, 579

Creative activity, 24

Creativity, 155

Cybernetics, xviii, xxi-xxiii, 52, 605, 610

Decision theory, 257

Definition, 177, 196, 209, 211, 213, 214, 222, 225, 349, 387, 568, 570

Diagram, 12, 83, 358f, 400, 401, 402, 405, 407, 410, 426, 427, 428, 430, 434-435, 437, 455, 469, 470, 476, 500, 518

Didactics, 67, 158, 377, 609, 611

Didactic task, 397, 400, 401, 402, 404, 406, 408, 409, 411, 441

Difficulty, 244, 255, 267

Directive instruction (*see* Prescription)

Education (*see also* Instruction, Pedagogy), 248

Elementary operators, 56

Entropy, 285

Excitability, 319, 326

Experimental classes, 345, 549f, 576, 579, 585f

Experimental instruction, 344, 539, 543, 577, 592f, 608

Feedback, ix, 68, 83, 515

Fragmentation (of description), 53

Generalization, 97, 163

Graph, 58, 62, 63, 71, 105

H (Information measure), 284, 292

Identifiability, 37

Identification, 106, 176, 227, 236, 259, 428

Independent activity, 23

Indicative feature(s), 108, 110, 181, 191, 194, 197f, 214, 288, 348, 352, 388f, 458, 566, 570, 609

compatibility of, 261
probabilities of, 262, 270
selection of, 243
structure of, 216, 220, 224, 232, 592, 614
time for verification, 262, 268, 270, 305
verification of, 202, 228f, 245f, 260, 295, 321, 428
Individual instruction, 65
Individual study, 65
Information, 280f
Information theory, 278f, 302, 306
Informativity, 287f, 291, 293f
Instruction, ix, 43, 75, 81, 107, 181, 202, 236, 238, 321, 329, 377, 396, 549f, 604f, 612
adaptation of, 66
algorithmization of, 72
automation of, 66
effectiveness of, 69
efficient strategies of, 81
experimental instruction, 344
individualization of, 44, 169, 531
of machines, 91
mass instruction, 65
self-instruction, 90, 101
Instructional procedure, 60, 308, 438
Instructional science and technology, viii

Knowledge, 178, 307, 377, 513, 570

Learning, 81, 90, 101
Logic diagram, 56, 61, 62, 63, 217, 400, 464, 492, 498-499, 502-503
Logical condition, 56, 62, 64
Logical conjunction (&), 127, 193, 203, 225, 228
Logical correctives, 62, 192, 204, 214, 216
Logical disjunction (V), 20, 195, 203, 207, 225, 229
Logical implication (⇒), 192
Logical necessity, 193, 203
Logical negation (⁻), 144, 193
Logical operator, 215
Logical structure, 182, 206, 214, 259, 307, 350, 465, 601
Logical sufficiency, 193, 203

Members, 56
Method, 9, 29, 611
algorithmic, 10
heuristic, 117
non-algorithmic, 21, 133, 136
quantitative, 243
of reasoning, 8, 29, 248, 359, 570, 591
Mistakes, 156, 348f, 450f, 521, 550f, 585f

Object, 77, 105, 109, 195, 258

Operations, 303, 332, 396, 513, 570
  advance knowledge of, 25
  automatic execution, 46, 48, 50, 602, 604
  conscious execution of, 49, 50
  elementary, 17, 37f, 42, 44, 50, 120, 157, 158, 197, 537f
  intellectual, 10, 43, 157, 514, 519, 523, 559, 601f
  on material content, 35, 44f
  structure of, 226
  system of, 9, 156, 439, 519, 538, 558, 565
Operations research, 142, 256
Operators, 62, 72, 87, 105, 398
Optimality, 258, 264
Outcome, 89, 125, 278, 286

Pedagogical operations, 398, 400, 401, 404, 406, 409
Pedagogy (*see also* Education, Instruction), 73, ˙131, 146, 603
Predicate logic, 208
Prescription(s), 17, 18, 21, 50, 120
  algorithmic, 121, 143, 158
  non-algorithmic, 121
  quasi-algorithmic, 33f,
    40, 47
  univocality of, 45
Probabilistic solution, 127
  advantages of, 128
Problems:
  algorithmically solvable, 149
  algorithmically unsolvable, 16
  classification of, 114
  of alternative choice, 258f
  of multiple choice, 259, 278f
Problem class, 11, 50, 51, 136
Problem solving, 7, 70, 113, 148, 601
Process (creative), 95, 104
Process (of learning and teaching), 33
Programmed instruction, xix-xxi, 40, 68, 72, 170, 612
Programmed textbooks, 65
Propositional logic (*see* Sentential logic)
Psychological operations, 397, 400, 401, 404, 406, 409, 443
Psychology, 73, 131, 158, 248, 302, 605, 609

Quasi-algorithm (meaning in), 36

Reaction time, 329
Readiness (to perceive), 326f
Reasoning methods (proce-

dures), xii, 8, 157, 341
Receptivity, 319f
Reinforcement, 322
Rule, 29, 51, 134, 166, 211, 568

Search, 94, 254
Sensitivity (*see also* Receptivity), 319f, 330
Sentential logic, 208
Sequence (of actions, operations), 64, 82, 116, 143, 165, 185, 225, 249, 333, 577
Simulation, 96
Skill, 170
State, 19, 20, 77, 87, 100
Stimulus, 318, 329
Strategy, 257, 265, 273
    of search and identification, 94, 250, 259, 289
    of teaching, 147
    optimal strategy, 263, 270, 275, 276, 291, 306, 332
Structural analysis, 59
Structure (of processes), 53, 98
Subject-matter, 154, 212, 608
System, 40, 48, 59, 74, 84
    hierarchical structure of, 75
    purposive system, 89
Systematization, 164

Teaching (*see* Instruction)

Teaching algorithm (algorithm of instruction), 60, 398
Teaching machines, 65, 526f
Theory of games, 257
Thinking (efficient methods of), 70
Transfer, 97, 163
Transformation (*see also* Transition), 19, 106, 176
    probability of, 87, 100
    univocal, 30, 45
Transition, 19
    probability of, 79
    stochastic, 79f
    strictly determined, 77f
Trial-and-error, 93, 102, 124, 135, 145, 156, 567
Turing machines, 59

Uncertainty, 278f, 291
    degree (value) of, 23, 283f
Univocality, 200, 226, 393

A Glossary of Logical Symbols
Used in This Book

Logical Function	Symbol	Ordinary Language Meaning
Biconditional*	⇔	if and only if . . ., then . . .
Conjunction	&	and
Disjunction**	V	or (possibly both)
Implication†	⇒	if . . ., then . . .
Negation	$\overline{a}$	non-a; not a; a is not true (where a represents some proposition)

* $p \Leftrightarrow q$ means that proposition q is true, if and only if proposition p is true and vice versa.

**   In the case of a non-strict disjunction, e.g., $p \lor q$, either or both propositions may be true; in the case of the strict disjunction (only tangentially used in this book) both cannot be true.

†   The truth of the premise is not guaranteed, e.g., $p \Rightarrow q$ does not mean that p is necessarily true when q is true.

About the Author, Editor and Translator

The Author

L.N. Landa was born in 1927 in Rostow, U.S.S.R. He is a graduate of the University of Leningrad and the Yaroslav Pedagogical Institute. His graduate work culminated in a thesis on the "Formation in Students of General Methods of Reasoning," in 1955. His ideas on this subject were at the heart of subsequent research at the Academy of the Pedagogical Sciences in Moscow, leading to the publication of the original edition of *Algorithmization in Learning and Instruction*, in 1966. Dr. Landa is the author of more than 80 publications, which have been translated into 12 languages.

The Scientific Editor

Felix F. Kopstein is a Senior Staff Scientist with the Human Resources Research Organization, a behavioral science research institution in Alexandria, Virginia. Trained as an experimental psychologist, with a Ph.D. from the University of Illinois, his interests have shifted from the experimental analysis of human learning processes to the synthesis of instructional models which can be implemented with the aid of computers. He is the current Secretary of the American Society for Cybernetics.

The Translator

Virginia Bennett received her graduate degree in Russian studies from the Sorbonne (diplôme breveté) and her doctorate from Princeton University. She is presently teaching Russian at Colgate University, and is one of the editors of the forthcoming English language edition of F.M. Dostoyevsky's letters.